Springer Proceedings in Earth and Environmental Sciences

The series Springer Proceedings in Earth and Environmental Sciences publishes proceedings from scholarly meetings and workshops on all topics related to Environmental and Earth Sciences and related sciences. This series constitutes a comprehensive up-to-date source of reference on a field or subfield of relevance in Earth and Environmental Sciences. In addition to an overall evaluation of the interest, scientific quality, and timeliness of each proposal at the hands of the publisher, individual contributions are all refereed to the high quality standards of leading journals in the field. Thus, this series provides the research community with well-edited, authoritative reports on developments in the most exciting areas of environmental sciences, earth sciences and related fields.

Debabrata Mazumder
Editor

Sustainable Advanced Technologies for Environmental Management

Proceedings of SATEM 2023

Springer

Editor
Debabrata Mazumder
Department of Civil Engineering
Indian Institute of Engineering Science and Technology, Shibpur
Howrah, West Bengal, India

SATEM-2023, International Conference on Sustainable Advanced Technologies for Environmental Management
December 20-22, 2023
Howrah, India

ISSN 2524-342X ISSN 2524-3438 (electronic)
Springer Proceedings in Earth and Environmental Sciences
ISBN 978-3-031-64005-6 ISBN 978-3-031-64006-3 (eBook)
https://doi.org/10.1007/978-3-031-64006-3

This Springer imprint is published by the registered company Springer Nature Switzerland AG
The registered company address is: Gewerbestrasse 11, 6330 Cham, Switzerland

Dedicated to
All Scientist, Engineers, and Technologists, who are actively involved in the compliance of Sustainable Environmental Management

Preface

In the advent of rapidly increasing environmental pollution, untoward changes in global climate now become constraints for socioeconomic development. Enormous utilization of non-renewable sources for meeting high energy demand even in domestic sector also poses a serious threat to environment. Indiscriminate use of natural resources without tangible recycling causes a huge burden on end-of-pipe waste management. To combat these issues, a variety of advanced technologies have been developed; but a few of them are compatible for simplistic and versatile application. It is also a great challenge to sustain a new technology over a wide time span due to lack of expertise. To disseminate state-of-the-art knowledge on this issue among the relevant stakeholders, the First Version of International Conference on *Sustainable Advanced Technologies for Environmental Management* (*SATEM-2023*) was organized by the Department of Civil Engineering, Indian Institute of Engineering Science and Technology (IIEST), Shibpur, West Bengal, India, during December 20–22, 2023, in hybrid mode. The Springer Nature AG, Switzerland, a premier internationally acclaimed publishing organization, has come forward to bring to light one Proceedings Book of the above International Conference. It is our extreme pleasure that *Sustainable Advanced Technologies for Environmental Management: Proceedings of SATEM 2023* has been published with some selected and peer-reviewed full-length papers, presented in the above conference.

There are 5 thematic parts containing a total 38 numbers of chapters on various aspects of advanced technologies for environmental management. The thematic parts are spread over the titles—*Sustainable Water Quality Management*, *Advanced Water Treatment Technologies*, *Innovative Techniques for Measuring Environmental Pollutants*, *Advanced Wastewater Treatment Technologies*, as well as *Solid and Hazardous Waste Management*. Utmost care has been taken to include the chapters in their appropriate part as far as possible.

In spite of our best possible effort some inadvertent errors might have occurred, for which we solicit valuable comments from the readers.

We sincerely believe that *Sustainable Advanced Technologies for Environmental Management: Proceedings of SATEM 2023* would be gladly received by the scientific and engineering community and will definitely motivate us to bring more such Proceedings Book in future.

Howrah, West Bengal, India
29 February 2024

Debabrata Mazumder

Acknowledgment

In the very beginning, we extend our sincere thanks and gratitude to Prof. Parthasarathi Chakrabarti, Honorable Director, IIEST, Shibpur, for his inspiring motivation in organizing *SATEM-2023*. We express our deep sense of gratitude to Prof. Ambarish Ghosh, Chairman, SATEM-2023 and the Former Head, Department of Civil Engineering, IIEST, Shibpur, for his continuous support at every moment. We are thankful to Prof. Chaitali Ray, the Head, Department of Civil Engineering, IIEST, Shibpur for her cordial effort to make SATEM-2023 a successful event. We would also like to express our heartiest thanks to all the faculty and staff members as well as master's and doctoral scholars of the Civil Engineering Department for their commendable role in organizing SATEM-2023. In this regard, we express our deep sense of gratitude to Dr. Chanchal Majumder and Dr. Asok Adak, two Co-Chairmen of the above international conference, for their outstanding contribution. We would also like to extend our sincere thanks to Dr. Dipankana Bhattacharjee, Co-ordinator of SATEM-2023, for rendering her excellent service throughout the conference.

We would like to acknowledge our sincere thanks and indebtedness to the Members of the Editorial Board of *Springer Nature AG* for accepting the proposal for publication of the Proceedings Book of SATEM-2023. In this context, we cordially acknowledge the kind inspiration and co-operation of Dr. Zachary Romano and all his editorial team members, who have served with monumental patients to materialize publication of this Proceedings Book. We are also extremely grateful to Tina Silva and her production team members for their relentless effort to make the present book a perfect one.

We convey our sincere thanks and heartiest gratitude to all the Reviewers who have extended their best level of expertise in assessing the technical papers. We cordially acknowledge with great admiration, the scholastic contribution of the authors who enthusiastically responded with their final version of articles.

At the last, but not the least, we deeply acknowledge the contribution of all the stakeholders of the SATEM-2023, whose names might have been omitted above inadvertently. It would have not been possible to publish the present Proceedings Book without their best effort and kind co-operation.

Howrah, West Bengal, India
29 February 2024

Debabrata Mazumder

Contents

About the Editor

Debabrata Mazumder received his bachelor's and master's degrees in Civil Engineering from Jadavpur University in 1993 and 1996, respectively. He received his Ph.D. degree in Civil Engineering (Environmental Engineering Specialization) from Bengal Engineering and Science University, Shibpur (*presently Indian Institute of Engineering science and Technology, Shibpur*) in 2004. He has a total 31 years of industrial, teaching, and research experience. He published more than 160 technical papers and 11 book chapters in various national and international journals and conference proceedings. He has co-authored one Reference Book and edited two Proceedings Books in the Springer Nature. He is the member of American Water Works Association (AWWA) and the member of International Water Association (IWA). He is also the Life Fellow of Institution of Engineers (India) (IE, India) and the Life Fellow of Institution of Public Health Engineers (India) (IPHE, India). Apart from these, he is the Life Member of Indian Water Works Association (IWWA). He has supervised 6 Ph.D. (Engg.) and 26 M.Tech. thesis in the field of Environmental Engineering. He has also supervised more than 45 B.Tech. Projects in the same field. He has adjudicated several Ph.D. thesis from a number of institutions of India and abroad.

He is as a Consultant in National Mission for Clean Ganga (NMCG) and National River Conservation Directorate (NRCD) under the Ministry of Environment, Forests and Climate Change, Government of India. Apart from that he has been serving as a Principal

Consultant in several government and private organizations. He has been granted Five Indian Patents in the field of wastewater treatment and disposal. He is presently serving as the Professor of Civil Engineering Department, Indian Institute of Engineering Science and Technology, Shibpur. He was also the Dean (Students' Welfare) in the same institute for 3 years since 2019. He has been regularly organizing International Conference on "Sustainable Advanced Technologies for Environmental Management" (SATEM) as well as International Conference on "Advanced Technologies for Industrial Pollution Control" (ATIPC) from 2017 and 2018, respectively, in the Department of Civil Engineering, IIEST, Shibpur.

Part I
Sustainable Water Quality Management

Sunlight-Assisted Photocatalytic Degradation of Methyl Red Using g-C_3N_4 as Metal-Free Photocatalyst

Subhalaxmi Sahoo, Guddeti Bhargava Teja Reddy, and Prateeksha Mahamallik

Abbreviations

AOP	Advanced oxidation process
BOD	Biochemical oxygen demand
COD	Chemical oxygen demand
FTIR	Fourier transform infrared spectroscopy
MR	Methyl red
SEM	Scanning electron microscopy
TSS	Total suspended solids
UV–Vis	Ultraviolet–visible
XRD	X-ray diffraction

1 Introduction

Dyes are introduced to the environment through different industrial sources such as textiles, pulp and paper, plastic, paints, and leather [1]. Even at minimal concentrations (1–2 mg/L), these pigmented components stay visible, impeding light penetration and posing a hazardous impact on the environment of water bodies [2, 3]. The categorization of dyes includes acidic, basic, azo, diazo, anthraquinonoid, and metal-based types, where basic and azo dyes exhibiting higher toxicity rates [4]. The dye effluents containing various contaminants are normally characterized by their COD, BOD, TSS, pH value, and so on [5]. The presence of dyes in water reduces

S. Sahoo · G. B. T. Reddy · P. Mahamallik (✉)
Department of Civil Engineering, National Institute of Technology, Rourkela, India
e-mail: mahamallikp@nitrkl.ac.in

D. Mazumder (ed.), *Sustainable Advanced Technologies for Environmental Management*, Springer Proceedings in Earth and Environmental Sciences,
https://doi.org/10.1007/978-3-031-64006-3_1

transparency, hindering light transmission, thereby disrupting photosynthesis and diminishing dissolved oxygen levels, adversely affecting aquatic plants and animals [6]. Methyl red (MR), an azo dye ordinarily used in textile industries and paper printing, contains single or many azo groups (–N=N–) in its structure, rendering it cancer-causing and harmful to human health. Beyond permissible limits, MR can induce severe irritations in the eyes, respiratory system, skin, and alimentary canal [7, 8]. The discharge of untreated dyes carries the potential to harm aquatic life and vegetation by obstructing sunlight penetration in the water they inhabit [9]. Various techniques, including adsorption, flocculation, biodegradation, ozonation, chemical oxidation, ion exchange, etc., have been employed to eradicate these dyes from wastewater [3]. However, conventional methods are often expensive, generating poisonous sludge and volatile components as secondary toxins. AOPs have emerged as efficient techniques for degrading recalcitrant toxins from aqueous solutions [10]. Among these, the photocatalytic technique stands out as a novel type of AOP with higher efficiency, environmentally friendly, cheap, and safe, garnering increased attention for the high-efficiency degradation of harmful organic pollutants [11]. In recent years, researchers have actively sought metal-free photocatalysts due to their cost-effectiveness, ready availability, and lack of environmental pollution after use. The introduction of graphite phase C_3N_4 (g-C_3N_4) in 2009 marked the first report of a metal-free photocatalyst [12, 13]. g-C_3N_4, a nonmetal semiconductor having a moderate band gap (~2.7 eV), has gained prominence for its potential practical applications in the photodegradation of organic contaminants [14].

In this study, our aim is to prepare the catalyst, analyze its characteristics, and explore the impact of several operational factors on the photocatalytic degradation of MR.

2 Materials and Methods

2.1 Chemicals

MR and urea (for the preliminary synthesis of g-C_3N_4) were purchased from Merck and Oxford respectively. NaOH and H_2SO_4 were used for pH adjustment. The solution for material preparation was made with deionized water.

2.2 Synthesis of g-C_3N_4

Two grams of urea was taken and then placed in a crucible. The ceramic cup is covered with a silver foil. Then the urea was calcined at 520 °C at the heating rate of 5 °C/min. This was done for 2 h. Then it was allowed to cool inside the furnace for a few hours. After cooling the obtained material is g-C_3N_4 [15].

2.3 *Characterization of g-C_3N_4*

The as-synthesized material was characterized through XRD, FTIR, and SEM.

2.4 *Experimental Procedures for MR Degradation*

Degradation of MR was carried out in a glass beaker in the presence of sunlight. The required MR concentrations were prepared by diluting the stock solution of 100 mg/L. A specific quantity of catalyst was mixed to 50 mL of sample solution. Then the solution was kept in the dark for 30 min to attain the equilibrium of adsorption. After the adsorption, the samples were taken for the photodegradation study. The photocatalytic degradation was initiated by exposure of sunlight to the dye sample. The intensity of sunlight (lux) was measured by the light meter at the start of each time interval. The experiments were conducted under a light intensity between 25,000 and 30,000 lux. The degraded samples were collected after a certain interval of time. To separate the supernatant from the material, the sample was centrifuged at 4000 rpm and then the absorbance reading was taken using UV–Vis spectrophotometer.

3 Results and Discussion

3.1 *Characterization of g-C_3N_4*

3.1.1 X-Ray Diffraction (XRD)

The peaks observed at approximately 13.3° and 28.85° were allocated to the (100) and (002) planes, indicating the in-plane structural arrangement and interlaminar stacking of aromatic units, respectively (Fig. 1) [16].

3.1.2 Fourier Transform Infrared Spectroscopy (FTIR)

FTIR analysis was done for g-C_3N_4, and the observations are presented in Fig. 2. The peak at 810 cm^{-1} represents the triazine units [17]. The band at 1237 cm^{-1} belongs to the C–O–C group's stretching vibration. The broad absorption peaks from 1200 to 1600 cm^{-1} belong to stretching vibrations of heptazine resultant repeating unit/CN heterocycles [15].

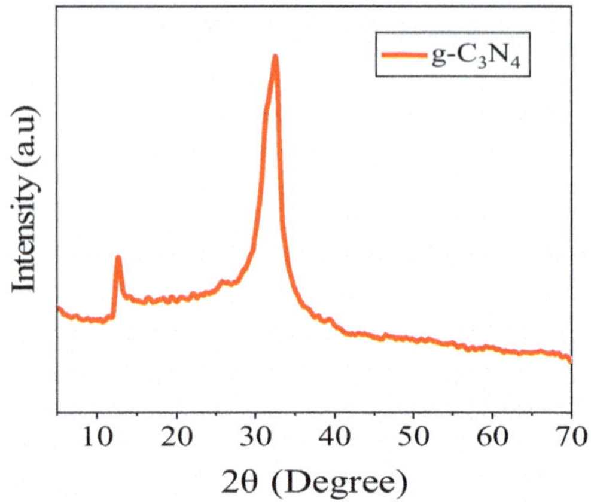

Fig. 1 XRD analysis graph of g-C_3N_4

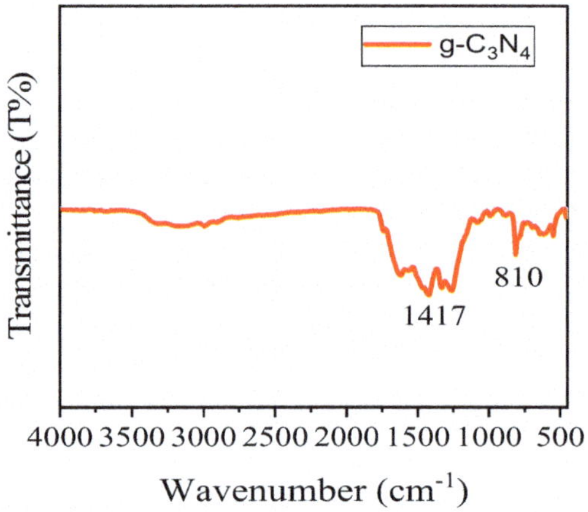

Fig. 2 FTIR analysis graph of g-C_3N_4

3.1.3 Scanning Electron Microscopy (SEM)

SEM analyses of g-C_3N_4 were made to gain an idea about the surface topography and composition of the material. SEM image of g-C_3N_4 is shown in Fig. 3. The structure of the pure g-C_3N_4 showing a layered sheets morphology [18].

3.2 Effect of Operating Factors on MR Degradation Using g-C_3N_4

Degradation of MR was carried out for different operational parameters including the effect of initial MR concentration, g-C_3N_4 dose, and pH under acidic, neutral, and basic range under sunlight irradiation (~30,000 lux). The dye was allowed to achieve adsorption–desorption equilibrium and it was found to be $<10\%$ in dark atmosphere.

Fig. 3 SEM image of g-C_3N_4

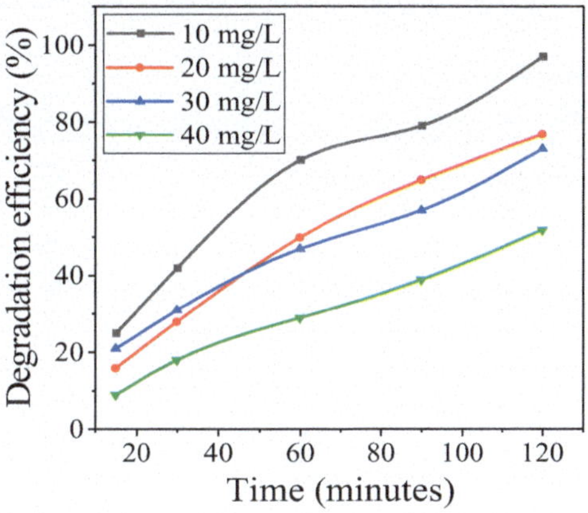

Fig. 4 Effect of initial MR concentration on degradation (g-C_3N_4 dose: 0.8 g/L)

3.2.1 Effect of Primary Concentration of MR

The reaction was observed with different MR concentrations (i.e., 10, 20, 30, 40 mg/L). It was observed that the degradation efficiency was 97% in 10 mg/L of MR concentration within 120 min (Fig. 4). It was noticed that with the increase in MR concentration the degradation rate decreased, which may be caused because at higher dye concentration, the availability of reactive sites gets cover by the dye molecules. Another reason might be the absorption of light by dye molecules becomes more, which hinders the photocatalytic process [19].

3.2.2 Effect of g-C_3N_4 Dose

The study was conducted with 10 mg/L of MR concentration and varying g-C_3N_4 dose (0.2, 0.4, 0.6, 0.8 g/L). It was perceived that the degradation efficiency was 96% within 120 min with 0.8 g/L of g-C_3N_4 dose (Fig. 5). With an increasing dose of catalyst, the availability of active radicals for the dye molecules becomes more, so the degradation also increases [19].

3.2.3 Effect of pH

The effect of the pH study was conducted under three initial pH conditions (acidic = 3, neutral = 7, and basic = 10). The MR concentration (20 mg/L) and g-C_3N_4 dose (0.8 g/L) were taken for this study. The degradation efficiency was 92% at initial pH 3 whereas only 41% at initial pH 10 within 120 min (Fig. 6). At lower pH, the availability of electrons in the conduction band reduced due to electrostatic

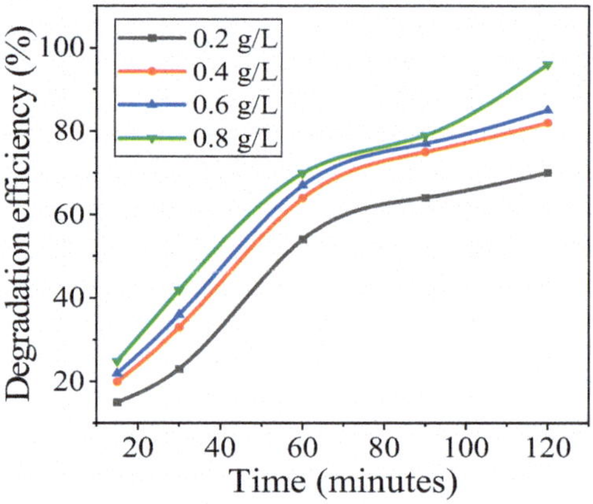

Fig. 5 Degradation efficiency of MR for different g-C_3N_4 dose (MR concentration: 10 mg/L)

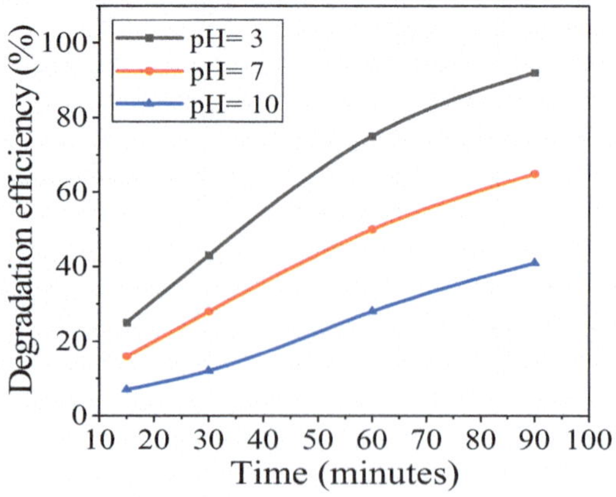

Fig. 6 Effect of pH on degradation of MR (concentration of MR: 20 mg/L and g-C_3N_4 dose: 0.8 g/L)

interactions of anions of dye and positive surface charge of the catalyst, resulting adsorption of dyes to the surface of the catalyst. Also in acidic regions, strong complex bonds were formed on the catalyst surface, which enhances the electron-transfer efficiency process [19].

4 Conclusions

The photocatalyst g-C_3N_4 was synthesized by the calcination process. The prepared material was characterized by XRD, FTIR, and SEM. Photocatalytic degradation was carried out using the as-prepared catalyst under sunlight irradiation. The MR degradation decreased with increasing primary concentration of dye, whereas with an increasing dose of g-C_3N_4, the degradation also increased. The maximum degradation efficiency was achieved 97% with 10 mg/L of MR concentration and 0.8 g/L g-C_3N_4, 96% with 0.8 g/L of g-C_3N_4 and 10 mg/L of MR concentration in 120 min, and 92% in acidic region (pH 3) with MR concentration 20 mg/L and g-C_3N_4 dosage 0.8 g/L.

References

1. M.A. Tariq, M. Faisal, M. Saquib, M. Muneer, Heterogeneous photocatalytic degradation of an anthraquinone and a triphenylmethane dye derivative in aqueous suspensions of semiconductor. Dyes Pigments **76**(2), 358–365 (2008)
2. N.K. Singh, S. Saha, A. Pal, Solar light-induced photocatalytic degradation of methyl red in an aqueous suspension of commercial ZnO: a green approach. Desalin. Water Treat. **53**(2), 501–514 (2015)
3. I. Ullah, M. Tariq, M. Muhammad, J. Khan, A. Raheem, A.Z. Abdullah, UV photocatalytic remediation of methyl red in aqueous medium by sulfate ($SO_4^{\bullet -}$) and Hydroxyl ($^{\bullet}OH$) radicals in the presence of Fe^{2+} and Co@ TiO_2 NPs photocatalysts. Available at SSRN 4546566 (2023)
4. Y. Badr, M.G. Abd El-Wahed, M.A. Mahmoud, Photocatalytic degradation of methyl red dye by silica nanoparticles. J. Hazard. Mater. **154**(1–3), 245–253 (2008)
5. N.N. Nyangiwe, B. Baatjie, C. Greyling, M. Khenfouch, M. Maaza, The decolourisation of Methyl Orange and textile effluent under UV using commercial and synthesized nano-TiO_2, in *Journal of Physics: Conference Series*, vol. 984, No. 1, (IOP Publishing, 2018, March), p. 012009
6. M.A. Lafta, S.H. Ammar, H.J. Khadim, Z.H. Jabbar, Improved photocatalytic degradation of methyl violet dye and pathogenic bacteria using g-C_3N_4 supported phosphotungstic acid heterojunction. J. Photochem. Photobiol. A Chem. **437**, 114506 (2023)
7. H. Lachheb, E. Puzenat, A. Houas, M. Ksibi, E. Elaloui, C. Guillard, J.M. Herrmann, Photocatalytic degradation of various types of dyes (Alizarin S, Crocein Orange G, Methyl Red, Congo Red, Methylene Blue) in water by UV-irradiated titania. Appl. Catal. B Environ. **39**(1), 75–90 (2002)
8. G. Mascolo, R. Comparelli, M.L. Curri, G. Lovecchio, A. Lopez, A. Agostiano, Photocatalytic degradation of methyl red by TiO_2: comparison of the efficiency of immobilized nanoparticles versus conventional suspended catalyst. J. Hazard. Mater. **142**(1–2), 130–137 (2007)

9. T. Gulzar, T. Huma, F. Jalal, S. Iqbal, S. Abrar, S. Kiran, S. Nosheen, W. Hussain, M.A. Rafique, Bioremediation of synthetic and industrial effluents by Aspergillus niger isolated from contaminated soil following a sequential strategy. Molecules **22**(12), 2244 (2017)
10. J. Khan, M. Tariq, M. Muhammad, M.H. Mehmood, I. Ullah, A. Raziq, F. Akbar, M. Saqib, A. Rahim, A. Niaz, Kinetic and thermodynamic study of oxidative degradation of acid yellow 17 dye by Fenton-like process: effect of HCO_3^-, CO_3^{2-}, Cl^- and SO_4^{2-} on dye degradation. Bull. Chem. Soc. Ethiop. **33**(2), 243–254 (2019)
11. Y.P. Sun, B. Zhou, Y. Lin, W. Wang, K.S. Fernando, P. Pathak, M.J. Meziani, B.A. Harruff, X. Wang, H. Wang, P.G. Luo, H. Yang, M.E. Kose, B. Chen, L.M. Veca, S.Y. Xie, Quantum-sized carbon dots for bright and colorful photoluminescence. J. Am. Chem. Soc. **128**(24), 7756–7757 (2006)
12. Z. Feng, L. Zeng, Q. Zhang, S. Ge, X. Zhao, H. Lin, Y. He, In situ preparation of g-C_3N_4/$Bi_4O_5I_2$ complex and its elevated photoactivity in Methyl Orange degradation under visible light. J. Environ. Sci. **87**, 149–162 (2020)
13. Y. Zhang, Z. Jin, Boosting photocatalytic hydrogen evolution achieved by NiSx coupled with g-C_3N_4@ ZIF-67 heterojunction. J. Phys. Chem. C **123**(30), 18248–18263 (2019)
14. F. Goettmann, A. Fischer, M. Antonietti, A. Thomas, Chemical synthesis of mesoporous carbon nitrides using hard templates and their use as a metal-free catalyst for Friedel–Crafts reaction of benzene. Angew. Chem. Int. Ed. **45**(27), 4467–4471 (2006)
15. H. Xie, Y. Zheng, X. Guo, Y. Liu, Z. Zhang, J. Zhao, W. Zhang, Y. Wang, Y. Huang, Rapid microwave synthesis of mesoporous oxygen-doped g-C_3N_4 with carbon vacancies for efficient photocatalytic H_2O_2 production. ACS Sustain. Chem. Eng. **9**(19), 6788–6798 (2021)
16. R. Chen, H. Yin, L. Wang, Z. Zhang, J. Ding, J. Zhang, H. Wan, G. Guan, Enwrapping g-C_3N_4 on In_2O_3 hollow hexagonal tubular for photocatalytic CO2 conversion: construction, characterization, and Z-scheme mechanism insight. J. Colloid Interface Sci. **631**, 122–132 (2023)
17. Y. Yang, Z. Bian, Oxygen doping through oxidation causes the main active substance in g-C_3N_4 photocatalysis to change from holes to singlet oxygen. Sci. Total Environ. **753**, 141908 (2021)
18. A. Modwi, B. Mustafa, A. Toghan, K.K. Taha, Scalable fabrication and characterization of Y_2O_3@ g-C_3N_4 nanocomposite for the enhancement of photocatalytic removal of Congo red dye under visible light. J. Mater. Sci. Mater. Electron. **34**(4), 298 (2023)
19. A. Rani, K. Singh, A.S. Patel, P. Sharma, Factors affecting photocatalytic degradation of methyl red by MoS_2 nanostructures prepared by hydrothermal technique. Bull. Mater. Sci. **46**(2), 94 (2023)

Fabrication of Cellulose Acetate-Based Electrospun Nanofiber Membranes and Its Application for Ciprofloxacin Removal

Sushma Das, Asok Adak, and Ananya Barui

Abbreviations

AA	Acetic acid
Ac	Acetone
CA	Cellulose acetate
CF	Chloroform
CIP	Ciprofloxacin
DCM	Dichloromethane
DMAc	Dimethyl acetamide
DMF	Dimethyl formamide
EDS	Electron dispersive spectroscopy
ENFM	Electrospun nanofiber membrane
FE-SEM	Field emission scanning electron microscope
HCl	Hydrochloric acid
MeOH	Methanol
MMT	Montmorillonite
NNMP	*N*,*N*-methyl pyrrolidine
PCL	Polycaprolactone
PEO	Polyethylene oxide
PVA	Polyvinyl alcohol
TFA	Trifluoroacetic acid

S. Das (✉)
Dream Institute of Technology (DIT), Thakurpukur, Kolkata, West Bengal, India

A. Adak
Department of Civil Engineering, IIEST, Shibpur, Howrah, India

A. Barui
Centre for health Care Science and Technology, IIEST, Shibpur, Howrah, India

D. Mazumder (ed.), *Sustainable Advanced Technologies for Environmental Management*, Springer Proceedings in Earth and Environmental Sciences,
https://doi.org/10.1007/978-3-031-64006-3_2

1 Introduction

The sustenance of human and animal life is dependent on the precious natural resource—water. Rapid population growth and urbanization have an impact on environmental resources and also lead to an increase in diseases, necessitating the increase in pharmaceuticals and antibiotics, which subsequently has a huge impact on the availability of clean water in the environment due to the release of these chemicals. In addition to pharmaceuticals and antibiotics, personal care products, industrial chemicals, steroid hormones, pesticides, and many other new compounds are entering the environment and unbalancing the ecosystem. These pollutants are present in water in trace amounts as nanograms per liter (ng/L) to micrograms per liter (μg/L).

A detailed review has been published on the presence of antibiotics in different environmental matrixes [1]. In India, many studies have been done to determine the extent of antibiotics in the wastewater and their removal efficiencies by various methods. Among different classes of antibiotics, fluoroquinolones and tetracyclines have been detected in a higher range than any other classes of antibiotics. Ciprofloxacin (CIP) was detected as the highly used antibiotic and most commonly detected in the effluent samples from hospitals [1–6]. A CIP concentration of around 31,000 μg/L was measured in the wastewater from Hyderabad, India [3]. Ampicillin was also present in a higher concentration of around 104.2 μg/L in the Yamuna River [2].

Due to the discharge of these antibiotics, measures need to be taken for the wastewater treatment with easy and effective methods for the removal of antibiotics. As the wastewater treatment plants cannot remove these antibiotics, various mechanisms have been implemented and various treatments have been established for the removal of antibiotics from wastewater, such as UV and UV-H_2O_2 degradation [7, 9], degradation by advanced oxidation process, electron beam, UV and ozone [8], advanced treatment by UV radiation [10], adsorption by carbon-based materials viz. activated carbon, carbon nanotubes and carbon xerogel [11], adsorption by ordered mesoporous carbon CMK3 and bamboo based carbon [12], adsorption by sawdust [13], and so on. The above mechanisms are under thorough research, but among them, the adsorption process is regarded as the most reliable method for the removal of antibiotics because of its simple design parameters, low cost, and less production of toxic components. So, it is of great importance to make such an adsorbent that satisfies all the above criteria and can lead to a new innovation in the field of adsorption.

In this study, an attempt was made to remove CIP from wastewater by using a nanofiber sorptive membrane made from different polymers such as cellulose acetate (CA), polyethylene oxide (PEO), polyvinyl alcohol (PVA), and polycaprolactone (PCL). Different polymer solutions were made with different solvents, and electrospun nanofiber membrane (ENFM) was formed by electrospinning. Very fine ENFMs were formed with a size of <100 nm (approximately) and were used as adsorbent. Various other adsorbents have been reviewed for the removal of antibiotics in wastewater, but not much research has been done on the fabrication of sorptive nanofiber membranes and using them for treatment purposes [14]. To enhance the adsorption capacity, surface-functionalized sorptive ENFM was fabricated by impregnating montmorillonite (MMT) with the polymer.

2 Materials and Methods

2.1 Reagents

CA was obtained from Lobachemicals (acetyl content: 45%; MW: 30,000 g/mol), and PCL and PEO were obtained from Hi-Media. The solvents such as acetic acid (AA), acetone (Ac), chloroform (CF), methanol (MeOH), dimethyl formamide (DMF), *N,N*-methyl pyrrolidine (NNMP), trifluoroacetic acid (TFA), dichloromethane (DCM), and dimethyl acetamide (DMAc) were obtained from Merck having purity more than 90%. Ciprofloxacin hydrochloride (CIP) injection (200 mg/100 mL)—purity higher than 99.6% and molecular weight 385.8 g/mol—was obtained from a normal medical shop Frank Ross Pharmacy, Sakher Bazar, India. Montmorillonite (MMT) clay was obtained from the Geotechnical Department of IIEST Shibpur, India.

2.2 Preparation of Different Combinations for Fabrication of ENFM by Electrospinning

CA, PEO, and PCL were used with different solvent proportions to fabricate ENFM. The solution was kept at a constant stirring of 1000 rpm overnight (12 h) in all the combinations to ensure the proper mixing of the polymer solutions. The combinations are shown in Table 1. MMT was also used for impregnation with polymer solutions. Electrospinning was used for ENFM fabrication, as shown in Fig. 1, and parameters were also changed depending on the combination of polymers used. For electrospinning, ESPIN-NANO (M/s. Peco, India) was used. High-voltage power supply up to 40 kV DC was used for fabrication. ENFMs formed after electrospinning were used for the adsorption studies to determine the CIP removal from wastewater. The CIP concentration was taken as 10 mg/L, maintaining a pH of 5.0. The shaking speed was kept at 100 rpm, and temperature was maintained at 30 °C with varying adsorbent dosages. The concentration of CIP was evaluated by a double-beam UV–VIS spectrophotometer. The absorbance was checked at a wavelength of 263.6 nm.

Table 1 Combination of different polymers with CA

Sl. no.	Combinations	Solvents used
1	CA (5%) and PEO (0.5%)	AA:H_2O:DMF (4:2:1)
2	CA (5%) and PCL (1%)	CF:MeOH (1:1)
3	CA (5%) and PCL (2%)	Ac:DMF (2:1)
4	CA (5%) and PCL (2%) with Fe nanoparticles	Ac:DMF (2:1)
5	CA (5%) and PCL (2%) solvents	Ac:NNMP (2:1)
6	CA (10%) with different solvents	Ac:DMAc (2:1), Ac:DMAc:DCM (2:1:0.5)
7	CA (10%) ultrasound	Ac:DMAc (2:1)
8	CA (10%) and MMT (5%)	Ac:DMAc (2:1)

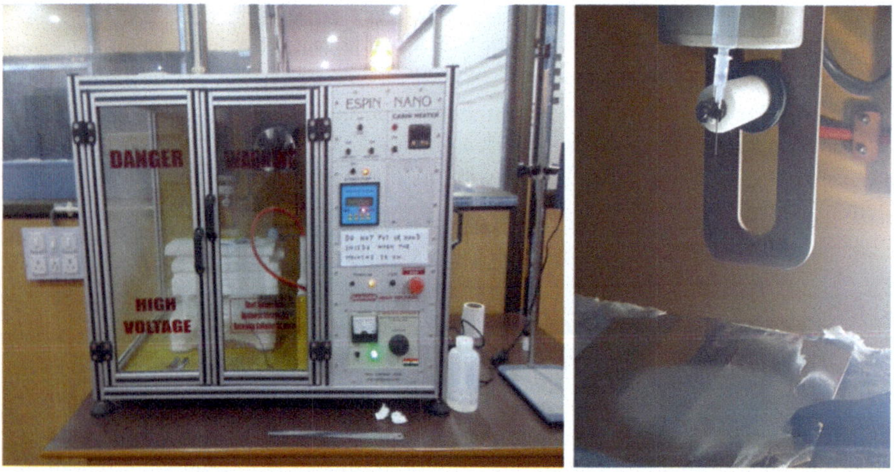

Fig. 1 Electrospinning of polymer solution

3 Results and Discussion

3.1 Different Combinations of Membrane Fabrication

Different combinations were attempted for CA and other polymers with different solvents, as shown in Table 1. The solution was fed at different flow rates using a syringe and at a different positive voltage at a collecting distance of 10 cm. Different types of ENFM fabricated are analyzed for CIP removal in the following sections.

3.1.1 Preparation of CA (5%) and PO (0.5%) ENFM

CA (5%) and PEO (0.5%) were mixed in different solvents with different proportions. AA:Water:DMAc at a different proportion of 4:2:1 and 2:1:0.5 and CF:MeOH in different proportions of 1:1 and 4:1 were used for membrane fabrication. The flow rates of 3 and 1.5 mL/h with a positive voltage of 15 and 12 kV, respectively, were taken as electrospinning parameters. ENFM, once formed, was crosslinked for 4 h with glutaraldehyde (15 mM) and 0.1 M HCl because PEO, being soluble in water, might come out from the membrane, hindering the adsorption process. The percentage removal of CIP obtained was about 5% for the first combination and about 17% for the second combination at a dosage of 3 g/L, respectively. Since the percentage removal was very low, different polymers were used in combination with varying solvent ratios.

3.1.2 Preparation of CA (5%) and PCL (1%) ENFM

CA (5%) and PCL (1%) were chosen as polymers with solvents CF and MeOH in the ratio of 1:1. The fabrication of the ENFM was hindered due to the less conducting nature of CF. The percentage removal of CIP was around 20 and 43% at an adsorbent dosage of 1 and 3 g/L, respectively.

3.1.3 Preparation of CA (5%) and PCL (2%) ENFM

CA (5%) and PCL (2%) were taken as a combination by changing the solvents to Ac: DMF (2:1). The membrane fabricated is shown in Fig. 2. The electrospinning parameters and adsorption parameters were the same as above. The percentage removal obtained was 15, 28, and 36% at an adsorbent dosage of 1, 3, and 5 g/L, respectively. Since this combination could not yield better removal of CIP, membrane modification by impregnating nanoparticles was also considered.

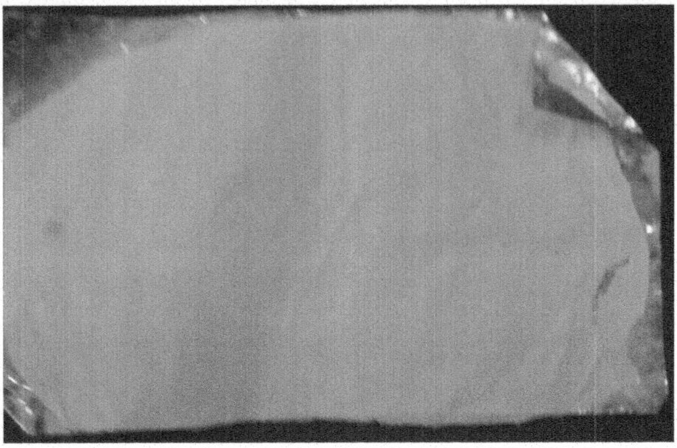

Fig. 2 Fabrication of ENFM with CA (5%) and PCL (2%) in solvents Ac: DMF (2:1)

3.1.4 Warning: Processing-instruction not allowed here!!!Preparation of CA (5%) and PCL (2%) ENFM with Solvents Ac:NNMP (2:1)

CA (5%) and PCL (2%) were used, and solvents were changed to Ac:NNMP (2:1). The electrospinning parameters were changed to a voltage of 20 kV, flow rate of 3 mL/h, and a collecting distance of 15 cm, respectively. ENFM formed was cut into small pieces for adsorption study, as shown in Fig. 3. The adsorbent dosages were taken as 1, 3, and 5 g/L. The adsorption parameters were the same as in the above experiments. However, the removal efficiency was less than 5%. So, this combination also did not produce the desired result, and hence, it was discarded.

3.1.5 Preparation of CA (10%) ENFM with Different Solvents

CA was mixed in solvents Ac:DMAc (2:1). The flow rate of 4 mL/h at a collecting distance of 15 cm and at a positive voltage of 20 kV were chosen as electrospinning parameters. The ENFM formed was Deacetylased by immersing in ethanol with the proportion of 0.1 M NaOH and ethanol 2 mL (Fig. 4). The adsorbent dosages were taken as 1, 3, and 5 g/L. The removal of CIP was found to be 56%.

In order to enhance the adsorption process and increase the removal of CIP, dichloromethane (DCM) was added to the above solvent. CA was dissolved in the solvents Ac:DMAc:DCM (2:1:0.5). The purpose of adding DCM was to fabricate a porous nanofiber membrane, which was possible due to the high vapor pressure of DCM. The adsorbent dosages were taken as 1, 3, and 5 g/L. However, this procedure resulted in a decrease of the adsorption process, and the percentage removal was lowered to 48%.

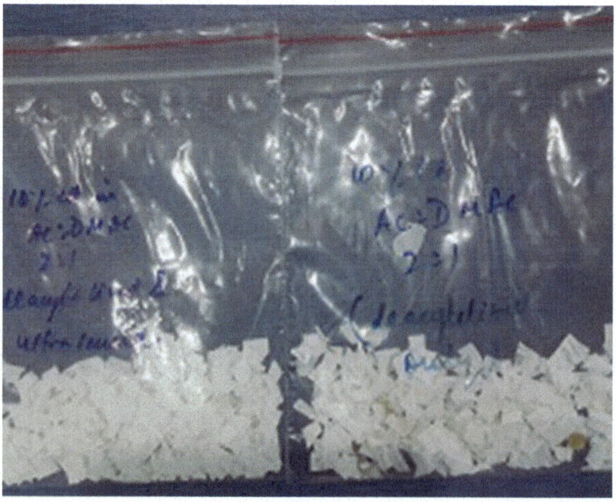

Fig. 3 Fabricated ENFM with CA(5%) and PCL(2%) in solvents Ac:NNMP (2:1) cut into pieces for adsorption study

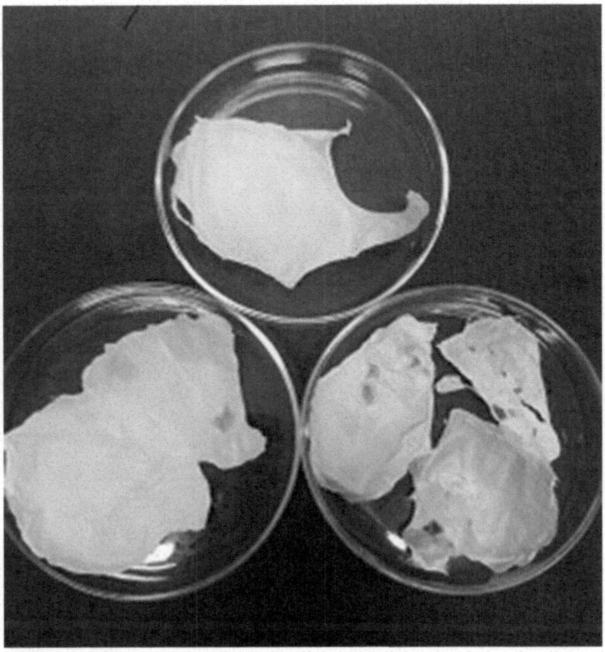

Fig. 4 Fabricated CA (10%) ENFM in solvents Ac:DMAc (2:1) after Deacetylation

3.1.6 Preparation of CA (10%) ENFM by Giving Ultrasound Mechanical Waves

Ultrasonic waves were passed through the membrane in order to increase the porosity of ENFM. The membrane was made as the above combination and ultrasound was given for 4–6 min, according to the literature review in order to increase the porosity of the ENFM. This experiment was conducted in triplicate but resulted in failure due to lower percentage removal of 20, 24, and 28% at an adsorbent dosage of 1, 3, and 5 g/L.

The lower CIP removal by CA-ENFM can be explained by the presence of unreacted functional groups in CA and its unsuitability for an adsorption process operating on the principle of affinity and compatibility [14]. Since the desired removal of CIP from synthetic wastewater was not achieved, surface functionalization was done to increase the adsorbing capacity of ENFM and the percentage removal of CIP.

3.1.7 Preparation of CA (5%) and PCL (2%) ENFM with Iron Nanoparticle Impregnation

CA (5%) and PCL (2%) were mixed in solvents Ac:DMF (2:1) to form ENFM. The iron nanoparticle was impregnated in the membrane to increase efficiency and enhance adsorption. The ENFM was immersed in $FeCl_3$ solution prepared by mixing $FeCl_3$ in ethylene glycol and adding ammonia solution to it as shown in Fig. 5a. The

Fig. 5 (**a**) Immersion of ENFM in the solution for impregnation of iron nanoparticle, (**b**) iron nanoparticle impregnated ENFM, (**c**) iron impregnated nanofiber membrane when dried

(a)

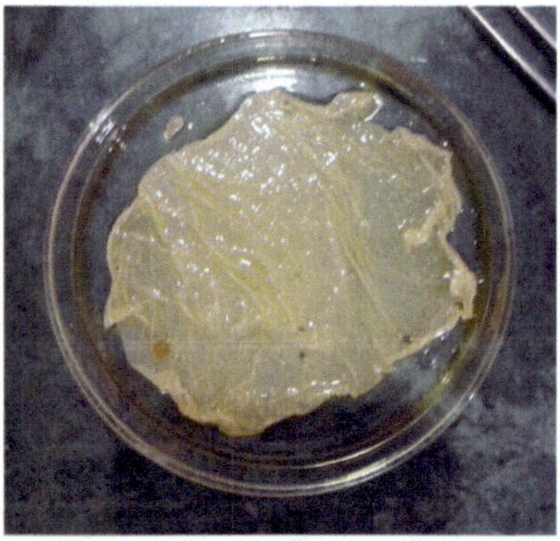

(b)

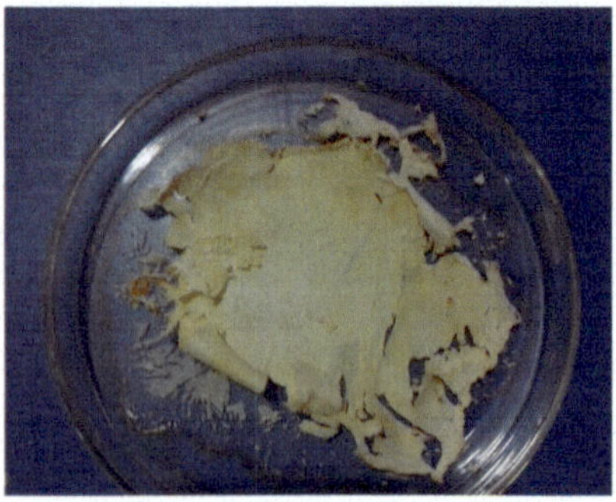

(c)

membrane was immersed in the solution of $FeCl_3$ for 2 h and then immersed in deionized water for 80 °C for 1 h in hot water bath (Fig. 5b). After 1 h the ENFM was taken out and then oven-dried at 60 °C to dry the membrane (Fig. 5c). Once the membrane was dried, it was used for adsorption process. The adsorbent dosages were taken as 1, 3, and 5 g/L, and the percentage removal of CIP was found to be 5, 16, and 24%, respectively.

The iron nanoparticles were impregnated in CA, but it did not achieve much removal of CIP. A case has been studied where clay was used with CA-Fe nanoparticles, which achieved better CIP removal due to increased surface functionalization [15]. According to the literature review, clay particles have been used for removal of CIP [16]. Fabrication of CA with MMT was reported earlier, but the membrane formed was not used for the treatment procedure [17]. By impregnating clay, the efficiency of ENFM can be increased. Hence, a trial for impregnating clay MMT as nanoparticle in cellulose acetate was done. Before using MMT for impregnation, it was undergone various treatments prior to its use. MMT was first ultrasound for 15 min and then washed with distilled water for 24 h to remove any impurities. It was then filtered and dried at 80 °C for 24 h. After drying in the oven, the MMT was crushed and ground into a fine powder for use in subsequent experiments.

3.1.8 Preparation of CA (10%) and MMT (5%) ENFM

CA (10%) with MMT (5%) was mixed in the solvents Ac:DMAc (2:1), and the ENFM was fabricated by electrospinning as shown in Fig. 6. The membrane formed was checked for adsorption study, and the removal efficiency of CIP was determined. The flow rate of 2 mL/h and a positive voltage of 20 kV were applied for electrospinning. The collecting distance of 12 cm was taken between the needle syringe and the collecting plate. ENFM formed was of good strength. The membrane was cut into smaller pieces in order to increase the surface area. An adsorption study was performed with adsorbent dosages of 1, 3, 5, 7, and 10 g/L. The CIP concentration was taken as 10 mg/L. The solution pH was conserved at 4.00 ± 0.05, and the time taken for adsorption was 5 h. The temperature was kept at 30 °C. The percentage removal was increased to 88% as shown in Fig. 7. Hence, after trial for different combinations and keeping in view of cost-effectiveness, the ENFM was formed with 5% CA and 15% MMT by electrospinning [18], and the ENFM was used for adsorption of ciprofloxacin from wastewater [19].

Field emission scanning electron microscope (FE-SEM; JSM-7610F JEOL) was used in characterization (morphological) of ENFM (Fig. 8). The average size of the nanofiber was around 100 nm. Energy-dispersive X-ray spectroscopy (EDS) (XMAX 50, OXFORD) was used to determine the composition of the ENFM where constituent elements were found to be oxygen (O; 40%), carbon (C; 28.5%), silicon (Si; 14.9%), aluminum (Al; 9.3%), iron (Fe; 2.68%), potassium (K; 1.89%), magnesium (Mg; 1.19%), zirconium (Zr; 0.68%), calcium (Ca; 0.34%), copper (Cu; 0.30%), and nickel (Ni; 0.15%).

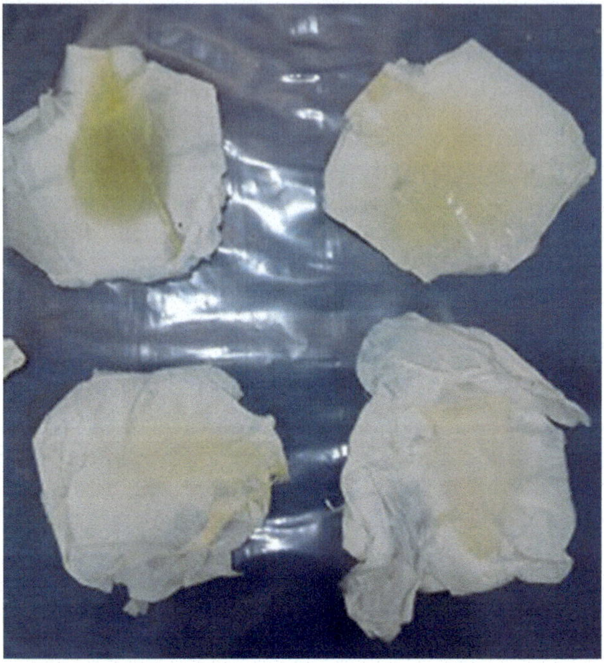

Fig. 6 Fabricated CA (10%) and MMT (5%) in solvents Ac:DMAc (2:1)

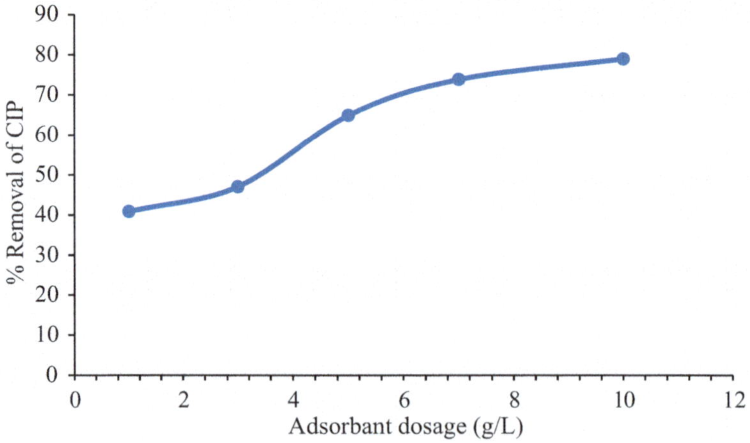

Fig. 7 Effect of CA-MMT ENFM dose on CIP removal (CIP: 10 mg/L, pH 6, temperature: 30 °C, shaker speed: 100 rpm, time of shaking:5 h)

Fig. 8 SEM analyses of CA (10%) and MMT (5%) in solvents Ac:DMAc (2:1)

4 Conclusion

Different polymers with various solvent ratios were used for the fabrication of the ENFM, but it did not achieve better removal of CIP. CA was combined with other polymers, such as PEO and PCL, at different solvent ratios for membrane fabrication. When CA and PEO were combined to form a membrane and used for dosage study, the percentage removal was around 13%, whereas when PCL replaced PEO, the percentage removal increased to 43%. CA was deacetylased and ultrasonicated, but the percentage removal of only 24% was achieved. The formed ENFM is not sufficiently suitable for CIP removal from wastewater due to the presence of nonfunctional groups and the reduced affinity of CIP molecules for the membrane surface. Hence, surface functionalization is required to increase the affinity and interaction of ENFM toward the adsorption of CIP from wastewater. MMT was impregnated with CA and ENFM was fabricated. This combination was found to achieve better removal of CIP as compared to other fabricated ENFMs. The CIP removal was increased to 88% at a dosage of 5 g/L. Hence, we can conclude that MMT impregnated with CA in a suitable combination yields higher removal efficiency of CIP.

References

1. R. Gothwal, T. Shashidhar, Antibiotic pollution in the environment: a review. Clean Soil Air Water **43**(4), 479–489 (2015)
2. P.K. Mutiyar, A.K. Mittal, Occurrences and fate of selected human antibiotics in influents and effluents of sewage treatment plant and effluent-receiving river Yamuna in Delhi (India). Environ. Monit. Assess. **186**(1), 541–557 (2014)
3. V. Diwan, A.J. Tamhankar, M. Aggarwal, S. Sen, R.K. Khandal, C.S. Lundborg, Detection of antibiotics in hospital effluents in India. Curr. Sci. **97**(12), 1752–1755 (2009)

4. V. Diwan, A.J. Tamhankar, R.K. Khandal, S. Sen, M. Aggarwal, Y. Marothi, C.S. Lundborg, Antibiotics and antibiotic-resistant bacteria in waters associated with a hospital in Ujjain, India. BMC Public Health **10**, 414 (2010)
5. P.K. Mutiyar, A.K. Mittal, Risk assessment of antibiotic residues in different water matrices in India: key issues and challenges. Environ. Sci. Pollut. Res. **21**, 7723–7736 (2014)
6. Y. Luo, W. Guo, H.H. Ngo, L.D. Nghiem, F.I. Hai, J. Zhang, X.C. Wang, A review on the occurrence of micropollutants in the aquatic environment and their fate and removal during wastewater treatment. Sci. Total Environ. **473–474**, 619–641 (2014)
7. A. Lopez, A. Bozzi, G. Mascolo, J. Kiwi, Kinetic investigation on UV and UV/H2O2 degradations of pharmaceutical intermediates in aqueous solution. J. Photochem. Photobiol. A Chem. **156**(1–3), 121–126 (2003)
8. T.H. Kim, S.D. Kim, H.Y. Kim, S.J. Lim, M. Lee, S. Yu, Degradation and toxicity assessment of sulfamethoxazole and chlortetracycline using electron beam, ozone and UV. J. Hazard. Mater. **227–228**, 237–242 (2012)
9. Y.J. Jung, W.G. Kim, Y. Yoon, J.W. Kang, Y.M. Hong, H.W. Kim, Removal of amoxicillin by UV and UV/H2O2 processes. Sci. Total Environ. **420**, 160–167 (2012)
10. L. Rizzo, A. Fiorentino, A. Anselmo, Advanced treatment of urban wastewater by UV radiation: effect on antibiotics and antibiotic-resistant E. coli strains. Chemosphere **92**(2), 171–176 (2013)
11. S.A.C. Carabineiro, T. Thavorn-Amornsri, M.F.R. Pereira, J.L. Figueiredo, Adsorption of ciprofloxacin on surface-modified carbon materials. Water Res. **45**(2011), 4583–4459 (2011)
12. Z. Liu, H. Dai, Adsorption behavior and mechanisms of ciprofloxacin from aqueous solution by ordered mesoporous carbon and bamboo-based carbon. J. Colloid Interface Sci. **460**(2015), 349–360 (2015)
13. S.K. Bajpai, M. Bajpai, N. Rai, Sorptive removal of ciprofloxacin hydrochloride from simulated wastewater using sawdust: Kinetic study and effect of pH. Water SA **38 No**, 673–682 (2012)
14. M.D. Islam, F.J. Uddin, T.U. Rashid, M. Shahruzzaman, Cellulose acetate-based membrane for wastewater treatment—A state-of-the-art review. Mater. Adv. **4**, 4054–4102 (2023)
15. G. Gopal, C. Natarajan, A. Mukherjee, Adsorptive removal of fluoroquinolone antibiotics using green synthesized and highly efficient Fe clay cellulose-acrylamide beads. Environ. Technol. Innov. **28**, 102783 (2022)
16. M.E.R. Jalil, M. Baschini, K. Sapag, Influence of pH and antibiotic solubility on the removal of ciprofloxacin from aqueous media using montmorillonite. Appl. Clay Sci. **114**, 69–76 (2015)
17. Q. Wu, Z. Li, H. Hong, K. Yin, L. Tie, Adsorption and intercalation of ciprofloxacin on montmorillonite. Appl. Clay Sci. **50**, 204–211 (2010)
18. S. Das, A. Adak, A. Barui, Fabrication of Montmorillonite impregnated cellulose acetate nanofiber membrane and its use for adsorption of Ciprofloxacin. J. Indian Chem. Soc. **96**, 478–479 (2019)
19. S. Das, A. Adak, A. Barui, Montmorillonite impregnated electrospun cellulose acetate nanofiber sorptive membrane for ciprofloxacin removal from wastewater. J. Water Process Eng. **37**(2020), 101497 (2020)

Water Quality Index in Arsenic Exposed Areas of West Bengal: Sustainability Toward Proactive Management for Future Risk

Sharmistha Majumder, Antara Das, Madhurima Joardar, and Tarit Roychowdhury

Abbreviations

ARP	Arsenic removal plants
BW	Body weight
CR	Carcinogenic risk
CSF	Cancer slope factor
DT	Deep tube well
EC	Electrical conductivity
EDI	Estimated daily intake
HQ	Hazard quotient
NCR	Noncarcinogenic risk
PSW	Pipeline supplied water
R_fD	Oral reference dose
SAMOE	Severity adjusted margin of exposure
SI	Subindex value
ST	Shallow tube well
TA	Total alkalinity
TDI	Tolerable daily intake
TDS	Total dissolved solid
TH	Total hardness
WQI	Water Quality Index

S. Majumder · A. Das · M. Joardar · T. Roychowdhury (✉)
School of Environmental Studies, Jadavpur University, Jadavpur, Kolkata, India

D. Mazumder (ed.), *Sustainable Advanced Technologies for Environmental Management*, Springer Proceedings in Earth and Environmental Sciences,
https://doi.org/10.1007/978-3-031-64006-3_3

1 Introduction

Arsenic (As), a perilous metalloid known for its carcinogenic properties, is predominantly found in Earth's crust within varying concentrations across diverse geographical regions. Its mobilization from the Earth's surface into groundwater, primarily facilitated by natural geological and hydrological processes, has led to widespread As contamination in groundwater [1]. Among the severely affected areas, South-East Asian countries, particularly West Bengal in India, have faced escalating challenges due to the contamination of groundwater by As and came to the point of attention as for the whole world [2]. Human exposure to As occurs predominantly through the consumption of contaminated drinking water, posing severe health hazards to affected populations [3]. The first detection of As contamination in groundwater from West Bengal traces back to the year 1984 [4], since then, over the past few decades, groundwater had become the greatest source of As contamination concerning the Indo-Bangladesh Gangetic delta. The upper alluvial plain of West Bengal exhibits a complex geological structure characterized by aquifers having three interlinked systems. These aquifers are stratified into distinct depths: (1) the shallowest aquifer, reaching depths of 12–15 m beneath the surface, (2) the intermediary aquifer extending to depths of 35–45 m, and (3) the deeper aquifer ranging from 70 to 150 m. Notably, the aquifers' depths vary across different districts of West Bengal, emphasizing the heterogeneous nature of the subsurface hydrogeology [5]. Following the implementation of domestic hand pumps, a consequential trend has emerged as the progressive escalation of human exposure to As through the utilization of groundwater sources. With each passing day, this trajectory intensifies the intricate interplay between anthropogenic activities and the inadvertent consequences associated with the extraction of groundwater, accentuating the risk of As exposure for the human population [6].

In addressing the perilous impacts of As contamination in groundwater mainly in shallow tube wells (STs), extensive research endeavors have been undertaken, focusing on the exploration of alternative secure sources. The primary objective has been to assess their feasibility in ensuring widespread access to safe water and thereby to fulfill the aims of sustainable development goals [7]. Simultaneously, the West Bengal government has initiated strategic interventions, such as the distribution of Ganges treated water in numerous endemic rural areas through a well-organized pipeline system, i.e., pipeline supplied water (PSW). Government also plants deep tube wells (DTs) in many places for serving as an integral component of community-level initiatives aimed at providing potable water to rural populations. Apart from this, several organizations are implemented various As removal plants (ARPs) by the use of different technologies. All of the mentioned procedures are very costly, even not able to provide the water all over throughout the rural areas and have limitations with less suitability for use in rural households of West Bengal where the situation is dominated by high poverty level [1].

This present work is a follow-up study to analyze the current contamination scenario and to evaluate the awareness concern regarding As toxicity among local populace after several previous studies and mitigation measures implemented in those endemic areas over time. Thus, this study focuses on the present scenario of As contamination in domestic level STs in different studied areas. Apart from that, it describes the As contamination level in alternative sources, which had been developed as mitigation plans by the government and different organizations and asses the quality of available water sources through Water Quality Index (WQI). Most importantly, this study contributes to understand the intricate dynamics of the impact of awareness campaigns on individuals. It delves into a comprehensive evaluation of the pivotal role played by governing authorities in upholding the quality of drinking water sources, particularly in the aftermath of multiple studies that have underscored their inefficacies. Lastly, this chapter also has few recommendations about some sustainable, eco-friendly, low-cost approaches by which As contamination could be eliminated in future. It might help to create more awareness among the rural population about the risk of consuming As through available drinking water sources and can create a concern to the responsible authorities to implement new approaches. Hence, this kind of studies are much required to knock the safety door toward sustainability and better management.

2 Materials and Methods

2.1 Study Area

Four distinct areas from three district namely North 24 Parganas (N 24 Pgs), Murshidabad, and Nadia in West Bengal were selected for this study. Specifically, the areas include Deganga and Gaighata from N 24 Pgs, Raninagar-II in the Murshidabad district, and all blocks within the Nadia district. A part of the Gangetic delta belongs to North 24 Parganas and Nadia, which located at the eastern side of the Hooghly River, whereas, Murshidabad is situated at the eastern side of the Bhagirathi River and are reportedly been found to be severely polluted with As [2, 8, 9].

2.2 Sample Collection and Preservation

A total of 580 samples of available drinking water (480 samples in general and 100 random samples in duplicates of those 480) were collected. The collected samples were carefully preserved in 50 mL containers with 0.1% concentrated nitric acid that serve as a preservative for subsequent analysis of As, calcium (Ca^{2+}) and iron (Fe). Furthermore, 250 mL containers were employed without any preservatives for the estimation of various other parameters for water quality. All the samples were kept at a temperature of 4 °C. Before sampling, meticulous preparation was undertaken to ensure the cleanliness and reliability of the collected samples. The sample containers were dipped in a 1:1 hydrochloric acid (HCl) solution for a complete day, followed by thorough cleaning with deionized water and washing with the sample's own filtrates. For getting electrically stable samples during collection, the STs and DTs were kept running for 10–15 min prior to sample collection. The detailed procedure was given elsewhere [10].

2.3 Sample Analysis

Various physicochemical parameters, i.e., pH, total alkalinity (TA), total hardness (TH), chloride (Cl^-), electrical conductivity (EC), total dissolved solids (TDS), Ca^{2+}, magnesium (Mg^{2+}), fluoride (F^-), Fe, As, nitrate (NO_3^-), and sulphate (SO_4^{2-}), have been assessed using established methods. The pH of the water samples was determined using a pH meter manufactured by Toshcon Industries Pvt. Ltd. Measurement of EC was accomplished through an auto digital conductivity meter (model- LT-16) from Labtronics and TDS was quantified gravimetrically. Fluoride and NO_3^- concentration in water samples were measured using Ion Selective Electrode (Thermo Scientific, Orion Star A214) combined with their respective electrode. Total hardness, TA, and Cl^- ions were assessed via titration methods, while Fe and SO_4^{2-} concentrations were determined using a UV–Vis spectrophotometer (Thermo Scientific, Orion Aquamate, 8000 from USA) and Ca^{2+} using Flame Photometer (HPG System, model number G-301 from Chandigarh, India). Arsenic in collected water samples was determined directly without any digestion. The HG-AAS method was employed for As estimation, utilizing a Varian Model AA140 Atomic Absorption Spectrophotometer (USA) coupled with Vapor Generation Accessory (VGA-77, Agilent Technologies, Malaysia). Detailed information of water sample analysis has been provided elsewhere [9].

2.4 Chemicals and Reagents

Chemicals of analytical grade were used in this study. The entirety of the analytical procedures was done using double-distilled water. For the estimation of As content using the HG-AAS technique, high-quality reagents such as 0.6% $NaBH_4$ (in 0.5% NaOH) and 5–10 M HCl sourced from Merck, Mumbai, India, were utilized. During sample preparation, concentrated HCl (8%) and 10% of KI solution were added in respective to the specific volume of water sample. The combined solution was then left for 45 min to settle before being tested for As. Detailed information about compounds and reagents was described elsewhere [11].

2.5 Data Analysis

2.5.1 Water Quality Index (WQI)

The WQI offers a comprehensive framework that encapsulates the holistic water quality assessment [12]. Parameters within this index are assigned specific weights (w_i) based on their significance in contributing to the overall quality of water intended for drinking purposes. The determination of relative weights is meticulously calculated as

$$\text{Relative weight } (W_i) = w_i / \sum_{i=1}^{n} w_i \tag{1}$$

where w_i denotes the weight assigned to each parameter, where n represents the total number of parameters involved in the assessment.

Thereafter, one quality rating scale (q_i) is given for each parameter and calculated with the help of the following equation:

$$q_i = (C_i / S_i) \times 100 \tag{2}$$

where C_i stands for concentration obtained for each parameter; S_i denotes the drinking water standard values.

WQI values were further obtained through calculating the subindex value (SI) and the equations were as follows [12]:

$$\text{SI}_i = W_i \times q_i \tag{3}$$

$$\text{WQI} = \sum \text{SI}_i \tag{4}$$

where the term SI_i refers to the subindex assigned to the ith parameter.

The calculated WQI values underwent categorization into distinct classes for interpretative clarity. Specifically, the classifications include: WQI $<$ 50, indicating excellent water quality; 50–100, denoting good water quality; 100–200, signifying poor water quality; 200–300, indicating very poor water quality; and WQI $>$ 300, suggesting water unsuitable for drinking [13].

2.5.2 Health Risk Assessment

2.5.2.1 Risk Assessment for Chemical Risk Characterization

The Swedish National Food Agency has developed a crucial analytical tool termed the "Risk thermometer," also referred to as the severity adjusted margin of exposure (SAMOE), to effectively categorize the potential human health risks associated with chemically hazardous components [14]. This sophisticated metric plays a pivotal role in assessing the toxicity exposure levels of individuals by comparing them with the tolerable daily intake (TDI), which serves as the health-based reference value for specific toxicants. The following Eq. (5) was implied to calculate the various risk levels of exposure among the people from different areas caused by consumption of drinking water contaminated with As [9]:

$$\mathrm{SAMOE} = \mathrm{TDI}/(\mathrm{AF_{BMR}} \times \mathrm{AF} \times \mathrm{SF} \times E) \tag{5}$$

where TDI is 3.0 μg/kg BW/day for As; AF_{BMR} designates the nonlinear range of the toxicant, defined as 1/10, with BMR representing the benchmark response; AF represents the assessment factor, set at 10; SF signifies the severity factor, fixed at 100; and *E* represents the diverse exposure factors associated with the chemical under consideration.

2.5.2.2 Carcinogenic and Noncarcinogenic Risk

Carcinogenic risk (CR) and noncarcinogenic risk (NCR) were calculated with the help of estimated daily intake (EDI) of As through available drinking water, cancer slope factor (CSF) and oral reference dose (R_fD). The EDI for three different studied groups (adult male, adult female, and children) was calculated [15] as:

$$\mathrm{EDI} = (C/\mathrm{BW}) \times \mathrm{IR} \tag{6}$$

where EDI is the estimated daily intake of As through the consumption of drinking water (mg/kg BW/day); *C* is the concentration of As in drinking water (mg/L); BW is the body weight (kg); IR is the ingestion rate of water (L/day).

Thereafter, CR was calculated as

$$\mathrm{CR} = \mathrm{EDI} \times \mathrm{CSF} \tag{7}$$

The NCR in terms of hazard quotient (HQ) was calculated as

$$\mathrm{HQ} = \mathrm{EDI}/\mathrm{R_fD} \tag{8}$$

where CSF = 1.5 per mg/kg BW/day and $R_fD = 3 \times 10^{-4}$ mg/kg BW/day [16].

Cancer risks beyond 1×10^{-4} are unacceptable [17], whereas $HQ > 1$ suggested the exposure to be harmful, and the threat of rising the noncarcinogenic effects increasing with higher HQ value [16].

2.6 Quality Control and Quality Assurance

To ensure the precision and reliability of the acquired data, a meticulous approach was adopted. Following the analysis of every ten collected water samples, standard As samples were systematically introduced to validate the accuracy of the results obtained. Furthermore, routine blank measurements and the analysis of spiked samples were incorporated into the analytical process. These additional measures served as essential quality control mechanisms, systematically verifying the reliability, accuracy, and overall quality of the analytical data [18]. Analysis of the samples were also performed in duplicates (±5% variation) and the mean value was used for calculation.

2.7 Statistical Analysis

We used statistical software Excel 2019 (Microsoft Office) to calculate descriptive statistics and to prepare required figures for data interpretation.

3 Results and Discussion

3.1 Arsenic Contamination Scenario in Available Drinking Water Sources

Most of the rural populations from the study areas are still reliant on domestic level STs. All of the study regions do not have access to the water supplied from the government supplied pipeline water service, DTs, or any other water treatment plants throughout all over the area. Therefore, in this study, DTs from Deganga and Gaighata, ARPs from Gaighata and government PSW, water from ARP, RO, and DW in Raninagar-II have been considered as an alternative source of water. No water sample from alternative sources have been collected from Nadia district as the randomly selected households mostly used water from the STs. The present As concentration in STs from As exposed studied areas showed a considerable risk to the population using these for drinking purposes. Table 1a shows the concentration of As in domestic level STs from three different studied areas. The mean As content in domestic level STs from Deganga, Gaighata, and Raninagar-II were found to be 580 ± 663 μg/L ($n = 62$; range: 61–2301 μg/L), 132 ± 48.2 μg/L ($n = 80$; range: 19.4–202 μg/L), and 83.5 ± 151 μg/L ($n = 121$; range: 4.45–857 μg/L), respectively. The mean As content in all the available water sources reported from Nadia district was 21.7 ± 23.4 μg/L ($n = 105$; range: <3–205 μg/L). Furthermore, the mean As concentration of all water samples from STs exceeds the WHO recommended value of As by 58, 13, 8, and 2 times higher in respect to Deganga, Gaighata, Raninagar-II,

Table 1a Arsenic concentration (μg/L) in domestic level shallow tube well from the studied areas

Study area				
District	North 24 Parganas		Murshidabad	Nadia
Block	Deganga	Gaighata	Raninagar-II	17 blocks
Number of samples (*n*)	62	80	121	105
Mean	580	132	83.5	21.7
SD[a]	663	48.2	151	23.4
Range	61–2301	19.4–202	4.45–857	<3–205

[a]*SD* standard deviation

Table 1b Percentage (%) of shallow tube well with different arsenic concentration range (μg/L)

		Arsenic concentration (μg/L)				
Study area	Number of samples (*n*)	<10 μg/L (%)	>10 μg/L (%)	10–50 μg/L (%)	>50 μg/L (%)	Highest concentration (μg/L)
Deganga	62	–	62 (100%)	–	62 (100%)	2301
Gaighata	80	–	80 (100%)	46 (57.5%)	34 (42.5%)	202
Raninagar-II	121	34 (28.1%)	87 (71.9%)	46 (38%)	41 (33.9%)	857
Nadia	105	51 (48.6%)	54 (51.4%)	34 (32.4%)	20 (19%)	205

and Nadia. Additionally, Table 1b shows the distribution of As in these STs from each of the studied areas, where the high As percentage (%) was more pronounced in Deganga, i.e., all samples ($n = 62$) with As concentration exceeding 50 μg/L and maximum As concentration was 2301 μg/L. Out of 80 groundwater samples from Gaighata block, all the samples (100%) had As concentration greater than 10 μg/L and among them 34 samples (42.5%) had As concentration more than 50 μg/L. From Raninagar-II, out of 121 groundwater samples, 34 samples (28.1%) have As concentration within the safe range, i.e., <10 μg/L and 87 samples (71.9%) with As content >10 μg/L. Apart from these, in Nadia district, 51.4% collected water samples have been found with As concentration higher than the safe limit.

The DTs in Deganga and Gaighata and ARPs from Gaighata showed an average of 38.4 ± 39.9 μg/L (range: <3–162 μg/L; $n = 30$), 34.6 ± 22.8 μg/L (range: <3–71.2 μg/L; $n = 35$), and 48.3 ± 26.1 μg/L (range: <3–98 μg/L; $n = 10$) As, respectively (Fig. 1a). About 33.3% and 31.4% of the analyzed water samples from DTs and 90% of ARPs have As concentration above permissible limit for Deganga and Gaighata respectively (Fig. 1b). Only one ARP (Surface water treatment plant) situated in Madhusudankati from Gaighata had been found to produce treated water with As concentration below the permissible range. Likewise, PSW, ARP, and RO water samples from Raninagar-II have been found with a mean As concentration of 22.3 ± 16.2 μg/L (range: <3–46.7 μg/L; $n = 15$), 15.3 ± 17.1 μg/L (range: <3–54.4 μg/L; $n = 14$), and 7.95 ± 4.36 μg/L (range: <3–18.5 μg/L; $n = 5$),

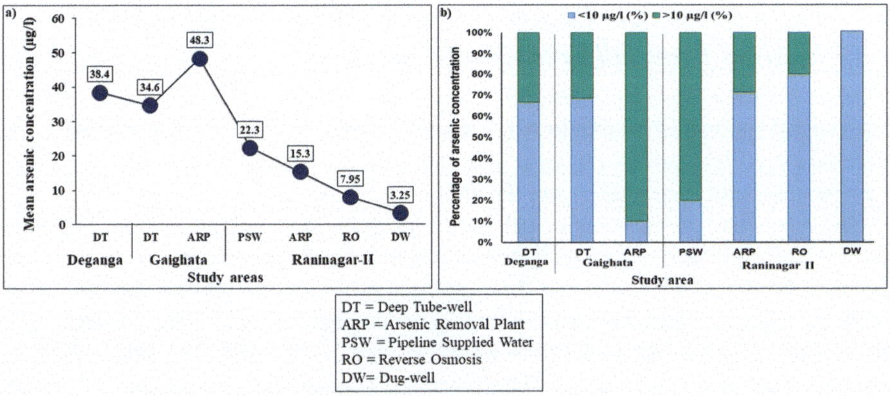

Fig. 1 (**a**) Variation in arsenic concentrations among alternative water sources. (**b**) Percentage of drinking water samples with respective to the acceptable limit of arsenic

respectively (Fig. 1a). Only about 20% of the analyzed samples from PSW were found with As concentration below permissible limit; therefore, it is not a safe alternative source for drinking water, while about 71.4% and 80% of the evaluated water samples from ARP and RO were found to be safe for consumption (Fig. 1b). Meanwhile, similar to the previous study, all the DW water samples are found to be safe from As contamination [18]. However, DW water is very prone to be polluted with anthropological activities and thus need precautions and treatment before consumption and this study did not consider harmful microbes and other pollutants. Therefore, among all these alternative sources from these study areas, treated water from RO plants was apparently safer to use. Another study found that groundwater samples from Deganga block is severely contaminated with As and had As level more than 10 µg/L in around 57% of total collected water samples [19]. Similarly, in our previous study, the groundwater from Deganga and Gaighata blocks had been found to contaminated with As [20]. Hand tube well water samples from Sagarpara (86.2%) and Raninagar-II blocks (67.6%) in the Murshidabad district were also detected with As content over 10 µg/L [21]. In our previous study, most of the groundwater samples from Murshidabad and Nadia districts were found to be contaminated with As concentration above the permissible limit [11, 18]. Despite the awareness and evaluation of the contamination scenario by these studies, people continue to drink contaminated water; thus, this kind of study is further required for continuous evaluation of the drinking water quality and creating more awareness to the local populace about the severity of As consumption.

3.2 Physicochemical Parameters of Available Drinking Water Sources from Different Study Areas

The overall quality of available water samples from four respective studied areas were analyzed by various physical and chemical parameters. The mean values of these water quality parameters are detailed in Table 2.

The respective mean value of pH for all available water samples from Deganga ($n = 92$), Gaighata ($n = 125$), Raninagar-II ($n = 158$), and Nadia ($n = 105$) are found to be 7.5, 7.45, 7.42, and 7.51, respectively, which signifies that collected available water samples from all the studied areas were basic in nature. Furthermore, the average values of total alkalinity were obtained as 458, 466, 271, and 378 mg/L for the respective study areas. These values surpass the recommended Indian Standard limit of 200 mg/L. The pH variability in groundwater is attributed to the mineral composition of the adjacent rocks near the water table and the constituents within the lithological unit of the groundwater recharge zones [22]. The alkalinity in the groundwater is presumed to originate from the presence of cationic species, contributing to the elevated total alkalinity levels observed in the samples from the study areas [23, 24].

The mean value of TH, i.e., 369, 435, 256, and 332 mg/L for respective studied areas were showed the value above permissible limit (200 mg mg/L). The higher mean value of TH in water samples signifies that the feature of groundwater is relatively hard. Groundwater hardness is primarily attributed to the presence of divalent cations. However, it is crucial to recognize that the abundance of cations in groundwater serves as more than a mere determinant of water hardness or softness; it also plays a pivotal role in influencing TDS and EC levels. The intricate interplay between these factors underscores the multifaceted nature of groundwater

Table 2 Different water quality parameters of all available drinking water sources from the studied areas

		Mean			
Parameters	Unit	Deganga	Gaighata	Raninagar-II	Nadia
pH	–	7.5	7.45	7.42	7.51
TDS	(mg/L)	2154	124	363	528
EC	(μS/cm)	108	954	728	785
Total hardness	(mg/L)	369	435	256	332
Total alkalinity	(mg/L)	458	466	271	378
Chloride	(mg/L)	58.9	86.1	39.2	77.2
Calcium	(mg/L)	84.7	97.8	2.13	54.4
Magnesium	(mg/L)	31.8	44.7	1.73	45.3
Fluoride	(mg/L)	0.17	0.26	0.31	0.17
Iron	(mg/L)	4.23	4.49	11.4	3.11
Arsenic	(μg/L)	403	96.8	67.7	21.7
Nitrate	(mg/L)	1.89	3.51	5.21	9.81
Sulphate	(mg/L)	8.52	27.4	57.6	27.3

composition, where the presence and interactions of cations extend beyond categorizing water solely in terms of hardness, contributing significantly to the broader physicochemical characteristics of the water body [25]. Electric conductivity and TDS are two such parameters that help to understand the total concentration of soluble salts in water. TDS defines the number of inorganic salts in water, whereas EC measures electrical current in water. These parameters have a strong relation between themselves and can be expressed by the following equation:

$$\mathrm{TDS}\left(\frac{\mathrm{mg}}{\mathrm{l}}\right) = k \times \mathrm{EC}\left(\frac{\mu\mathrm{S}}{\mathrm{cm}}\right) \tag{9}$$

where the factor "k" represents the TDS/EC ratio, which mostly ranges between 0.5 and 1.00 for freshwater samples [26].

However, the obtained mean for EC and TDS were 108, 954, 728, and 785 μS/cm and 2154, 124, 363, and 528 mg/L, respectively. The ratio between EC and TDS of water samples from Raninagar-II and Nadia were found within this range, whereas weak correlation was found for Deganga and Gaighata. Similarly, in a South African study, this ratio was fall between 0.25 and 1.34 [27]. The relationship between TDS and EC in natural waters is often complex and nonlinear due to the diverse composition of the natural aqueous systems [27]. TDS encompasses dissolved substances, including ions and undissociated species. EC measures a solution's ability to conduct electricity, influenced by ion concentration and activity. The interaction between specific ions, ion pairing, and complexation further complicates the relationship and different electrolytes affect the EC in a different fashion [28]. These factors lead to nonlinear trends where changes in TDS may not proportionally affect EC. Understanding these complexities is crucial for accurate interpretation of water quality and conductivity data in environmental monitoring and analysis. The mean TDS values observed in the drinking water samples from Deganga and Nadia were above the limit prescribed by the Bureau of Indian Standards (BIS) (500 mg/L), which signifies the presence of inorganic salts and minerals in the studied areas. Whereas for Gaighata and Raninagar-II, the TDS value observed is within the permissible limit, which signifies that the presence of inorganic salts and minerals is lesser in this studied area.

Moreover, the mean Cl^- values, i.e., 58.9 (Deganga), 86.1 (Gaighata), 39.2 (Raninagar-II), and 77.2 mg/L (Nadia), are also within the recommended value (250 mg/L). The presence of calcium, magnesium, and Fe ions denotes the hardness of drinking water. The mean Ca^{2+} ion concentration is higher than mean Mg^{2+} ion in all available drinking water sources. However, very little amount is found in case of drinking water samples from Raninagar-II (Ca^{2+}: 2.13 mg/L; Mg^{2+}: 1.73 mg/L). The observed mean Fe and As concentrations from all the studied areas (4.23 and 0.403, 4.49 and 0.097, 11.4 and 0.068, 3.11 and 0.022 mg/L, respectively) are greater than the recommended values (0.3 and 0.01 mg/L, respectively), indicating that the groundwater is Fe-enriched and contaminated with As. The observed mean concentrations of NO^{3-}, F^-, and SO_4^{2-} ions in drinking water samples are found to be 1.89, 0.17, and 8.52 mg/L for Deganga, 3.51, 0.26, and 27.4 mg/L for Gaighata,

5.21, 0.31, and 57.6 mg/L for Raninagar-II, and 9.81, 0.17, and 27.3 mg/L for Nadia. These observed concentrations of NO^{3-}, F^-, and SO_4^{2-} ions for all available drinking water samples of the studied Deganga block are lower than the recommended values of 45, 1.5, and 200 mg/L, respectively. Overall, the mean anionic concentration of the water samples from the studied areas showed the order of the ions as $Cl^- > SO_4^{2-} > NO^{3-}$, which justifies that the water quality of Deganga block is basic in nature.

3.3 *Evaluation of WQI*

A comprehensive assessment of water quality for drinking purposes involves the allocation of weight (w_i) to different studied parameters, each reflective of their respective importance. This weight allocation signifies the impact of these parameters on the overall quality of drinking water. Table 3 shows a detailed of these parameters and their respective W_i.

The estimated quality (in percentage) of all the available waters with respective to their calculated categories were depicted in Fig. 2.

All available water from STs from Deganga, approximately 80% in Gaighata, were found to be unsuitable for drinking purpose. Whereas in Raninagar-II and Nadia, 3% and 7% were designated as unsuitable: 21% and 19% as very poor, 33% and 37% as poor, and only 38% and 37% were found to be in good quality. No water sample from STs was found in excellent quality in case of Nadia; however, only 5% of collected water samples from Raninagar-II can be labeled as excellent. Similarly, DTs are also not safe as only 27% and 40% provide a good quality of drinking water.

Table 3 Physicochemical parameters with their allotted relative weight

Parameters	Unit	Weight (w_i)	Relative weight (W_i)	Standards
pH	–	3	0.088235294	6.5–8.5
TDS	(mg/L)	2	0.058823529	500–2000
Fluoride	(mg/L)	4	0.117647059	1.5
Chloride	(mg/L)	3	0.088235294	250–1000
Nitrate	(mg/L)	4	0.117647059	45
Sulphate	(mg/L)	2	0.058823529	200–400
Total hardness	(mg/L)	2	0.058823529	200–600
Total alkalinity	(mg/L)	2	0.058823529	200–600
Iron	(mg/L)	3	0.088235294	0.3
Calcium	(mg/L)	2	0.058823529	75–200
Magnesium	(mg/L)	2	0.058823529	30–150
Arsenic	(μg/L)	5	0.147058824	10
Sum (∑)		34	1	

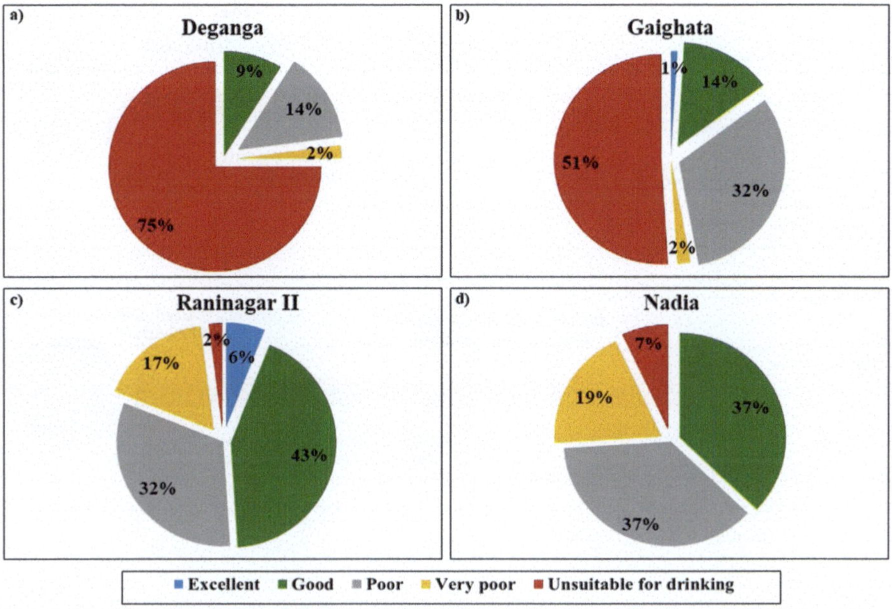

Fig. 2 Classification of all the available water sources (in percentage) based on WQI value for (**a**) Deganga, (**b**) Gaighata, (**c**) Raninagar-II, and (**d**) Nadia

Among other alternative sources out of five RO treated water, one was found to be excellent and four with good quality, whereas water from only 8 PSW out of 15 were found to be in good quality. Moreover, water from 50% ARPs in Gaighata and 64% ARPs in Raninagar-II can be used for drinking purpose. As a whole, approximately 91%, 85%, 51%, and 63% of water samples from Deganga, Gaighata, Raninagar-II, and Nadia, respectively, cannot be recommended for drinking purpose. This result is somewhat similar to previous studies, and the situation still has not changed after so many years [11, 18, 20, 29].

3.4 Risk Assessment

3.4.1 Risk Categorization of Available Drinking Water Sources Through Risk Thermometer

Risk of As exposure through the available drinking water ingestion by the population from particular areas has been estimated using risk thermometer (Fig. 3).

SAMOE value and risk level of drinking water sources from different studied areas have been shown in Table 4. It is observed that As exposure in adult (male and female) and children was high from drinking water from domestic level STs (class 5, SAMOE: 0.00062, 0.00071, and 0.00078 in Deganga; 0.00273, 0.00313, and

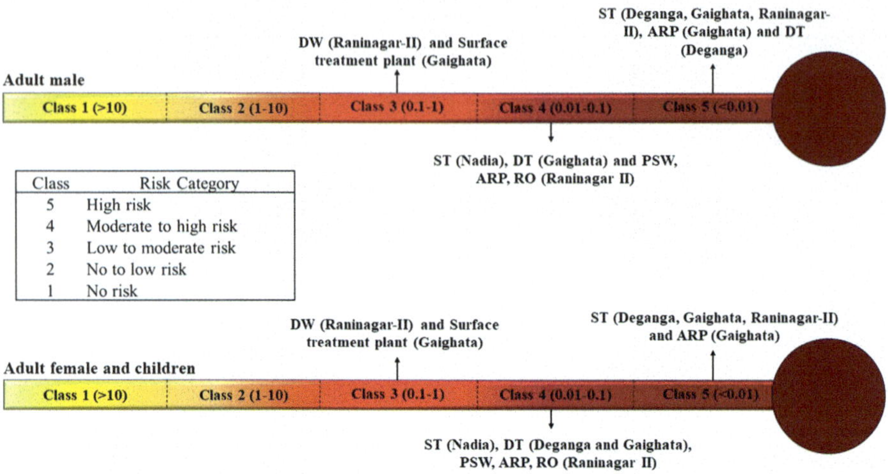

Fig. 3 Risk classification of arsenic exposure among all available drinking water sources from the studied areas through risk thermometer

0.0034 in Gaighata, respectively) followed by ARPs from Gaighata (class 5, SAMOE: 0.00745, 0.00854, and 0.00932, respectively, for three groups) except one surface treatment plant (risk class 3) and DTs from both study areas (class 5, SAMOE: 0.00938 for adult male and class 4, SAMOE 0.01074 and 0.01172 for adult female and children in Deganga; class 4, SAMOE: 0.01040, 0.01192, and 0.01301, in Gaighata). However, in Deganga, the exposure level is much higher for both the sources. Similar result was observed in Raninagar-II for domestic level STs (class 5, SAMOE: 0.00431 for adult male, 0.00494 for female, and 0.00539 for children, respectively) among other available sources and the remaining available drinking water sources those were fall in class 4 with moderate–high risk except DW. Dug well was observed to have class 3, i.e., low–moderate risk with SAMOE value 0.11077, 0.12692, and 0.13846 for respective population groups. Since, there were no alternative water sources studied from Nadia, the STs are fall under the risk category 4 for respective population groups (0.01659, 0.01901, and 0.020737, respectively) and all the mean values were taking into consideration to describe the level of exposure of As in all drinking water sources together regardless of their areas.

3.4.2 Assessment of Carcinogenic and Noncarcinogenic Risk

The CR and HQ of three studied population groups from four respective areas are shown in Table 5.

The carcinogenic risk among all were higher than USEPA recommended limit (1×10^{-4}) [17] in Deganga; all were at more risk from STs water (7.3×10^{-2}, 6.3×10^{-2}, and 5.8×10^{-2}, respectively, for adult male, female, and children).

Table 4 SAMOE values of arsenic toxicity in available drinking water from the studied areas

			Daily arsenic intake through available drinking water (μg/day)			Daily exposure or "E" (μg/kg BW/day)					
			[Adult male = 5 L/day; Adult female = 4 L/day; Children = 2 L/day]			[$BW_{Adult\ male}$ = 60 kg; $BW_{Adult\ female}$ = 55 kg; $BW_{children}$ = 30 kg]		SAMOE value			
Study area	Source of water	Mean arsenic concentration (μg/L)	Adult male	Adult female	Children	Adult male	Adult female	Children	Adult male	Adult female	Children
Deganga	ST	580	2900	2320	1160	48.3	42.2	38.7	0.00062	0.00071	0.00078
	DT	38.4	192	153.6	76.8	3.20	2.79	2.6	0.00938	0.01074	0.01172
Gaighata	ST	132	660	528	264	11.0	9.60	8.8	0.00273	0.00313	0.00341
	DT	34.6	173	138.4	69.2	2.88	2.52	2.3	0.01040	0.01192	0.01301
	ARP	48.3	241.5	193.2	96.6	4.03	3.51	3.2	0.00745	0.00854	0.00932
Raninagar-II	ST	83.5	417.5	334	167	6.96	6.07	5.6	0.00431	0.00494	0.00539
	PSW	22.3	111.5	89.2	44.6	1.86	1.62	1.5	0.01614	0.01850	0.02018
	ARP	15.3	76.5	61.2	30.6	1.28	1.11	1.0	0.02353	0.02696	0.02941
	RO	7.95	39.75	31.8	15.9	0.66	0.58	0.5	0.04528	0.05189	0.05660
	DW	3.25	16.25	13	6.5	0.27	0.24	0.2	0.11077	0.12692	0.13846
Nadia	ST	21.7	108.5	86.8	43.4	1.81	1.58	1.4	0.01659	0.01901	0.020737

Table 5 Carcinogenic and noncarcinogenic risk of arsenic among studied population groups

		CR			HQ		
Study area	Source of water	Adult male	Adult female	Children	Adult male	Adult female	Children
Deganga	ST	7.3×10^{-2}	6.3×10^{-2}	5.8×10^{-2}	161.1	141	129
	DT	4.8×10^{-3}	4.2×10^{-3}	3.84×10^{-3}	10.7	9.31	8.5
Gaighata	ST	1.7×10^{-2}	1.4×10^{-2}	1.32×10^{-2}	36.7	32.0	29.3
	DT	4.3×10^{-3}	3.8×10^{-3}	3.46×10^{-3}	9.61	8.39	7.7
	ARP	6×10^{-3}	5.3×10^{-3}	4.83×10^{-3}	13.4	11.7	10.7
Raninagar-II	ST	1×10^{-2}	9.1×10^{-3}	8.35×10^{-3}	23.2	20.2	18.6
	PSW	2.8×10^{-3}	2.4×10^{-3}	2.23×10^{-3}	6.19	5.41	4.96
	ARP	1.9×10^{-3}	1.7×10^{-3}	1.53×10^{-3}	4.25	3.71	3.40
	RO	9.9×10^{-4}	8.7×10^{-4}	7.95×10^{-4}	2.21	1.93	1.77
	DW	4.1×10^{-4}	3.5×10^{-4}	3.25×10^{-4}	0.90	0.79	0.72
Nadia	ST	2.7×10^{-3}	2.4×10^{-3}	2.17×10^{-3}	6.03	5.26	4.82

Likewise, each population group from all studied areas was under serious NCR as all of them exceeded the threshold for HQ (HQ > 1) except for DW water (0.90, 0.79, and 0.72, respectively, for three studied groups). However, raw water from DW is not recommended for consumption before treatment due to the probability of being polluted through anthropogenic activities over time. Despite having both CR and HQ values greater than threshold level (9.9×10^{-4}, 8.7×10^{-4}, 7.95×10^{-4}, and 2.21, 1.93, 1.77, respectively, for three studied population groups), RO treated water from Raninagar-II can be used with least risk among other collected available water sources. The risk was found to be higher in male followed by female and children. One study from West Bengal also found higher risk in adult than children [30].

3.5 *Future Risk and Further Recommendation*

Several health hazards, including lung cancer, prostate cancer, high blood pressure, and diabetes, can be caused due to As exposure in drinking water [31]. Cancer in several organs is considered to be the most severe future risk from long-term As exposure. Increased and medium levels of As exposure are associated with high mortality and lung cancer rates in humans [32]. Arsenic contamination of groundwater can be reduced by two distinct, but complementary, methods: avoiding future As contamination of the physical environment and treating previously affected populations. Finding efficient As treatment and implementing those treatment facilities at the community scale were common recommendations [33]. Another proposal is to replace As-contaminated shallow tube wells with DTs at a depth of 50–100 m and to monitor the background soil quality for 5 years [34]. Several studies try to draw attention to government PSW [34]. However, in this follow-up study, the observed As contamination scenario and risk of As exposure through the alternative drinking water sources revealed that implementation of the above-mentioned mitigation strategies could not have the full potency to supply As-free water as found in our previous study. Still, after so many years, the situation has not changed and the local populace are still not aware of As-exposure and its toxic severity. Water from DTs is not safe for human consumption, while government PSW is not accessible to all the people throughout the rural area. Although RO plants and ARP facilities have an approximate As removal efficacy, but they still require constant monitoring. The local population may not receive free drinking water from ARPs that are managed by some individuals or nonprofit groups. Apart from these, the economic status and lack of awareness of the population move them toward easily available water sources, i.e., domestic level STs, which cause severe health risk. This type of follow-up studies are required in future to evaluate the contamination scenario at several time intervals to understand the present situation and to assess the effectiveness of these types of studies.

According to the risk calculated from all different studied sources of drinking water, the safest source with the lowest As exposure level was found for DW. It is another form of groundwater storage system. Similar result was observed in a study conducted by Chakraborty et al. [35]. They found that DW can be opted as alternative safe source for drinking water in Bangladesh; however, treatment is required before using DW water for drinking [36]. Other options include the use of surface water from ponds and rainwater collection, which might be simple, cost-effective, and safe to develop. Similar to this study, our previous study found surface-treated drinking water to be a safe source, and consumption of this safe water can be helpful in suppressing the symptoms and exposure toxicity level of As in arsenicosis patients [37]. Therefore, immediate prohibition of As-exposed groundwater is required with the help of color-marking method, so that the rural populace can easily understand the risk for particular sources. Therefore, for better management of the drinking water quality, immediately prohibiting the utilization of groundwater are encouraged for the rural belt in order to preserve life. It is strongly advised to make DW, increase well sharing, well switching, use of well shaded surface water, gradual and sand-filtered pond water, and treated rainwater collection to reduce As toxicity.

4 Conclusion

Being a state of developing country, West Bengal has also been facing several issues related to sustainable development for environmental safety. Arsenic contamination in STs and DTs and other alternative sources in economically deprived rural As-exposed communities is one of the major public health concerns. Carcinogenic potentiality and noncarcinogenic effect of As can affect several organs severely to exposed population. However, growth in new technologies and their implementations are increasing nowadays, but their monitoring and maintenance are required for proper functioning. Apart from this, economically poor rural population on most of the cases could not afford the cost of new technologies to get rid of exposure. There is a need for policy upgrading and strategy development for mitigating As in the light of current scientific knowledge for developing water sources free from As and a proper functional organization should be required to safeguard the availability of As safe and good quality of water all over the exposed areas.

Acknowledgment This research work is supported by the research fellowship program of University Grant Commission (UGC), Government of India.

References

1. R. Kumar, S. Kumar, A. Ghosh, Current scenario of groundwater arsenic contamination in West Bengal and its mitigation approach, in *Arsenic Toxicity: Challenges and Solutions*, (Springer, Singapore, 2021), pp. 193–216
2. D. Chakraborti, B. Das, M.M. Rahman, U.K. Chowdhury, B. Biswas, A.B. Goswami, B. Nayak, A. Pal, M.K. Sengupta, S. Ahamed, A. Hossain, Status of groundwater arsenic contamination in the state of West Bengal, India: a 20-year study report. Mol. Nutr. Food Res. **53**(5), 542–551 (2009)
3. T.K. Dey, P. Banerjee, M. Bakshi, A. Kar, S. Ghosh, Groundwater arsenic contamination in West Bengal: current scenario, effects and probable ways of mitigation. Int. Lett. Nat. Sci. **13**(1), 45–58 (2014)
4. D. Chakraborti, M.M. Rahman, K. Paul, U.K. Chowdhury, M.K. Sengupta, D. Lodh, C.R. Chanda, K.C. Saha, S.C. Mukherjee, Arsenic calamity in the Indian subcontinent: what lessons have been learned? Talanta **58**(1), 3–22 (2002)
5. D. Stüben, Z. Berner, D. Chandrasekharam, J. Karmakar, Arsenic enrichment in groundwater of West Bengal, India: geochemical evidence for mobilization of As under reducing conditions. Appl. Geochem. **18**(9), 1417–1434 (2003)
6. R.T. Nickson, J.M. McArthur, P. Ravenscroft, W.G. Burgess, K.M. Ahmed, Mechanism of arsenic release to groundwater, Bangladesh and West Bengal. Appl. Geochem. **15**(4), 403–413 (2000)
7. A. Ahmad, P. Bhattacharya, Environmental arsenic in a changing world. Groundw. Sustain. Dev. **8**, 169–171 (2019)
8. N.R. Chowdhury, S. Ghosh, M. Joardar, D. Kar, T. Roychowdhury, Impact of arsenic contaminated groundwater used during domestic scale post harvesting of paddy crop in West Bengal: arsenic partitioning in raw and parboiled whole grain. Chemosphere **211**, 173–184 (2018)
9. M. Joardar, A. Das, N.R. Chowdhury, D. Mridha, A. De, K.K. Majumdar, T. Roychowdhury, Health effect and risk assessment of the populations exposed to different arsenic levels in drinking water and foodstuffs from four villages in arsenic endemic Gaighata block, West Bengal, India. Environ. Geochem. Health **43**(8), 3027–3053 (2021)
10. P. Bhattacharya, A.C. Samal, J. Majumdar, S.C. Santra, Arsenic contamination in rice, wheat, pulses, and vegetables: a study in an arsenic affected area of West Bengal, India. Water Air Soil Pollut. **213**(1), 3–13 (2010)
11. A. Das, S.S. Das, N.R. Chowdhury, M. Joardar, B. Ghosh, T. Roychowdhury, Quality and health risk evaluation for groundwater in Nadia district, West Bengal: an approach on its suitability for drinking and domestic purpose. Groundw. Sustain. Dev. **10**, 100351 (2020)

12. Q. Meng, J. Zhang, Z. Zhang, T. Wu, Geochemistry of dissolved trace elements and heavy metals in the Dan River Drainage (China): distribution, sources, and water quality assessment. Environ. Sci. Pollut. Res. **23**(8), 8091–8103 (2016)
13. S.M. Yidana, A. Yidana, Assessing water quality using water quality index and multivariate analysis. Environ. Earth Sci. **59**, 1461–1473 (2010)
14. S. Sand, R. Bjerselius, L. Busk, H. Eneroth, J.S. Färnstrand, R. Lindqvist, *The Risk Thermometer–A Tool for Risk Comparison* (Swedish National Food Agency, 2015)
15. A.C. Samal, P. Bhattacharya, P. Biswas, J.P. Maity, J. Bundschuh, S.C. Santra, Variety-specific arsenic accumulation in 44 different rice cultivars (O. sativa L.) and human health risks due to co-exposure of arsenic-contaminated rice and drinking water. J. Hazard. Mater. **407**, 124804 (2021)
16. S. Majumder, M. Joardar, A. Das, A. De, D. Mridha, S. Ghosh, U. Lama, A. Dey, N.R. Chowdhury, A. Majumdar, T. Roychowdhury, Arsenic toxicity, biomarkers of exposure and risk assessment among different aged young population from endemic areas of West Bengal, India. Groundw. Sustain. Dev. **23**, 101022 (2023)
17. USEPA, *Exposure Factors Handbook: 2011 Edition* (National Center for Environmental Assessment, Washington, DC, 2011)
18. A. Das, M. Joardar, A. De, D. Mridha, N.R. Chowdhury, M.T.B.K. Khan, P. Chakrabartty, T. Roychowdhury, Pollution index and health risk assessment of arsenic through different groundwater sources and its load on soil-paddy-rice system in a part of Murshidabad district of West Bengal, India. Groundw. Sustain. Dev. **15**, 100652 (2021)
19. M.M. Rahman, B.K. Mandal, T.R. Chowdhury, M.K. Sengupta, U.K. Chowdhury, D. Lodh, C.R. Chanda, G.K. Basu, S.C. Mukherjee, K.C. Saha, D. Chakraborti, Arsenic groundwater contamination and sufferings of people in North 24-Parganas, one of the nine arsenic affected districts of West Bengal, India. J. Environ. Sci. Health A **38**(1), 25–59 (2003)
20. M. Joardar, N.R. Chowdhury, A. Das, D. Mridha, T. Roychowdhury, Groundwater quality indexing for drinking purpose from arsenic prone areas, West Bengal, in *Toxic Metals Contamination: Generation, Disposal, Treatment and Valuation*, (CRC Press, 2022), p. 10
21. M.M. Rahman, M.K. Sengupta, S. Ahamed, U.K. Chowdhury, D. Lodh, M.A. Hossain, B. Das, D. Chakraborti, K.C. Saha, I. Kaies, A.K. Barua, Status of groundwater arsenic contamination and human suffering in a Gram Panchayet (cluster of villages) in Murshidabad, one of the nine arsenic affected districts in West Bengal, India. J. Water Health **3**(3), 283–296 (2005)
22. K. Nakagawa, H. Amano, M. Persson, R. Berndtsson, Spatiotemporal variation of nitrate concentrations in soil and groundwater of an intensely polluted agricultural area. Sci. Rep. **11**(1), 2598 (2021)
23. M. Chowdhury, P.K. Paul, Quantification of groundwater resource of Kandi subdivision of Murshidabad district, West Bengal. Environ. Dev. Sustain. **22**, 5849–5871 (2020)
24. M. Kumar, P. Kumar, A.L. Ramanathan, P. Bhattacharya, R. Thunvik, U.K. Singh, M. Tsujimura, O. Sracek, Arsenic enrichment in groundwater in the middle Gangetic Plain of Ghazipur District in Uttar Pradesh, India. J. Geochem. Explor. **105**(3), 83–94 (2010)
25. S. Mahato, A. Mahato, P.K. Karna, N. Balmiki, Investigating aquifer contamination and groundwater quality in eastern Terai region of Nepal. BMC. Res. Notes **11**(1), 1–7 (2018)
26. V.H. McNeil, M.E. Cox, Relationship between conductivity and analysed composition in a large set of natural surface-water samples, Queensland, Australia. Environ. Geol. **39**, 1325–1333 (2000)
27. E. Hubert, C. Wolkersdorfer, Establishing a conversion factor between electrical conductivity and total dissolved solids in South African mine waters. Water SA **41**(4), 490–500 (2015)
28. M. Siosemarde, F. Kave, E. Pazira, H. Sedghi, S.J. Ghaderi, Determine of constant coefficients to relate total dissolved solids to electrical conductivity. World Acad. Sci. Eng. Technol. **46**, 258–260 (2010)
29. Ghosh, A., & Chowdhury, A. (2019). Assessing Spatial Variability of Groundwater Quality Using GIS Techniques in Nadia District, West Bengal

30. A. De, D. Mridha, M. Joardar, A. Das, N.R. Chowdhury, T. Roychowdhury, Distribution, prevalence and health risk assessment of fluoride and arsenic in groundwater from lower Gangetic plain in West Bengal, India. Groundw. Sustain. Dev. **16**, 100722 (2022)
31. S.H. Lamm, I.J. Boroje, H. Ferdosi, J. Ahn, Lung cancer risk and low (≤50 μg/L) drinking water arsenic levels for US counties (2009–2013) – a negative association. Int. J. Environ. Res. Public Health **15**(6), 1200 (2018)
32. K.M. Khan, R. Chakraborty, J. Bundschuh, P. Bhattacharya, F. Parvez, Health effects of arsenic exposure in Latin America: an overview of the past eight years of research. Sci. Total Environ. **710**, 136071 (2020)
33. S.V. Flanagan, R.B. Johnston, Y. Zheng, Arsenic in tube well water in Bangladesh: health and economic impacts and implications for arsenic mitigation. Bull. World Health Organ. **90**, 839–846 (2012)
34. R. Johnston, S.J. Hug, J. Inauen, N.I. Khan, H.J. Mosler, H. Yang, Enhancing arsenic mitigation in Bangladesh: findings from institutional, psychological, and technical investigations. Sci. Total Environ. **488**, 477–483 (2014)
35. M. Chakraborty, A. Mukherjee, K.M. Ahmed, A review of groundwater arsenic in the Bengal Basin, Bangladesh and India: from source to sink. Curr. Pollut. Rep. **1**(4), 220–247 (2015)
36. B. Sefa-Ntiri, B. Kwakye-Awuah, R. Edziah, B. Anderson, F.A. Armah, P. Mensah-Amoah, S.S. Sackey, F. Sam, A.G. Akyea, S. Yunus, F. Kumi, Dissolved organic matter in hand-dug well water as groundwater quality indicator: assessment using laser-induced fluorescence spectroscopy and multivariate statistical techniques. SN Appl. Sci. **2**, 1–15 (2020)
37. M. Joardar, A. Das, N.R. Chowdhury, D. Mridha, J. Das, A. De, S. Majumder, K.K. Majumdar, T. Roychowdhury, Impact of treated drinking water on arsenicosis patients with continuous consumption of contaminated dietary foodstuffs: a longitudinal health effect study from arsenic prone area, West Bengal, India. Groundw. Sustain. Dev., 100786 (2022)

Defluoridation by Sustainable Utilization of Iron Oxide Nanoparticles-Enriched Waste Tea Residue

Urvashi Lama, Shuvankar Halder, and Tarit Roychowdhury

Abbreviations

CTR	Crushed tea residue
EDX-FESEM	Energy-dispersive X-ray spectroscopy-field emission scanning electron microscopy
FeONP-TR	Iron oxide nanoparticles-enriched tea residue
FTIR	Fourier transform infrared spectroscopy
HCl	Hydrochloric acid
HSAB	Hard acid soft base
NaF	Sodium fluoride
NaOH	Sodium hydroxide
NPs	Nanoparticles
TISAB-III	Total ionic strength adjustment buffer-III

1 Introduction

Fluorine possesses the highest electronegativity among all elements. It is commonly prevalent in aqueous environment as F^- [1]. Fluoride enrichment in groundwater has been an epidemic global concern. Although geogenic sources are more common, anthropogenic activities also contribute towards its contamination in the environment [2]. In India, the groundwater of 20 different states is already overexposed with it causing F^- toxicity to the large population dependent on water [3–5]. The acceptable concentration of F^- in drinking water is 1.5 mg L^{-1} and to some extent

U. Lama · S. Halder · T. Roychowdhury (✉)
School of Environmental Studies, Jadavpur University, Kolkata, India

D. Mazumder (ed.), *Sustainable Advanced Technologies for Environmental Management*, Springer Proceedings in Earth and Environmental Sciences,
https://doi.org/10.1007/978-3-031-64006-3_4

it is beneficial to the human body [6]. It prevents dental cavities when present in an optimum limit of 0.5 mg L^{-1}. However, excess F^- (>1.5 mg L^{-1}) can lead to dental, skeletal and crippling fluorosis [6].

Therefore, removal of surplus F^- from water is a pressing necessity as it poses serious health risks in humans. Reverse osmosis, membrane filtration technology, electrocoagulation, electrochemical ion exchange and nanofiltration are some defluoridation techniques. These techniques often include significant maintenance costs, excessive energy consumption and secondary pollution such as the production of hazardous sludge [7]. Among these, the adsorption method has drawn the most attention since it is inexpensive, easy to use, efficient and responds quickly. Adsorption is a well-researched and efficient defluoridation method, and a number of adsorbents have been created for defluoridation [8–12]. Defluoridation of water and waste water is commonly accomplished using iron oxide and their mixtures with other adsorbents. Compared to other metal oxides, iron oxide is the most affordable, easily accessible and highly efficient due to its surface morphology, electromagnetic and chemical characteristics that aid in F^- removal [13]. Studies have reported the induced increase in adsorption capability of adsorbents on modifications with different agents [14–16].

Based on the literature, very little work has been done on metal oxide nanoparticles produced through green synthesis using organic waste material. Thus, this present study seeks to increase the F^- removal efficiency by synthesizing iron oxide nanoparticles utilizing waste tea residues. Black tea extract was recently employed for the synthesis of an iron nanoparticle-incorporated adsorbent, which when applied at a dose of 2.5 g L^{-1} to 4.0 mg L^{-1} concentration of F^- exhibited 90% defluoridation in 25 min at 25 °C and pH 7 [17]. This study aims for a lower adsorbent dosage, i.e., 1 g L^{-1}, which is kept fixed. Analytical techniques were used to examine characteristics of the synthesized NPs and a number of variables that impacted the defluoridation process were evaluated. Adsorption isotherm models were studied. The present study promotes a step towards a sustainable approach of defluoridation using a common household and tea shops' waste, i.e., tea residue. Proper waste management is a challenging problem worldwide and this study has tried to upgrade the waste tea residue for defluoridation of water.

2 Materials and Methods

2.1 Chemicals and Reagents

Analytical grade chemicals, reagents, distilled water and double distilled water were used for the experimental study. Ferric chloride anhydrous ($FeCl_3$) was purchased from Merck.

For estimation of F^-, 100 mg L^{-1} standard F^- solution (Orion 940911, Thermo Scientific, USA) was procured. Total Ionic Strength Adjusting Buffer (TISAB-III) was purchased from Thermo Fisher Scientific (USA, Orion 940911). Plastic containers were used to analyse the F^- containing samples.

2.2 Collection of Waste Tea Residue

Tea preparation results to the tea residue, which is generally discarded away as waste. The waste tea residues generated were collected from households and local tea shops near Jadavpur University, West Bengal, India. The collected waste tea residue was thoroughly washed with tap water followed by distilled water. After drying, a portion of it was ground, which was termed as crushed tea residue (CTR), and another portion was processed for iron oxide nanoparticles-enriched tea residue (FeONP-TR) synthesis.

2.3 Synthesis of Iron Oxide Nanoparticles-Enriched Tea Residue (FeONP-TR)

Iron oxide nanoparticles-enriched tea residue (FeONP-TR) was synthesized following the process mentioned in an earlier study [18] employing some modifications. The initial steps included a thorough wash with tap water followed by distilled water. The tea residue was then oven dried. In a beaker, 20 g of anhydrous $FeCl_3$ was dissolved in 200 mL of distilled water to which 40 g of dried tea residue was added. It was stirred for 4 h and left overnight. It was then filtered and the filtrate was collected and dried in oven. It was further thermally treated in a muffle furnace at 450 °C for 6 h. A black dry char was obtained, which was then washed and dried. This dry residue was then homogenized with the help of a mortar and pestle and stored for future experiments.

2.4 Characterization of FeONP-TR

The physicochemical properties of FeONP-TR were characterized by EDX-FESEM and FTIR analysis. EDX-FESEM was used to analyse CTR.

2.5 Adsorption Experiment

The adsorption of F^- on FeONP-TR and CTR was assessed through batch experiments. Fluoride concentrated solutions (5, 10, and 20 mg L^{-1}) were prepared and transferred to plastic Erlenmeyer flasks, which were used for the experimental work. FeONP-TR and CTR were administered at the specified dose of 1 g L^{-1}. The study was based upon parameters such as different initial F^- concentrations, different pH of the fluoridated solutions and a range of contact time up to 24 h. The pH was adjusted using 0.1 M NaOH or 0.1 M HCl to 5, 7, and 9.

The defluoridation capacities or removal efficiencies of the FeONP-TR and CTR were calculated as:

$$\text{Removal Efficiency} = \frac{C_\text{i} - C_\text{e}}{m} \times v \tag{1}$$

where C_i is the initial concentration of F^- (in mg L^{-1}), C_e is the equilibrium concentration of F^- (in mg L^{-1}) after adsorption, v is volume of the adsorbate solution (in L) and m is the given dosage of adsorbent (in g).

2.6 F^- Analysis

An ion-selective electrode meter (Model: Orion Star A214, Thermo Scientific) along with F^- ion-selective electrode (Orion, Model no. 9609BNWP) was used for analysis of F^- following the recommended protocol by the American Society for Testing and Materials [19]. From the standard stock solution of 100 mg L^{-1} NaF, solutions of 0.1, 1, and 10 mg L^{-1} were prepared through serial dilutions. TISAB-III was added to the F^- aliquots in a 1:10 ratio and the mixture solution was used for calibration. Prior to the measurement of F^-, the ion meter was calibrated to a slope of -59.1 ± 2.0 mV dec^{-1}.

3 Results and Discussion

3.1 Characterization of CTR and FeONP-TR

The surface morphology and elemental composition of FeONP-TR and CTR was studied from EDX-FESEM as shown in Fig. 1a, b, respectively. It was observed that the size of the CTR particles was larger, while the particles of FeONP-TR lied in the range of 180–250 nm. The magnetic particles of FeONP-TR tend to cluster and agglomerate as they are magnetic in nature. There may be a possibility of exposure of the adsorbent to moisture during sample preparation and handling, which also cannot be ruled out.

Analysis for the presence of any chemical groups in the synthesized nanoparticles was performed by FTIR as presented in Fig. 2. The scanning wave number range was 500–4000 cm^{-1}, and the relative functional groups were marked in the graph. The FTIR of FeONP-TR shows broad and strong bands at 3734 and 3324 cm^{-1} due

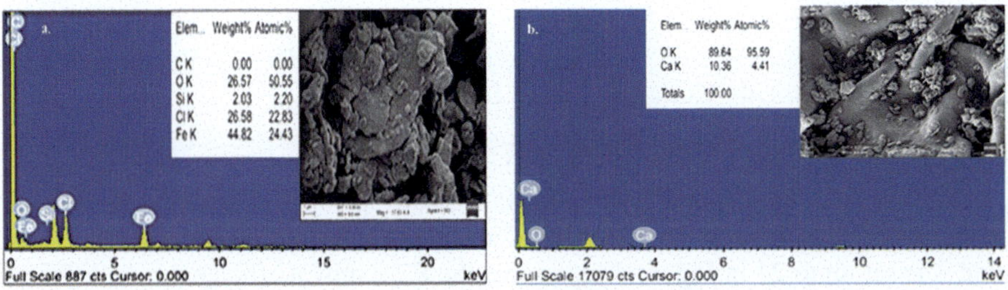

Fig. 1 (**a**) EDX-FESEM analysis of FeONP-TR and (**b**) EDX-FESEM analysis of CTR particles

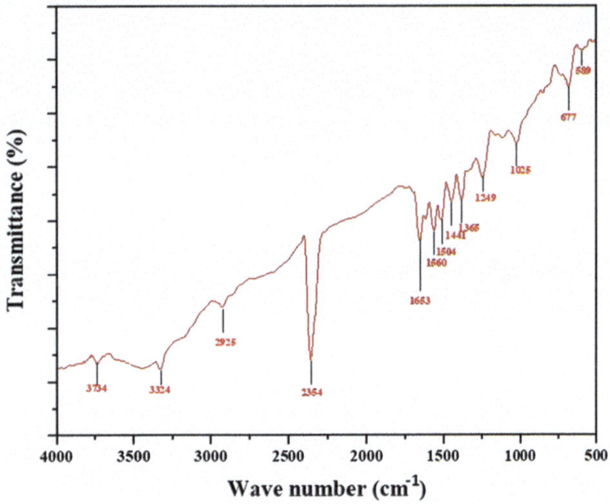

Fig. 2 Characterization of synthesized FeONP-TR by FTIR

to O–H stretch, which may be due to absorption of OH– by the nanoparticles. The peaks at 2925 and 2354 cm^{-1} correspond to the C–H stretch of aliphatic carbon and O=C=O groups, respectively. The observed peak at 1441 cm^{-1} indicates in-plane bending vibration of OH– groups. The peak observed at 1249 cm^{-1} indicates C–O stretching, while the observation at 1025 cm^{-1} is attributed to Fe–OH structural vibration. The peaks at 1653 cm^{-1} suggests C=N stretching, and at 1560 and 1504 cm^{-1} may be due to N–H vibrations. The peaks at 1441 and 1365 cm^{-1} suggest O–H bending and C–H vibrations, respectively, which indicate residual carbon of tea in FeONP-TR [20]. The peaks at 677 and 589 cm^{-1} are attributed to the Fe–O bond vibration of Fe_3O_4, confirming the synthesized particles to be iron oxide nanoparticles [20, 21].

3.2 *Batch Adsorption Studies*

3.2.1 Initial Fluoride Concentration Effects

The initial F^- solution concentration exerts an influence upon the adsorption capacity. The initial F^- concentrations were 5, 10, and 50 mg L^{-1} under different pH conditions (5, 7, and 9), and the amount of adsorbent was 1 g L^{-1} and the adsorption time was set up to 24 h. Figure 3 represents the F^- removal efficiencies at different initial F^- concentrations. In case of CTR, the F^- removal efficiency increased from 45.8% to 57.3% when the initial F^- concentration increased from 5 to 20 mg L^{-1} in acidic condition of pH 5. Similar trend was noticed with increasing initial F^- concentration in other pH conditions. FeONP-TR also exhibited increase in F^- removal efficiency with increasing F^- conditions, the highest being ranging from 53.6% to 74.85% under pH 7. However, there was a decrease in removal efficiency by FeONP-TR in alkaline condition when the initial F^- concentration reached 20 mg L^{-1}. This may be possibly due to the saturation of available sites on its surface and the high pH condition.

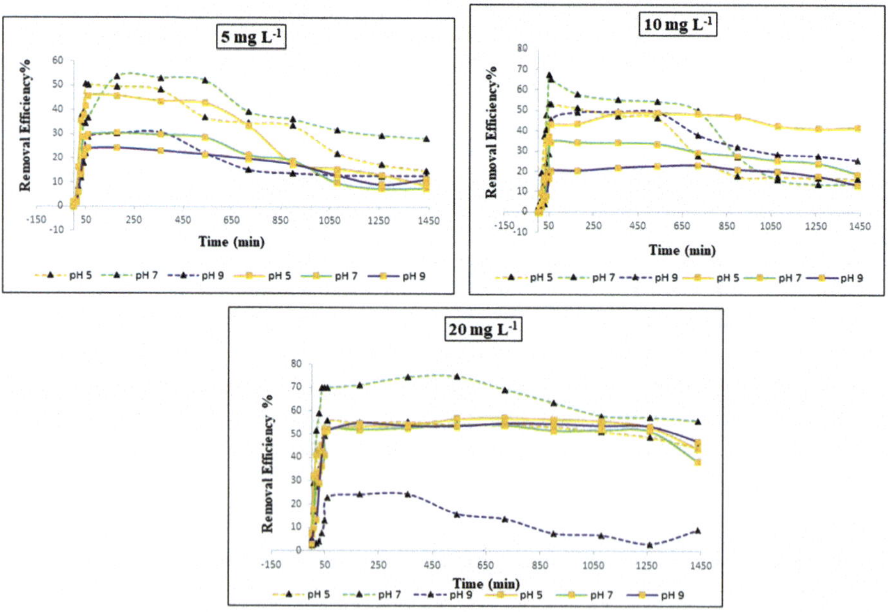

Fig. 3 Removal efficiencies of F^- by FeONP-TR (represented by ----) and CTR (represented by ····) at different pH and contact time

3.2.2 Influence of Contact Time

The effect of contact time with the defluoridation capacities of both CTR and FeONP-TR under different pH conditions is presented in Fig. 3. The adsorption ability of the adsorbents increased with time in the beginning. This may be attributed to the sufficient active sites on the surface of the adsorbent, which is filled up with time in the process of adsorption. FeONP-TR, however, showed faster adsorption rate compared to CTR. As the contact period was further increased and the adsorbent was fully saturated, desorption was noticed, which was confirmed with the slow increase in F^- concentration.

3.2.3 Influence of pH

The effect of pH on the adsorption of F^- is shown in Fig. 3. It was observed that the adsorption of F^- was better in lower pH conditions. CTR showed maximum defluoridation capacities in solutions of pH 5, while FeONP-TR demonstrated its best ability in pH 7. The surface of adsorbent is positively charged in acidic conditions, which forms an electrostatic bond with the F^- ions contributing towards the enhanced adsorption of F^- by CTR and FeONP-TR. At lower pH, the protonation of –OH groups is facilitated, which aids to ligand exchange. It results to a strong attraction between the F^- ions and the positive charged surface of CTR and FeONP-TR. The adsorption of F^- ions may occur via ion exchange at moderate pH conditions. The reduced removal efficiencies in alkaline conditions may be due to the competition for active sites by the excessive OH^- produced in basic pH conditions [22].

3.3 *Adsorption Isotherm*

Adsorption is influenced by characteristics of the adsorbent and the adsorbate and the chemistry between them, which was indicated by isotherm study. The adsorption mechanism involved was understood through three isotherm models namely Langmuir, Freundlich and Temkin adsorption isotherms. While Langmuir isotherm defines monolayer adsorption, Freundlich isotherm describes the creation of heterogeneous layers on the adsorbent surface as a result of adsorption.

The Langmuir adsorption model can be represented in linear form as follows [23]:

$$1/q_e = (1/K_L q_{max}) \times (1/C_e) + (1/q_{max}) \tag{2}$$

where q_e represents adsorbent equilibrium F^- concentration (mg g^{-1}); C_e is the F^- concentration in the solution at equilibrium (mg g^{-1}); q_{max} indicates the maximum quantity of F^- adsorbed per unit weight of adsorbent, sufficient to establish a fully saturated monolayer on the surface at high C_e (mg g^{-1}) and K_L is Langmuir adsorption constant. The parameters for Langmuir isotherm were obtained by plotting $1/C_e$ vs. $1/q_e$.

The Freundlich isotherm model is derived from Langmuir isotherm model, which assumes the existence of sites on the adsorbent with different affinities for each specific adsorbent. The model further elucidates the individual site behaviour in accordance with the principles outlined in the Langmuir isotherm. The Freundlich adsorption model can be represented as [24]:

$$\log q_e = \log K_f + (1/n) \log C_e \tag{3}$$

where K_f indicates the ability of the adsorbent referred as the Freundlich adsorption capacity and n indicates the heterogeneous nature of the system, which reflects the variation in the attraction towards the adsorbate with changes in the biosorption density. The parameters for Freundlich adsorption isotherm were obtained by plotting $\log C_e$ vs. $\log q_e$.

The adsorbent–adsorbate interaction is explained by the Temkin isotherm [25]:

$$q_e = (RT/b_T) \ln A_T + (RT/b_T) \ln C_e \tag{4}$$

where A_T refers to Temkin isotherm equilibrium binding constant (L g^{-1}), b_T indicates Temkin adsorption isotherm constant, R represents universal gas constant (8.314 J mol^{-1} K^{-1}) and T is the temperature at 298 K. The parameters for Temkin isotherm were obtained by plotting $\ln C_e$ vs. q_e.

Table 1 lists the fitting parameters for the Langmuir, Freundlich and Temkin adsorption isotherms. Considering the best defluoridation capacity shown in pH 5, the study results fit the Freundlich model best. Overall, Freundlich and Temkin isotherms show better fit than Langmuir isotherm as per the batch adsorption study. Fluoride may be adsorbed in multilayers, which can be explained by the higher regression coefficients. Because of multilayer adsorption and greater surface area, FeONP-TR has a comparatively higher adsorption capacity when compared to CTR. The adsorption may further be explained by different types of van der Waals contact and the electrostatic attraction force between positive and negative sites of the adsorbent and the adsorbate, respectively.

3.4 Fluoride Removal Mechanism

The defluoridation mechanism of FeONP-TR may be attributed mainly to the HSAB (hard acid soft base) principle, which states that hard bases and acids form favourable strong bonds. Based on the principle, the negatively charged F^- (hard base) binds with the positively charged Fe^{3+} (hard acid) through electrostatic attraction. In addition, Fe^{3+} ions have a great propensity to use the complexation mechanism to form complexes with F^- ions. Under acidic conditions, the amine

Table 1 Adsorption isotherm parameters for F^- adsorption by CTR and FeONP-TR

	Langmuir isotherm	q_{max} **(mg g^{-1})**	K_L **(L mg^{-1})**	R^2
CTR	pH 5	22.3864	0.03439	0.99821
	pH 7	4.27369	0.07034	0.99131
	pH 9	5.91051	0.04256	0.68195
FeONP-TR	pH 5	33.0469	0.02906	0.99979
	pH 7	7.04424	0.12614	0.80713
	pH 9	320.5128	0.001466	0.41167
	Freundlich isotherm	n	K_F **(L mg^{-1})**	R^2
CTR	pH 5	0.827328	0.685472	0.99999
	pH 7	0.516809	0.112717	0.91121
	pH 9	0.532555	0.081779	0.39215
FeONP-TR	pH 5	0.859202	0.884403	0.99939
	pH 7	0.599215	0.758124	0.8917
	pH 9	1.553808	0.691274	0.06532
	Temkin isotherm	B_T **(J mol^{-1})**	K_T **(L mg^{-1})**	R^2
CTR	pH 5	6.5005	0.486116946	0.91373
	pH 7	8.46035	0.299737052	0.61942
	pH 9	7.32688	0.274596298	0.08236
FeONP-TR	pH 5	6.77765	0.545739218	0.90584
	pH 7	11.86917	0.541458204	0.9999242
	pH 9	1.74715	1.115290428	0.07629

group (–NH_2) protonates to (NH_3^+), while –OH protonates to –OH_2^+ that binds electrostatically with F^-. This is why the defluoridation capacity was more in acidic and moderate pH conditions, which decreased significantly in higher pH water solutions.

Although FeONP-TR demonstrated better F^- removal efficiency, the removal efficiency of F^- by simple CTR is also acknowledgeable the reason being tea a good accumulator of F^-. The presence of minute Ca^{2+} in the CTR also seems to have aided in adsorption of F^- due to formation of ionic bonds. There is a possibility that in cases where tea particles were used to make milk tea, these ions might have been absorbed by them, which remained in the waste tea residue. Although the collected waste tea residue was washed, some part of it might have been retained by the tea residue.

4 Conclusion

The removal efficiency by FeONP-TR was evidently higher compared to that of CTR. For 5, 10, and 20 mg L^{-1} F^- contaminated water at pH 5, CTR showed defluoridation capacity of 45.8%, 49.10%, and 57.3%, respectively, while FeONP-TR showed 50.8%, 53.1%, and 56% for the same. FeONP-TR showed maximum F^-

removal efficiency of 74.85% under pH 7 with an initial F^- concentration of 20 mg L^{-1} compared to 54.05% shown by CTR under same condition. It can be said that modification of tea waste aided in enhancing defluoridation of water. The present study demonstrates an effective way to utilize tea residue, a waste generally discarded after preparing tea, in removal of F^- from water. Application of waste tea residue promises as a sustainable method to remove F^- from F^- contaminated water.

Acknowledgement The authors acknowledge the contributions of researchers from School of Environmental Studies, Jadavpur University, India, for this research study.

References

1. A. Ghosh, K. Mukherjee, S.K. Ghosh, B. Saha, Sources and toxicity of fluoride in the environment. Res. Chem. Intermed. **39**, 2881–2915 (2013)
2. D.K. Nordstrom, P.L. Smedley, *Fluoride in Groundwater* (The Groundwater Project, Guelph, 2022)
3. CGWB-AQUIM Report, Central Groundwater Board, Ministry of Water Resource, Govt. of India, Nagpur, Maharashtra (2015)
4. N. Subba Rao, D. Marghade, A. Dinakar, I. Chandana, B. Sunitha, B. Ravindra, T. Balaji, Geochemical characteristics and controlling factors of chemical composition of groundwater in a part of Guntur district, Andhra Pradesh, India. Environ. Earth Sci. **76**(21), 1–22 (2017)
5. A. De, A. Das, M. Joardar, D. Mridha, A. Majumdar, J. Das, T. Roychowdhury, Investigating spatial distribution of fluoride in groundwater with respect to hydro-geochemical characteristics and associated probabilistic health risk in Baruipur block of West Bengal, India. Sci. Total Environ. **886**, 163877 (2023)
6. G. WHO, *Guidelines for Drinking-Water Quality*, vol 216 (World Health Organization, Geneva, 2011), pp. 303–304
7. P. Pillai, S. Dharaskar, S. Sasikumar, M. Khalid, Zeolitic imidazolate framework-8 nanoparticle: a promising adsorbent for effective fluoride removal from aqueous solution. Appl Water Sci **9**, 1–12 (2019)
8. D. Khandare, S. Mukherjee, A review of metal oxide nanomaterials for fluoride decontamination from water environment. Mater. Today Proc. **18**, 1146–1155 (2019)
9. G.J. Millar, S.J. Couperthwaite, L.A. Dawes, S. Thompson, J. Spencer, Activated alumina for the removal of fluoride ions from high alkalinity groundwater: new insights from equilibrium and column studies with multicomponent solutions. Sep. Purif. Technol. **187**, 14–24 (2017)
10. C. Shen, L. Wu, Y. Chen, S. Li, S. Rashid, Y. Gao, J. Liu, Efficient removal of fluoride from drinking water using well-dispersed monetite bundles inlaid in chitosan beads. Chem. Eng. J. **303**, 391–400 (2016)
11. S. Dubey, M. Agarwal, A.B. Gupta, Experimental investigation of Al-F species formation and transformation during coagulation for fluoride removal using alum and PACl. J. Mol. Liq. **266**, 349–360 (2018)
12. A. Mohammad, C.B. Majumder, Removal of fluoride from synthetic waste water by using "bio-adsorbents". Int. J. Res. Eng. Technol **3**(4), 776–785 (2014)
13. M.K. Adak, S. Chakraborty, S. Sen, D. Dhak, Synthesis and optimization of low-cost and high efficient zirconium-aluminium modified iron oxide nano adsorbent for fluoride removal from drinking water. Adv. Mater. Proc. **2**(11), 716–724 (2017)
14. K. Mukhopadhyay, A. Naskar, U.C. Ghosh, P. Sasikumar, One-pot synthesis of β-cyclodextrin amended mesoporous cerium(IV) incorporated ferric oxide surface towards the evaluation of

fluoride removal efficiency from contaminated water for point of use. J. Hazard. Mater. **384**, 121235 (2020)
15. P. Pillai, Y. Lakhtaria, S. Dharaskar, M. Khalid, Synthesis, characterization, and application of iron oxyhydroxide coated with rice husk for fluoride removal from aqueous media. Environ. Sci. Pollut. Res. **27**, 20606–20620 (2020)
16. L.Y. Kuang, Y.Y. Liu, D.D. Fu, Y.P. Zhao, FeOOH graphene oxide nanocomposites for fluoride removal from water: acetate mediated nano FeOOH growth and adsorption mechanism. J. Colloid Interface Sci. **490**, 259–269 (2017)
17. I. Ali, Z.A. Alothman, M. M., Sanagi., Green synthesis of iron nano-impregnated adsorbent for fast removal of fluoride from water. J. Mol. Liq. **211**, 457–465 (2015)
18. S. Lunge, S. Singh, A. Sinha, Magnetic iron oxide (Fe_3O_4) nanoparticles from tea waste for arsenic removal. J. Magn. Magn. Mater. **356**, 21–31 (2014)
19. D. ASTM, *1179-10. Standard Test Methods for Fluoride Ion in Water* (American Society for Testing and Materials, 2010)
20. R.M. Dhoble, S. Lunge, A.G. Bhole, S. Rayalu, Magnetic binary oxide particles (MBOP): a promising adsorbent for removal of As (III) in water. Water Res. **45**(16), 4769–4781 (2011)
21. J. Sun, S. Zhou, P. Hou, Y. Yang, J. Weng, X. Li, M. Li, Synthesis and characterization of biocompatible Fe3O4 nanoparticles. J. Biomed. Mater. Res. A **80**(2), 333–341 (2007)
22. C. Li, N. Chen, Y. Zhao, R. Li, C. Feng, Polypyrrole-grafted peanut shell biological carbon as a potential sorbent for fluoride removal: Sorption capability and mechanism. Chemosphere **163**, 81–89 (2016)
23. I. Langmuir, The adsorption of gases on plane surfaces of glass, mica and platinum. J. Am. Chem. Soc. **40**(9), 1361–1403 (1918)
24. H. Freundlich, Über die adsorption in lösungen. Z. Phys. Chem. **57**(1), 385–470 (1907)
25. M.I. Temkin, Adsorption equilibrium and the kinetics of processes on nonhomogeneous surfaces and in the interaction between adsorbed molecules. Zh. Fiz. Chim. **15**, 296–332 (1941)

Mechanistic Insight and Optimal Remediation Strategy of Methylene Blue Using Waste Banana Peel in a Circular Economy Approach

Lohit Paul and Surabhi Chaudhuri

Abbreviations

AD	Adsorption dose
ANOVA	Analysis of variance
ATR-FTIR	Attenuated total reflectance-Fourier transform infrared spectroscopy
B1	Banana peel before adsorption
B2	Banana peel after adsorption
BBD	Box–Behnken design
BET	Brunauer–Emmett–Teller analysis
BP	Banana peel
CCD	Central composite design
CM	Contaminant concentration
CP	Cucumber peel
FE-SEM	Field emission scanning electron microscope
MB	Methylene blue
PP	Potato peel
RPM	Rotation per minute
RSM	Response surface methodology

L. Paul
Department of Earth and Environmental Studies, National Institute of Technology Durgapur, Durgapur, West Bengal, India
e-mail: schaudhuri.bt@nitdgp.ac.in

S. Chaudhuri (✉)
Department of Biotechnology, National Institute of Technology Durgapur, Durgapur, West Bengal, India
e-mail: schaudhuri.bt@nitdgp.ac.in

D. Mazumder (ed.), *Sustainable Advanced Technologies for Environmental Management*, Springer Proceedings in Earth and Environmental Sciences,
https://doi.org/10.1007/978-3-031-64006-3_5

1 Introduction

Water pollution is defined as the polluting of water bodies such as lakes, rivers, seas and groundwater. It can happen either spontaneously or as a result of human activities. Industrial waste, agricultural runoff and sewage frequently contribute to water contamination [1]. Polluted water may have a harmful influence on the ecosystem, including the extinction of aquatic species, as well as on human health. To prevent water pollution, it is critical to enforce legislation and best practices in businesses and agriculture, as well as to promote individual activities such as safe waste disposal and water resource conservation [2].

Water pollution caused by synthetic dyes is now acknowledged as one of the most pressing environmental issues [3]. By lowering the quantity of light that reaches the water and harming photosynthetic organisms such as algae and plants, cationic dyes can seriously disrupt aquatic ecosystems when they are dumped into rivers. Additionally, hazardous to aquatic life, cationic dyes can do so through interfering with their metabolic processes or destroying their tissues. In addition to having an immediate impact on the environment, cationic dyes can also interact with other chemicals to generate potentially dangerous compounds. For instance, chlorinated aromatic compounds, which are extremely hazardous and persistent in the environment, and can be created when cationic dyes react with chlorine or other oxidizing agents [4]. Because of their toxicity, carcinogenicity and mutagenicity, the presence of 70% of all industrial dyes in wastewater effluents and their use in production could lead to serious disorders [5]. A total of 7×10^5 tonnes of dye compounds are produced industrially per year [6].

Popular cationic dyes are methylene blue, safranin, haematoxylin, methyl green and malachite green.

Biosorption is an environmentally benign and sustainable method that removes heavy metals and other pollutants from wastewater and contaminated environments by using biological materials such as microbes, algae or plant-based compounds. Biosorption, as opposed to traditional chemical approaches, is more cost-effective and often more efficient in binding and sequestering pollutants via diverse biological mechanisms such as ion exchange, complexation and surface adsorption. This developing technique has a lot of potential for reducing water and soil pollution, especially in sectors with high metal concentrations [7].

Both adsorption and biosorption involve the accumulation of molecules on the surface of a solid or liquid. The main distinction between the two is that the adsorbent in adsorption is a non-living substance such as activated carbon or clay minerals, whereas the adsorbent in biosorption is an organic material such as bacteria or plant materials [8]. When compared to chemical or physical adsorption, biosorption has the extra benefit of being eco-friendly and biodegradable, making it a more sustainable option.

Resources such as plants, algae and microorganisms can be used to create biosorbents, effectively removing dye molecules from water. Biosorbents such as chitosan and peat moss are efficient and eco-friendly [9].

Applying circular economy principles to biosorbents can minimize waste and promote resource reuse. Utilizing waste materials such as agricultural waste for biosorbent manufacturing makes it economical and environmentally responsible. For instance, agricultural waste such as rice husks, sawdust and maize cobs, which are plentiful and frequently disposed of as waste, can be used to make biosorbents. The manufacturing of biosorbents can be economical and environmentally responsible by employing waste materials as a source of raw materials because it lowers waste generation and the need for virgin ingredients [10].

Banana peels have been discovered to be a very efficient biosorbent for the removal of dyes from water. They are high in natural components including cellulose, lignin and pectin, making them ideal for this use. The cellulose and lignin in banana peels provide a wide surface area for dye molecules to cling to, while pectin aids in the binding of dye molecules to the peel. Banana peels have been found in studies to successfully remove colours such as methylene blue [11], Rhodamine-B [12], orange-G [5] and malachite green [13] from water. Furthermore, banana peels are widely available, affordable and biodegradable, making them a sustainable and ecologically beneficial colour removal option. In this study, we have used banana peel in its purest form, i.e., without any further chemical or biological treatment, thereby reducing the cost and time of production of biosorbent and achieving maximum removal.

2 Materials and Methods

2.1 *Materials*

The banana peels were obtained from fresh bananas, which were collected from a local shop in Durgapur in West Bengal, India. The methylene blue dye was obtained from Merck Life Science Private Limited, Mumbai, through a local distributor.

2.2 *Preparation of Biosorbent*

After the consumption of the bananas, the waste peels were collected and stored in a clean container and were transported to the laboratory. The peels were washed thoroughly under running water, followed by distilled water. They were then cut into small pieces (~1 × 1 cm). The cut pieces were then transferred onto a petri dish and were spread evenly to allow maximum evaporation. The cut peels were then placed in a 24-h hot air oven set at 50 °C. The completely dried-up peels then were ground in a grinder until a powdery mass was obtained [14]. The process diagram is illustrated in Fig. 1. The banana peel powder obtained was then stored in an amber bottle at room temperature for future use.

Fig. 1 Biosorbent preparation [15]

2.3 Preliminary Dye Adsorption Determination

The methylene blue dye was made as a stock solution at 100 mg/L and stored in a conical flask. For the preliminary treatment, several batch tests were run to obtain an optimum level for the maximum level of adsorption. The pH was varied from 5 to 8, time was considered at 24 h, the adsorbent dose was taken as 200, 400, 600, 800, and 1000 mg/L, and the concentration of the dye was taken as 10, 20, 30, 40, and 50 mg/L. The initial and the final absorbance was measured under a UV-spectrophotometer at 660 nm. The adsorption percentage (P_{abs}) was then calculated using the formula (Eq. 1) given as follows [16]:

$$P_{\text{abs}} = [(m_{\text{i}} - m_{\text{f}})/m_{\text{i}}] \times 100 \quad (1)$$

where m_{i} and m_{f} are the absorbance before and after treatment, respectively. Before measuring the m_{f}, the specimen was centrifuged for 15 min at 10,000 rpm to get rid of the biosorbent particles.

2.4 Procedure for Dye Removal

The stock dye solution was prepared by mixing 100 mg in 1 L distilled water (100 mg/L). A total of 5 mL of the stock solution was diluted with 50 mL of distilled water to create a 10 mg/L solution, which was then put into a 50 mL conical flask. The initial absorbance was recorded at 660 nm in a spectrophotometer. The pH was altered from 7. Then, 20 mg of banana peel powder was added to 50 mL of the contaminant. The solution was then stirred at 120 rpm for 24 h. The final absorbance was recorded at 660 nm for methylene blue in a spectrophotometer. The colour change after treatment is displayed in Fig. 2. The process for colour change is depicted in Fig. 3.

Fig. 2 Colour change after treatment

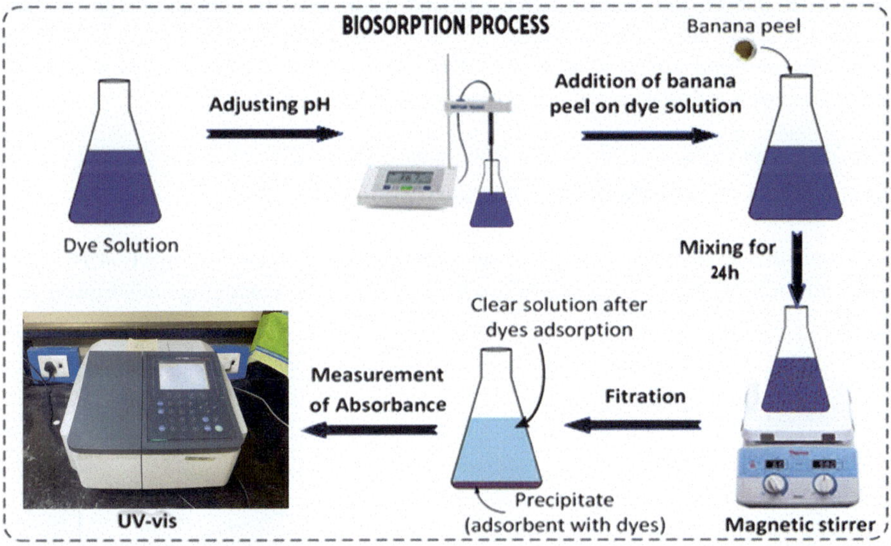

Fig. 3 Biosorption process [15]

3 Results and Discussion

3.1 Effect of Different Factors

Several factors are responsible for the amount of adsorption. Such factors, their effect on adsorption and the optimum levels are described in the following subsections:

3.1.1 Effect of pH on Adsorption

It has been observed that the biosorbents' surface charge density and their ability to adsorb are significantly influenced by the pH level of the dye solution at the beginning [17, 18]. In this study, 5–8 pH was considered with a minimum contaminant dosage of 10 and 400 mg/L as adsorbent dosage. The mixture was agitated at 120 rpm for 1 h. Maximum adsorption was achieved at pH 7 with 89% MB removal. It is generally known that the dye's hue will start to fade itself around pH above 8 and below 4, due to the development of new species [17]. Hence, for further investigation, pH 7 was used. The graph in Fig. 4 shows the variation in removal percentage with changes in pH levels. Lower pH levels resulted in an increase in positively charged adsorbent surface sites at the expense of negatively charged surface sites. At a lower pH, the dyes' carboxylic groups were protonated and possessed a significant amount of positive charge [19]. As a result, the capacity of the dyes to adsorb to the adsorbent decreased because with increasing solution pH, the electrostatic attraction between the positively charged dye molecule and its positively charged surface increased. Additionally, the competition between H+ ions and cationic dye molecules lowered the adsorption [20, 21].

3.1.2 Effect of Adsorbent Dose on Adsorption

As the adsorbent dosages were increased starting from 200 to 1000 mg/L at 10 mg/L concentration of dye on an equilibrium time of 24 h, the percentage adsorption and

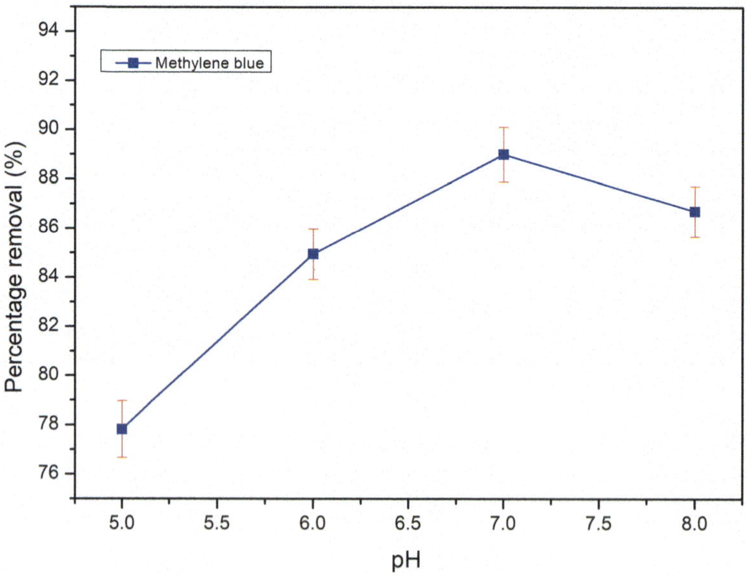

Fig. 4 Percentage removal vs pH

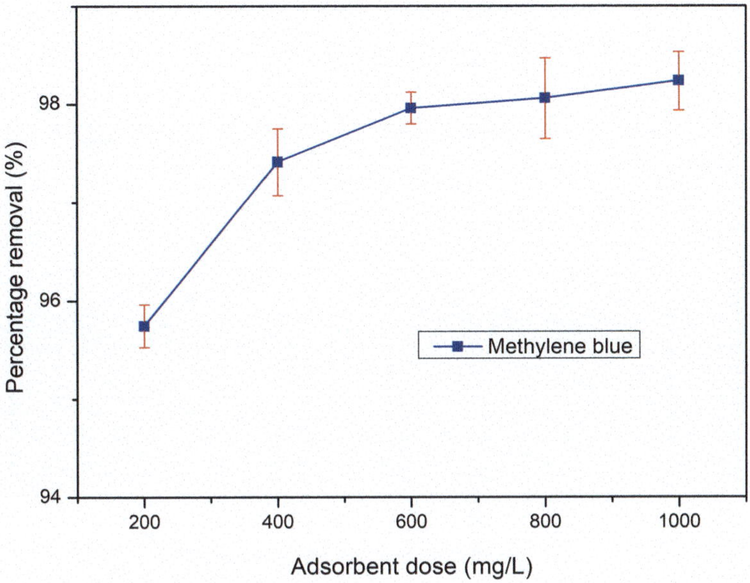

Fig. 5 Percentage removal vs. adsorbent dose

amount adsorbed increased. There was 12.35 mg/g of adsorbed material detected, with 96.5% removal of MB dye. Figure 5 clearly shows that increasing the adsorbent dosage increases the number of available adsorption sites as well as the surface area, which in turn increases the amount of dye adsorbed. The ideal adsorbent dosage was determined to be 400 mg/L because there was no significant increase in the rate of dye removal after that dosage.

Figure 5 shows the variation of the percentage of removal with different levels of adsorbent dose.

3.1.3 Effect of Contaminant Concentration on Adsorption

Figure 6 clearly shows that the lowest contaminant concentration of 10 mg/L resulted in the highest adsorption of 95.9%. The amount of adsorption was almost constant till 20 mg/L but then reduces drastically signifying the increase of contaminant but less fully active bare surface. The adsorption capacity for BP rose from 23.99 to 114.19 mg/g when the MB concentration in the test solution increased from 10 to 50 mg/L.

The greatest amount of dye was removed from the solution within 24 h of the start of each experiment. For BP to reach equilibrium, it took around 24 h. In addition to supplying the necessary energy to get past the obstacles to dye mass transfer, the dye and adsorbent interact better when the concentration is higher. As a result, when dye

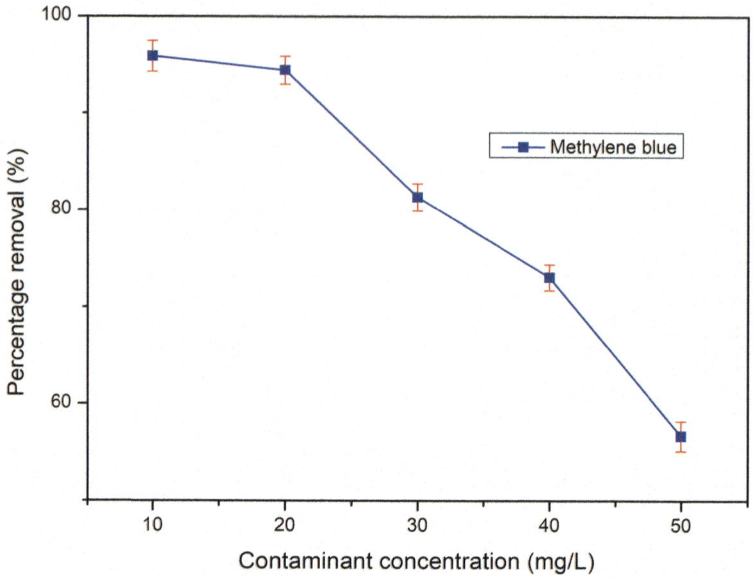

Fig. 6 Percentage removal vs. contaminant concentration

concentration increased, so did the rate of adsorption and, consequently, dye uptake [17].

3.2 *Adsorption Optimization by RSM*

A statistical approach called response surface methodology (RSM) was utilized in planning and evaluation of the trials in order to optimize the process's parameters. RSM can be used to design adsorption experiments that maximize the efficiency of the process of adsorption. Using mathematical equations and statistical analysis, the method entails modelling the adsorption process. The response of the process to changes in the experimental parameters, such as, in this case, pH, the adsorbent dose and the concentration of the contaminant, was then predicted using the model. Resources can be used more effectively, and fewer tests are needed to optimize the process with the help of RSM.

Box–Behnken design (BBD) and central composite design (CCD) are the two most used experiment designs in response surface methodology (RSM). Compared to CCD, a more intricate design that comprises a full factorial and a centre point, BBD is a three-level factorial design that necessitates fewer experimental runs. BBD has an edge over CCD since it offers a good balance between the quantity of experiments and the calibre of the outcomes. BBD minimizes the possibility of confusing or redundant outcomes while enabling effective exploration of the design space. Furthermore, BBD only needs a small number of experimental runs, which

Table 1 Response and the independent variables for the experiment

		Coded levels		
Experimental variables	Factors	−1	0	1
pH	A	6	7	8
AD: Adsorbent dose (mg/L)	B	200	400	600
CM: Contaminant concentration (mg/L)	C	10	20	30

can save time and resources while still producing precise and dependable results [22].

Every factor level in BBD needs to be limited to three levels. The formula $N = 2k(k - 1) + \text{cp}$ calculates the number of experiments in BBD. The factor number in this formula is k and the centre point is cp. In this case, the removal percentage was provided as a response to this investigation and the pH, the concentration of contaminant (CM) and the adsorbent dose (AD) were chosen as independent factors, as shown in Table 1. Design Expert 13 was used to calculate the influence of these independent factors on the removal percentage. According to the conditions listed in Table 1, three sets of experiments were carried out, and their mean values were then displayed [23].

Analysis of variance (ANOVA) is a statistical approach used in response surface methodology (RSM) to analyse experimental data and identify the significance of input variables on process response. ANOVA is used to divide response variable variability into multiple sources of variation, such as experimental error, pure error and error owing to lack of fit. Analysing the ANOVA table allowed us to ascertain the significant effect of input components on the process's response and construct an appropriate model of the process. Three-dimensional (3D) surface plots were created in the quadratic model while maintaining a variable constant. Finally, the best experimental conditions for removing the most MB were identified.

The experimental and projected methylene blue removal percentage are shown in Table 2 following 17 adsorption runs, indicated by BBD. The range of adsorption efficiencies was 69–98%. The combined interactive effects of the components were investigated in addition to the second-order and linear effects of the three independent variables on the percentage of removal. Table 3 provides the ANOVA findings on the proposed quadratic model's statistical significance. The relevance of the model is highlighted by the model F-value of 35.95 and p-value of <0.0001, which effectively indicates that an F-value this large has only a 0.01% probability of occurring due to noise.

It is obvious from the required p-values provided by the linear (B, C), interactive effects (BC) and quadratic (C^2) terms are significant. The below-mentioned polyfunctional model equation (Eq. 2), which depicts the adsorption process, makes it evident which variables have a positive or negative impact, and the numerical coefficient relates to the magnitude of this impact:

Table 2 Experimental and predicted values of the response of BBD for the percentage removal of methylene blue

Std	Run	A: pH	B: Adsorbent dose (mg/L)	C: Contaminant concentration (mg/L)	Percentage removal (%)	
					Experimental	Predicted
1	17	6	200	20	93.0198	90.9174
2	16	8	200	20	88.1022	87.3225
3	11	6	600	20	97.2079	97.9876
4	2	8	600	20	97.3856	99.4881
5	6	6	400	10	97.8024	98.2585
6	1	8	400	10	98.1882	97.3215
7	10	6	400	30	80.5343	81.401
8	4	8	400	30	80.6998	80.2437
9	15	7	200	10	95.589	97.2354
10	5	7	600	10	98.1955	96.9598
11	12	7	200	30	69.1385	70.3743
12	7	7	600	30	91.5321	89.8857
13	8	7	400	20	92.3844	92.3844
14	14	7	400	20	92.3844	92.3844
15	13	7	400	20	92.3844	92.3844
16	9	7	400	20	92.3844	92.3844
17	3	7	400	20	92.3844	92.3844

Table 3 ANOVA for quadratic model

Source	Sum of squares	df	Mean square	*F*-value	*p*-value
Model	945.22	9	105.02	35.95	<0.0001
A: pH	2.19	1	2.19	0.7507	0.415
B: Adsorbent dose	185.01	1	185.01	63.33	<0.0001
C: Contaminant concentration	575.8	1	575.8	197.09	<0.0001
AB	6.49	1	6.49	2.22	0.1797
AC	0.0121	1	0.0121	0.0042	0.9504
BC	97.88	1	97.88	33.5	0.0007
A^2	5.27	1	5.27	1.8	0.2213
B^2	0.7643	1	0.7643	0.2616	0.6248
C^2	74.16	1	74.16	25.38	0.0015
Residual	20.45	7	2.92		
Lack of fit	20.45	3	6.82		
Pure error	0	4	0		
Cor total	965.67	16			

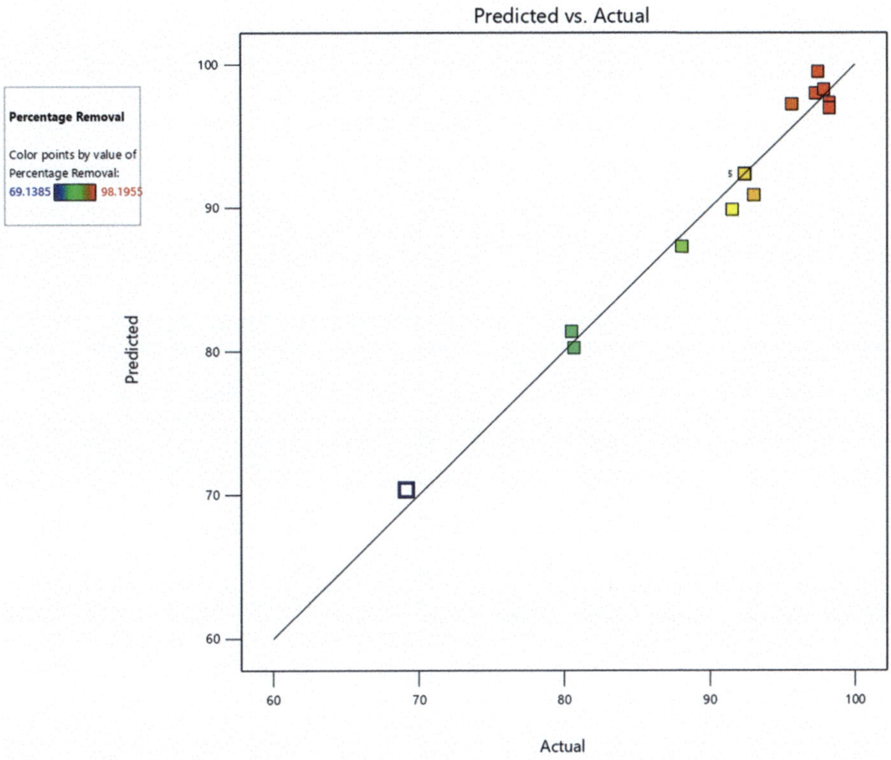

Fig. 7 Correlation between the actual and the predicted values of adsorption yield

$$\begin{aligned}\text{Percentage Removal\%} = & +92.38 - 0.5236\text{A} + 4.81\text{B} - 8.48\text{C} + 1.27\text{AB} \\ & - 0.0551\text{AC} + 4.95\text{BC} + 1.12\text{A}^2 + 0.4260\text{B}^2 - 4.20\text{C}^2\end{aligned} \tag{2}$$

It is clear right away that the contaminant concentration (C) was crucial to the outcome of the experiment. The effect of the MB concentration was inversely proportionate.

The employed model's R^2, adjusted R^2 and predicted R^2 coefficients of the model were calculated to be, respectively, 0.9788, 0.9516, and 0.6612. These R^2 figures represent the degree of agreement between empirically determined percentage of removal data and model predictions (Fig. 7).

The signal-to-noise ratio measurement's sufficient precision is 22.209, which is substantially higher than the lowest permissible value of 4. This simply means that, within the constraints of the previously defined variables, the suggested model can be used to consistently navigate the design space.

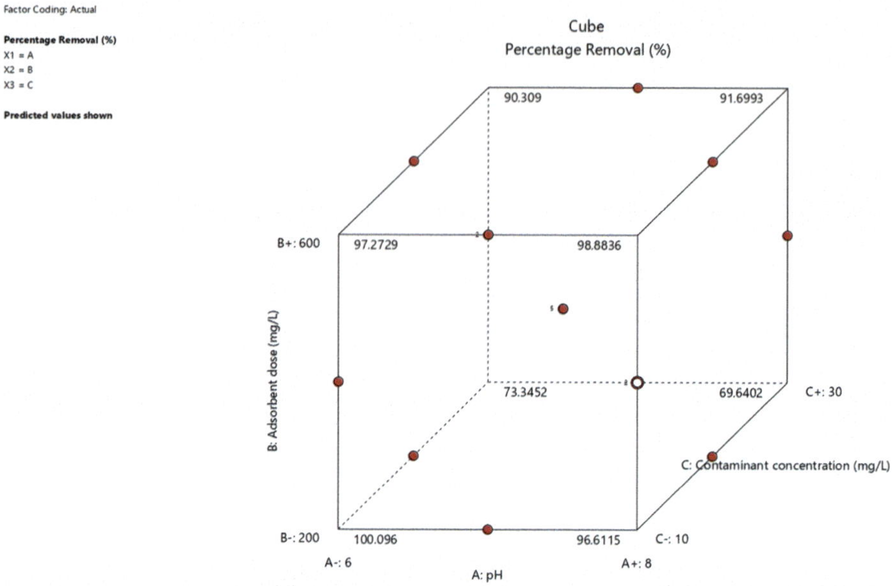

Fig. 8 Cube plot of BBD adsorption models

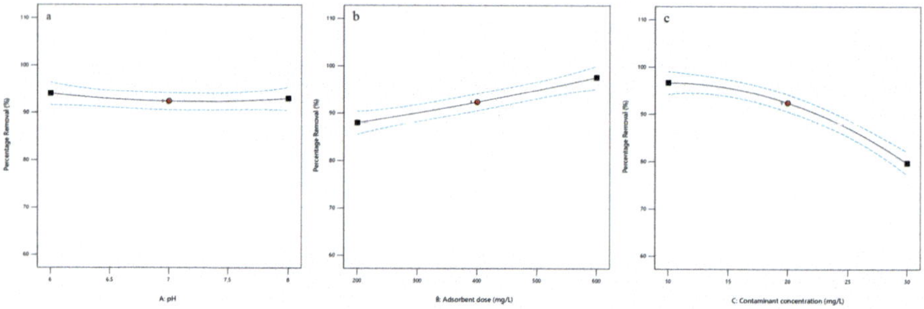

Fig. 9 Effect of independent variables on the response. (**a**) pH vs percentage removal (%), (**b**) adsorbent dose (mg/L) vs percentage removal (%), (**c**) contaminant concentration (mg/L) vs percentage removal (%)

The highest MB adsorption, 98.8836%, was attained in the cubical representation shown in Fig. 8 at a pH (A) of 8, adsorbent dose (B) of 600 mg/L and contaminant concentration (C) of 10 mg/L.

Figure 9 displays the impact of all of the three independent variables on the response. pH 7.25 was chosen as the equilibrium pH (Fig. 9a). This number falls within the average 6–8 range observed in previous methylene blue dye adsorption studies.

The effect of the adsorbent dose Fig. 9b is essentially linear, which signifies the adsorbent's active surface sites are easily accessible for interaction with the methylene blue dye throughout the adsorption process.

The influence of contaminant concentration Fig. 9c on removal percentage is negligible between 10 and 16.5 mg/L. Beyond that, the adsorption reduction is more pronounced, and the slope gets slightly sharper. This might be caused by greater MB concentrations destabilizing adsorbed MB molecules and saturating the adsorbent pores.

While Eq. (2) and Fig. 9 examined the separate effects of the independent variables on methylene blue adsorption, Fig. 10 (3D contour graphs) depicts the cumulative influence of the independent components.

Figure 10a, b represents the effect of pH and adsorbent dose. The graph shows very negligible variation caused due to pH but significant changes has been observed due to the changes in level of adsorbent dose, removal of methylene blue dye increases with increase in adsorbent dose. There has also been a relationship between pH levels and adsorbent dose—at 200 mg/L adsorbent dosage, maximum removal has been noticed at pH 6 and at 600 mg/L adsorbent dosage, pH 8 shows maximum removal. As the contaminant concentration level changes from 30 to 10 mg/L, the removal percentage increases from 70 to 97% at 200 mg/L adsorbent dosage and from 89% to 96% at 600 mg/L adsorbent dosage. The effect of pH is almost linear and parallel to the x-axis, all throughout.

The combined effect of pH and contaminant concentration is demonstrated in Fig. 10c, d. As mentioned earlier, pH levels are almost linear and parallel to the x-axis but shows a dip at pH 7. There is a drastic increase in 70–89% removal of MB dye at contaminant level 30 mg/L when the adsorbent dosage is varied from 200 to 600 mg/L but remains almost the same in 10 mg/L contaminant level with a slight increase from 96% to 98% at pH 8. Equation (2) shows that the overall interaction effect of AC ($\beta_{AC} = -1.75$), on the adsorption efficiency, was somewhat negative.

Finally, the effect of adsorbent dose and contaminant concentration is shown in Fig. 10e, f which, according to Eq. (2), has the maximum impact ($\beta_{BC} = +4.95$). Significant changes have been observed in removal percentage from 70% to 89% at pH 7 and contaminant concentration at 30 mg/L. Removal percentage change was minimal at 10 mg/L of contaminant level with pH variation from 6 to 8.

The experimental findings conformed well to the model, and BP demonstrated varied patterns of extractability, with significant variance of the independent variables in the linear, quadratic and interaction effects. The optimized conditions were pH 7, 400 mg/L adsorbent dose and 20 mg/L contaminant concentration. The corresponding predicted value is 92.3844%. From the experimentation the value of removal has been found to be 92.3844%. So, under optimized conditions the experimental values agree well with the values predicted by the proposed models.

The final optimization was done by the solutions received after entering the lower and upper limits (according to Table 1) and giving equal weightage to all. The following conditions were suggested with Desirability 1: pH 7.572, adsorbent dose 225.008 mg/L and contaminant concentration 10.122 mg/L. Figure 11 shows the

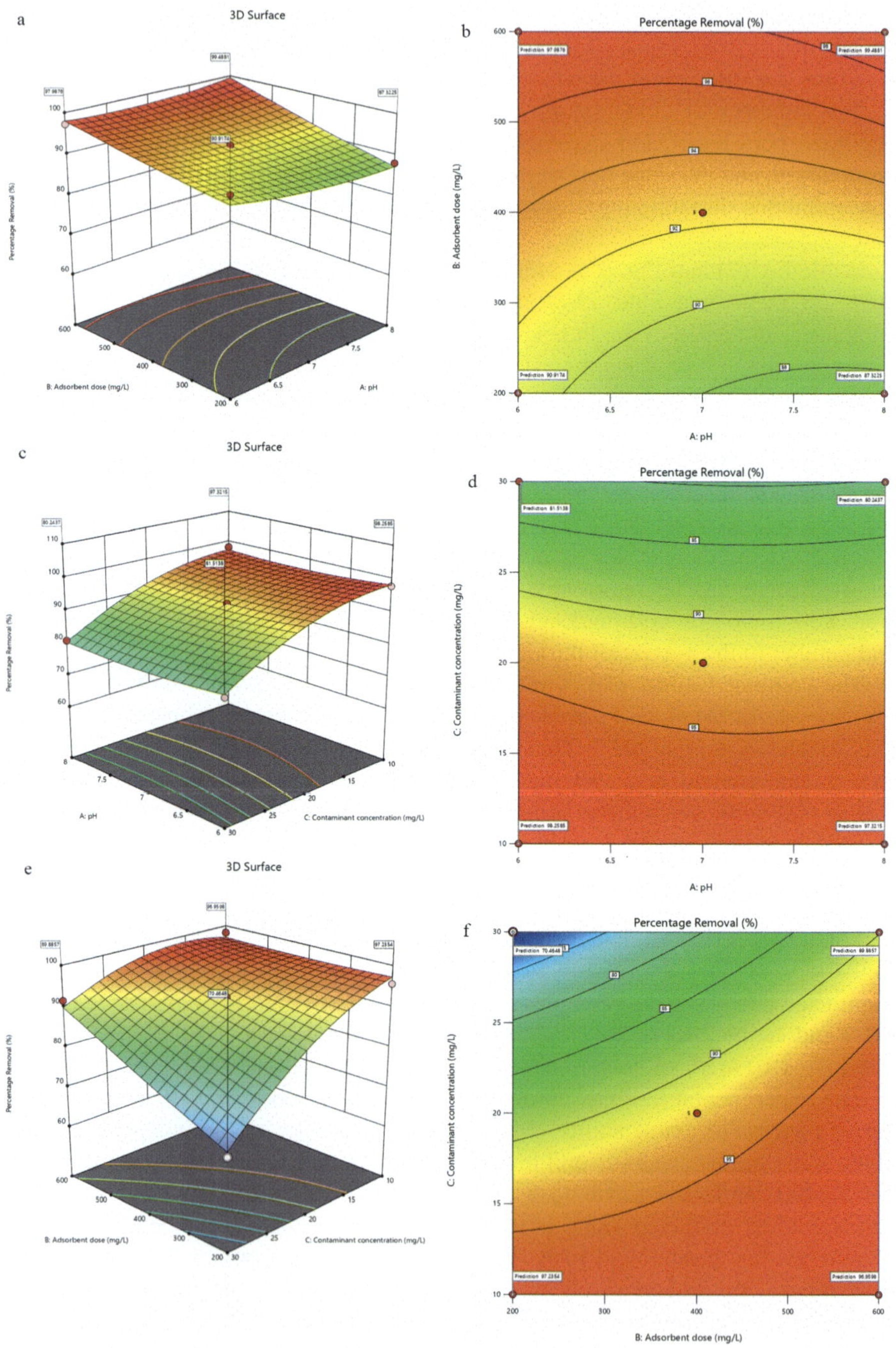

Fig. 10 3D and contour plots demonstrate the mutual interaction effects of pH and AD (**a**, **b**), pH and CM (**c**, **d**) and AD and CM (**e**, **f**)

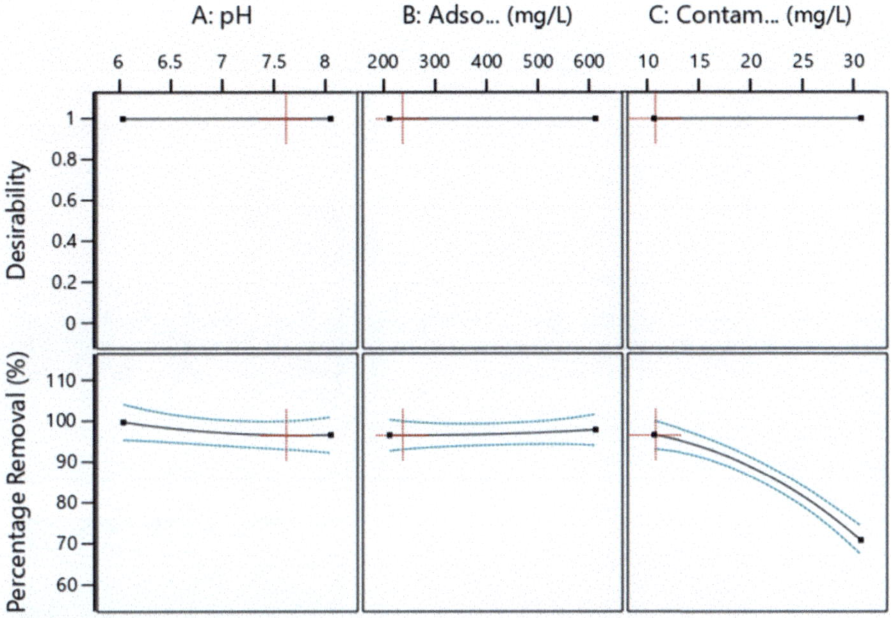

Fig. 11 Desirability of the independent variables as suggested by RSM

variation of the independent variables and the optimized conditions suggested by RSM. The pH is slightly on the alkaline side and the adsorbent dose, and the contaminant concentration are on their lowest levels. Under the following conditions, 96.524% removal of methylene blue dye was achieved. Under the above-mentioned conditions, a treatment was performed to check for the correctness of the model. The removal percentage turned out to be 97.257%, which is at par with our model. Hence, the model values are justified.

Figure 12 shows the graphic representation of the interactive effects that was produced after the optimal adsorption conditions were chosen based on the desirability ratings.

3.3 Characterization

3.3.1 FTIR Analysis

ATR-FTIR spectroscopy was performed on Shimadzu IRAffinity1S to ascertain the BP's surface chemistry both before to and following MB adsorption. The FTIR analysis graph is shown in Fig. 13. Several peaks have been identified under the wavenumber 3850, 3726, 3618, 2900, 2350, 1650–1700, 1523, 1037, 524, and 459 cm^{-1}. The two peaks at ~3850, ~3726 cm^{-1} and the formation of another peak at ~3618 cm^{-1} after adsorption indicate the presence of hydroxyl (O–H)

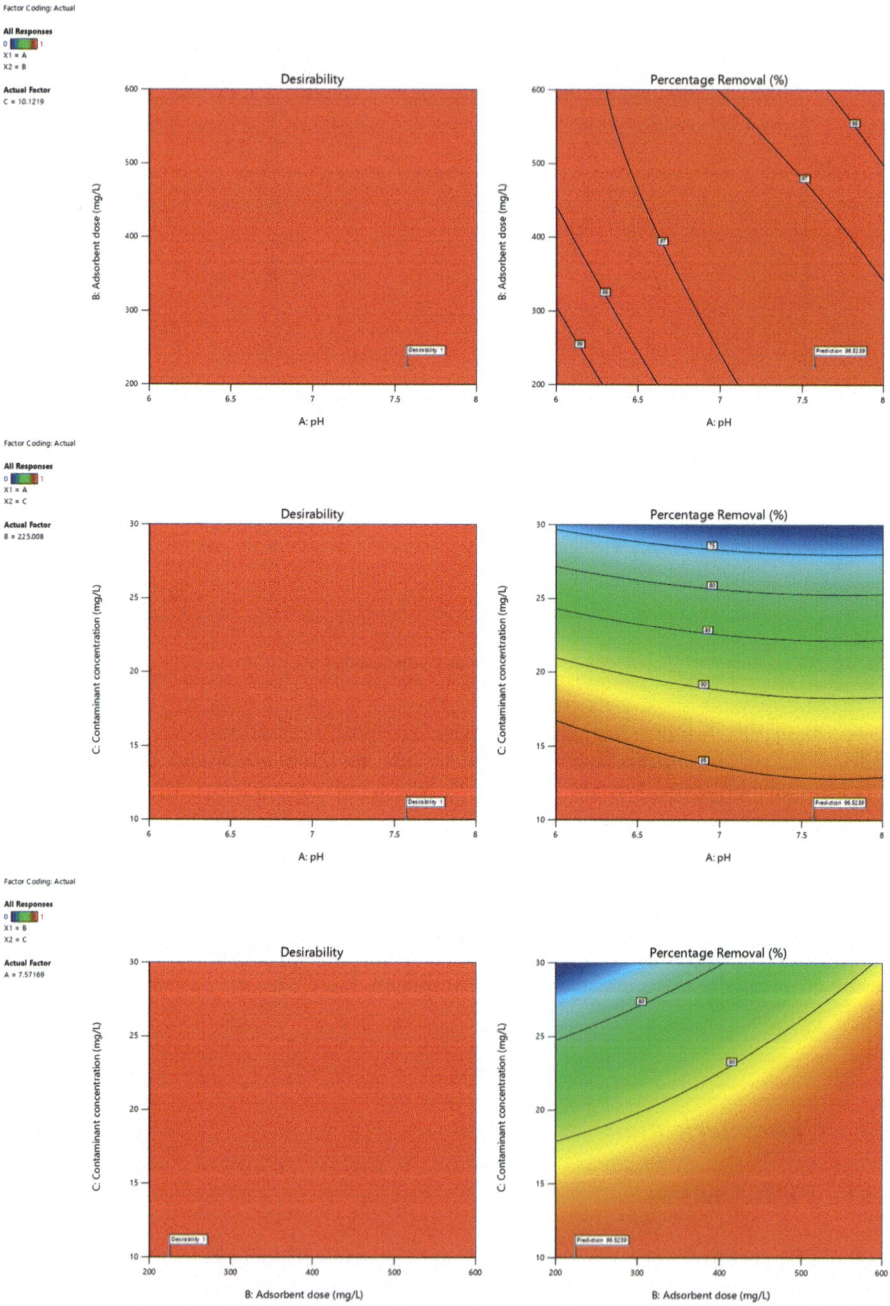

Fig. 12 Recommended optimum adsorption conditions and their desirability

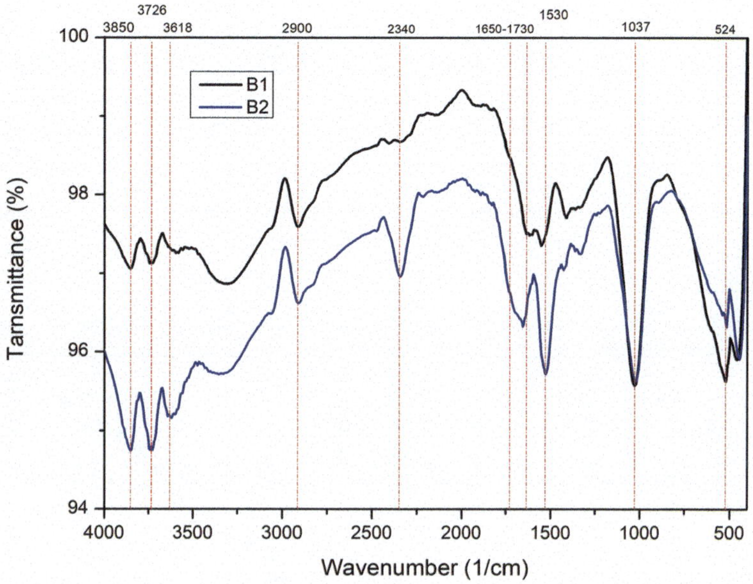

Fig. 13 FTIR spectra of banana peel before adsorption (B1) and after adsorption (B2) of methylene blue

stretching, which is found on the surface of the biosorbent and in lignin, cellulose and hemicellulose as well as adsorbed water [5]. The peak at ~2900 cm^{-1} corresponds to C–H methylene bond. This peak does not match with the aromatic C–H stretching since there was no carbonization or activation performed [23]. The formation of a peak at ~2340 cm^{-1} in the post-treatment graph is close to the asymmetric stretching vibration of the C=O bond in CO_2 [24]. A portion of the curve from 1650 to 1730 cm^{-1} and at ~1530 cm^{-1} deepens in samples B2, signifying the presence of ester (C=O) stretching and benzenoid rings, respectively. The C–OH stretching vibration of the alcohol group is what causes the broad peak at ~1040 cm^{-1} [25]. The peak at ~524 cm^{-1} clearly signifies the formation of halo compound (C–Cl) with the help of the chlorine ion from the dyes.

ATR spectra clearly demonstrate the presence of several functional groups, and when the dye was adsorbed onto the surface of the biosorbents, some absorption peaks shifted, disappeared or resurfaced, indicating that the aforementioned functional groups were engaged in the adsorption process [5].

3.3.2 BET Analysis

The surface area of the biosorbent before and after MB adsorption was determined by using BET (Brunauer–Emmett–Teller) analysis on Quantachrome Nova 100e. The specific surface area of the banana peel is 3.548 m^2/g, which upon adsorption becomes 32.253 m^2/g for methylene blue. The detailed results are mentioned in

Table 4 BET measurements

		B1	B2
	Surface area	2.421 m^2/g	12.298 m^2/g
Adsorption	Pore volume	0.004 cc/g	0.019 cc/g
	Pore radius Dv (*r*)	16.026 Å	17.252 Å
	Surface area	2.121 m^2/g	12.506 m^2/g
Desorption	Pore volume	0.003 cc/g	0.018 cc/g
	Pore radius Dv (*r*)	15.764 Å	15.944 Å
Surface area		3.548 m^2/g	32.253 m^2/g

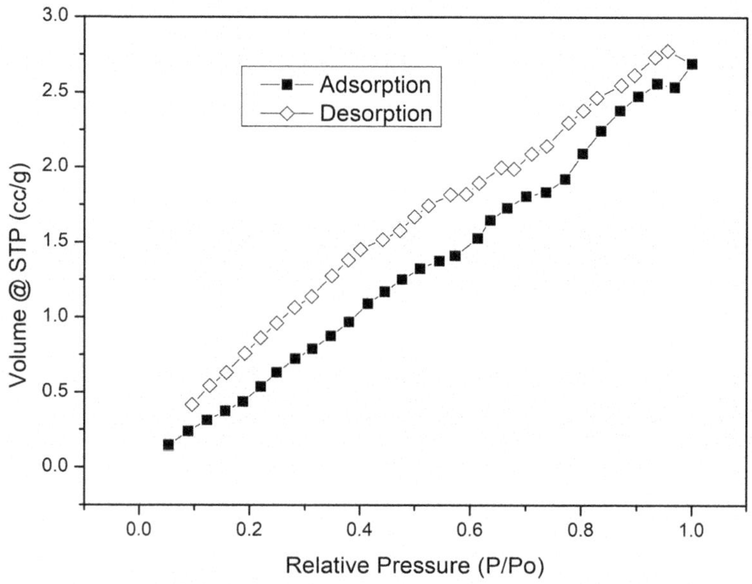

Fig. 14 BET isotherm before treatment

Table 4. In Table 4, B1 represents banana peel before adsorption and B2 represents banana peel after adsorption of methylene blue.

This outcome demonstrates that the dye molecules for methylene blue can easily diffuse into the inner accessible pore of the biosorbent [26].

The method of choice for analysing the porosity and distribution of pore sizes in porous materials is N_2 gas adsorption–desorption. The procedure entails subjecting the sample to nitrogen gas at low temperatures and high pressures, allowing the gas to adsorb onto the material's surface and into its pores. Afterwards, a pressure-dependent measurement of the amount of gas adsorbed is made. Calculating the surface area and pore size distribution of the material involves examining the amount of gas adsorbed at various pressures. To remove the adsorbed gas from the material, the pressure is gradually reduced during the desorption process [27].

As shown in Figs. 14 and 15, the nitrogen adsorption–desorption isotherm was attained at 77 K for banana peel before treatment and after the adsorption of

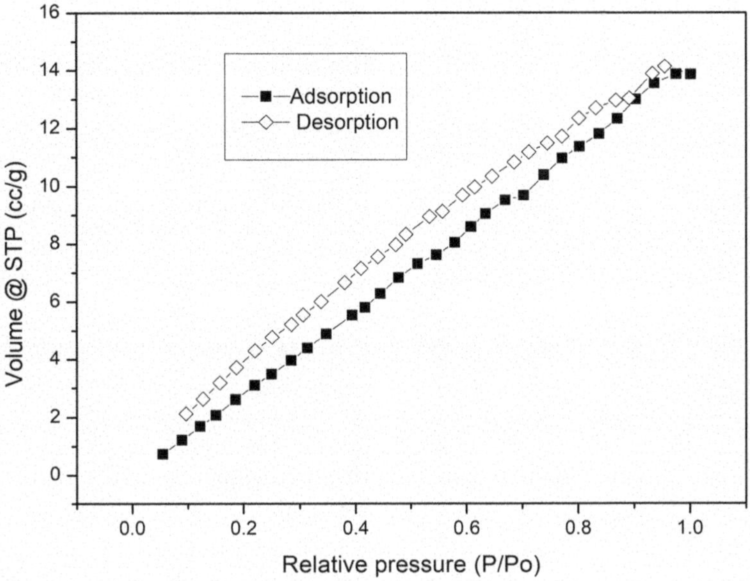

Fig. 15 BET isotherm after methylene blue adsorption

methylene blue. The isotherm curves are classified as Type VI by IUPAC classification. It demonstrated that the biosorbent surface was macroporous.

3.3.3 FE-SEM Analysis

The before-and-after adsorption of the MB by the banana peel was examined under scanning electron microscope (SEM) by Zeiss Sigma. The images of the morphology (Fig. 16) shows that prior to treatment, the particles were clear and the porosity was high, which suggests the presence of void spaces between them. The void diminishes and the particles appear rough, indicating the diffusion of the molecules of the dye among the biosorbent's porous structures.

3.4 Comparison with Existing Work

This section presents a comprehensive comparison of relevant scholarly articles focused on various biosorbents used for MB removal from wastewater. The objective of this analysis was to identify the most effective and efficient biosorbent, refrain from using any chemicals and evaluate its potential for practical applications in wastewater treatment.

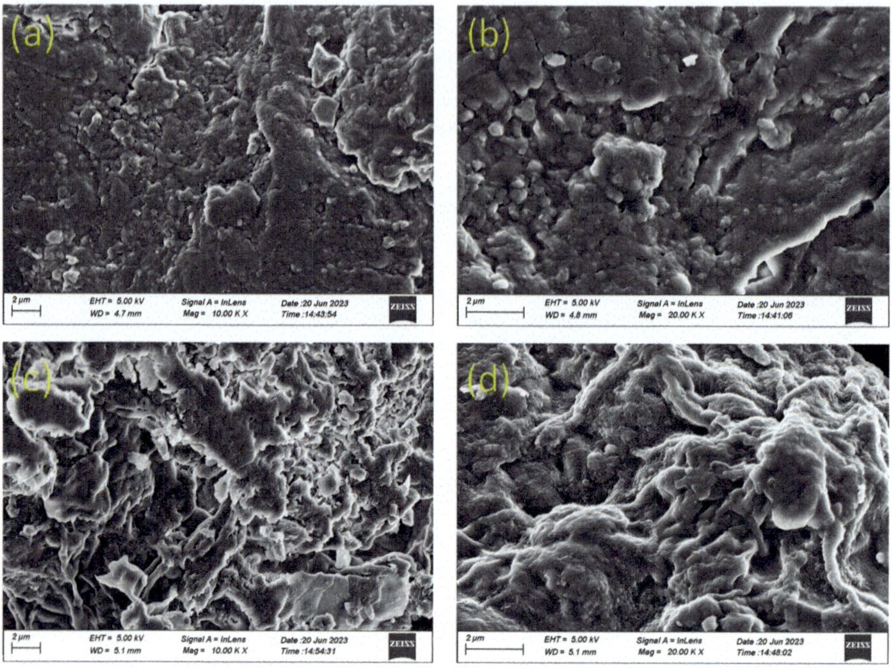

Fig. 16 Representative FE-SEM images of the BP at (**a**) 10× and (**b**) at 20× magnification before adsorption and (**c**) at 10× and (**d**) at 20× magnification after MB adsorption

Table 5 Comparison between the different biosorbents and their effect on the methylene blue

Biosorbent	Contaminant	Contaminant concentration (mg/L)	pH	Adsorbent dose (mg/L)	Removal percentage	Reference
Terminalia catappa shell	Methylene blue	800	5	4000	90.56	[28]
Pine needles	Methylene blue	50	9	500	56.00	[29]
Banana, cucumber and potato peel	Methylene blue	10	6	1000	64.6, 81.1 and 96.8 for CP, PP and BP	[5]
Banana peel	Methylene blue	10	7	400	97.41	This study

The biosorbent preparation was undertaken in a controlled environment, without the use of any chemicals or chemical treatment so as to get the maximum biological remediation possible, along with the aim to create an environment-friendly adsorbent. The current study is conducted at room temperature when compared to other adsorbents (used in their pure and untreated form) for MB removal (Table 5).

According to Ref. [28], pH 5, adsorbent dosage at 4 g/L and an initial dye concentration of 800 mg/L were the ideal conditions for TC shell absorption of MB with a maximum removal capacity of 90.56%.

According to Ref. [29], 56% of the removal of MB was achieved at 9 pH, with 0.5 g/L adsorbent in 45 min with an initial dye concentration of 50 mg/L. Although the pine needle adsorbent was treated with a mixture of strong and weak acids, the higher removal levels were not achieved.

In the study [5], it was stated that at 6 pH, and with an adsorbent dosage of 1.0 g/L, the removal percentage was obtained as 64.6%, 81.1%, and 96.8% for cucumber peel, potato peel and banana peel, respectively. The adsorbent was used in an acidic medium and 2.5 times higher than this study to achieve the desired results.

All the above-mentioned studies portray the fact that higher levels of removal can be achieved through the use of sufficiently large quantities of biosorbent, at non-neutral pH levels, or pre-treatment with acids.

The removal efficiency achieved is maximum with the minimum amount of biosorbent used. This information leads one to the conclusion that, when compared to the stated adsorbents, BP is the most cost-effective, efficient and effective biosorbent. The current study addresses the problem of water contamination in the country and provides a way to use solid waste. Despite the widespread use of functionalized polymers, biosorption is preferred since it is an environmentally friendly method. Adsorption's inability to efficiently handle highly concentrated effluents and/or a huge volume of wastewater stream is its drawback as a water treatment technology [28].

4 Conclusion

In this study, it was demonstrated that a solid waste material can be turned into a high-value product using a straightforward and economical process. The obtained biosorbent showed positive and favourable results in terms of the adsorption of methylene blue dye. FTIR analysis showed the appearance of various organic and non-metallic bondings, which were formed after the adsorption. BET analysis revealed that the biosorbent was macroporous, which resulted in the dye molecules for methylene blue easily diffusing into the inner accessible pore of the biosorbent. The SEM images helped to study the morphology of the biosorbent both before and after the adsorption, clearly indicating the diffusion of dye molecules into the void spaces of the biosorbent. The response surface methodology and the Box–Behnken design (RSM-BBD) were used to optimize the methylene blue adsorption conditions on the banana peel adsorbent. Under the optimum conditions of pH 7.572, adsorbent dose 225.008 mg/L and contaminant concentration 10.122 mg/L, a removal efficiency of 96.524% was achieved. BP adsorbent can be mass-produced due to the cheap and environmentally friendly nature of the raw materials. Thus, these BPs can be used for the purification of natural water.

Acknowledgement The authors are obliged to the Director, National Institute of Technology Durgapur, West Bengal, India; the Department of Biotechnology, National Institute of Technology Durgapur, West Bengal, India; and The Centre of Excellence in Advanced Materials, National

Institute of Technology Durgapur, West Bengal, India, for providing essential support to carry out the present research work.

References

1. H. Zeng, W. Qi, L. Zhai, F. Wang, J. Zhang, D. Li, Magnetic biochar synthesized with waterworks sludge and sewage sludge and its potential for methylene blue removal. J. Environ. Chem. Eng. **9**(5), 105951 (2021). https://doi.org/10.1016/j.jece.2021.105951
2. R. Afroz, M.M. Masud, R. Akhtar, J.B. Duasa, Water pollution: challenges and future direction for water resource management policies in Malaysia. Environ. Urban. ASIA **5**(1), 63–81 (2014). https://doi.org/10.1177/0975425314521544
3. J.K. Bediako et al., Characterization of the residual biochemical components of sequentially extracted banana peel biomasses and their environmental remediation applications. Waste Manag. **89**, 141–153 (2019). https://doi.org/10.1016/j.wasman.2019.04.009
4. A.P. Bhat, P.R. Gogate, Degradation of nitrogen-containing hazardous compounds using advanced oxidation processes: a review on aliphatic and aromatic amines, dyes, and pesticides. J. Hazard. Mater. **403**(2020), 123657 (2021). https://doi.org/10.1016/j.jhazmat.2020.123657
5. A. Stavrinou, C.A. Aggelopoulos, C.D. Tsakiroglou, Exploring the adsorption mechanisms of cationic and anionic dyes onto agricultural waste peels of banana, cucumber and potato: Adsorption kinetics and equilibrium isotherms as a tool. J. Environ. Chem. Eng. **6**(6), 6958–6970 (2018). https://doi.org/10.1016/j.jece.2018.10.063
6. V. Katheresan, J. Kansedo, S.Y. Lau, Efficiency of various recent wastewater dye removal methods: a review. J. Environ. Chem. Eng. **6**(4), 4676–4697 (2018). https://doi.org/10.1016/j.jece.2018.06.060
7. Q. Maqsood, N. Hussain, M. Mumtaz, M. Bilal, H.M.N. Iqbal, Novel strategies and advancement in reducing heavy metals from the contaminated environment. Arch. Microbiol. **204**(8), 1–18 (2022). https://doi.org/10.1007/s00203-022-03087-2
8. M. Fomina, G.M. Gadd, Biosorption: current perspectives on concept, definition and application. Bioresour. Technol. **160**, 3–14 (2014). https://doi.org/10.1016/j.biortech.2013.12.102
9. H. Zohoorian, M. Shourian, H. Ahmadzadeh, M. Molazadeh, S. Lyon, *Microalgal Bioremediation of Heavy Metals and Dyes* (Elsevier Inc., 2020). https://doi.org/10.1016/B978-0-12-818305-2.00041-3
10. M. Madeła, M. Skuza, Towards a circular economy: analysis of the use of biowaste as biosorbent for the removal of heavy metals. Energies **14**(17) (2021). https://doi.org/10.3390/en14175427
11. A.H. Hashem, E. Saied, M.S. Hasanin, Green and ecofriendly bio-removal of methylene blue dye from aqueous solution using biologically activated banana peel waste. Sustain. Chem. Pharm. **18**(July), 100333 (2020). https://doi.org/10.1016/j.scp.2020.100333
12. S. Singh, N. Parveen, H. Gupta, Environmental Technology & Innovation Adsorptive decontamination of rhodamine-B from water using banana peel powder: a biosorbent. Environ. Technol. Innov. **12**, 189–195 (2018). https://doi.org/10.1016/j.eti.2018.09.001
13. U. Maheshwari, R.V. Thakur, D. Deshpande, S. Ghodke, Efficiency evaluation of orange and banana peels for dye removal from synthetic industrial effluent. Mater. Today Proc. **76** (2023). https://doi.org/10.1016/j.matpr.2022.11.023
14. I.K. Da Silva Correia, P.F. Santos, C.S. Santana, J.B. Neris, F.H.M. Luzardo, F.G. Velasco, Application of coconut shell, banana peel, spent coffee grounds, eucalyptus bark, piassava (Attalea funifera) and water hyacinth (Eichornia crassipes) in the adsorption of Pb2+ and Ni2+

ions in water. J. Environ. Chem. Eng. **6**(2), 2319–2334 (2018). https://doi.org/10.1016/j.jece.2018.03.033

15. M. Akter, F.B.A. Rahman, M.Z. Abedin, S.M.F. Kabir, Adsorption characteristics of banana peel in the removal of dyes from textile effluent. Text **1**(2), 361–375 (2021). https://doi.org/10.3390/textiles1020018
16. A. Prasad, K.V.B. Rao, Physico chemical analysis of textile effluent and decolorization of textile azo dye by Bacillus endophyticus strain VITABR13. IIOAB J. **2**(2), 55–62 (2011)
17. A.S. Sartape, A.M. Mandhare, V.V. Jadhav, P.D. Raut, M.A. Anuse, S.S. Kolekar, Removal of malachite green dye from aqueous solution with adsorption technique using Limonia acidissima (wood apple) shell as low cost adsorbent. Arab. J. Chem. **10**, S3229–S3238 (2017). https://doi.org/10.1016/j.arabjc.2013.12.019
18. R. Zhou et al., An efficient bio-adsorbent for the removal of dye: adsorption studies and cold atmospheric plasma regeneration. J. Taiwan Inst. Chem. Eng. **68**, 372–378 (2016). https://doi.org/10.1016/j.jtice.2016.09.030
19. V.K. Garg, R. Gupta, A.B. Yadav, R. Kumar, Dye removal from aqueous solution by adsorption on treated sawdust. Bioresour. Technol. **89**(2), 121–124 (2003). https://doi.org/10.1016/S0960-8524(03)00058-0
20. M.H. Baek, C.O. Ijagbemi, O. Se-Jin, D.S. Kim, Removal of Malachite Green from aqueous solution using degreased coffee bean. J. Hazard. Mater. **176**(1–3), 820–828 (2010). https://doi.org/10.1016/j.jhazmat.2009.11.110
21. X. Pan, D. Zhang, Removal of malachite green from water by Firmiana simplex wood fiber. Electron. J. Biotechnol. **12**(4), 1–10 (2009). https://doi.org/10.2225/vol12-issue4-fulltext-4
22. S.L.C. Ferreira et al., Box-Behnken design: an alternative for the optimization of analytical methods. Anal. Chim. Acta **597**(2), 179–186 (2007). https://doi.org/10.1016/j.aca.2007.07.011
23. F. Çatlıoğlu et al., Preparation and application of Fe-modified banana peel in the adsorption of methylene blue: process optimization using response surface methodology. Environ. Nanotechnol. Monit. Manag. **16**(June) (2021). https://doi.org/10.1016/j.enmm.2021.100517
24. E.M. Ait Ben Hassi, Fourier transform. Theory Distrib., 205–262 (2023). https://doi.org/10.1002/9781394236411.ch8
25. K. Aziz, F. Aziz, R. Mamouni, L. Aziz, N. Saffaj, Engineering of highly Brachychiton populneus shells@polyaniline bio-sorbent for efficient removal of pesticides from wastewater: optimization using BBD-RSM approach. J. Mol. Liq. **346**, 117092 (2022). https://doi.org/10.1016/j.molliq.2021.117092
26. A. Bukhari et al., Removal of Eosin dye from simulated media onto lemon peel-based low cost biosorbent. Arab. J. Chem. **15**(7), 103873 (2022). https://doi.org/10.1016/j.arabjc.2022.103873
27. Z.A. Alothman, A review: fundamental aspects of silicate mesoporous materials. Materials **5**, 2874–2902 (2012). https://doi.org/10.3390/ma5122874
28. L. Hevira, J.O. Ighalo, H. Aziz, R. Zein, Terminalia catappa shell as low-cost biosorbent for the removal of methylene blue from aqueous solutions. J. Ind. Eng. Chem. **97**, 188–199 (2021). https://doi.org/10.1016/j.jiec.2021.01.028
29. D. Pandey, A. Daverey, K. Dutta, V.K. Yata, K. Arunachalam, Valorization of waste pine needle biomass into biosorbents for the removal of methylene blue dye from water: kinetics, equilibrium and thermodynamics study. Environ. Technol. Innov. **25**, 102200 (2022). https://doi.org/10.1016/j.eti.2021.102200

Risk Assessment for Arsenic in Groundwater in the Ganges Delta of West Bengal, India

Dipanwita Das and Asok Adak

Abbreviations

AAS	Atomic absorption spectroscopy
DWLR	Digital water level recorder
GIS	Geographic information system
HQ	Hazard Quotient
IDW	Inverse distance weightage
IS	Indian Standard
LULC	Land use/land cover
PHED	Public Health Engineering Department
SWID	State Water Investigation Directorate
WRIDD	Water Resources Investigation and Development Department

1 Introduction

Arsenic contamination in groundwater may be due to various geological, hydrological, and anthropogenic factors. It can naturally leach into groundwater from arsenic-rich geological formations influenced by redox conditions, pH, and mineral compositions [1]. Arsenic exists in groundwater as As(III) and As (V) oxidation states, where As(III) is much difficult to remove and is oxidized to As(V) for removal from groundwater by various methods [2]. Excessive groundwater pumping for agricultural, domestic, and industrial uses can lead the arsenic-

D. Das (✉) · A. Adak
Civil Engineering Department, Indian Institute of Engineering Science and Technology, Howrah, West Bengal, India
e-mail: 2022cem001.dipanwita@students.iiests.ac.in; asok@civil.iiests.ac.in

D. Mazumder (ed.), *Sustainable Advanced Technologies for Environmental Management*, Springer Proceedings in Earth and Environmental Sciences,
https://doi.org/10.1007/978-3-031-64006-3_6

contaminated water from deeper to shallow, previously uncontaminated aquifers. Industrial discharges, mining activities, use of arsenic-rich pesticides, and improper disposal of arsenic-containing wastes can expose arsenic-bearing substances to the environment, thereby contaminating groundwater [3]. Hydrological conditions such as flow patterns and characteristics of aquifer and subsoil strata variations influence the movement and distribution of arsenic in groundwater [4]. As per IS 10500:2012, the acceptable limit for arsenic in drinking water in India is 0.01 mg/L, whereas the permissible limit in the absence of alternate sources is 0.05 mg/L, and arsenic contamination can have serious health, agricultural, and domestic implications. As groundwater is the major source of drinking water, its contamination with higher arsenic levels beyond standard values can cause arsenic poisoning, leading to abdominal problems, cancer, skin lesions, and respiratory problems on long-term consumption [5]. High levels of arsenic in irrigation water can accumulate in crops, affecting food safety or causing soil degradation, leading to reduced agricultural productivity. Arsenic-contaminated water poses a risk of being used in domestic households for cooking, washing, and so on, and aquatic ecosystems and biodiversity may also be affected to a great extent [6]. There are various methods to evaluate arsenic risk in groundwater, such as groundwater sample collection from various locations and depths to analyze arsenic content by equipment [7]. Geographical information systems (GIS) maps can be used to visualize the spatial distribution of arsenic-contaminated areas and identify the high-risk zones by incorporating geological formations, land use/land cover, and demographic factors [8].

Das et al. studied the high groundwater arsenic concentrations in the gram panchayats of Raninagar II block, Murshidabad, and its detrimental effects on agriculture and the health of the local population [9]. Rahman et al. observed that majorly 17 blocks in Nadia district have groundwater arsenic concentrations of more than 0.05 mg/L, and the village populations using this water for consumption and agriculture were substantially exposed to chronic As toxicity, skin lesions, and neuropathy disorders [10]. Singh et al. studied the hazard and cancer risk due to high arsenic concentration in groundwater and irrigated rice consumed in villages of Nadia and North 24 Parganas [11]. Ghosh et al. prepared GIS-based arsenic concentration maps by interpolation and studied the chronological arsenic infestation levels in tube wells and their spatial distribution in North 24 Parganas district with the scope for trend analysis and land use comparisons [12]. Chaudhuri et al. used GIS maps to locate the arsenic-rich areas at the village level in Baruipur, South 24 Parganas, and its impact on children [13]. Shah and Chatterjee studied by geospatial analysis that the Ganges Delta region has arsenic-rich Holocene aquifers that contain iron oxides, creating reducing conditions, and concentrations of arsenic were found to be greater at shallow depths overall, endangering the region with high arsenic contamination [14].

There requires a study to comprehensively put together all the factors that cause variation in groundwater arsenic throughout the Gangetic Delta plain over time, along with the risk associated with the health of the general population of the region as a robust approach. Also, the sporadic nature of arsenic, which makes it very difficult to predict the concentrations over a large area, could be represented geospatially, and through data interpolation techniques, an average value for the region at large could be obtained. The objective of this chapter, therefore, is to allow a better understanding of the extent of arsenic contamination in groundwater with respect to the groundwater table depth, cropping pattern and hydrogeomorphology of the study area and support the development of effective risk reduction and mitigation strategies for the benefit of livelihood and health of the local population habituating in the Ganges Delta of West Bengal.

2 Materials and Methods

The study area considered includes the districts of Murshidabad, Nadia, North 24 Parganas, and South 24 Parganas in the West Bengal Gangetic Delta region. The groundwater arsenic concentration data over 5 years (2018–2022) duration for various blocks in the districts of Murshidabad, Nadia, North 24 Parganas, and South 24 Parganas have been collected from Public Health Engineering Department (PHED), Govt. of West Bengal. The pre- and post-monsoon groundwater table data were collected from the State Water Investigation Directorate (SWID), Water Resources Investigation and Development Department (WRIDD), Govt. of West Bengal. As the data is collected from respective government departments, for planning and prediction purposes, blocks are considered to be micro-level administrative units instead of natural boundaries. So, blockwise data have been collected from the departments as conventional reports. The arsenic concentrations obtained from PHED, West Bengal, were determined by physical and chemical sampling and testing by atomic absorption spectroscopy (AAS) based on Indian Standard code IS 3025(Part 3):1987. The groundwater table data were obtained from SWID, WRIDD, West Bengal through permanent hydrograph stations where the water level data is collected four times a year manually in January, April, August, and November using a digital water level recorder (DWLR) that has been recently installed in good numbers to get the real-time water level data.

The presence of arsenic in groundwater, being sporadic in nature, could be assessed by the principle that closer points have a stronger influence on the estimation of the arsenic concentration of the surrounding region and can be typically interpolated in GIS by assigning higher weightage to nearby locations and lower weightage to distant ones. The methodology used here is Inverse Distance Weightage (IDW) and can be considered suitable in this regard due to its simplicity, adaptability, computational efficiency, and ability to establish controlled spatial relationships in data typically when estimating values at unsampled locations, thereby aligning well with the real-world spatial patterns due to various geological factors [15]. The data interpolation was done with the help of geostatistical tools and the identification of high-risk arsenic zones using geospatial analysis in ArcGIS 10 software. The maps on arsenic concentration, groundwater table depth, land use/land cover, and hydrogeomorphology were prepared for the area under review for groundwater arsenic risk assessment in a common coordinate system based on pre-monsoon and post-monsoon seasons with general characteristics of the study region.

After obtaining the pre-monsoon and post-monsoon arsenic concentrations and groundwater table data for all the blocks in each district, the data was interpolated on the basis of statistically determined numerical weighted average value in Microsoft Excel, 2016 for every block positioned at their central coordinate as the arsenic concentrations were collected from tube wells at various depths throughout each block and due to its sporadic nature, it varies with location, depth and time. This was done to ensure which blocks in the district were mostly affected by groundwater arsenic contamination and to what extent the surrounding areas were affected as well. After preprocessing the geospatial data obtained from satellite imagery and mosaicking, land use/land cover (LULC) map of the region was prepared, and the features were classified on the basis of supervised classifications by the maximum likelihood method to assign land cover classes for different zones [16] to understand the cropping pattern and identify the agricultural areas where groundwater extraction could be a possible factor to analyze arsenic contamination. Lineament density and geomorphology maps were overlayed using the Weighted Index Overlay method [17] to identify the geological landforms and understand their significance in the arsenic contamination of groundwater in that region.

On obtaining all the maps, the variations in groundwater table depth over the years with respect to pre-monsoon and post-monsoon, the cropping patterns from the land use/land cover, and the general hydrogeomorphology of the regions were analyzed to find out their impact on the variation of arsenic concentrations over time. Spatial visualizations were created to understand the distribution of arsenic concentrations and identify potential risk zones. The actual measurements in the region validated interpolated arsenic concentrations by identifying combined geological, hydrological, and anthropogenic factors such as the presence of arsenic in rock formations, movement of groundwater, and agricultural/industrial activities, respectively, along with risk factors such as Hazard Quotient (HQ), i.e., the ratio of exposure concentration in mg/L to the reference concentration of arsenic in drinking water (0.05 mg/L as per IS 10500: 2012) to determine the toxicological effect due to arsenic contamination. HQ values can be classified into three categories: HQ greater than 1 indicates hazardous conditions, HQ lesser or equal to 1 has non-adverse effects, while HQ lesser than 0.5 implies negligible toxicological effects on health due to arsenic exposure [18]. Finally, after the identification of the potential risk zones, suitable risk reduction and mitigation strategies were sought to eradicate the various problems, including agricultural, domestic, and health issues that are most likely to prevail among the local population in the region.

3 Results and Discussion

The interpolation of the blockwise data obtained for arsenic concentration and groundwater table depth over 5 years was done and mapped for all four districts considered in the Gangetic Delta region of West Bengal. In Nadia district, it was observed that the majority of the district suffers from groundwater arsenic contamination, with Karimpur I and Karimpur II blocks being most affected with an average pre- and post-monsoon concentration of 0.0664, 0.0769 mg/L (pre-monsoon) and 0.0628, 0.0513 mg/L (post-monsoon), respectively, which is beyond the permissible limit as per IS 10500:2012. The other blocks also have arsenic concentrations ranging from about 0.02–0.04 mg/L, which is beyond the acceptable limit of IS 10500:2012, i.e., 0.01 mg/L. So, Nadia district can be considered highly susceptible to groundwater arsenic contamination; the pre-monsoon and post-monsoon arsenic level maps are shown in Fig. 1a, b. The overall groundwater table depth on average for Nadia district in pre-monsoon season was observed to be ranging between 4.17 and 6.64 m and in post-monsoon season between 3.26 and 5.86 m with Chakdah, Tehatta-II, and Krishnaganj blocks having more deep and Ranaghat-II, Krishnanagar, and Nabadwip blocks having more shallow water tables (Fig. 2a, b). Over the past decade, there has been significant depreciation in groundwater table depth in Nadia at the rate of 1.6 cm/year (pre-monsoon) and 6.5 cm/year (post-monsoon). The trendline is shown in Fig. 3.

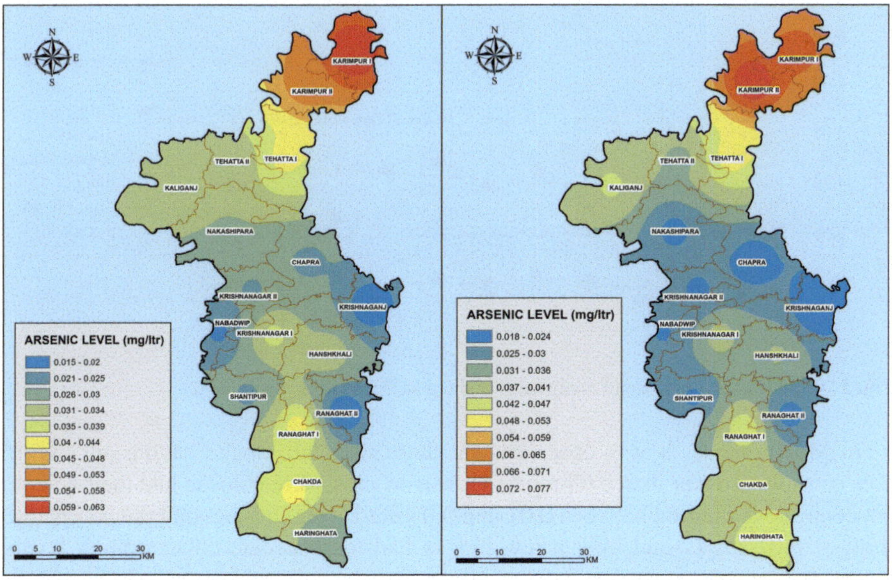

(a) (b)

Fig. 1 Arsenic concentration in groundwater in Nadia: (**a**) pre-monsoon and (**b**) post-monsoon

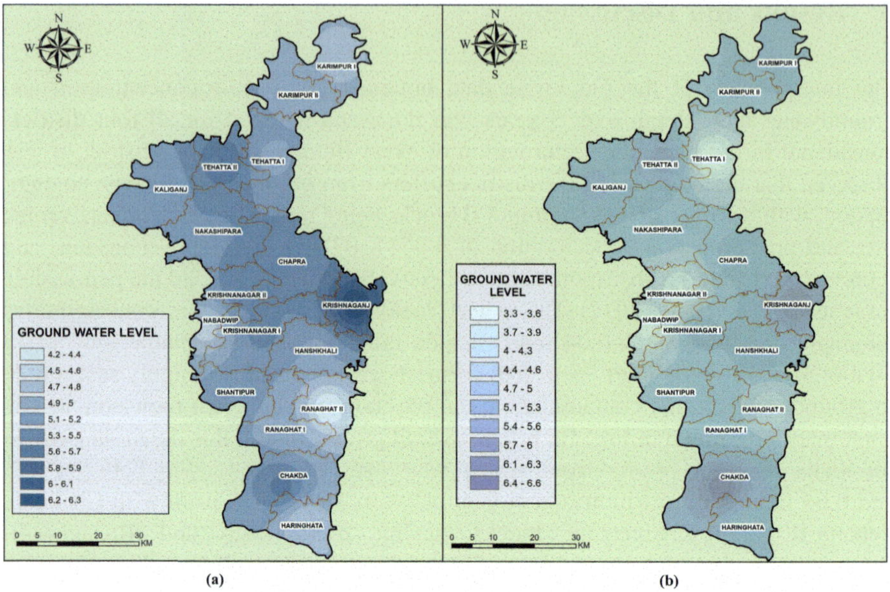

Fig. 2 Groundwater table depth in Nadia: (**a**) pre-monsoon and (**b**) post-monsoon

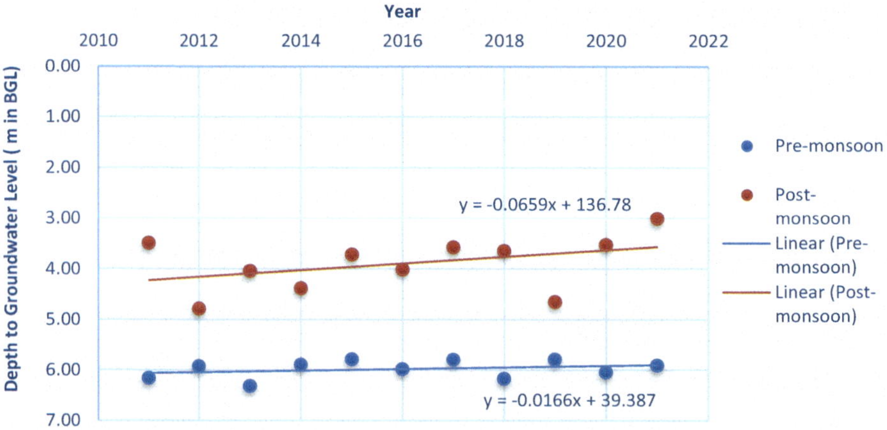

Fig. 3 Groundwater level trend analysis over the past decade for Nadia

In Murshidabad, it was observed that there were no blocks having an arsenic concentration greater than 0.05 mg/L. However, most of the blocks had their arsenic concentrations ranging between 0.01 and 0.05 mg/L, which is beyond the acceptable limit of 0.01 mg/L, and only a few blocks had their arsenic concentration levels below 0.01 mg/L, ensuring safe drinking water quality as per IS 10500:2012. Domkal, Hariharpara, Suti-II (post-monsoon), and Raninagar-II blocks were found to have highly arsenic-contaminated groundwater with an average concentration of

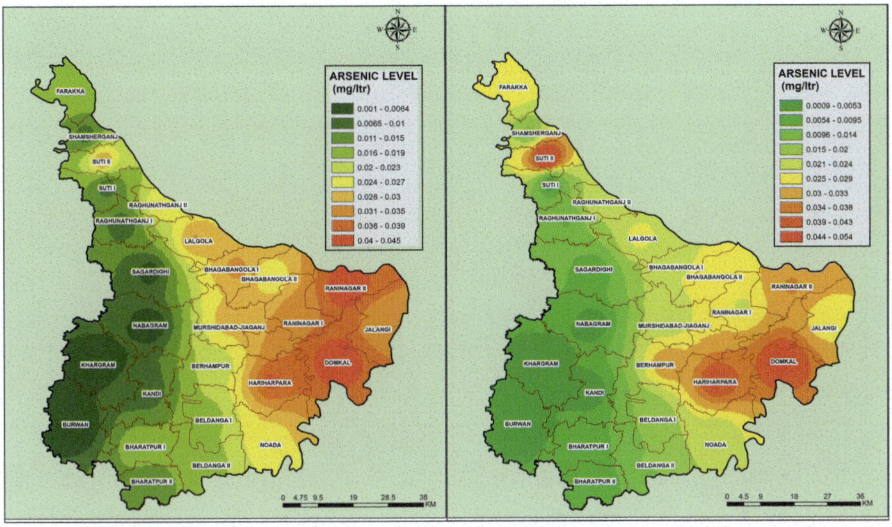

(a) (b)

Fig. 4 Arsenic concentration in groundwater in Murshidabad: (**a**) pre-monsoon and (**b**) post-monsoon

about 0.0481, 0.0453, and 0.0449 mg/L (pre-monsoon) and 0.0455, 0.0406, 0.0543, and 0.0338 mg/L (post-monsoon), respectively. The pre-monsoon and post-monsoon arsenic level maps of Murshidabad are shown in Fig. 4a, b. The overall groundwater table depth for the district was found to range between about 4.16 and 23.78 m in pre-monsoon and between 1.88 and 17.06 m in post-monsoon season, with Nabagram, Khargram, Kandi, Barwan, and Bharatpur blocks having deeper, while Lalgola, Beldanga-I, and Raghunathganj-II blocks having shallower groundwater table, the variation of which is shown in Fig. 5a, b. The groundwater table in Murshidabad has been depleting predominantly over the past decade at the rate of 2.9 cm/year (pre-monsoon) and 11.6 cm/year (post-monsoon), the trendline for which is given in Fig. 6.

In North 24 Parganas, the groundwater is quite arsenic-contaminated throughout, with Gaighata, Habra-I (pre-monsoon), and Basirhat-I blocks having maximum arsenic concentrations of about 0.0665, 0.0574, 0.0834 (pre-monsoon) and 0.0592, 0.0807 mg/L (post-monsoon), respectively, which is beyond the permissible limits as per IS 10500:2012. While few blocks have their groundwater arsenic concentrations in safe ranges, most of the remaining blocks were observed to have values ranging between 0.0127 and 0.0339 mg/L, more than the acceptable limit of 0.01 mg/L. The distribution of arsenic is shown in Fig. 7a, b. The average groundwater table depth was seen to be varying between 5.02 and 12.62 m in pre-monsoon and 1.40 and 8.72 m in post-monsoon months, with Amdanga, Barrackpore, and Barasat-I blocks having deeper and Gaighata, Hasnabad, and Basirhat-II having shallower groundwater table as observed in Fig. 8a, b. Overall, the declination trend

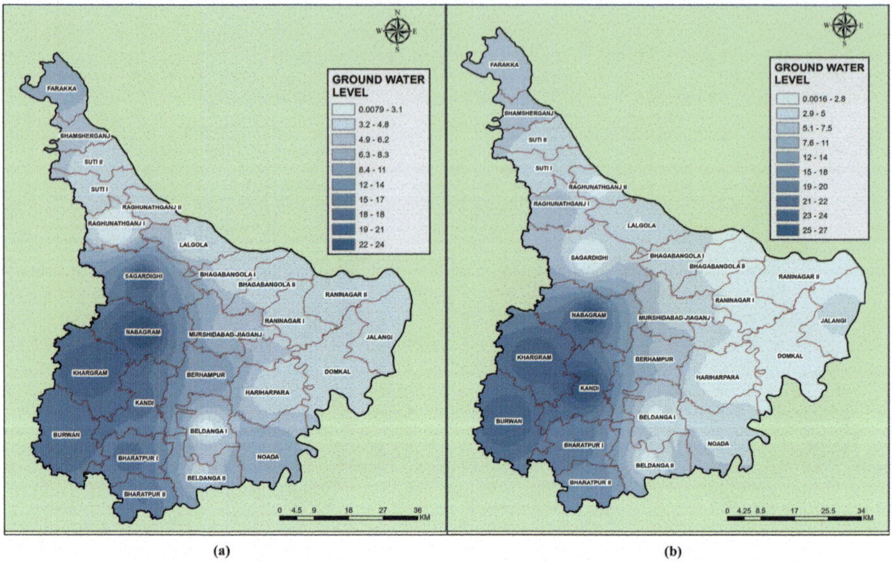

Fig. 5 Groundwater table depth in Murshidabad: (**a**) pre-monsoon and (**b**) post-monsoon

Fig. 6 Groundwater level trend analysis over the past decade for Murshidabad

of the groundwater table over the last decade was found to be at the rate of 8.3 cm/year (pre-monsoon) and 8.6 cm/year (post-monsoon) and shown in Fig. 9.

In South 24 Parganas, there was no significant arsenic contamination found in most of the blocks, while some blocks such as Baruipur, Bhangore-I, II, and Mograhat-II were found to bear arsenic concentrations in groundwater in very low ranges between 0.0006 and 0.0062 mg/L, which are below the acceptable limit (Fig. 10a, b). The groundwater table depth was observed to lie between about 5.87

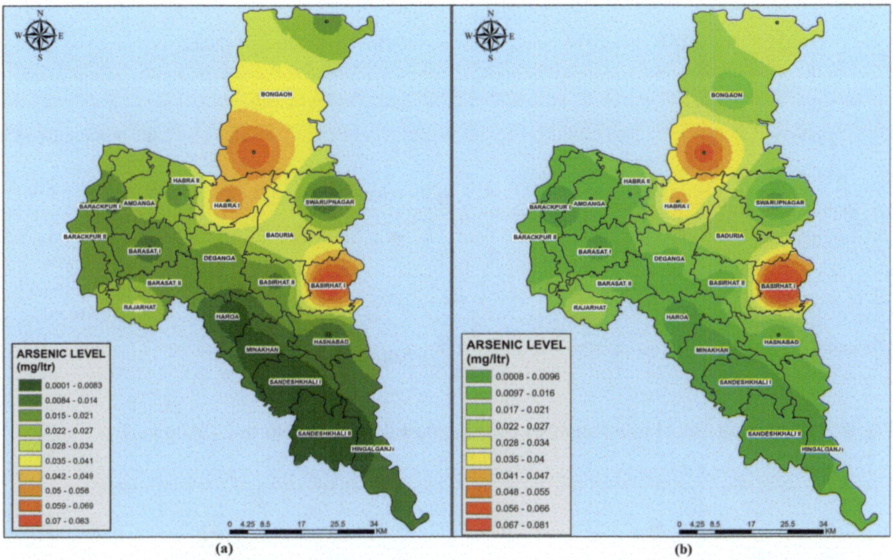

Fig. 7 Arsenic concentration in groundwater in North 24 Parganas: (**a**) pre-monsoon and (**b**) post-monsoon

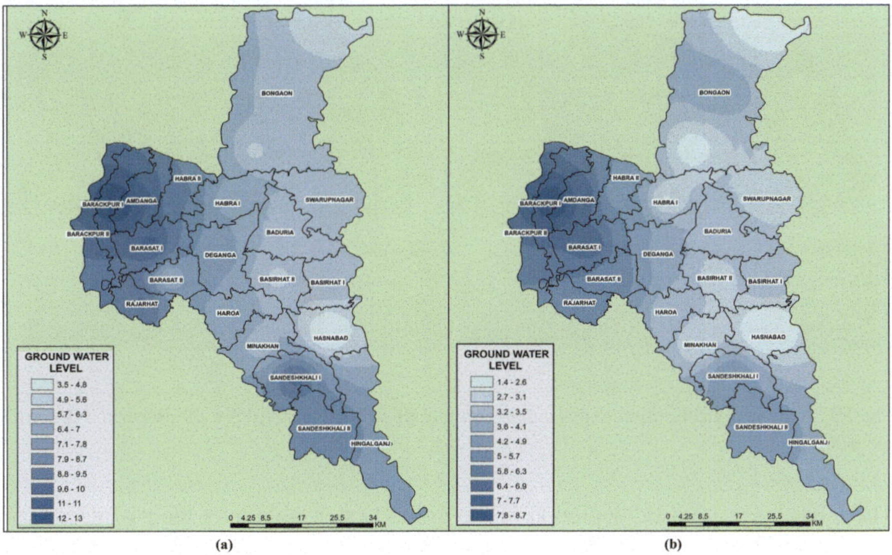

Fig. 8 Groundwater table depth in North 24 Parganas: (**a**) pre-monsoon and (**b**) post-monsoon

and 16.87 m pre-monsoon and 3.34 and 9.94 m post-monsoon, with most of the blocks having very shallow water tables except Kulpi, Sonarpur, Thakurpukur, Maheshtala, and Bishnupur blocks bearing deeper ones as shown in Fig. 11a, b.

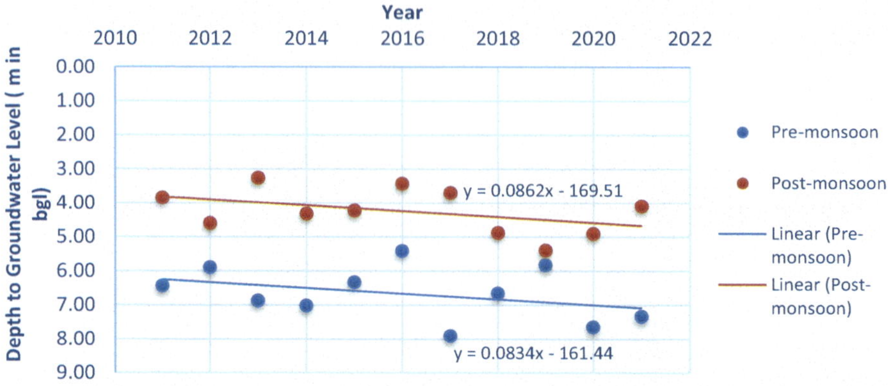

Fig. 9 Groundwater level trend analysis over the past decade for North 24 Parganas

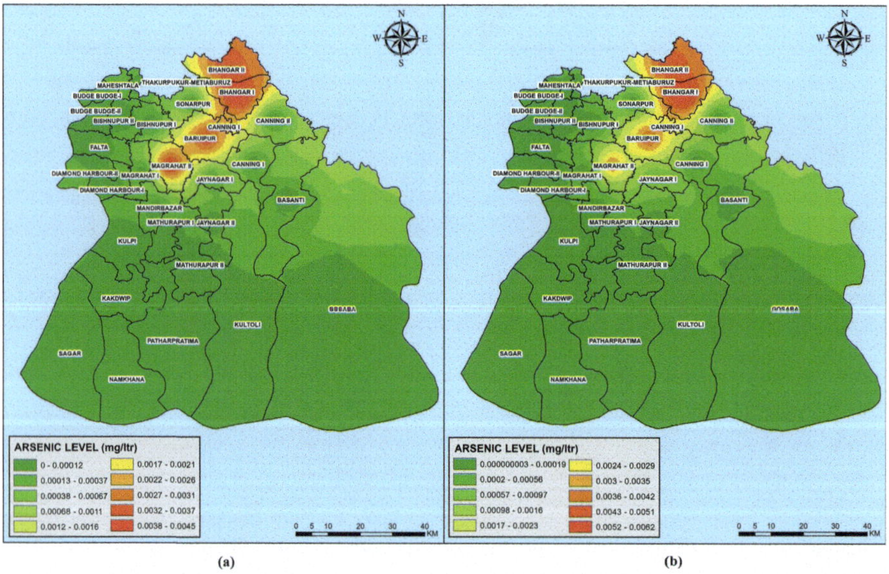

Fig. 10 Arsenic concentration in groundwater in South 24 Parganas: (**a**) pre-monsoon and (**b**) post-monsoon

There has been a drastic depletion of the groundwater table over the past decade at the rate of 39 cm/year (pre-monsoon) and 34 cm/year (post-monsoon), expressed as a trendline in Fig. 12.

After analyzing all the data for the four districts in the Gangetic Delta region of West Bengal, it could be said that the study region is quite affected by groundwater arsenic contamination, and its variation with water table depth was observed in context to pre-monsoon and post-monsoon seasons. In general, the groundwater table depth of the region in the pre-monsoon period was found to be more than in the

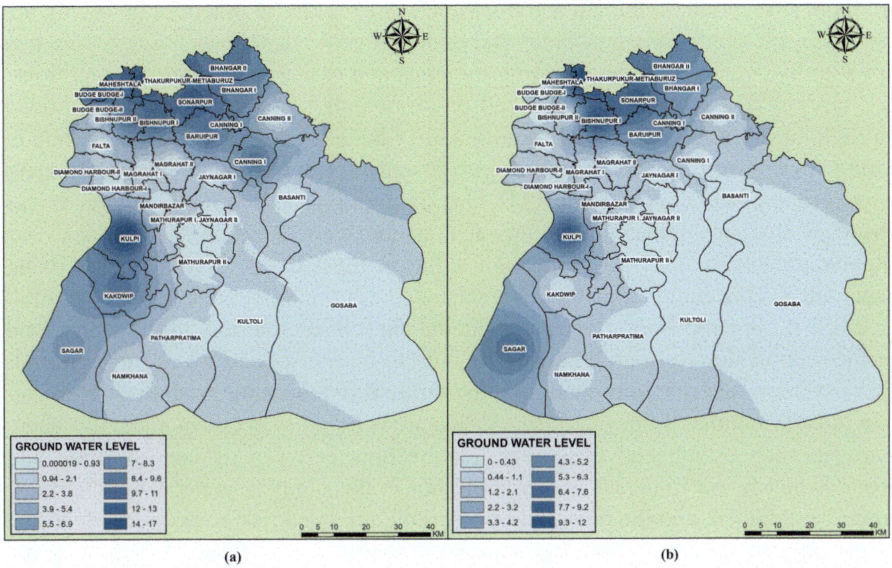

Fig. 11 Groundwater table depth in South 24 Parganas: (**a**) pre-monsoon and (**b**) post-monsoon

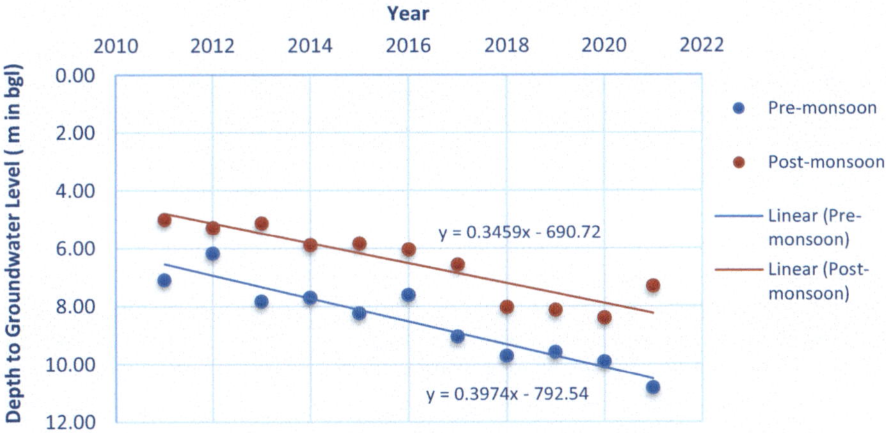

Fig. 12 Groundwater level trend analysis over the past decade for South 24 Parganas

post-monsoon period because groundwater recharge is limited due to lower precipitation before monsoon arrives. Increased rainfall during monsoon causes the replenishment of groundwater, causing the water table to rise in the later months, and then, due to groundwater extraction and subsurface runoff, the water table again declines gradually [19]. It was also mostly observed that the arsenic concentrations in areas of shallow water tables in the districts were greater than the deeper ones. Arsenic could be often found more in shallow water table depths as shallow aquifers are more susceptible to human activities such as agricultural practices involving the use of arsenic-containing pesticides and other geogenic factors such as the presence of naturally occurring arsenic-bearing minerals and oxidation–reduction conditions, i.e., arsenic is less soluble in reduced environments compared to when oxidized through air infiltration in shallow aquifers, they become more soluble and mobile in nature. In contrast, deeper aquifers limit arsenic release and mobility, leading to lower groundwater concentrations [4]. On comparing arsenic level variations, it was observed that pre-monsoon groundwater arsenic concentration for a block was slightly greater than its post-monsoon value in most cases, even with the rising water table. This could be justified by enhanced dilution effect due to heavy rainfall received during monsoon, thereby reducing the concentrations, the flushing, and surface runoff due to rainfall causing arsenic to be transported away from groundwater systems and due to reduction in agricultural practices by groundwater extraction in later months of the year [20]. As arsenic is found to occur sporadically, there were no such established consistent relationships between its concentration and groundwater levels to depict its variation, but in this study, a generalized overview could be obtained, and the potential arsenic-contaminated risk zones were identified in the area of interest. There has also been observed significant depletion in the groundwater table over the past decade in this delta region as obtained from groundwater estimation data by SWID, West Bengal, which, for obvious reasons, is due to high groundwater extraction for agricultural, domestic, and industrial activities along with subsurface runoff due to increased hydraulic gradient [21]. The groundwater table data that was obtained over the decade was not consistent enough to establish a clear relation, so a straightforward analysis was done by linear fitting to easily identify the overall trend and also determine the declination rate of the water table. Although it was so, there were no major changes in arsenic concentrations observed over the 5-year time period in regards to the decline in the water table.

Upon analyzing the land use/land cover obtained for all districts, about 70% of the deltaic region was agricultural lands among other land cover classes such as built-up areas, forests, wastelands, and water bodies as shown in Fig. 13. This showed that extensive agricultural activities practiced in these districts can contribute to the elevated arsenic levels in groundwater through pesticides, groundwater extraction, and arsenic-rich organic matter in the soil, causing arsenic mobility from soil to groundwater. The presence of meandering rivers in this region can influence arsenic concentrations flowing through arsenic-rich geological formations and large sediment interactions that have been carried away by the glacier-melted rivers, Ganga, and its tributaries from the Himalayas supposedly containing arsenic deposits, thereby transporting the arsenic downstream to the delta plains. In contrast, rivers can also dilute the concentration by combining fresh water with lower arsenic levels in its due course [22]. Hydrogeomorphologically, no significant geological variations were found, and the area was mainly plain terrain with recent alluvium, as shown in Fig. 14. Hence, arsenic level variations in groundwater, with its respect, could not be satisfactorily defined in the delta region. Another probable theory regarding the presence of arsenic in groundwater associated with geomorphic

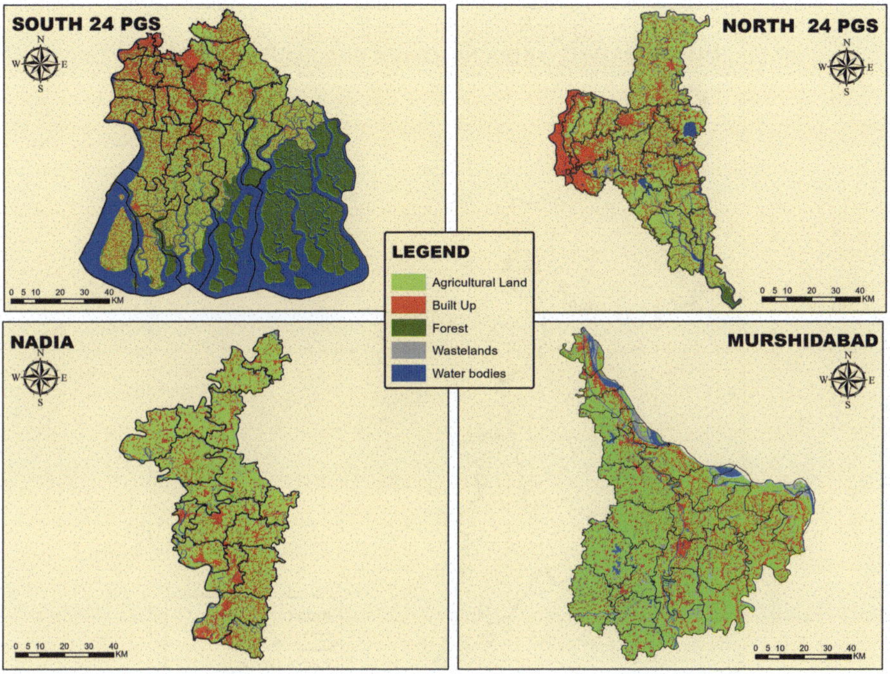

Fig. 13 Land use and land cover maps for Gangetic Delta districts (Murshidabad, Nadia, South 24 Parganas, and North 24 Parganas)

features of the region could be the presence of natural levees and backswamps in the recent alluvial flood plains of Gangetic Bengal plains, which contain mostly coarse-grained sand and fine-grained clayey soil, respectively [23]. In a generalized overview, clayey soils could be found bearing arsenic content greater than sandy soils, which could lead to variations in the level of arsenic contamination in groundwater accordingly [24]. The study area mainly consists of confined aquifers between predominantly thick sandy layers from which groundwater extraction is made for various purposes. Due to its proximity toward the Bay of Bengal, South 24 Parganas district bears salinity in the upper strata, and the suitable groundwater for usage is obtained from very deep aquifers present there, as a result of which there is relatively low groundwater arsenic contamination compared to the other districts in the study region.

The Hazard Quotient was calculated for all the most arsenic-susceptible blocks found in the deltaic districts by determining the ratio of their exposure concentrations and reference concentration of arsenic in drinking water, i.e., 0.05 mg/L as per IS 10500:2012. It was found that Karimpur I and Karimpur II blocks in Nadia district had HQ greater than 1, thereby indicating hazardous conditions in the area, and a similar situation was found to be prevailing in the blocks of Gaighata, Basirhat I, and Habra I in North 24 Parganas as well in both pre-monsoon and post-monsoon seasons. Domkal, Hariharpara, Suti II, and Raninagar II blocks in Murshidabad

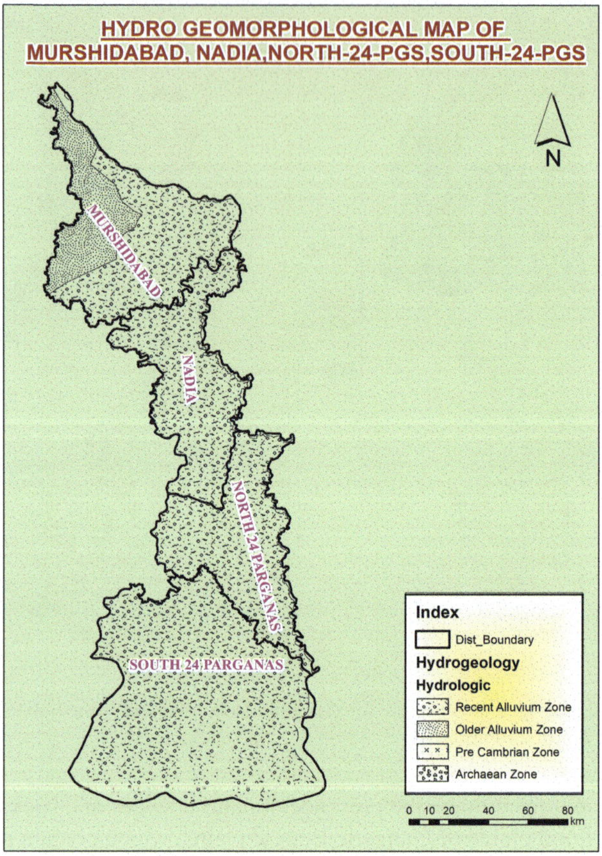

Fig. 14 Hydrogeomorphology map for Gangetic Delta region of West Bengal

district were found to possess HQ less than 1, which indicated that adverse effects are not likely to occur. As South 24 Parganas district was already found to be quite insignificant in groundwater arsenic contamination, the arsenic-containing blocks such as Baruipur, Bhangore, and Mograhat were found to have HQ lesser than 0.5, which further implies negligible hazardous effects in the region. The HQ calculated for all the arsenic-rich blocks for pre-monsoon and post-monsoon of the Gangetic Delta are shown districtwise in Table 1. These values help in determining the risk assessment of the toxicological impact on the health of the inhabitants on exposure to arsenic-contaminated groundwater over time.

Table 1 Hazard Quotient for high arsenic contaminated blocks of the districts in the Gangetic Delta of West Bengal

Districts	Blocks	HQ (pre-monsoon)	Comments	HQ (post-monsoon)	Comments
Nadia	Karimpur I	1.328	Hazardous	1.256	Hazardous
	Karimpur II	1.538	Hazardous	1.026	Hazardous
Murshidabad	Domkal	0.962	Non-adverse	0.91	Non-adverse
	Hariharpara	0.906	Non-adverse	0.812	Non-adverse
	Raninagar II	0.898	Non-adverse	0.676	Non-adverse
	Suti II	–	–	1.086	Hazardous
North 24 Parganas	Gaighata	1.33	Hazardous	1.184	Hazardous
	Basirhat I	1.668	Hazardous	1.614	Hazardous
	Habra I	1.148	Hazardous	–	–
South 24 Parganas	Baruipur	<0.5	Negligible	<0.5	Negligible
	Bhangore	<0.5	Negligible	<0.5	Negligible
	Mograhat	<0.5	Negligible	<0.5	Negligible

4 Conclusion

This study highlights the effectiveness of utilizing data interpolation techniques combined with geospatial analysis to assess groundwater arsenic risk. The districts of Nadia and North 24 Parganas overall pose the major risk toward groundwater arsenic contamination with Karimpur I and Karimpur II (Nadia) and Gaighata, Habra I, and Basirhat I (North 24 Parganas) blocks on the verge of severe consequences. In the Murshidabad district, there are no blocks under the threat of surpassing the permissible limits of the Indian Standard. Other blocks have values beyond the acceptable limit as per the IS code, such as Domkal, Hariharpara, Suti II, and Raninagar II blocks being potential high-risk zones. Due to close proximity to the sea and the presence of aquifers at greater depths, South 24 Parganas district had almost no groundwater arsenic contamination, with little traces found in Baruipur, Bhangore, and Mograhat blocks that were within safe limits. The level of arsenic concentration variation in comparison to the groundwater table depth showed that, on average, mostly shallow water tables were more susceptible to arsenic contamination than deeper ones due to more anthropogenic activities and local geogenic factors. The groundwater table has been depleting for the past 10 years due to increased groundwater exploration activities and subsurface runoff. The pre-monsoon values compared to post-monsoon arsenic level variations in each block are greater in most cases due to dilution, surface runoff, and reduced groundwater extraction. The general hydrogeomorphology of the region did not vary much to determine arsenic level changes, majorly being a plain terrain having recent alluvium. The region being agriculturally rich as obtained by the land use/land cover map with a dense population makes it all the more susceptible to be under the threats of arsenic contamination and its consequences on the health and agriculture sectors of these districts under study.

References

1. C.K. Jain, S.K. Sharma, S. Singh, Physico-chemical characteristics and hydrogeological mechanisms in groundwater with special reference to arsenic contamination in Barpeta District, Assam (India). Environ. Monit. Assess. **190**(7) (2018). https://doi.org/10.1007/s10661-018-6781-5
2. M.T. Al Samman, S. Sotelo, J. Sánchez, B.L. Rivas, Arsenic oxidation and its subsequent removal from water: An overview. Sep. Purif. Technol. **309**, 123055 (2023). https://doi.org/10.1016/j.seppur.2022.123055
3. M.C. Teixeira, A.C. Santos, C.S. Fernandes, J.C. Ng, Arsenic contamination assessment in Brazil – Past, present and future concerns: A historical and critical review. Sci. Total Environ. **730** (2020). https://doi.org/10.1016/j.scitotenv.2020.138217
4. S. Fendorf, H.A. Michael, A. Van Geen, Spatial and temporal variations of groundwater arsenic in South and Southeast Asia. Science AAAS **328**(1123) (2010). [Online]. Available: www.sciencemag.org
5. D. Chakraborti et al., Groundwater arsenic contamination in the ganga river basin: A future health danger. Int. J. Environ. Res. Public Health **15**(2) (2018). https://doi.org/10.3390/ijerph15020180. MDPI.
6. M.A. Rahman, A. Rahman, M.Z.K. Khan, A.M.N. Renzaho, Human health risks and socio-economic perspectives of arsenic exposure in Bangladesh: A scoping review. Ecotoxicol. Environ. Saf. **150**, 335–343 (2018). https://doi.org/10.1016/j.ecoenv.2017.12.032. Academic Press.
7. K.F. Akter, Z. Chen, L. Smith, D. Davey, R. Naidu, Speciation of arsenic in ground water samples: A comparative study of CE-UV, HG-AAS and LC-ICP-MS. Talanta, 406–415 (2005). https://doi.org/10.1016/j.talanta.2005.09.011. Elsevier.
8. N.I. Khan, G. Owens, D. Bruce, R. Naidu, Human arsenic exposure and risk assessment at the landscape level: A review. Environ. Geochem. Health **31**(Suppl. 1), 143–166 (2009). https://doi.org/10.1007/s10653-008-9240-3
9. A. Das et al., Pollution index and health risk assessment of arsenic through different groundwater sources and its load on soil-paddy-rice system in a part of Murshidabad district of West Bengal, India. Groundw. Sustain. Dev. **15** (2021). https://doi.org/10.1016/j.gsd.2021.100652
10. M.M. Rahman et al., Status of groundwater arsenic contamination in all 17 blocks of Nadia district in the state of West Bengal, India: A 23-year study report. J. Hydrol. (Amst) **518**(PC), 363–372 (2014). https://doi.org/10.1016/j.jhydrol.2013.10.037
11. S. Singh, A. Shukla, S. Srivastava, G.S. Kamble, P.K. Patra, V.P. Venugopalan, An evaluation of arsenic contamination status and its potential health risk assessment in villages of Nadia and North 24 Parganas, West Bengal, India. Environ. Sci. Pollut. Res. (2023). https://doi.org/10.1007/s11356-023-28542-5
12. M. Ghosh, D.K. Pal, S.C. Santra, Spatial mapping and modeling of arsenic contamination of groundwater and risk assessment through geospatial interpolation technique. Environ. Dev. Sustain. **22**(4), 2861–2880 (2020). https://doi.org/10.1007/s10668-019-00322-7
13. P. Chaudhuri, P. Aitch, A. Dutta, *Identification of Arsenic Hazard Locations and Impact on Children—A Case Study on Baruipur Block, South 24 Parganas, West Bengal* (2020), pp. 427–449. https://doi.org/10.1007/978-981-32-9527-8_24

14. B.A. Shah, S. Chatterjee, Geospatial assessment of groundwater arsenic contamination in the Holocene aquifers of Bengal Delta from western parts of the Bhagirathi River in West Bengal, India. Arab. J. Geosci. **14**(3) (2021). https://doi.org/10.1007/s12517-020-06375-7
15. A. Setianto, T. Triandini, Comparison of kriging and inverse distance weighted (IDW) interpolation methods in lineament extraction and analysis. J. Appl. Geol. **5** (2013). https://doi.org/10.22146/jag.7204
16. F. SunarErbek, C. Özkan, M. Taberner, Comparison of maximum likelihood classification method with supervised artificial neural network algorithms for land use activities. Int. J. Remote Sens. **25**(9), 1733–1748 (2004). https://doi.org/10.1080/0143116031000150077
17. S. Murasingh, R. Jha, S. Adamala, Geospatial technique for delineation of groundwater potential zones in mine and dense forest area using weighted index overlay technique. Groundw. Sustain. Dev. **7**, 387–399 (2018). https://doi.org/10.1016/j.gsd.2017.12.001
18. M. Goumenou, A. Tsatsakis, Proposing new approaches for the risk characterisation of single chemicals and chemical mixtures: The source related Hazard Quotient (HQS) and Hazard Index (HIS) and the adversity specific Hazard Index (HIA). Toxicol. Rep. **6**, 632–636 (2019). https://doi.org/10.1016/j.toxrep.2019.06.010. Elsevier Inc.
19. S. Shekhar, M. Jha, Groundwater level prediction of Varanasi wells during pre-monsoon and post-monsoon using intelligence approach. Arab. J. Geosci. **15**(1) (2022). https://doi.org/10.1007/s12517-021-08875-6
20. M. Shamsudduha, L.J. Marzen, A. Uddin, M.K. Lee, J.A. Saunders, Spatial relationship of groundwater arsenic distribution with regional topography and water-table fluctuations in the shallow aquifers in Bangladesh. Environ. Geol. **57**(7), 1521–1535 (2009). https://doi.org/10.1007/s00254-008-1429-3
21. A. Mukherjee, S.N. Bhanja, Y. Wada, Groundwater depletion causing reduction of baseflow triggering Ganges river summer drying. Sci. Rep. **8**(1) (2018). https://doi.org/10.1038/s41598-018-30246-7
22. S. Chakraborty, P.K. Sikdar, Understanding the control of geology, geomorphology and landuse/landcover on arsenic distribution in groundwater of Bengal Basin using RS, GIS and PCA. Survey (London, England:1961) **52**(3 & 4), 70–85 (2012). [Online]. Available: https://www.researchgate.net/publication/274952113
23. K.R. Kidd, C.A. Copenheaver, W.M. Aust, Sediment accretion rates and radial growth in natural levee and backswamp riparian forests in southwestern Alabama, USA. For. Ecol. Manag. **358**, 272–280 (2015). https://doi.org/10.1016/j.foreco.2015.09.025
24. A.F. Gerdelidani et al., Arsenic geochemistry and mineralogy as a function of particle-size in naturally arsenic-enriched soils. J. Hazard. Mater. **403** (2021). https://doi.org/10.1016/j.jhazmat.2020.123931

Assessment of Water Quality and Evaluation of Water Quality Index of Loktak Lake

Sudhakar Ningthoujam and Potsangbam Albino Kumar

Abbreviations

BOD	Biochemical oxygen demand
COD	Chemical oxygen demand
DO	Dissolved oxygen
EC	Electrical conductivity
TDS	Total dissolved solids
WQI	Water Quality Index

1 Introduction

Due to population growth and unplanned development, natural water bodies such as lakes, ponds, wetlands and rivers are threatened by pollution, encroachment, displacement, eutrophication and overdevelopment [1]. Natural and anthropogenic sources continue to contribute to many forms of pollution in lakes and other freshwater bodies, particularly in developing countries such as India. Many countries have implemented a variety of regulations and recommendations to improve surface water quality and water resource management in order to revert this downtrend [2]. Water Framework Directive has already been adopted on member states of the European Union (EU) in the year 2000 and it has proven to be an excellent tool for managing water quality and ecosystems [3]. Water quality assessment is one of the key steps in defining water quality guidelines, regulations and policies as well as

S. Ningthoujam · P. A. Kumar (✉)
Civil Engineering Department, National Institute of Technology, Manipur, India
e-mail: albino@nitmanipur.ac.in

D. Mazumder (ed.), *Sustainable Advanced Technologies for Environmental Management*, Springer Proceedings in Earth and Environmental Sciences,
https://doi.org/10.1007/978-3-031-64006-3_7

analysing efforts made to follow these frameworks. The primary objective of analysing water quality is to ensure that the water source is suitable for various applications such as drinking water supply, irrigation, bathing, industrial cooling and so on [4, 5]. Many methods for analysing water quality are being proposed and implemented. Using relevant indices is one of the most effective techniques for studying water quality. The benefit of indices is that they can provide the general public and decision-makers about the state of the ecosystem.

The Water Quality Index (WQI) is one of the numerical scales that represents the quality of water for a specific purpose such as drinking, swimming or supporting aquatic life. It offers the public an easy way to understand complex information about water quality [6–9]. The index determines a single value or grade to indicate the overall quality of the water by considering multiple parameters related to water quality. This chapter aims to determine the extent of pollution in the Loktak Lake waters a Ramsar site at different areas by assessing the physical, chemical and biological characteristics of the water. It also aims in providing valuable information about the present lake water quality for decision-making by developing WQI for different location of the lakes.

2 Materials and Methods

2.1 Study Area

Loktak Lake, a fresh-water wetland system in Manipur, Northeast India, is the present study area. Loktak Lake is famous for one and only floating National Park in the world, Keibul Lamjao Nation Park. Locally referred to as Sangai, brow-antlered deer inhabits Keibul Lamjao National Park. Loktak Lake is a distinctive natural habitat with multiple national and international significances as recognized by the Ramsar Convention [10, 11]. It is located between 93°46′E–93°55′E and 24° 25′N–24°42′N, which is at the Bishnupur district of Manipur (Fig. 1), the lake is oval in shape with an area of 246.72 km^2 and has direct catchment area of 980 km^2. The lake receives an annual rainfall of 1400 mm on average. It has a maximum length of 26 km and width of 13 km; the depth of the lake varies from 0.5 to 4.58 m with an average depth of 2.7 m [10, 12]. Loktak Lake is surrounded by 12 towns and 52 settlements, home to 220,017 people, which is approximately 9% of Manipur state's total population [1].

2.2 Data Collection

Sixteen physicochemical parameters, namely temperature, pH, total dissolved solids (TDS), electrical conductivity (EC), turbidity, hardness, alkalinity, dissolved oxygen (DO), biochemical oxygen demand (BOD), chemical oxygen demand (COD), chloride (Cl), magnesium (Mg), sodium (Na), phosphate (PO_4^{2-}), nitrate (NO_3) and

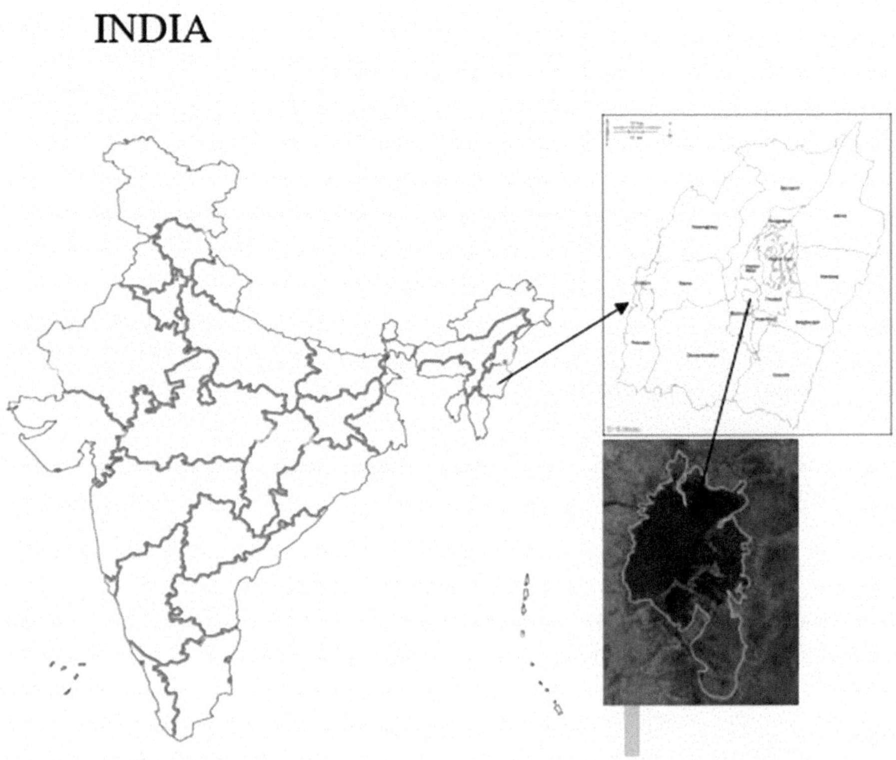

Fig. 1 Loktak Lake location

Table 1 Various coordinates of sampling sites at Loktak Lake

Sl. no	Sites	Latitude	Longitude
1	Mayang Imphal	24°34′38.26″N	93°49′59.44″E
2	Omba	24°30′23.88″N	93°47′46.78″E
3	Sendra	24°30′54.01″N	93°47′17.92″E
4	Thanga	24°31′43.44″N	93°49′39.14″E
5	Karang	24°32′46.91″N	93°49′49.71″E
6	Ithing	24°31′19.76″N	93°47′53.70″E
7	KLNP	24°30′35.77″N	93°50′7.81″E
8	Ningthoukhong	24°34′30.23″N	93°47′14.42″E

nitrite (NO_2) were taken into account in this study in order to calculate WQI for each location. The samples were collected from eight different locations from the lake. The sampling locations were selected based on the settlement, drainage of river and tourist hotspot in and around the lake. The sampling sites are shown in Table 1. Every sampling location was chosen to include important sites that reflect the lake's water quality. In this study the samples were collected bellow 0.5 m from the surface. The samples were collected during August 2022–April 2023 for different season, i.e. post-monsoon, winter and pre-monsoon. Parameters such as pH, TDS, EC, temperature and turbidity were measured in situ using portable water analyser. At each site, triplicate water samples were collected by employing random sampling. The collected water samples are kept in the pre-treated polyethylene bottles at 4 °C until they are brought to the lab for analysis. The samples were filtered using 0.45 μm filters prior to analysis. The samples were then analysed for parameters such as dissolved oxygen (DO), chemical oxygen demand (COD), biochemical oxygen demand (BOD), hardness, alkalinity and the major ions using the American Public Health Association (APHA) standard technique [13].

2.3 Water Quality Index

WQI was calculated using the Weighted Arithmetic Index method. In essence, the Weighted Arithmetic Index is an ascending scale index, with higher scores denoting higher pollution. WQI value ranging from 0 to 25 indicates as excellent water quality and graded as A, 26–50 indicates as good water quality and graded as B, 51–75 as poor water quality and graded as C, 76–100 as very poor water quality and graded as D and WQI value above 100 in unsuitable for drinking purpose. The primary goal of the Weighted Arithmetic Index is to evaluate the quality of water according to whether or not it is suitable for human consumption [14]. The index is quite flexible because the user can select which water quality parameters to include in the process and define boundary concentration values for each of them [6, 9, 15, 16]. The following formulas were used to compute the WQI. Each water quality parameter's unit weight (W_n) was computed using Eq. (1):

$$W_n = K/S_n \tag{1}$$

where W_n stands for the unit weight of nth parameters, K is a proportionality constant and S_n is the standard value of each parameter.

$$K = 1/\sum_{n=1}^{n}(1/S_n) \tag{2}$$

Each parameter's quality rating scale (Q_n) was calculated using Eq. (3):

$$Q_n = 100 * [(V_n - V_0)/(S_n - V_0)] \tag{3}$$

where S_n is the standard value, V_n is the observed value, V_0 is the ideal value. In most cases $V_0 = 0$ except in certain parameters such as pH and dissolved oxygen (DO) where the ideal value are 7 and 14.6 mg/L, respectively. Quality rating calculation for pH and DO are evaluated using Eqs. (4) and (5) respectively as follows:

$$Q_{\mathrm{pH}} = 100(V_{\mathrm{pH}} - 7.0)/(V_s - 7.0) \tag{4}$$

$$Q_{\mathrm{DO}} = 100(V_{\mathrm{DO}} - 14.6)/(V_s - 14.6) \tag{5}$$

WQI was calculated using the following equation:

$$\mathrm{WQI} = \sum\nolimits_{n=1}^{n} Q_n * W_n / \sum\nolimits_{n=1}^{n} W_n \tag{6}$$

3 Results and Discussion

3.1 Surface Water Quality

Sixteen physiochemical characteristics of Loktak Lake's surface water in Manipur were assessed for this study, and compared with the drinking water quality standards given by Bureau of Indian Standards (BIS) [17] and further WQI was determine from these results. The water temperature for different location throughout Loktak Lake and throughout the study period ranges from 22.1 to 16.3 °C (Table 2); this wide variation is most likely because of the influence of solar radiation or thermal circulation caused by interacting land [18]. Turbidity ranges from 3 to 8 NTU and with the lowest being recorded at Thanga and Ithing of 3 NTU during winter. Also, the highest turbidity was observed at Omba Ching area in all the three respective seasons, which is 7.6 NTU during post-monsoon, 8 NTU during winter and 5.6 NTU during pre-monsoon. All three values are above the standard limit given by the BIS and WHO. This is due to the fact that Omba hillock is a tourist spot and with more inhabitants and tourist, pollution is higher at this site. From Table 2, we can also see that some of the sites has recorded higher turbidity as compared to the recommendation given by the WHO and BIS. For TDS, lowest value was observed at Karang area, which is 67 mg/L during winter season and highest was recorded at Ningthoukhong, which is 89 mg/L during pre-monsoon. Such wide variation is also observed in EC, which ranges from 125 to 169 μS/m.

Important chemical parameters of the lake such as pH ranges from 6.6 to 7.5, which are within the limits given by the WHO and BIS. The range of BOD, an essential parameter and widely used indicator of organic matter pollution in water, is 1.8–6 mg/L (Table 3). From Table 3, we can see the highest BOD was observed at Omba, i.e. 6 mg/L during the post-monsoon and the reason is due to the interference from the tourist and the inhabitants. To examine the degree of purity of the lake, DO is an important parameter and its value ranges from 3.5 to 7.2 mg/L (Table 3). The low DO indicates poor water quality at the lake. The COD vales ranges from 24 to 51.7 mg/L (Table 3), most of the sampling location exceed the COD values given by the BIS and WHO. The highest COD was observed at Mayang Imphal area, high COD values around the Mayang Imphal area indicates the presence of inorganic matter on the lake, which is probably due to the runoff of chemical fertilizers and

Table 2 In situ water quality of different location of Loktak Lake during post-monsoon, winter and pre-monsoon

Parameters		Temperature (°C)			pH			TDS (mg/L)			EC (μS/m)			Turbidity (NTU)		
Sl. no.	Location	Post-monsoon	Winter	Pre-monsoon	Post-monsoon	Winter	Pre-monsoon	Post-monsoon	Winter	Pre-monsoon	Post-monsoon	Winter	Pre-monsoon	Post-monsoon	Winter	Pre-monsoon
1	Mayang Imphal	20.4	16.6	16.3	7.08	7.01	7.3	73	85	84	146	163	162	4.6	4.19	3.2
2	Omba	20.1	16.3	16.7	7.5	7.36	7.8	76	86	83	155	164	160	7.6	8	5.6
3	Sendra	20.2	17.2	17.8	7.2	7.1	7.2	78	84	81	144	158	152	4.9	3.95	4.9
4	Thanga	20	18.5	18.2	7	6.9	7.4	77	86	84	142	157	158	4	3	3.6
5	Karang	21.9	18.1	18.6	7.4	7.2	7.3	79	67	70	128	136	134	3.9	3.1	3.2
6	Ithing	22	17.4	18.6	7.3	7.3	7.2	84	76	79	125	157	153	3.7	3	3.1
7	KLNP	20.9	18.3	17.9	6.9	6.6	7.7	69	84	88	146	165	160	3.8	3.22	3.5
8	Ningthoukhong	22.1	18.1	17.9	7.1	7.06	7.6	73	89	89	155	169	162	4.2	4	4.4
Standard deviation (±)		0.9	0.8	0.8	0.2	0.2	0.2	4.5	7.2	5.9	11.1	10.1	9.3	1.3	1.7	0.9
BIS					6.5–8.5			1000			250			5		

Table 3 Water quality of different locations of Loktak Lake during post-monsoon, winter and pre-monsoon

Parameters		Alkalinity (mg/L)			Hardness (mg/L)			BOD (mg/L)			COD (mg/L)			DO (mg/L)		
Sl. no.	Location	Post-monsoon	Winter	Pre-monsoon	Post-monsoon	Winter	Pre-monsoon	Post-monsoon	Winter	Pre-monsoon	Post-monsoon	Winter	Pre-monsoon	Post-monsoon	Winter	Pr[illegible] mo[illegible]
1	Mayang Imphal	120	100	115	80	78	80	5.4	2.7	2.9	51.7	35	29	4.9	5.6	4.[illegible]
2	Omba	102	78	96	46	40	46	6	2.5	4.3	44.7	30.3	30	3.5	5.2	5
3	Sendra	107	78	97	68	62	68	5.6	2.5	2.7	47	31.2	32	6.1	6.3	6.[illegible]
4	Thanga	98	70	93	38	30	38	4.3	2.2	2.4	43.7	29.7	30	4.2	4.5	4.[illegible]
5	Karang	100	85	95	40	34	40	5.5	1.9	2.1	41.9	23.7	28	7	7.2	7
6	Ithing	95	80	87	37	32	37	5.3	1.8	2.3	37.6	21.6	32	6.7	6.9	6.[illegible]
7	KLNP	92	68	86	35	29	35	5.3	2.3	2.6	49.3	28.7	30	4.3	4.8	4.[illegible]
8	Ningthoukhong	121	107	113	47	42	47	5.5	2.5	2.8	39	29.5	24	4.2	4.6	4.[illegible]
Standard deviation (±)		10.9	13.7	10.8	16.4	17.6	16.4	0.5	0.3	0.7	4.9	4.2	2.6	1.3	1.1	1.[illegible]
BIS		200–600			100			5			10			5		

insecticides from the nearby agricultural lands. The elevated concentration of physicochemical parameters around the region is brought about by the two significant rivers, the Nambol and Nambul, which discharge contaminated water straight into the lake and is the cause of Loktak Lake's poor quality [19]. The overall impact on the lake is due to the dumping of solid waste, direct disposal of sewage and chemical fertilizers runoff from the nearby agricultural runoff.

3.2 Determination of Major Ions

Measured concentration of major ions such as Na, Mg, PO_4^{3-}, Cl, NO_2 and NO_3 for post-monsoon, winter and pre-monsoon is displayed in Figs. 2, 3 and 4, respectively. During the post-monsoon, it was noticed that the concentrations of Na and Mg remained predominant in the water. Na concentration was observed highest at Omba during the post-monsoon and the lowest was observed at Keibul Lamjao National Park (KLNP). The concentration of Mg was found to be lowest in Ningthoukhong (6.3 mg/L) during the post-monsoon and to be highest at Thanga (17.41 mg/L) during the pre-monsoon. The reason for the elevated level of Mg at Thanga is because of its periphery of the hills that contains rocks and minerals such as magnesium. Concentration of PO_4^{3-} shows the least among all the other parameters, which ranges from 0.001 to 0.01 mg/L and below standard limits. This results that

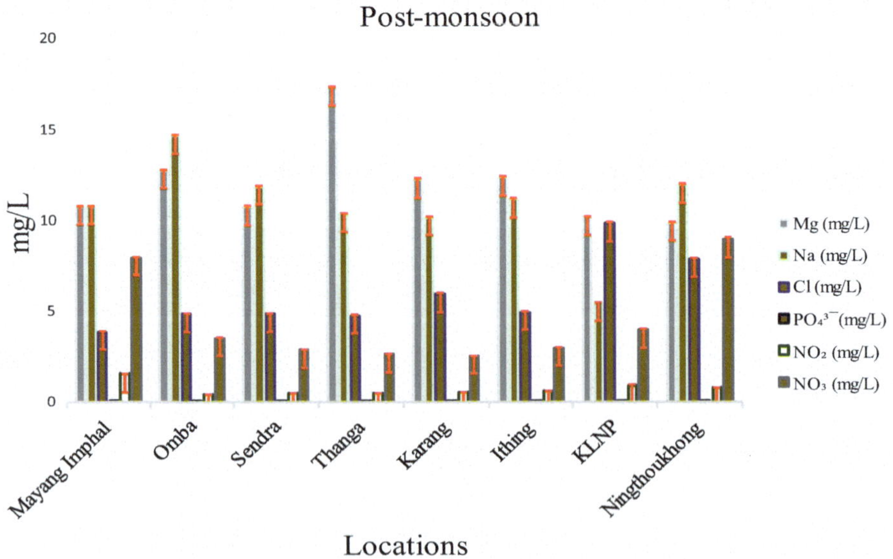

Fig. 2 Concentration of ions at different location of Loktak Lake during post-monsoon

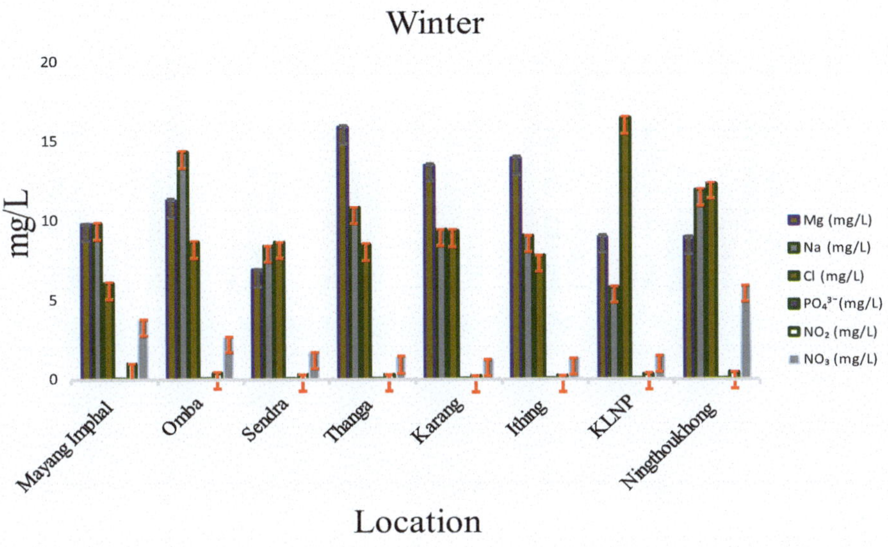

Fig. 3 Concentration of ions at different location of Loktak Lake during winter

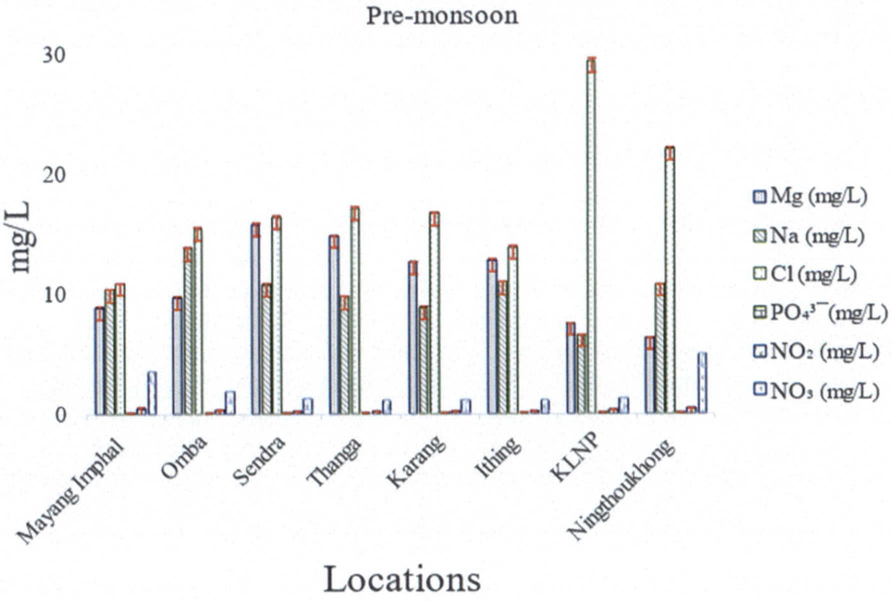

Fig. 4 Concentration of ions at different location of Loktak Lake during pre-monsoon

the concentration of PO_4^{3-} is below the standard limit. The concentration of NO_2 is higher than the permissible limit of 0.1 (WHO), which ranges from 0.15 to 1.75 mg/L. NO_3 value varies seasonally, reaching a maximum of 8.98 mg/L in Ningthoukhong during the post-monsoon and a minimum of 1.1 mg/L at

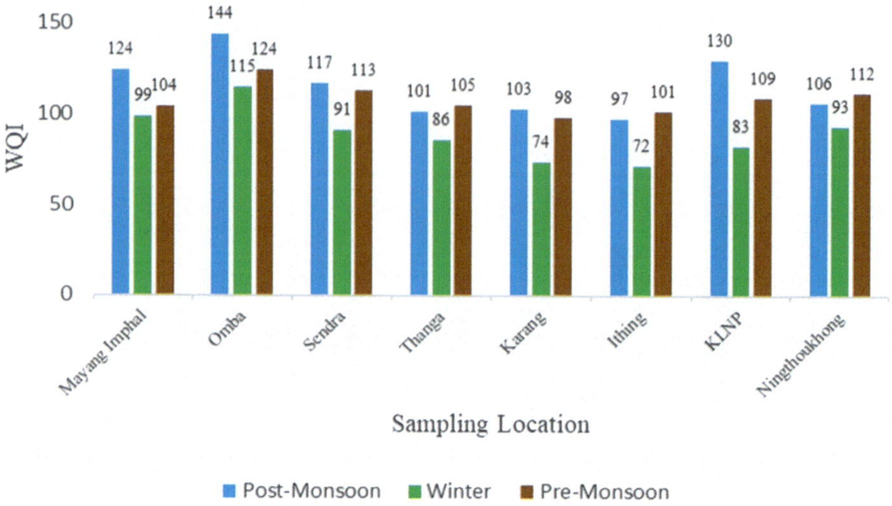

Fig. 5 Water Quality Index for different sites during post-monsoon, winter and pre-monsoon

Ningthoukhong during the post-monsoon. High concentration of NO_3 and NO_2 at Ningthoukhong is due to the runoff of inorganic nitrogen fertilizers and manures from the aquaculture and agricultural farm.

3.3 Water Quality Index

Figure 5 shows the comparison of WQI for different seasons of different locations of Loktak Lake. WQI during the three seasons, i.e. post-monsoon, winter and pre monsoon, ranges from 97 to 144, 72 to 115 and 98 to 124, respectively. The findings indicate that the lake's water quality varies depending on the location and season. The result indicates that in comparison to other seasons, the water quality during winter is better, which is due to the cold temperature that slows down all the microbial activity inside and around the lake and the oxidation caused by the wind blowing during winter season. Among the various sites of Loktak Lake, Omba recorded the highest WQI value for all the three seasons compared to other sampling locations, which results from an increase of human activity around the hillock of Omba and due to the decay of vegetation, e.g. lotus plant, phumdis, water hyacinth, and so on inside and around the region. It can be seen that during the post-monsoon, the water quality of the lake is deteriorates as compared to other seasons, which is due to the decay of aquatic plants and phumdis before the monsoon and starts to spread across the lake after the monsoon.

4 Conclusion

A thorough investigation was carried out for the Loktak Lake by considering eight different locations to study the impact of pollution and its quality assessment. Based on the findings, certain parameters such as pH, EC, BOD and PO_4^{3-} were within the permissible limit. Other parameters such as turbidity, COD and NO_2 were due to high degree of anthropogenic activities such as waste disposal practices and chemical discharge from agriculture and aquaculture. The WQI investigation demonstrates that the water quality of the Loktak Lake is polluted and unfit for drinking, although the locals depend on it. The high WQI values for several locations were due to low DO and high values of subindices of turbidity, EC and COD. The study suggests constant monitoring and identifying pollution sources to prevent additional contamination and use of treated water.

References

1. R. Eliza Khwairakpam, A. Khosa, A.N. Gosaln, Monitoring and modelling water quality of Loktak Lake. SN Appl. Sci. **1**, 1–15 (2019)
2. M.G. Uddin, S. Nash, A. Rahman, A.I. Olbert, A comperhensive method for improvement of water quality index (WQI) models for costal water assessment. Water Res. **219**, 1–20 (2022)
3. P.C. von der Ohe, A. Prüß, R.B. Schäfer, M. Liess, E. de Deckere, W. Beack, Water quality indices across Europe-A comparison of the good ecological status of five river basins. J. Environ. Monit. **9**, 970–978 (2017)
4. M. Tripathi, S.K. Singal, Allocation of weights using factor analysing for development of a novel water quality index. Ecotoxicol. Environ. Saf. **183**, 1–7 (2019)
5. R. Kumar, S. Parvaze, M.B. Huda, S.P. Allaie, The changing water quality of lakes—A case study of Dal Lake, Kashmir Valley. Environ. Monit. Assess. **194**, 228 (2022). pp. 1–16
6. I. Zotou, V.A. Tsihrintzis, G.D. Gikas, Water quality evaluation of a lacustrine water body in the mediterranean based on different water quality index (WQI) methodologies. J. Environ. Sci. Health A **5**, 537–548 (2020)
7. K.K. Birawat, T. Hymavathi, M.C. Nachiyar, N.A. Mayaja, C.V. Srinivasa, Impact of urbanisation on lakes—A study of Bengaluru lakes through water quality index (WQI) and overall index of pollution (OIP). Environ. Monit. Assess. **193**, 1–20 (2021)
8. I. Zotou, V.A. Tsihrintzis, G.D. Gikas, Performance of seven water quality indices (WQIs) in a Mediterranean river. Environ. Monit. Assess. **191**, 1–14 (2019)
9. D.D. Patel, D.J. Mehta, H.M. Azamathulla, M.M. Shaikh, S. Jha, U. Rathnayake, Application of the weighted arithmetic water quality index in assessing groundwater quality: A case study of the South Gujarat Region. Water **15**, 1–10 (2023)
10. Loktak Development Authority, vol 1. (1999)
11. R. Roy, M. Majumder, Assessment of water quality trend in Loktak Lake, Manipur, India. Environ. Earth Sci. **78**, 1–12 (2019)
12. R.J. Laishram, G. Yumnam, W. Alam, Assessment of ecohydrogeochemical status of freshwater Loktak Lake of Manipur, India. Environ. Monit. Assess. **194**, 1–31 (2022)
13. American Public Health Association (APHA), *Standard Method for the Examination of Water and Wastewater*, 23rd edn. (2023)
14. Z. Menberu, B. Mogesse, D. Reddythota, Evaluation of water quality and eutrophication status of Hawassa Lake based on different water quality Indices. Appl. Water Sci. **11**, 1–10 (2021)

15. S.P. Prasood, M.V. Mukesh, K.S. Sajinkumar, K.P. Thrivikramji, Covid-19 pandemic lockdown modulation of physio-chemical parameters of surface water, Karmana river basin, Southwest India: A weighted arithmetic index and geostatistical perspective. Total Environ. Res. Themes **6**, 1–11 (2023)
16. R.D. Kangabam, S.D. Bhoominathan, S. Kanagaraj, M. Govindaraju, Development of a water quality index (WQI) for Loktak Lake in India. Appl. Water Sci. **7**, 2907–2918 (2017)
17. BIS, *Drinking Water Specification*. (2012)
18. F.G. Avila, P.L. Lajo-Suco, C. Siguenza-Jeton, M. Jimenez-Ordonez, L. Valdiviezo-Gonzales, R. Cabello-Torres, A. Aviles-Anazco, Evaluation of the water quality of a high Andean lake using different quantitative approaches. Ecol. Indic. **154**, 1–21 (2023)
19. B. Mayanglambam, S.S. Neelam, Physicochemistry and water quality of Loktak Lake water, Manipur, India. Int. J. Environ. Anal. Chem. **102**, 1638–1661 (2020)

Water Quality Assessment of Purulia District, West Bengal, India Under Changed Climate Scenario

Prasun Mukherjee, Gourab Banerjee, Nilanjan Saha, and Asis Mazumdar

Abbreviations

APHA	American Public Health Association
BOD	Biochemical oxygen demand
COD	Chemical oxygen demand
IMD	India Meteorological Department
IPCC	Intergovernmental Panel on Climate Change
TDS	Total dissolved solids

1 Introduction

Climate change and global warming have become increasingly relevant and impactful in recent years. One of the most visible and concerning effects of these changes is the shifting of seasons [1]. As global temperatures rise, seasons are becoming less distinct, with warmer temperatures occurring earlier in the year and colder temperatures arriving later [2]. This can cause a range of ecological and societal issues, including disruptions to agricultural cycles, changes in animal migration patterns and altered weather patterns. It has a significant impact on the ecology and economy of every part of the world, regardless of urban or rural land use pattern [3]. The United Nations' Intergovernmental Panel on Climate Change (IPCC) reported that more than 40% the global population is under 'highly vulnerable' status and future climate change impacts could be even more dangerous according to the predictions

P. Mukherjee (✉) · G. Banerjee · N. Saha · A. Mazumdar
School of Water Resources Engineering, Jadavpur University, Kolkata, West Bengal, India
e-mail: gourabb.swre@jadavpuruniversity.in; nilanjans.rs.swre@jadavpuruniversity.in

D. Mazumder (ed.), *Sustainable Advanced Technologies for Environmental Management*, Springer Proceedings in Earth and Environmental Sciences,
https://doi.org/10.1007/978-3-031-64006-3_8

[4]. The vulnerability of populations to climate change, especially in developing nations such as India, is a major concern, with potential severe consequences for the environment and the economy [3].

The environment is experiencing paradigm change in relation to temperature change and change in rainfall pattern. Attempts at investing the causes and impacts have gathered considerable improvement over the past decades resulting in development of much sensitive data procurement and analysis tools. The two most fundamental and important parameters that help in determining the climatological health of a region are temperature and precipitation [5]. In a developing country such as India, regardless of the growth in industrialization, the economy is still dependent on agriculture [6]. Majority of the irrigation water comes from natural rainfall and/or canal irrigation as its main source of water, augmented with groundwater [7]. Ground water is also vulnerable due to overexploitation [8], saltwater intrusion and lowering of ground water table [9]. Surface water quality is also greatly impacted by changes in the amount of water available.

Studies carried out in Finland [10, 11] demonstrate how changes in water volume are closely related to changes in surface water quality, especially with regard to eutrophication and nutrient transfer. According to a study, variations in stream flow volume are directly related to the nitrogen loadings that reach coastal zones [12]. Prolonged dry spells have a noticeable impact on water quality as well, influencing elements including temperature, biological oxygen demand, chloride and ammonia content [13, 14]. Certain compounds may have higher levels of peak concentration as the amount of water decreases. These temporal variables such as precipitation and temperature act as an indirect indicator of air pollution in a region [15, 16]. This paper focuses on the intensive analysis of climatic data in Purulia, West Bengal, India, and the surface water quality in the region aiming to understand the temporal variability, status of water quality scenario and correlation between rainfall and temperature.

2 Materials and Methods

2.1 Study Area

Purulia is an administrative district within the state of West Bengal in the East of India having a total area of 6275 km^2. The area lies in between 22.60°N 85.75°E and 23.50°N 86.65°E as shown in Fig. 1. It has a semi-arid climate and majority of the precipitation happens during the monsoon period [17]. Physiologically, the region is part of the upper catchment of the river basins of the Damodar, Dwarakeswar and Sali.

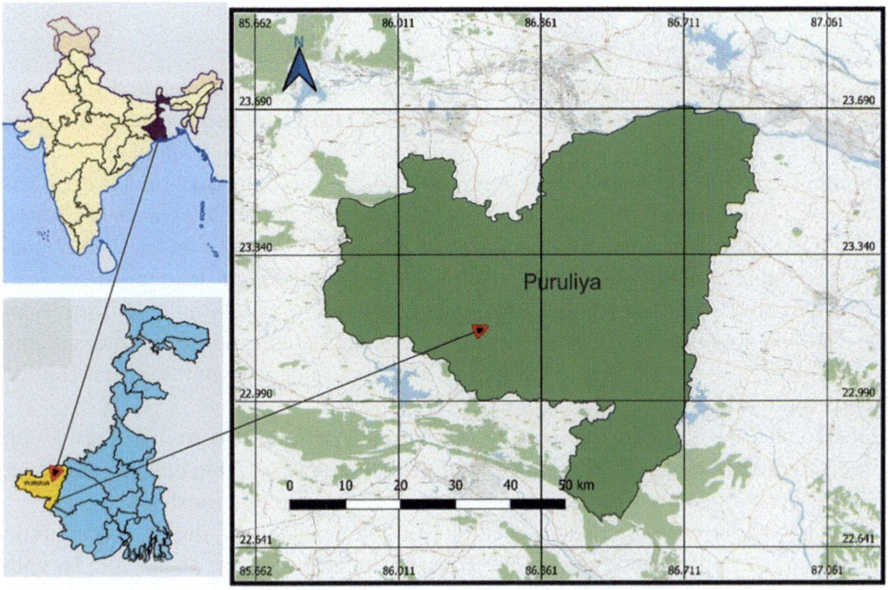

Fig. 1 Study area (Purulia, West Bengal, India)

Extremely hot summers accompanied with extreme heat wave causes major life-threatening situations in the region [18]. Droughts are a recurring phenomenon in this area. With insufficient water supplies to satisfy the region's needs, agriculture is mostly dependent on rain, which causes agricultural and hydrological drought almost every year in the area [19]. The relative humidity in the region broadly varies from 20% to 85% [19]. Lying at an average elevation of 115 m above mean sea level, the general inclination lies to the east [19]. The region is majorly lateritic soil and hard rock undulating, with gravelly loam and sandy skeletal type soil spread across the region [17]. Owing to the undulated topographical characteristics, more than 50% of the precipitation results in surface runoff [19]. The district being majorly dependent on agriculture has only one major town Purulia, which is the district administrative headquarter.

The district has several forest reserves namely Ayodhya Hill Reserve Forest, Baghmundi Reserve Forest, Murguma Reserve Forest, Matha Reserve Forest, Piyasala-Bandwan Reserve Forest, Turga Reserve Forest, Joypur Reserve Forest and Barabazar Reserve Forest. Sericulture and lac industries are also a great contributor to the economics of the district. In spite of such industrial and agricultural developments, the district still lies in the list of 250 most backward districts in India according to the Ministry of Panchayati Raj. Out of the 11 districts in the state of West Bengal, the district also receives grants from the Backward Regions Grant Fund for its development [20].

2.2 Methodology

The daily rainfall, maximum and minimum temperature were deciphered from grid data provided by Indian Metrological Department (IMD) for the period of 10 years (2012–2022). IMD provides gridded rainfall data with a very high spatial resolution, rainfall at 0.25° × 0.25° and temperature at 1° × 1°. The required gridded data was then identified using QGIS 3.22.14 software, and the necessary files were converted to simplified readable format (.csv) using plugin QGIS python console. An annual data file contains of 365/366 data points (daily) according to non-leap year/leap year.

For this study the leap day has not been considered to maintain the uniformity of the data and to simplify the numerical dataset. The dataset, sourced from satellite intelligence, was monitored, eliminating missing data as satellite data can be misinterpreted [21].

Trend analysis fundamentally is the extraction of definite pattern in a dataset with respect to time. Usually, the relationship of the dependent variable is associated over a longer period of time to establish a definitive and stable relationship [22].

The most common statistical method for the analysis of the temporal trend assessment of rainfall and temperature trend is linear regression analysis [23], with the determination coefficient (r^2), which was done using Minitab software. The standard deviation, mean and the coefficient of variability of all parameters were analysed to derive a conclusive indication of the nature and severity of climate change in the region. Furthermore, water quality analysis of 20 surface water bodies in study area was monitored. The physical parameters such as water temperature, pH and total dissolved solids (TDS) were measured in situ using HACH Multimeter. The chemical parameters such as phosphate, nitrate, ammonia, sulphate and iron were tested in the laboratory using powder pillow method (as stated in IS 10500: 2012) in HACH DR 2800 spectrophotometer. To prevent unexpected changes in water quality, the samples were collected in sterile dark plastic bottles from locations distributed all over the district as shown in Fig. 2 and transported to the laboratory in an icebox within 24 h. All of the containers were carefully cleansed three times with sample water before the collecting process as per standard procedure (as stated in APHA, 2005).

The average annual meteorological and water quality data of the region were then analysed using Pearson correlation. A fundamental statistical approach for determining the extent of relationship between many research variables is correlation analysis. To display the extent of a linear relationship between any two parameters, a correlation matrix was created and the degree of correlation was represented as correlation coefficient (r). The strongly linked and associated water quality parameter or climatic variable that may have an impact on the surface water quality was identified using the R-value. R has a value that ranges between +1 and −1; a value equal to 1 or close to 1 denotes the largest positive correlation among the two parameters under comparison, while a value of −1 or close to negative indicates an inverse linear correlation.

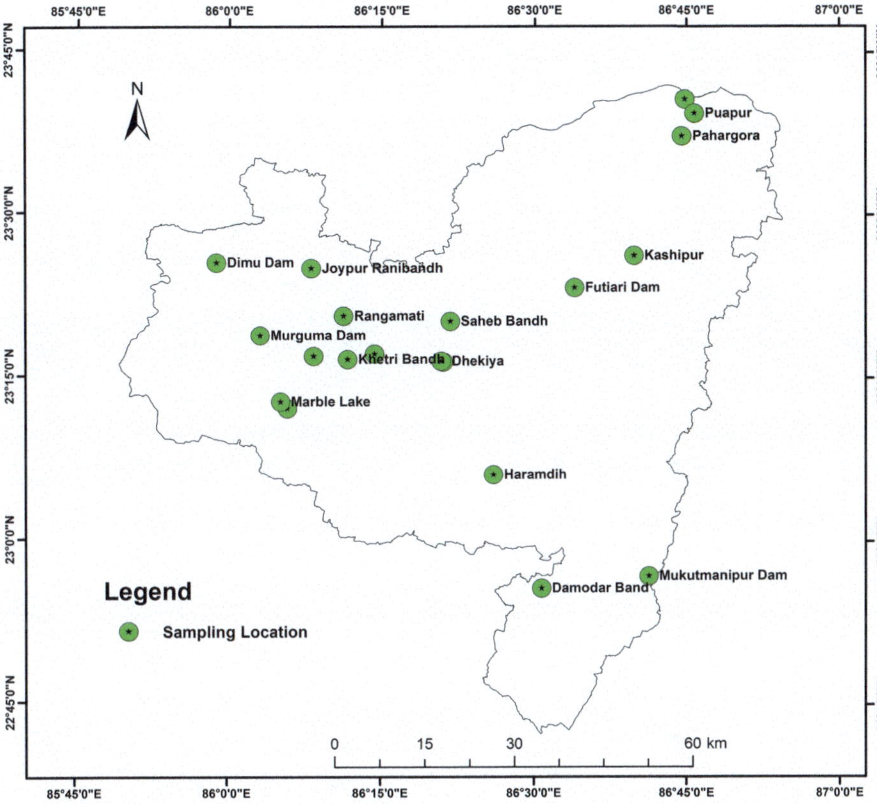

Fig. 2 Sampling points spread around the Purulia district

3 Results and Discussion

The average annual temperature (°C) and rainfall (in mm) data of the region for a 10-year period, ranging from 2012 to 2022, provide insightful knowledge into the climatic behaviour of the area and environmental changes. The clear decreasing trend in the yearly rainfall data indicates a continuous decline in precipitation over the study period. The area had a significant quantity of precipitation in 2012 (1656.35 mm) but by 2022, that amount had decreased by 35.64%. The region experienced an average rainfall of 1386.17 mm with a standard deviation of 256.54 during the study period. At the same time, Fig. 3 shows that average temperatures have been steadily rising. Starting in 2012 at 25.9 °C, the temperature rises each year until reaching a maximum of 26.8 °C in 2021. Even while the temperature drops somewhat to 26.6 °C in 2022, the general rising trend is a sign of changing weather patterns. The region experienced an average of 26.38 °C with a standard deviation of 0.309. The normalized temporal variation-wise analysis (*Z* score) of the water quality variables is shown in Figs. 3 and 4.

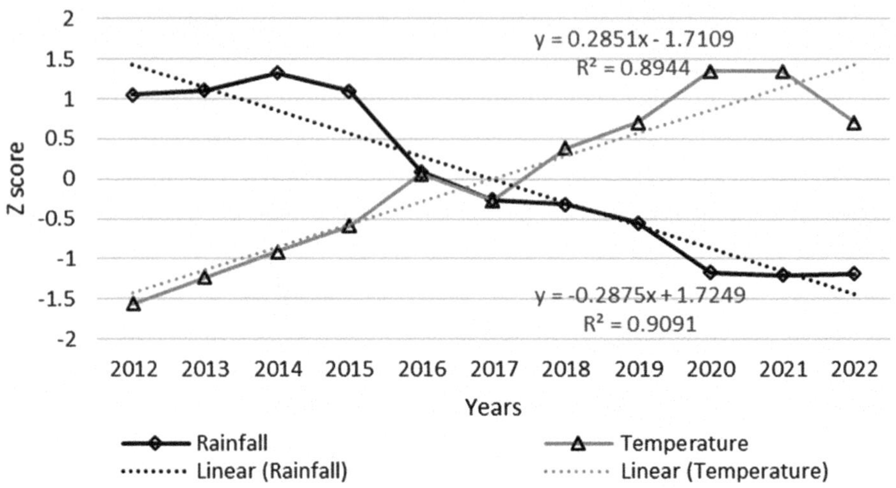

Fig. 3 Trend analysis of rainfall and temperature from 2012 to 2022

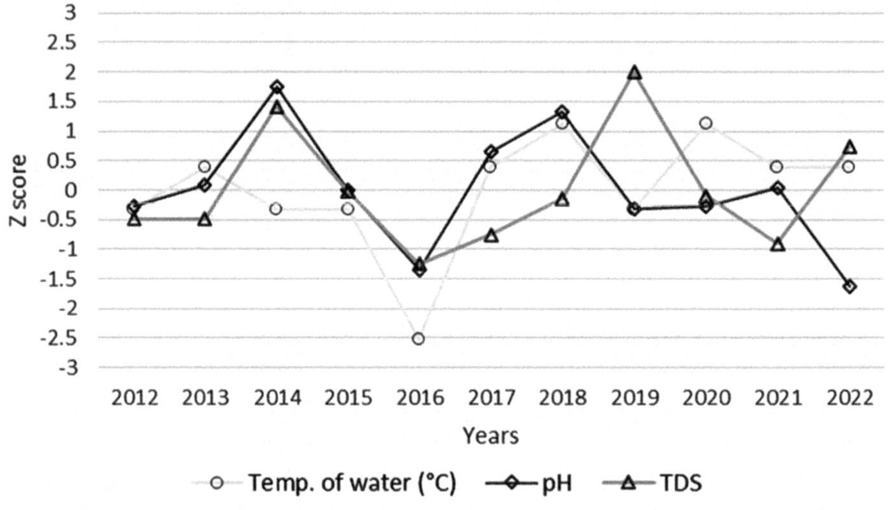

Fig. 4 Temporal variation of water temperature, pH and TDS

The apparent negative link between temperature and rainfall is one of the most interesting features. The average annual temperatures tend to rise when the amount of yearly rainfall decreases. This association suggests that temperature fluctuations and precipitation patterns interact intricately. Although general patterns are apparent, certain years show notable variations. For example, the year 2016 is unusual in that it had much less rainfall, indicating a pattern abnormality. The year 2020 also displays a lower temperature than nearby years with comparable rainfall totals, suggesting the existence of distinct climatic occurrences. These annual variations highlight how crucial it is to take short-term variations into account in the context of larger climate patterns. There are many ramifications for the patterns of rising temperatures and decreased rainfall that have been documented. Reduced precipitation might affect the amount of water available, which would be problematic for ecosystems and agricultural methods that rely on steady moisture. Increased temperatures have the potential to affect many ecological processes, such as plant growth, water quality and ecological processes.

The water quality data of physical parameters did not show much change in the region during the period of study. The average annual water temperature varied from 21 °C (2016) to 26 °C (2018 and 2020). The average water temperature over the period was around 24.45 °C, with a moderate standard deviation of 1.37 °C. This indicates a relatively stable water temperature with periodic fluctuations, such as the notable temperature drop in 2016 and peaks in 2018 and 2020.

The pH scale runs from 7.15 (2022) to 8.38 (2014), with the neutral to slightly alkaline range being the most common. The dataset reflects a consistent pH range around 7.74, suggesting a slightly acidic to neutral environment. The low standard deviation of 0.36 indicates minimal variability in acidity levels, contributing to the overall stability of the aquatic system.

TDS (total dissolved solids) varies from 206 mg/L (2016) to 376 mg/L (2019), with 2019 showing the greatest concentration, suggesting that dissolved solids are at an increased level but as the samples are of surface water, it is fairly within a range much lower than 500 mg/L. Average TDS concentrations was around 271.45, with a standard deviation of 52.39. This signifies also a low variability. The trend analysis of the physiochemical water parameters have been shown in Fig. 4.

The range of biochemical oxygen demand (BOD) from 2.45 to 7.35 mg/L in the timeframe of 2013–2020 indicates variations in the concentrations of organic pollutants. The average BOD is approximately 4.57 mg/L, exhibiting a moderate standard deviation of 1.95 mg/L. This rise in BOD levels signifies a notable shift in the quantity of biodegradable organic matter present within the aquatic ecosystem. Such an elevation suggests a heightened rate of organic matter decomposition, potentially leading to a surge in microbial activity. Furthermore, the increase in BOD underscores the mounting pressure on the ecosystem's capacity to effectively process and assimilate organic pollutants. This influx of organic material can overwhelm the natural processes of nutrient cycling and decomposition, thereby exacerbating ecological stress within the aquatic habitat.

The value of chemical oxygen demand (COD), which represents the amount of oxygen needed for chemical processes, varies greatly; in 2014, it reached its highest point of 38.63 mg/L. Average COD values stand at around 23.60 mg/L, with a standard deviation of 11.73 mg/L. Industrial activities, characterized by the discharge of effluents containing complex organic compounds and chemical by-products, contribute to the elevation of COD levels. Similarly, agricultural runoff, laden with fertilizers, pesticides and other agrochemicals, adds to the burden of organic matter and nutrient loading in water bodies, thereby raising the COD levels. Furthermore, domestic grey water discharge, encompassing household

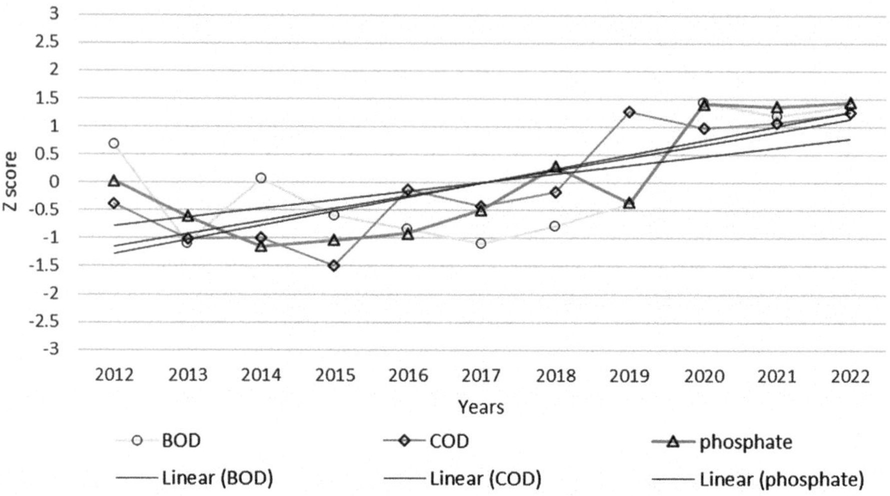

Fig. 5 Temporal trend analysis of standardized values of BOD, COD and phosphate

wastewater from sinks, showers and laundry, introduces organic matter, detergents and other contaminants into aquatic ecosystems. The cumulative effect of these anthropogenic inputs on COD levels underscores the potential implications for oxygen availability and aquatic life within the affected water bodies.

The phosphate levels range from 0.04 mg/L in 2014 to a maximum of 0.264 mg/L in 2022. The average phosphate concentration was 0.139 mg/L, with a relatively low standard deviation of 0.09 mg/L, indicating a consistent pattern with limited deviation from the average. An evident trend emerges, indicating an increase in phosphate concentrations over the years, with a significant rise from 2014 to 2022. This upward trend could be associated to changes in rainfall patterns, human activities and alterations in land use patterns. Figure 5 denotes the temporal trend of BOD, COD and phosphate on a normalized scale.

Nitrate concentrations ranges from a minimum of 0.07 mg/L in 2015 to a peak of 0.724 mg/L in 2013. These diverse values suggest a sensitivity to environmental influences, potentially influenced by factors such as agricultural practices, industrial effluents and domestic discharge, which affect the ecological processes of a region. The average nitrate concentration is 0.280 mg/L, with a standard deviation of 0.217. The sulphate concentrations, also exhibit substantial variability, ranging from 5.13 mg/L in 2014 to a high of 30.21 mg/L in 2022. The average sulphate concentration of 17.357 mg/L, coupled with a standard deviation of 8.214 mg/L.

Ammonia concentrations, with values ranging from a low of 0.083 mg/L in 2013 to a peak of 1.77 mg/L in 2020, showcase a distinct temporal pattern. The average ammonia concentration is 0.778 mg/L, with a standard deviation of 0.568, indicating a moderate degree of variability. Ammonia levels are often linked to agricultural runoff, wastewater effluents and natural decomposition processes of organic matter in surface water bodies. The highest concentrations in 2013 for nitrate, 2022 for

Table 1 Annual average concentration of water quality parameters (in mg/L)

Year	Nitrate	Sulphate	Ammonia
2012	0.604	26.677	0.273
2013	0.724	27.092	0.083
2014	0.27	5.13	0.76
2015	0.07	10.42	0.74
2016	0.103	14.59	0.24
2017	0.102	11.88	0.715
2018	0.111	23.86	0.281
2019	0.282	13.82	0.852
2020	0.129	15.84	1.77
2021	0.344	11.404	1.16
2022	0.346	30.21	1.68

sulphate and 2020 for ammonia suggest potential environmental events or localized influences during those periods. Table 1 denotes the specific water quality, which were found during the years.

The correlation matrix provides insights into the relationships between various water quality parameters and environmental variables. Notably, time (in years) have a strong positive correlation with water quality parameters such as COD (0.85) and atmospheric temperature (0.95), and fairly strong correlation with phosphate (0.76) and ammonia (0.78). The rainfall over the years have a very strong negative correlation with years (−0.95), indicating the increase of temperature and the decrease of rainfall over the years.

The physical parameters do not show any correlation with any of the parameters, depicting their independencies. The BOD values show a positive correlation to COD (0.63) and ammonia (0.78). The strong correlation between BOD and phosphate (0.81) indicates a rise of agricultural runoff such as pesticides in the surface water bodies of the region. The COD values have a positive correlation with phosphate (0.78) and ammonia (0.66), depicting correlation between the release of ammonia and phosphate from different sources. The strong negative correlation of rainfall with COD (−0.9), phosphate (−0.82) and ammonia (0.7) indicates the reduction in concentration of these parameters on the event of precipitation. Table 2 shows the correlation matrix between the parameters.

4 Conclusion

In conclusion, the comprehensive analysis of the datasets has yielded insights into the temporal variations of water quality parameters within the studied environment. Temperature exhibited fluctuations over the years, indicating a responsive interplay with environmental conditions. Rainfall significantly impacted water temperature, phosphate levels and atmospheric temperature. The negative correlation between rainfall and water temperature suggested cooling effects during rainy periods. pH

Table 2 Correlation matrix for water quality parameters including climate data

	Year	[a]Water temp.	pH	TDS	BOD	COD	[a]Phos.	[a]Nitr.	[a]Sulph.	[a]Ammo	rain	[a]Atm. temp.
Year	1.00											
[a]Water temp.	0.35	1.00										
pH	−0.33	0.39	1.00									
TDS	0.15	0.13	0.18	1.00								
BOD	0.52	0.31	0.32	0.13	1.00							
COD	0.85	0.21	−0.47	0.22	0.63	1.00						
[a]Phos.	0.76	0.58	−0.33	−0.09	0.81	0.78	1.00					
[a]Nitr.	−0.39	0.10	−0.11	0.00	0.13	−0.06	0.06	1.00				
[a]Sulph.	0.01	0.26	−0.41	−0.14	0.15	0.18	0.39	0.56	1.00			
[a]Ammo	0.78	0.40	−0.29	0.30	0.78	0.66	0.71	−0.28	−0.10	1.00		
rain	−0.95	−0.35	0.37	0.02	−0.51	−0.90	−0.82	0.32	−0.10	−0.70	1	
[a]Atm. temp.	0.95	0.28	−0.27	0.07	0.48	0.83	0.71	−0.47	−0.13	0.71	−0.93	1

[a]Water temperature, phosphate, nitrate, sulphate, ammonia, atmospheric temperature

levels displayed variability influenced by atmospheric temperature and rainfall. Higher atmospheric temperatures correlated with elevated pH, while rainfall contributed to a decrease in pH. TDS demonstrated a mild correlation with atmospheric temperature. However, overall, TDS levels exhibited variability without strong associations with other parameters. BOD and COD showcased correlations, revealing shared influencing factors related to organic pollution sources. COD, notably, exhibited a robust positive correlation with water temperature. Chemical parameters such as phosphate, nitrate, sulphate and ammonia presented diverse patterns, indicating varied sources and residual decomposition rates. Phosphate concentrations, in particular, strongly correlated with temperature. This analysis underscores the complex and multifaceted nature of water quality dynamics, emphasizing the need for ongoing research, statistical scrutiny and contextualized interventions to ensure the sustainable water management.

Acknowledgement This study was supported by Start-up Research Grant (SRG) scheme under Science and Engineering Research Board, Department of Science and Technology, for the project entitled 'Development of sustainable drinking water treatment through optimization of biological filtration step in Multistage Filtration Process by aerated slow sand filtration to mitigate emerging contaminant in rural sector under change climate scenario'.

References

1. J. Wang et al., Changing lengths of the four seasons by global warming. Geophys. Res. Lett. **48**(6), e2020GL091753 (2021). https://doi.org/10.1029/2020GL091753
2. J. Guo, S. Hu, Y. Guan, Regime shifts of the wet and dry seasons in the tropics under global warming. Environ. Res. Lett. **17**(10), 104028 (2022). https://doi.org/10.1088/1748-9326/ac9328
3. M. Zeleňáková, P. Purcz, H. Hlavatá, P. Blišťan, Climate change in urban versus rural areas. Proc. Eng. **119**, 1171–1180 (2015). https://doi.org/10.1016/j.proeng.2015.08.968
4. J. Tollefson, Climate change is hitting the planet faster than scientists originally thought. Nature (2022). https://doi.org/10.1038/d41586-022-00585-7
5. O. Singh, P. Arya, B.S. Chaudhary, On rising temperature trends at Dehradun in Doon valley of Uttarakhand, India | SpringerLink. J. Earth Syst. Sci. **122**, 613–622 (2013). https://doi.org/10.1007/s12040-013-0304-0
6. K. Smolińska, The role of agriculture in the national economy of China and India. Ann. PAAAE **XXI**(3), 432–440 (2019). https://doi.org/10.5604/01.3001.0013.3515
7. P. Mukherjee, S. Das, A. Mazumdar, Introducing winter rice cropping by using non-saline tidal water influx in western basins of South 24 Parganas, India. Sci. Rep. **11**(1), Art. no. 1 (2021). https://doi.org/10.1038/s41598-020-80797-x
8. B. John, P. Roy, S. Das, Analysing the influence of groundwater exploitation on its quality in Kolkata, in *Sustainability in Environmental Engineering and Science*, Lecture Notes in Civil Engineering, ed. by S. Kumar, A. Kalamdhad, M.M. Ghangrekar, (Springer, Singapore, 2021), pp. 83–89
9. S. Chakraborty, C. Rajak, S. Das, The contamination of groundwater and possible remedial measures in the coastal zone of Purba Medinipur, India. J. Coast. Conserv. **27**(4), 29 (2023). https://doi.org/10.1007/s11852-023-00960-0
10. T. Frisk, A. Bilaletdin, K. Kallio, M. Saura, Modeling the effects of climatic change on lake eutrophication. Boreal Environ. Res. **2**, 53–57 (1997)

11. K. Kallio, S. Rekolainen, P. Ekholm, K. Granlund, Y. Laine, H. Johnsson, Impacts of climate change on agricultural nutrient losses in Finland. Boreal Environ. Res. **2**, 33–42 (1997)
12. R.B. Alexander, P.S. Murdoch, R.A. Smith, Streamflow-induced variations in nitrate flux in tributaries to the Atlantic Coastal Zone. Biogeochemistry **33**(3), 149–177 (1996). https://doi.org/10.1007/BF02181070
13. M.A. Mimikou, E. Baltas, E. Varanou, K. Pantazis, Regional impacts of climate change on water resources quantity and quality indicators. J. Hydrol. **234**(1), 95–109 (2000). https://doi.org/10.1016/S0022-1694(00)00244-4
14. H.A.J. Senhorst, J.J.G. Zwolsman, Climate change and effects on water quality: A first impression. Water Sci. Technol. **51**(5), 53–59 (2005). https://doi.org/10.2166/wst.2005.0107
15. A. Dilawar et al., Investigating the potential climatic effects of atmospheric pollution across China under the National Clean Air Action Plan. Remote Sens. **15**(8), Art. no. 8 (2023). https://doi.org/10.3390/rs15082084
16. F. Giorgi, F. Meleux, Modelling the regional effects of climate change on air quality. Compt. Rendus Geosci. **339**(11), 721–733 (2007). https://doi.org/10.1016/j.crte.2007.08.006
17. A. Bera, S. Das, Water resource management in semi-arid Purulia District of West Bengal, in the context of sustainable development goals, in *Groundwater and Society: Applications of Geospatial Technology*, ed. by P.K. Shit, G.S. Bhunia, P.P. Adhikary, C.J. Dash, (Springer International Publishing, Cham, 2021), pp. 501–519. https://doi.org/10.1007/978-3-030-64136-8_23
18. Draft District Disaster Management Plan – Purulia. (Office of the District Magistrate, Purulia Disaster Management Section, 2021, 2020)
19. S. Hazra, S. Roy, S. Mitra, *Enhancing Adaptive Capacity and Increasing Resilience of Small and Marginal Farmers of Purulia and Bankura Districts, West Bengal to Climate Change* (Development Research Communication and Services Centre, 2019). [Online]. Available: https://www.drcsc.org/CCA/3/Docs/Vulnerability%20Index.pdf
20. A Note on the Backward Regions Grant Fund Programme (National Institute of Rural Development and Panchayati Raj (NIRD&PR), 2009). Accessed 16 Mar 2023. [Online]. Available: https://web.archive.org/web/20120405033402/http://www.nird.org.in/brgf/doc/brgf_BackgroundNote.pdf
21. R. Dasgupta, S. Das, G. Banerjee, A. Mazumdar, Revisit hydrological modeling in ungauged catchments comparing regionalization, satellite observations, and machine learning approaches. HydroResearch **7**, 15–31 (2024). https://doi.org/10.1016/j.hydres.2023.11.001
22. R. Mahmood, S. Jia, W. Zhu, Analysis of climate variability, trends, and prediction in the most active parts of the Lake Chad basin, Africa. Sci. Rep. **9**(1), Art. no. 1 (2019). https://doi.org/10.1038/s41598-019-42811-9
23. N. Altman, M. Krzywinski, Simple linear regression. Nat. Methods **12**(11), Art. no. 11 (2015). https://doi.org/10.1038/nmeth.3627

Part II
Advanced Water Treatment Technologies

Study the Fluoride Removal by Batch Mode Adsorption Method Using Laterite Soil

Barnali Mandal

Abbreviations

BET	Brunauer–Emmet–Teller
FTIR	Fourier transform infrared spectroscopy
SEM	Scanning electron microscopy
XRD	X-ray diffraction analysis

1 Introduction

Fluoride has beneficial as well as detrimental effects on human health. Fluoride is an essential element for proper formation of teeth, bone and skeletal system in human body [1]. However, the consumption of higher fluoride concentration through drinking water can cause skeletal and dental fluorosis [2]. The World Health Organization (WHO) suggested a fluoride concentration standard in drinking water of 1–1.5 mg/L [3].

Several techniques such as chemical precipitation, electro-coagulation [4], electro-dialysis, reverse osmosis, nano-filtration [5], ion exchange [6] and adsorption were used for the excess fluoride removal from drinking water. Among the techniques, adsorption process is very popular due to its low operating cost, higher efficiency, generation of less quantity sludge and adsorbent regeneration. Several adsorbents such as activated alumina [7], bauxite [8], hydrous ferric oxide [9] and goethite [10] have been applied for the adsorption of fluoride.

B. Mandal (✉)
Department of Chemical Engineering, University of Calcutta, Kolkata, India
e-mail: bmchemengg@caluniv.ac.in

D. Mazumder (ed.), *Sustainable Advanced Technologies for Environmental Management*, Springer Proceedings in Earth and Environmental Sciences,
https://doi.org/10.1007/978-3-031-64006-3_9

Under the study, the percent removal of fluoride from drinking water has been compared between untreated and Zr-treated laterite soil as low-cost adsorbent in batch mode process. FTIR, XRD, BET and SEM have been used for the characterization of the mentioned adsorbents before and after adsorption. Adsorption isotherm studies have also been conducted through Langmuir and Freundlich model. The effect of pH on the removal efficiency of fluoride has also been investigated.

2 Materials and Methods

2.1 Adsorbent Preparation

Powdery form of laterite soil was prepared by a ball mill to be used as adsorbent. The soil has been treated with zirconium oxide chloride octahydrate ($ZrOCl_2 \cdot 8H_2O$). The treatment process has been reported in the previous study [11].

2.2 Analytical Method

SPADNS method has been applied for the determination of fluoride concentration. A calibration curve has been developed by different known concentration of fluoride against respective absorbance measured by a UV–Vis spectrophotometer.

2.3 Batch Studies

A number of batch processes have been carried out in a 250 mL container. The samples were collected at equilibrium condition. After filtering, the residual fluoride concentration has been measured. The effect of initial fluoride concentration, initial adsorbent dose and contact time on fluoride removal from drinking water investigated in an optimization study based on response surface methodology has been reported previously [11]. The optimum value of contact time, initial adsorbent dose and initial fluoride concentration were obtained as 90 min, 3 g/L and 40 mg/L, respectively.

2.4 Effect of pH

The effect of pH on the fluoride removal from aqueous solution has been studied with varying pH in the range of 4–10, respectively, for both untreated and Zr-treated laterite soil at contact time of 90 min with a constant adsorbent dose of 3 g/L and initial fluoride concentration of 40 mg/L (based on previous study). Different pH value of fluoride samples has been maintained by adding 1 N NaOH and 1 N HCl solution.

2.5 Batch Isotherm Study

The equilibrium data were fitted with two isotherm models, namely Langmuir and Freundlich.

Langmuir isotherm model The distribution of adsorbate concentration in both liquid and solid phases in equilibrium condition can be explained by the Langmuir isotherm [12]. The Langmuir model represents the following equation:

$$Q = Q_{\max} \frac{K_L C_e}{1 + K_L C_e}$$

The above equation can be linearly presented as

$$\left(\frac{1}{Q}\right) = \left(\frac{1}{Q_{\max}}\right) + \left(\frac{1}{Q_{\max} K_L C_e}\right) \tag{1}$$

where C_e is fluoride concentration (mg/L) at equilibrium, Q is fluoride per gram of adsorbent (mg/g), $Q_{\max}$ is maximum capacity of adsorbent (mg/g) and K_L is Langmuir isotherm constant.

The values of $Q_{\max}$ and K_L have been determined from the Langmuir plot of $1/Q$ versus $1/C_e$ [13].

R_L, an equilibrium parameter, can be expressed as dimensionless constant:

$$R_L = \frac{1}{1 + (1 + K_L C_i)}$$

where C_i is initial fluoride concentration, and the value of R_L indicates the sorption nature. It is unfavourable if R_L is greater than 1, linear if R_L is 1, irreversible if R_L is 0, favourable if R_L is greater than 0 but less than 1.

Freundlich isotherm model The Freundlich isotherm describes the characteristics of adsorption process [14]. The Freundlich model represents the following equation:

$$Q_e = K_f C_e^{\frac{1}{n}}$$

where Q_e is fluoride adsorbed per gram of the adsorbent at equilibrium (mg/g), K_f is Freundlich constant (mg/g), C_e is the equilibrium concentration of fluoride (mg/L) and n is adsorption intensity.

Freundlich equation can be linearly presented as:

$$\log Q_e = \log K_f + \frac{1}{n} \log C_e \tag{2}$$

The values of Q_e and K_f have been determined from the Freundlich plot of log Q_e vs. log C_e. K_f is an approximate adsorption capacity indicator and $1/n$ is a strength function of adsorption process [15]. The two phases (solid and liquid) are independent of concentration, when the value of n is 1. For normal adsorption, $1/n$ is always below 1. If $1/n$ is greater than 1, it indicates cooperative adsorption [16].

2.6 Characterization of Adsorbent

Several techniques have been used for the characterization of untreated and Zr-treated laterite soil.

FTIR spectroscopy analysis FTIR spectroscopy technique has been applied to identify the chemical groups present in both adsorbents, namely untreated and Zr-treated laterite soil in the spectra range of 4000 cm^{-1} to 400 cm^{-1}.

XRD XRD was used to study the crystalline structure of untreated and Zr-treated laterite soil.

BET analysis The specific surface area of both untreated and Zr-treated laterite soil was measured by Brunauer–Emmett–Teller (BET) surface analyser including the pore size distributions.

Scanning electron microscope (SEM) The morphology of adsorbent of both untreated and Zr-treated laterite soil in the case of before and after adsorption has been investigated.

3 Results and Discussion

3.1 Effect of pH

The performance of adsorption process is highly dependent on the pH of the solution. Removal of fluoride from water has been studied in varying pH at 5, 6, 7, 8 and 9 for both untreated and Zr-treated laterite soil. The highest adsorption efficiency was observed at pH 5 as shown in Fig. 1. The percent removal of fluoride has been decreased with the increasing pH of the solution. At the end of batch experiments, the removals of fluoride were observed as 97.42% and 81.3% for treated and untreated soil, respectively.

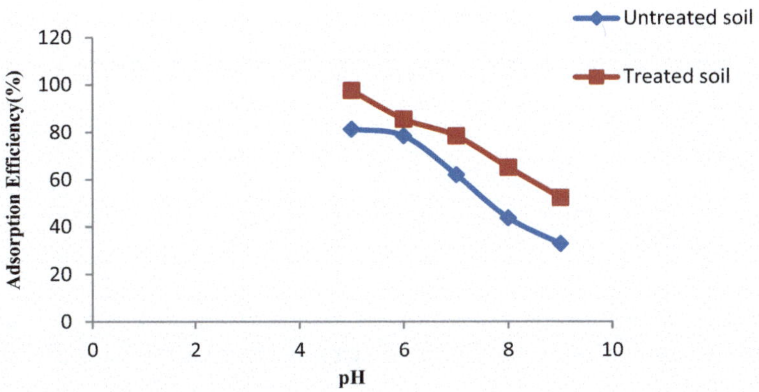

Fig. 1 Effect of pH on fluoride removal

3.2 Batch Isotherm Study

The experimental results obtained from batch studies have been analysed by employing Langmuir and Freundlich isotherm models.

Langmuir isotherm model According to Eq. (1), the graph was plotted for $1/Q$ vs. $1/C_e$, as shown in Fig. 2. The high value of R^2 (0.942) indicated that the data of sorption were well-fitted to Langmuir model. The value of constants ($Q_{\max}$, K_L and R_L) calculated has been given in Table 1. The R_L value of 0.0648 indicates the perfect adsorption performance.

Freundlich isotherm model According to Eq. (2), log Q_e vs. log C_e was plotted, as shown in Fig. 3. The value of R^2 is 0.900, implying that the data of sorption were fitted to Freundlich model. The whole set of constants was calculated, as shown in Table 2. The value of $1/n$ is 0.363, which is below 1 and indicates the normal adsorption.

Among the two isotherm models, the Langmuir equation was best adjusted with equilibrium data.

3.3 Characterization of Adsorbent

FTIR for Untreated Laterite Soil FTIR spectra of untreated laterite soil is shown in Fig. 4. The functional groups are present as O–H and ArO–H vibration stretching at bandwidth at 3621 cm^{-1}. The peaks at 3442 cm^{-1} are due to amides stretching vibrations in the R–CONHR groups. The asymmetric stretching vibrations of C=N bridges were observed at 1631 cm^{-1}. The peaks at around 1039 cm^{-1} indicate the presence of Si–OR groups. The peaks at 775 cm^{-1} indicate the presence of aromatics with the structure of 1,2,3-trisub. The peaks at around 694 cm^{-1} are due to the

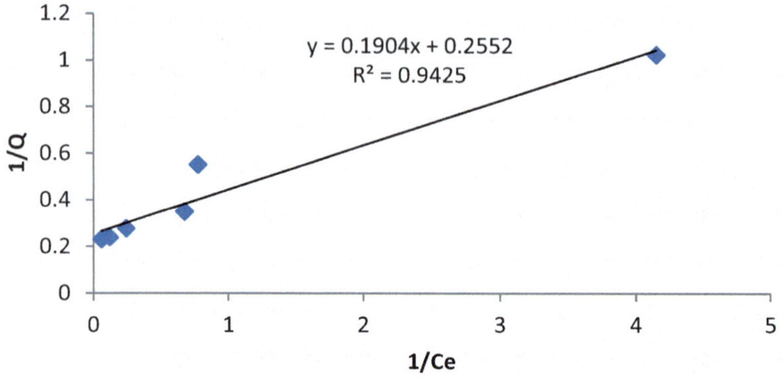

Fig. 2 Data fitted to Langmuir isotherm

Table 1 Langmuir isotherm constant

Q_{max} (mg/g)	K_L (L/mg)	R_L	R^2
3.921	1.342	0.0648	0.942

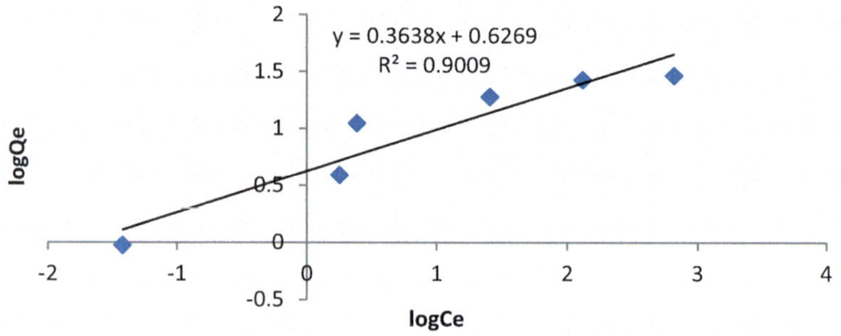

Fig. 3 Data fitted to Freundlich isotherm

Table 2 Freundlich isotherm constant

$1/n$	n	K_f	R^2
0.363	2.754	1.870	0.900

presence of aromatics groups with the structure of mono- and meta disub. The peaks at 561 cm^{-1} indicate the presence of alkyl halides with the structure of R–Br. The functional groups are 3,3 disulfide vibration stretching at bandwidth from 501 cm^{-1} to 427 cm^{-1}.

FTIR for Treated Laterite Soil The FTIR spectra of treated laterite soil (Fig. 5) has been investigated. The functional groups are O–H vibration stretching at bandwidth 3624 cm^{-1}. The peaks at around 3435 cm^{-1} are attributive to the amide groups with the structure of R–CONH–R′, which is N–H stretch free. The peaks at around 2359 cm^{-1} indicate the presence of Si–H silane structure. The peaks at

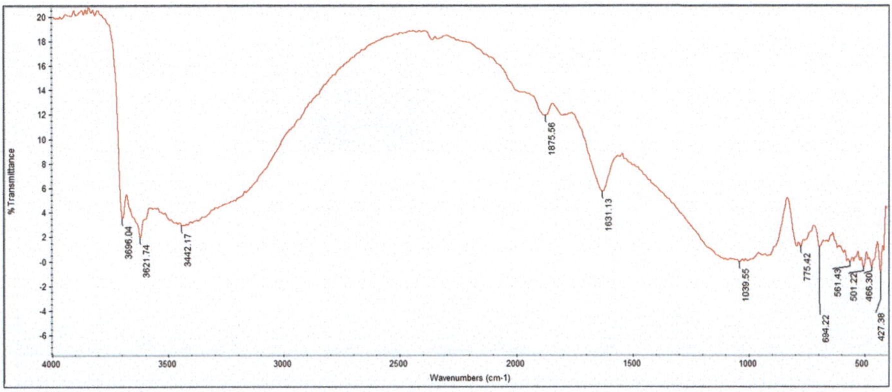

Fig. 4 FTIR spectra for untreated laterite soil

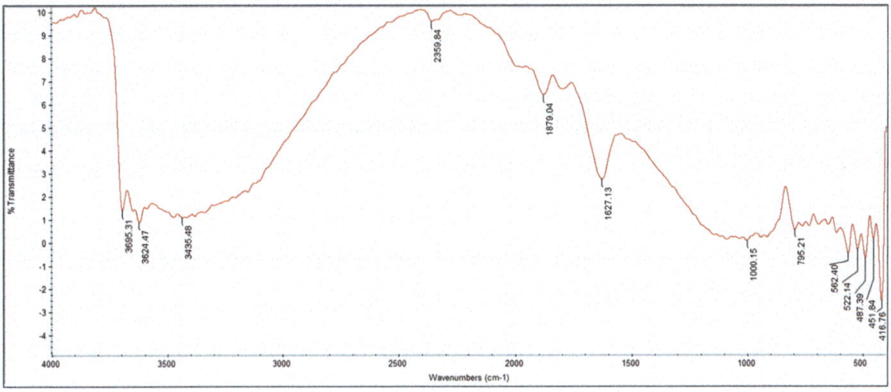

Fig. 5 FTIR spectra for treated laterite soil

around 1627 cm^{-1} is attributed to the alkene groups with the structure of Ar–CH=CH–R. The peaks at around 1000 cm^{-1} reveal the presence of carboxylic acids with the structure of R–COOH. The functional groups of alkenes with the structure of R_2C=CHR are vibrated at bandwidth 795 cm^{-1}. The peaks at around 562 cm^{-1} are attributive to alkyl halides with the structure of R–Br. There are some others from around 522 cm^{-1} to 416 cm^{-1} can be assigned for S–S disulfide.

X-Ray Diffraction (XRD) X-ray powder diffraction has been used to investigate the crystalline structure of untreated and treated laterite soil. The comparison of X-ray diffraction patterns between untreated laterite soil and treated laterite soil is given in Fig. 6. The XRD data of the untreated and treated soil exhibited the evidence of considerable modification over the crystal cleavages by indicating some peaks.

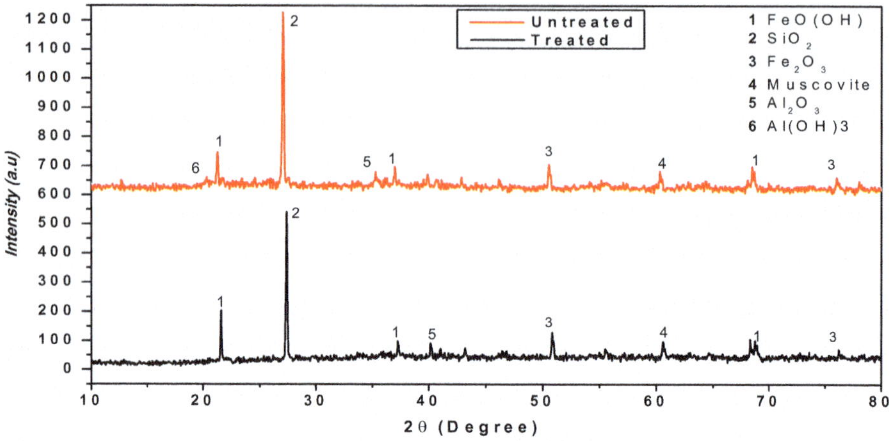

Fig. 6 XRD spectra of both untreated and treated laterite soil

Peak 1 (Fig. 6) indicates the presence of FeO(OH), which is present in both untreated and treated soil; however, from the figure it is clear that the intensity for the treated soil is higher than the untreated soil, which means the purity of FeO(OH) for treated soil is modified. The peak around 20.4Θ is attributive to the presence of $Al(OH)_3$, while for the treated soil there is no peak around 20.4 Θ. At the peak 3 indicate the presence of Fe_2O_3, which is present in both untreated and treated soil; however, from the figure it clear that the intensity for the treated soil is higher than the untreated soil, which means the purity of Fe_2O_3 for treated soil is modified. The peak around 35.3Θ is attributive to the presence of Al_2O_3, while for the treated soil there is no peak around 20.4 Θ; however, it was found at around 40.3 Θ, so the crystalline structure of the material is modified. The materials present in both treated and untreated soil are shown in Table 3.

BET Analysis The specific surface area of both untreated and treated laterite soil was also estimated by Brunauer–Emmett–Teller (BET) surface analyser, including the pore size distributions. The BET surface area analysis of untreated soil is shown in Fig. 7. BET summary was found from the figure, which indicates that slope is 11.096, correlation coefficient (*r*) is 1.000 and specific surface area is 311.303 m^2/g.

The BET surface area analysis of treated soil is shown in Fig. 8. BET summary was found specific surface area of 2631.184 m^2/g, pore volume of 11.097 cc/g and pore radius of 31.193 Å. The specific surface area of treated soil was higher than untreated soil.

Scanning Electron Microscope (SEM) The morphology of both untreated and treated laterite soil in the case of before and after adsorption were investigated as shown in Figs. 9, 10, 11, and 12. Figure 9a, b represents SEM and EDAX for untreated soil before adsorption. Similarly, Fig. 10a, b represents SEM and EDAX for treated soil before adsorption. Figure 11a, b represents SEM and EDAX for

Table 3 List of materials present in both treated and untreated soil

Sl. no.	Material	FeO (OH)	SiO_2	Fe_2O_3	Muscovite	Al_2O_3	$Al(OH)_3$
1	Untreated laterite soil	Present	Present	Present	Present	Present	Present
2	Treated laterite soil	Present	Present	Present	Present	Present	Not present

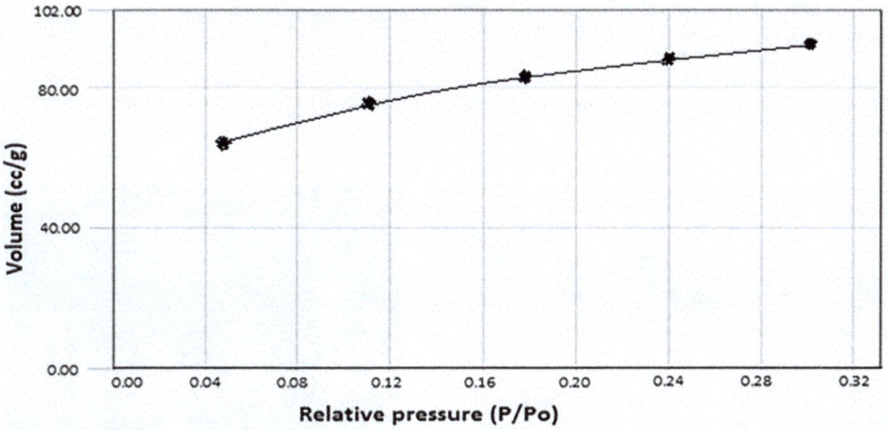

Fig. 7 BET analysis for untreated soil

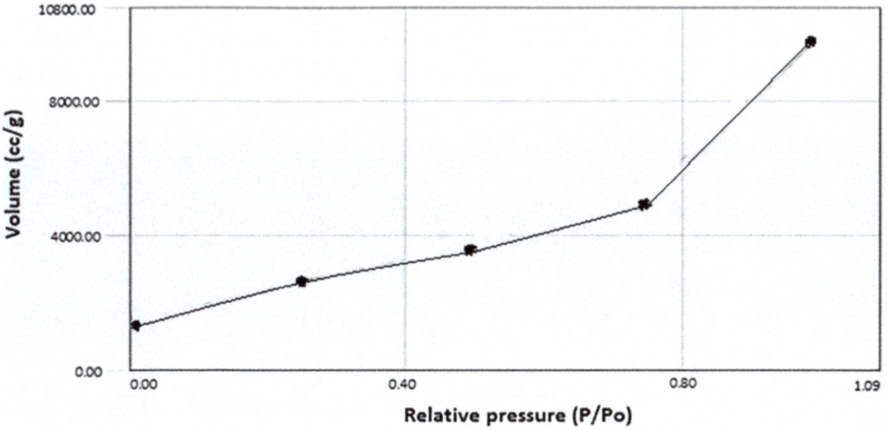

Fig. 8 BET analysis for treated soil

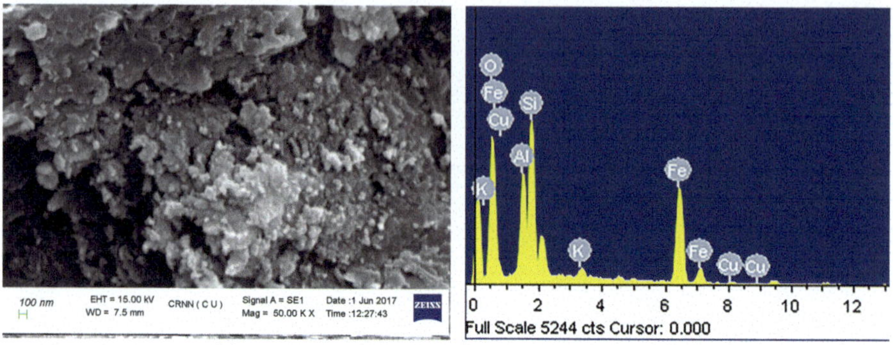

(a) (b)

Fig. 9 (**a**) SEM for untreated soil before adsorption and (**b**) EDAX for untreated soil before adsorption

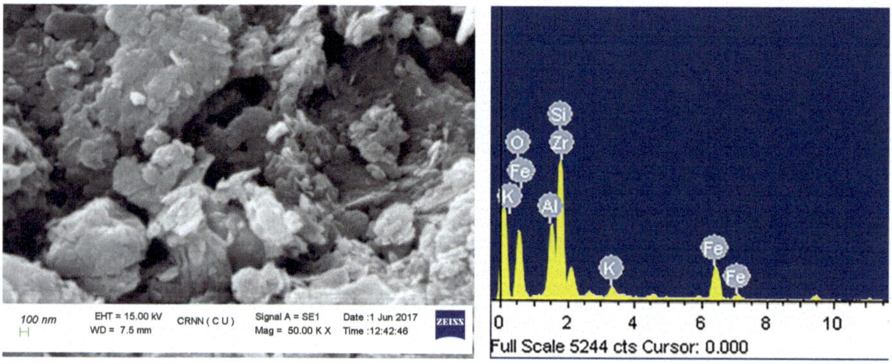

(a) (b)

Fig. 10 (**a**) SEM for treated soil before adsorption and (**b**) EDAX for treated soil before adsorption

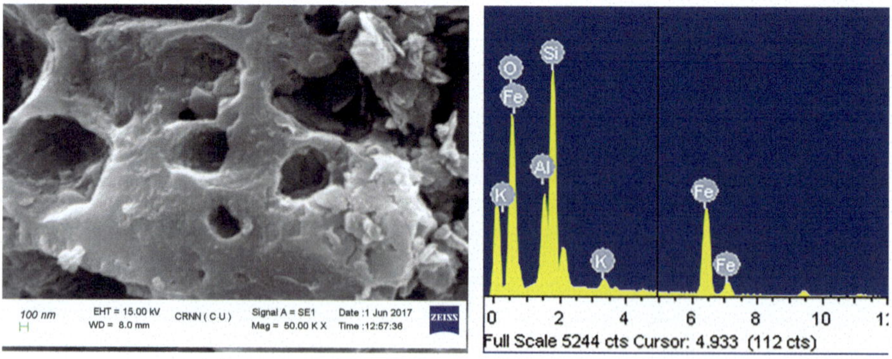

(a) (b)

Fig. 11 (**a**) SEM for untreated soil after adsorption and (**b**) EDAX for untreated soil after adsorption

(a)

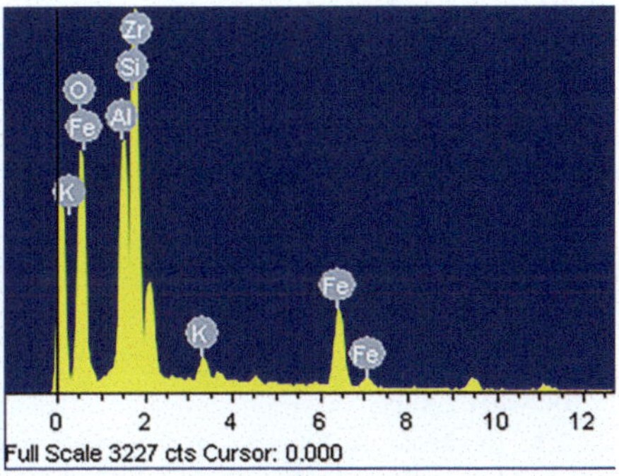

(b)

Fig. 12 (**a**) SEM for treated soil after adsorption and (**b**) EDAX for treated soil after adsorption

untreated soil after adsorption and Fig. 12a, b represents SEM and EDAX for treated soil after adsorption. All the figures of SEM and EDAX are given below.

In EDAX (for treated laterite soil before adsorption) (Fig. 10a), much higher pore volume is clearly seen compared with EDAX (for untreated laterite soil before adsorption) (Fig. 9a). Also, significant changes have been observed between Fig. 12a (EDAX for treated laterite soil after adsorption) and Fig. 11a (EDAX for untreated laterite soil after adsorption).

For treated soil, the Zr-band height in Fig. 12b (after adsorption) is much higher compared with Zr-band in Fig. 10b (before adsorption). This indicates that fluoride removal is maximized for Zr-treated soil. Also, Si-band, Al-band and Fe-band are increased in Fig. 12b than Fig. 10b. The fluoride has co-precipitation effects on Al, Fe and Si. On the other hand, for untreated soil, in Fig. 11b (after adsorption) and 9b (before adsorption), the co-precipitation effects of fluoride with Al, Fe and Si has been observed clearly.

Laterite soil has been used as adsorbent for fluoride removal in many studies [17, 18]. Zirconium has also been applied as nano-particle with other adsorbents for fluoride removal [19–21]. However, Zr-treated laterite soil has been used for the first time in the investigation [11].

4 Conclusion

The objective of the study was to compare the adsorptive removal of fluoride from drinking water with untreated and Zr-treated laterite soil, as well as to investigate the characteristics of the adsorbents before and after adsorption. The effect of pH on the removal of fluoride has also been investigated. The optimum value of pH is 5. Isotherm studies such as Langmuir and Freundlich has also been conducted. The equilibrium data were well-fitted with Langmuir isotherm model. The removal efficiency of fluoride was found to be 97.42% and 81.3% for Zr-treated and untreated laterite soil, respectively. Zr-treated laterite soil can be utilized as suitable adsorbent for large-scale fluoride removal from drinking water.

References

1. J.V. Kumar, M. Moss, Fluorides in dental public health programs. Dent. Clin. N. Am. **52**, 387–401 (2008)
2. A.K. Susheela, *A Treatise on Fluorosis*, vol 1 (Fluorosis Research and Rural Development Foundation, New Delhi, 2000)
3. WHO, *Guidelines for Drinking-Water Quality, Volume 1: Recommendations*, 3rd edn. (WHO, Geneva, 2004)
4. G. Chae, S. Yun, B. Mayer, K. Kim, S. Kim, J. Kwon, K. Kim, Y. Koh, Fluorine geochemistry in bedrock groundwater of South Korea. Sci. Total Environ. **385**, 272–283 (2007)
5. M. Tahaikt, R. El Habbani, A. Ait Haddou, I. Achary, Z. Amor, M. Taky, A. Alamib, A. Boughriba, M. Hafsi, A. Elmidaou, Fluoride removal from groundwater by nanofiltration. Desalination **212**, 46–53 (2007)
6. S. Meenakshi, N. Viswanathan, Identification of selective ion-exchange resin for fluoride sorption. J. Colloid Interface Sci. **308**, 438–450 (2007)
7. B. Shimelis, F. Zewge, B.S. Chandravanshi, Removal of excess fluoride from water by aluminum hydroxide. Bull. Chem. Soc. Ethiopia **20**, 17–34 (2006)
8. N. Das, P. Pattanaik, R. Das, Defluoridation of drinking water using activated titanium rich bauxite. J. Colloid Interface Sci. **292**, 1–10 (2005)
9. M. Streat, K. Hellgardt, N.L.R. Newton, Hydrous ferric oxide as an adsorbent in water treatment, part 2. Adsorption studies. Process Saf. Environ. Prot. **86**, 11–20 (2008)
10. K.B.P.N. Jinadasa, C.B. Dissanayake, S.V.R. Weerasooriya, A. Senaratne, Adsorption of fluoride on goethite surfaces-implications on dental epidemiology. Environ. Gcol. **21**, 251–255 (1993)
11. B. Mandal, S. Mallick, Removal of fluoride from drinking water by adsorption on Zr-treated laterite soil using response surface methodology. J. Indian Chem. Soc. **95**, 377–382 (2018)
12. T.H. Vermeulan, K.R. Vermeulan, L.C. Hall, Fundamental. Ind. Eng. Chem. **5**, 212–223 (1966)
13. I. Langmuir, The adsorption of gases on plane surfaces of glass, mica and platinum. J. Am. Chem. Soc. **40**, 1362–1403 (1918)
14. N.D. Hutson, R.T. Yang, Theoretical basis for the Dubinin - Radushkevitch (D - R) adsorption isotherm equation, Adsorption. J. Colloid Interface Sci. **3**, 189–195 (2000)
15. E. Voudrias, F. Fytianos, E. Bozani, Sorption description isotherms of dyes from aqueous solutions and waste waters with different sorbent materials. Global Nest Int. J. **4**, 75–83 (2002)
16. S. Mohan, J. Karthikeyan, Removal of lignin and tannin color from aqueous solution by adsorption on to activated carbon solution by adsorption on to activated charcoal. Environ. Pollut. **97**, 183–187 (1997)
17. A. Iriel, S.P. Bruneel, N. Schenone, A.F. Cirelli, The removal of fluoride from aqueous solution by a lateritic soil adsorption: Kinetic and equilibrium studies. Ecotoxicol. Environ. Saf. **149**, 166–172 (2018)
18. M. Sarkar, A. Banerjee, P.P. Pramanick, Kinetics and mechanism of fluoride removal using laterite. Ind. Eng. Chem. Res. **45**(17), 5920–5927 (2006)
19. Y. Gao, M. Li, Y. Ru, J. Fu, Fluoride removal from water by using micron zirconia/zeolite molecular sieve: Characterization and mechanism. Groundw. Sustain. Dev. **13**, 100567 (2021)

20. A. Savari, A. Hamidi, S. Farjadfard, M. Omidvar, B. Ramavand, Zirconium-based materials for fluoride removal from aqueous environments: A literature review and scientometric analysis. Colloids Interface Sci. Commun. **55**, 100722 (2023)
21. A. Mohamed, E.P. Valadez Sanchez, E. Bogdanova, B. Bergfeldt, A. Mahmood, R.V. Ostvald, T. Hashem, Efficient fluoride removal from aqueous solution using zirconium-based composite nanofiber membranes. Membranes **11**, 147 (2021)

Batch and Column Adsorption of Hexavalent Chromium on Household Filter Waste Activated Charcoal (HFWAC)

Kapilas Das, Sanjib Moulick, and Sanjoy Kumar Maji

Abbreviations

BOD	Biochemical oxygen demand
EBCT	Empty bed contact time
EPA	Environmental Protection Agency
HFWAC	Household filter waste activated charcoal
Sat	Saturation

1 Introduction

Chromium, the 3d block transitional element, is a significant toxic material because of its nonbiodegradability. Cr(VI) is considered as human carcinogen [1]. It causes damage of liver, congestion of the lungs, edema, and skin irritation [2]. Oxidation states of chromium affect its physical characteristics, bioavailability, and toxicity [3]. Cr(VI) species are substantially hazardous, mutagenic, carcinogenic, and teratogenic, and have increased solubility and mobility in aqueous systems. Cr(VI) is prevalent in the form of chromate ($CrO_4{}^{2-}$), dichromate ($Cr_2O_7{}^{2-}$), and hydrogen chromate ($HCrO_4{}^{-}$). The forms of Cr(VI) species depend on the pH of the solution and its total concentration [4, 5]. In general, the Cr(VI) concentration in industrial wastewater ranges between 0.5 mg/L and 2.70×10^5 mg/L [6]. As per the Environmental Protection Agency (EPA) notification, the discharge concentration of Cr

K. Das · S. K. Maji (✉)
Department of Chemistry, School of Applied Sciences, KIIT, DU, Odisha, Bhubaneswar, India

S. Moulick
School of Civil Engineering, KIIT, DU, Odisha, Bhubaneswar, India

D. Mazumder (ed.), *Sustainable Advanced Technologies for Environmental Management*, Springer Proceedings in Earth and Environmental Sciences,
https://doi.org/10.1007/978-3-031-64006-3_10

(VI) into the inland surface water should not exceed 0.1 mg/L, whereas the acceptable limit of total chromium (as Cr) in potable water is 0.05 mg/L (IS 10500 2021) [7]. For the removal of Cr(VI) from wastewater, a variety of physicochemical procedures including chemical precipitation, electrochemical processes, electrodialysis, reverse osmosis, ultrafiltration, ion exchange, and adsorption have been studied [8]. Activated carbon-based adsorption has been successfully employed for Cr (VI) removal [9]. Except adsorption, none of the available techniques are economically viable due to low efficiency, high reagent costs, high energy consumptions, and the production of hazardous sludge or other waste that needs to be removed or treated further [10]. Activated carbon is still the most often employed adsorbent, despite the fact that several different adsorbents have been tried to extract chromium from wastewater [11, 12]. Temperature and pH are the two most significant parameters for its remediation. In the present study, household filter waste activated charcoal (HFWAC) was prepared and utilized to remove Cr(VI) from the spiked contaminated synthetic wastewater and pond water using both batch (varying different experimental conditions) and column study [13].

2 Materials and Methods

A stock solution 200 mg/L concentration of $K_2Cr_2O_7$ (Loba Chemicals) was prepared and diluted as required. HCl, NaCl, and NaOH were purchased from Loba Chemicals. UV–Vis spectrophotometer (UV-1800, Shimadzu, Japan) was utilized for spectral measurement of the concentration of the Cr(VI) before and after treatment. The four decimals sensitive Sartorius electronic weighing balance was used to weigh the analytes.

Household filter waste activated charcoal (HFWAC) was collected at no-cost from the disordered household water guard filter, crushed to get the appropriate geometric mean-sized particles of 106 μ and 150 μ (in the first case, the crushed HFWAC dust was allowed to pass through sieve with opening size of 150 μ and retain on sieve with opening size of 75 μ. Thus, the geometric mean size was calculated as $(150\ \mu \times 75\ \mu)^{1/2} = 106\ \mu$. Similarly, in the second case, opening size of the two sieves were 300 μ and 75 μ and the geometric mean size of the particles was determined as $(300\ \mu \times 75\ \mu)^{1/2} = 150\ \mu$), washed in double distilled water followed by tap water to remove unwanted powder/impurities, and dried in hot air oven for overnight at 100 °C. At room temperature, the particles were separately stored in airtight glass bottles wrapped with black paper and characterized by FESEM/EDX (Zeiss, Gemini SEM 450, Fig. 1) and powder XRD (Fig. 2, Broker D8 Advance, wavelength 1.5406 Å, scan range 10–80θ, step size 0.01, CuKα) before and after treatment with Cr(VI).

These particles were utilized as the adsorbent for Cr(VI) removal from wastewater. EDX and XRD analyzed data revealed the presence of chromium after the treatment.

Fig. 1 The FESEM/EDX of HFWAC (**a**) prior to treatment (**b**) posttreatment of Cr(VI)

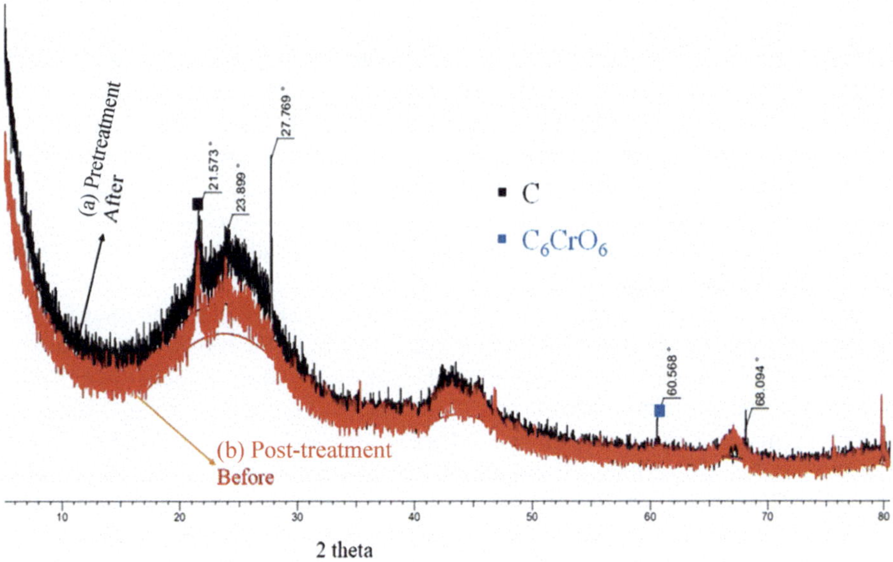

Fig. 2 The XRD pattern of HFWAC (**a**) pretreatment (**b**) posttreatment of Cr(VI)

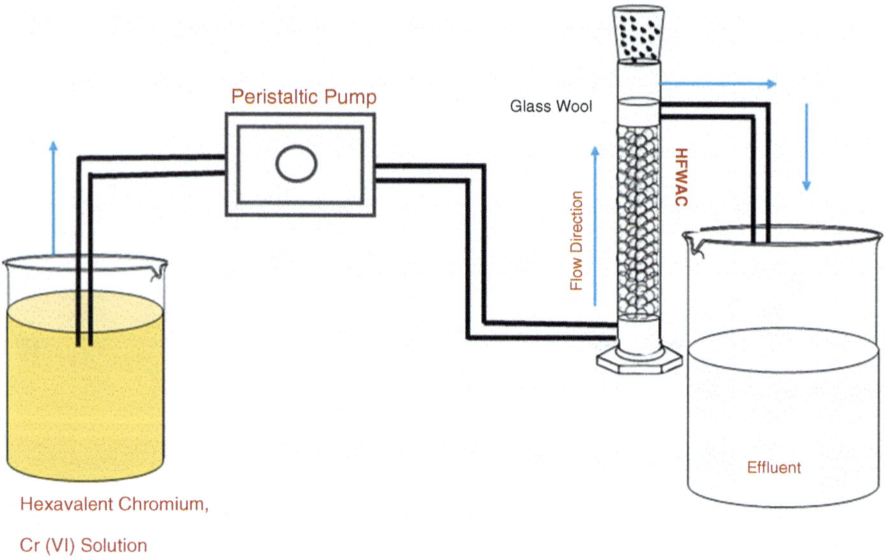

Fig. 3 Schematic sketch of the column operation

Desired concentration of synthetic 50 mL Cr(VI) contaminated solution was taken in the 100 mL plastic bottles (Tarson) with the administered dose of HFWAC and shaken in the BOD incubator shaker at 150 ± 5 rpm, 25 ± 2 °C and at pH 3.12. After completion of aging (6 h), the sample bottles were collected, settled for 5 min, and filtered. The eluent was measured for the residual concentration of Cr (VI) using the standard calibration curve, Abs (y) = $0.0314x + C\ (=0)$, $R^2 = 0.999$ at 350 nm. Lower Cr(VI) removal efficiency occurred in case of the bigger particles of 150 μ with lower surface area, whereas it is far better in case of smaller particles of 106 μ at the same experimental conditions. Hence, the smaller particle (106 μ) was used throughout the study.

A column of 55 cm height (2 cm i.d.) with the inlet and outlet was used for the continuous mode study. The outlet was 5 cm below from the top. The column was packed with 72.0 g HFWAC with the bed height of 20 cm (as a model bed depth) with two supporting glass wool layers (2 cm each) to control over the floating of adsorbent material. The column was charged (Peristaltic pump, Miclins India, Mumbai, India) with the batch-optimized Cr(VI) concentration of 12 mg/L with pH of 3.12, particle size 106 μ. To get the adequate quantity of effluents, the moderate upflow rate of 8 mL/min was maintained. The breakthrough point ($C/C_o = 0.0083$) reached after 40 h (volume of treated water 19.20 L) and bed-exhaust point (~90% influent concentration) reached after 65 h (volume of treated water 31.20 L). The effluents were collected within an interval of 1 h and the concentrations were measured. The obtained data was fitted in the logit model to understand the column design parameters with the column efficiency. The schematic column design is shown in Fig. 3.

3 Results and Discussion

3.1 Dosage and Concentration Variation Study

In the batch study, dosage was varied in order to understand the maximum (%) adsorption of Cr(VI) onto the HFWAC surfaces with an initial concentration of 12 mg/L, pH 3.12, and at the room temperature of 25 ± 2 °C. It is obvious that increasing the fresh precursor sites enhances the adsorption of the solutes on its surfaces. Hence, the definite and positive results have been achieved to ~97% removal of contaminant on HFWAC with a dose of 1.5 g/L and has been considered f throughout the study (Fig. 4, here presented in terms of removal as mg/L).

Similarly, the increment of the solute molecules, i.e., the chemical moieties with a fixed and definite number adsorbent surfaces, with other fixed experimental conditions, suggested the slowed down of the solute taking capability of the adsorbent, as the number of solute molecules became greater with respect to the sorbent particles. However, initial 12 mg/L Cr(VI) was found to remove ~97% from the water medium and was used throughout the study.

3.2 Time Variation Study

In the aqueous media, the heterogeneous combination of solute and sorbent come close to each other due to the weak van der Waals forces of attraction—solute molecules move toward the sorbent surfaces and get adsorbed. Hence, the process is time dependent.

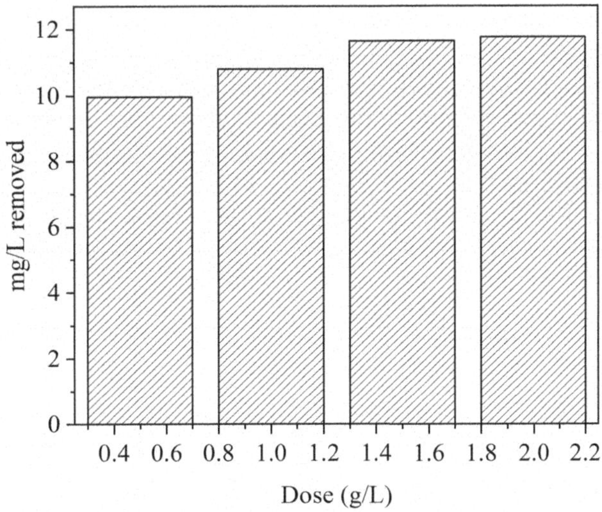

Fig. 4 Effective dose of HFWAC on Cr (VI) adsorption

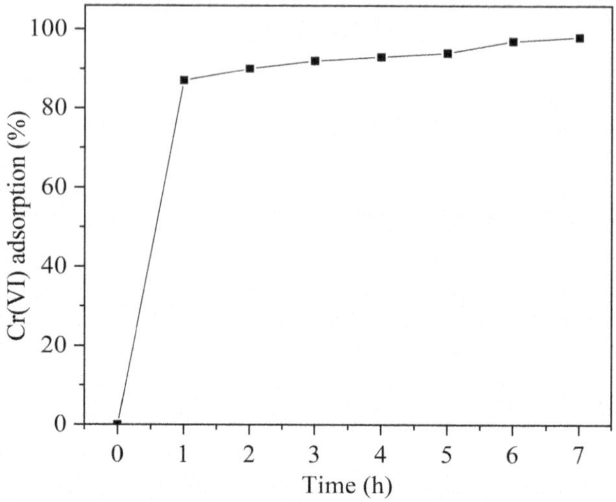

Fig. 5 Percentage adsorption of Cr(VI) on HFWAC with time

Table 1 Linear equations of kinetic models for Cr(VI) adsorption on HFWAC

Kinetic models	Linear equations	R^2
First order	$\ln C_t = -0.4382t + 1.5281$	0.766
Pseudo-first order	$\ln(q_1 - q_t) = -0.4728t + 0.8870$	0.727
Second order	$1/C_t = 0.3544t + 0.0741$	0.836
Pseudo-second order	$t/q_t = 0.1294t + 0.0126$	0.999

Figure 5 suggests that within 1 h of contact time, ~87% Cr(VI) was adsorbed, which may be due to the greater surface area ($S_{BET} = 452\ m^2/g$) of the smaller particles of 106 μ of the macropores HFWAC with respect to the bigger particles of 150 μ (352 m^2/g, after 6 h aging ~85%). However, after 6 h of agitation, ~97% Cr (VI) adsorption was observed and was used throughout the study. Again, the solute uptake capacity is time dependent; hence, the mass transfer to the adsorbent sites from the suspension with time must follow some kinetic models. Hence, to understand the adsorption kinetics models, various reaction models were examined. A comparison of the best-fit adsorption mechanism (Table 1) suggested pseudo-second order model for adsorption (Fig. 6).

The linearized equation of the four kinetic models are as follows [14]:

$$\ln(C_0/C_t) = k_1 t \tag{1}$$

$$\ln(q_1 - q_t) = \ln q_1 - ks_1 t \tag{2}$$

$$1/C_t - 1/C_0 = k_2 t \tag{3}$$

$$1/h = t/q_t - (1/q_e)\ t \tag{4}$$

Where, $h = k q_e^2$

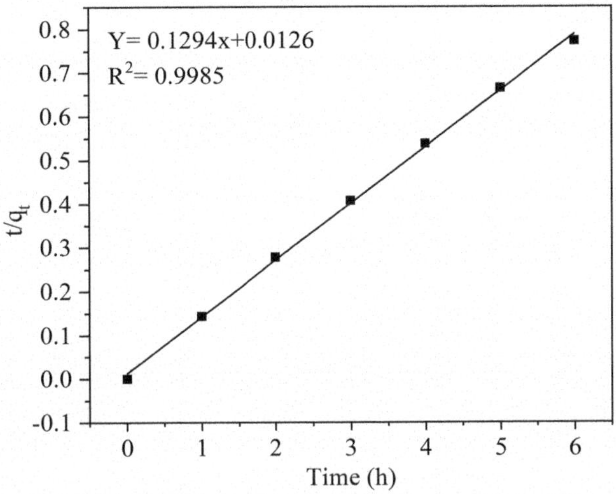

Fig. 6 Curve of pseudo-second order reaction model of Cr(VI) adsorption on HFWAC

Approximately, 97% Cr(VI) was adsorbed on HFWAC within 1 h of contact time and then slowed down, which may be due to the presence of insufficient numbers of active sorbent sites to accommodate the contaminant molecules. Hence, to understand the type of adsorption in terms film diffusion (D_f) or pore diffusion (D_p) followed, i.e., the rate limiting step of the adsorption process, the overall first-order kinetic data has been used. The D_p and D_f values were calculated to be 0.2130×10^{-7} and 0.500×10^{-6} cm^2/s, respectively. The D_f value clearly suggested the adsorption of the solute on to the adsorbent surfaces followed the film diffusion model, i.e., the rate limiting step ($10^{-6} \leq D_f \leq 10^{-8}$ cm^2/s) and not the pore diffusion model ($10^{-11} \leq D_p \leq 10^{-13}$ cm^2/s) [15], which again meets the Freundlich isotherm concept (multilayer physisorption).

3.3 *Isotherm Study*

Because the activated carbon/charcoal is a macropores material with larger surface area, it is considered as the superior adsorbents than any other sorbents for any contaminant, so isotherm studies, namely Langmuir and Freundlich, were carried out at the room temperature to understand its adsorptive capacity. When a solute and sorbent interact in aqueous media, a dynamic equilibrium is established between adsorption and desorption. In this theoretical scenario, the maximum solute uptake capacity is employed to comprehend isotherm studies. The investigation indicates that adsorption conforms more effectively to the Freundlich isotherm model, representing multilayer adsorption in contrast to the monolayer chemosorption of the Langmuir isotherm. The linearized expressions for the Langmuir and Freundlich isotherm models can be represented as [16]:

Table 2 Different Isotherm constants

	Langmuir constants				Freundlich constants			
[Cr]	Temperature (K)	Q_{max}(mg/g)	b (L/mg)	R^2	k_f	n	R^2	R_L
12 mg/L	298	417.50	0.6655	0.9602	171.09	2.3025	0.9768	0.111

$$(C_e/q_e) = C_e/Q_{max} + (1/bQ_{max}) \tag{5}$$

$$\ln\ (q_e/k_f) = (1/n) \times (\ln\ C_e) \tag{6}$$

The equation following Langmuir isotherm is $C_e/q_e = 0.0036 + 0.0024 \times C_e$ ($R^2 = 0.9602$), with the maximum adsorptive capacity of HFWAC (Q_{max}) was found to 417.50 (mg/g). The equation for Freundlich isotherm obtained is ln ($q_e/C_e^{0.4343}$) = 5.1422 (R^2 = 0.9768). The R^2 value suggested that the adsorption process followed Freundlich isotherm a better way than the Langmuir, which is again satisfied by the lower Langmuir isotherm constant value "b" and the higher adsorption intensity value "n" (Table 2, $b < 1$, $n > 1$). The lower calculated value (0.111) of dimensionless equilibrium parameter ($R_L < 1$) also confirms the favorable adsorption of Cr(VI) onto the HFWAC surfaces [16].

3.4 Study with pH of the Media

In aqueous media, Cr(VI) exists in the dimeric form of chromate or dichromate. i.e., CrO_4^{2-} or $Cr_2O_7^{2-}$. Hence, the adsorption of this anion on any adsorbent surfaces will depend on the surface charge of the used material. The pH_{ZPC} (zero-point charge) of the used adsorbent HFWAC was determined [17] experimentally and was found to be 6.50. At lower pH of the media, i.e., <pH_{ZPC}, due to the attachment of the proton on the HFWAC surfaces, it makes the surfaces positive in nature, whereas above it, it is negative in nature due to the presence of the hydroxyl radicles on it. Hence, it causes the adsorption and found good at lower pH (~97%, 3.12). The pH_{ZPC} is 6.50, which caused the surface charge of the adsorbent to be zero. So, at a higher pH than pH_{ZPC}, surface charge is negative. In this case, there was a chance to have the repulsion between the negative surface charge of the adsorbent with that of the anions of the solute. This may cause the suppression removal of Cr (VI) oxy-anions on HFWAC to ~31%. Again, the pH_{ZPC} is fairly well close to the neutral pH of 6.8 ± 0.2. However, at the neutral pH, due to the week van der Waals forces of attraction between the anionic solute to the sorbent, the adsorption may take place, and hence the result obtained. However, at higher pH than the pH_{ZPC}, due to the repulsion between the negative surface charge of the adsorbent and the Cr (VI) anions in the suspension, the suppression removal was achieved (Fig. 7).

The study was conducted in identical experimental conditions spiking the 12 mg/L of Cr(VI) onto the collected pond water (pH 7.86, no detection of Cr(VI)). The

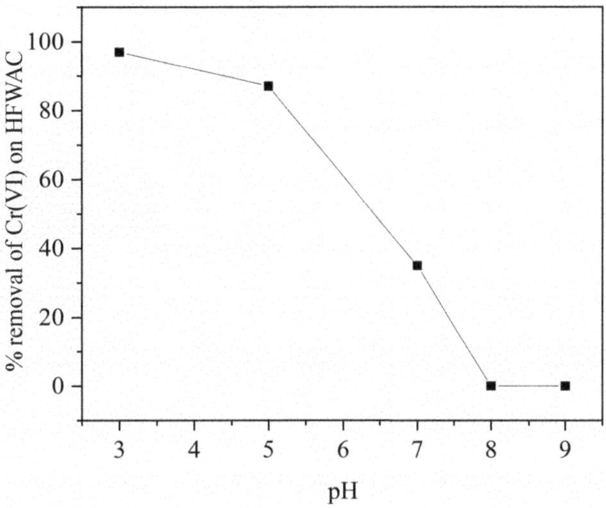

Fig. 7 Cr(VI) adsorption (%) on HFWAC vs. pH

Table 3 Pond water quality

Parameters of pond water	Quantitative values Before treatment	After treatment
pH	7.86	3.52
Chloride (mg/L)	200	195.0
Sulfate (mg/L)	180	187.0
Nitrate (mg/L)	110	120.0
Ferrous (mg/L)	1.2	0.45
[Cr(VI) added]	12 mg/L	85% (removed)
[Cr(VI) added] pH = 3.12 maintained	12 mg/L	92% (removed)
Hardness (mg/L)	150	133

adsorption results (Table 3) suggested that at lower pH (3.12, maintained by dilute HCl), Cr(VI) removed ~92%, which may be the presence of different coexisting ions in the water medium, which is again followed well the pH_{ZPC}, i.e., surface charge of the adsorbent.

HFWAC is a biodegradable/biocompatible material. It is composed of different lignin, cellulose, and hemicellulose with distinct hydroxyl radicles and having the ability to attract/bind the chromium ions on its surfaces forming the verities of complexes or may be adsorbed by the exchange of different cations present on the HFWAC. Again, it may form different precipitate with the anions present on HFWAC surfaces. However, HFWAC may behave as an electron donor (as Lewis base) to Cr(VI) and may reduce to low toxic Cr(III). The probable and postulated Cr (VI) adsorption mechanism on HFWAC is shown in Fig. 8.

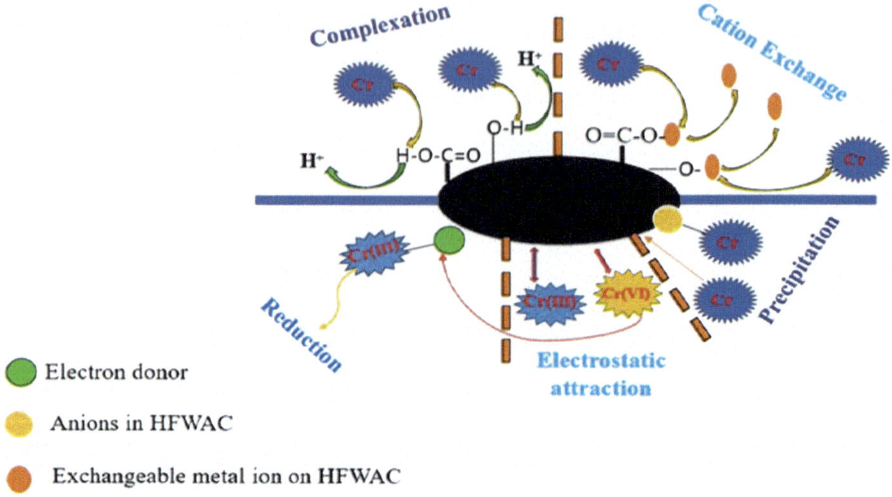

Fig. 8 Probable Cr(VI) adsorption mechanism on HFWAC

3.5 *Column Study*

The breakthrough curve was constructed with the elapsed time on the *X*-axis and C/C_0 of Cr(VI) on the *Y*-axis [18]. The point of breakthrough, indicating when the Cr (VI) concentration reaches the permissible level of 0.1 mg/L (at $C/C_o = 0.0083$), was marked on the curve. The "point of exhaustion" was defined as reaching approximately 90% of the influent concentration [19]. In this study, a 20 cm bed height HFWAC column was employed to investigate the removal of Cr(VI). A synthetic solution contaminated with Cr(VI) at 12 mg/L was passed through the HFWAC column in an upflow mode to prevent unnecessary column clogging [20]. The volumetric flow rate was maintained at 8 mL/min, determined as a practical flow rate for modeling purposes [21]. The breakthrough and bed exhaustion times are illustrated in Fig. 9, and the values of the column design parameters are detailed in Table 4.

In this study, the 20 cm bed-depth column was considered using the logit method [21–23]. The linearized logit equation is expressed as follows:

$$\ln\left[\frac{C/C_o}{1-C/C_o}\right] = -\frac{KNx}{V} + KC_o t \tag{7}$$

where C is Cr(VI) concentration at time t, C_o is initial Cr(VI) concentration, V is approach velocity (~1.533 m/h), x is bed depth (20 cm), K is adsorption rate constant (L/mg h), and N is the adsorption capacity of the column (mg/L). Equation (7) may be rewritten as follows:

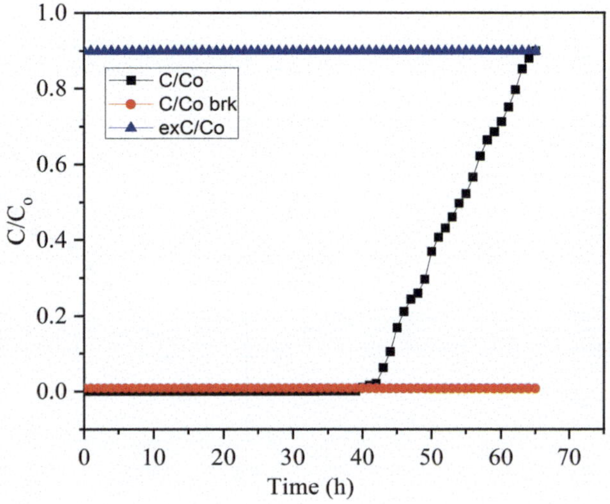

Fig. 9 Breakthrough and bed-exhaust curve of column adsorption with time for Cr(VI) adsorption

Table 4 Logit model parameters for 20 cm bed-depth column

Cr(VI), bed height (cm)	t_z (h)	K (L/mg h)	h_z (cm)	N (mg/L)	U_z (cm/h)	EBCT (min)	x_o (cm)	Saturation (%)
20	25.0	0.0135	7.99	5088.79	3.13	8.45	10.70	96.12

$$\ln\left[\frac{C}{C_o - C}\right] = -\frac{KNx}{V} + KC_o t \tag{8}$$

$$x_0 = \frac{V}{KN}\ln\left(\frac{C_o}{C_B} - 1\right) \tag{9}$$

where x_0 is minimum bed-depth, C_B (breakthrough concentration) = 0.1 mg/L.

The linear variation of $\ln[C/(C_o - C)]$ with time t is presented in Fig. 10.

As can be noted from Eq. (8), the slope of the straight line is represented by KC_o and the intercept by KNx/V. The values of K and N were determined as 0.0135 L/mg h and 5088.79 mg/L, respectively, for the minimum bed-depth of 10.70 cm (Table 4). There was no substantial leaching observed.

4 Conclusion

Household filter waste activated charcoal (HFWAC) was collected, crushed, and sieved in order to obtain two different geometric mean-sized particle of 106 μ and 150 μ particle. The HFWAC was utilized to eliminate Cr(VI) from wastewater, starting with an initial concentration of 12 mg/L. Both batch and column studies were conducted at room temperature and a pH of 3.12. The outcomes demonstrated

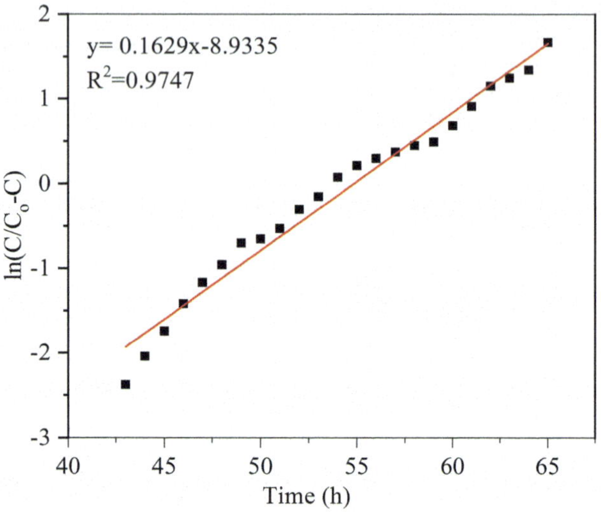

Fig. 10 The logit model

an impressive ~97% adsorption of Cr(VI) when 1.5 g/L of HFWAC was applied. The effectiveness of Cr(VI) adsorption at low pH was elucidated by the pH_{ZPC}. The Langmuir isotherm revealed an adsorptive capacity of 417.50 for the material. A column study with a 20 cm bed-depth of HFWAC was carried out, employing an upflow rate of 8 mL/min. Column design parameters were determined using the logit model. For the minimum bed-depth (x_0) of 10.70 cm, the calculated K and N values were 0.0135 L/mg h and 5088.79 mg/L, respectively. The designed HFWAC column demonstrated successful application in removing Cr(VI) from spiked pond water.

References

1. G.M. Cieslak, Toxic and magnetic effects of chromium (VI). Polyhedron **21**, 3665 (1996)
2. C. Raji, T.S. Anirudhan, Batch Cr(VI) removal by poly acrylamide-grafted sawdust: Kinetics and thermodynamics. Water Res. **32**, 3772 (1998)
3. E. Aranda-García, E. Cristiani-Urbina, Environ. Sci. Pollut. Res. **26**(4), 3157–3173 (2019). https://doi.org/10.1007/s11356-017-0248-z
4. L. Khezami, R. Capart, Removal of chromium (VI) from aqueous solution by activated carbons: Kinetic and equilibrium studies. J. Hazard. Mater. **123**(1–3), 223–231 (2005)
5. A. Boulett, K. Roa, E. Oyarce, L.P. Xiao, R.C. Sun, G.D.C. Pizarro, J. Sánchez, Reusable hydrogels based on lignosulfonate and cationic polymer for the removal of Cr (VI) from wastewater. Colloids Surf. A Physicochem. Eng. Asp. **656**, 130–359 (2023)
6. J.W. Patterson, *Industrial Waste Water Treatment Technology*, 25th edn. (Butterworth-Heineman, London, 1985)
7. M.D. Vu, V.A. Hoang, D.B. Ninh, D.L. Vu, Low-cost material derived from coconut husk for the removal of chromium (VI) from water environment. J. Solid Waste Technol. Manag. **49**(4), 376–383 (2023)

8. S. Babel, T.A. Kumiawan, Cr(VI) removal from synthetic waste water using coconut shell charcoal and commercial activated carbon modified with oxidizing agents and/or chitosan. Chemosphere **54**, 951 (2004)
9. A.J.M. Barros, S. Prasad, V.D. Leite, A.G. Souza, The process of biosorption of heavy metals in bioreactors loaded with sanitary sewage sludge. Braz. J. Chem. Eng. **23, 2**, 153 (2006)
10. I.J. Bithi, M.A. Mottalib, M.S. Miran, M.F. Ehsan, M.M. Rahman, Removal of Cr (VI) from wastewater by impregnated activated carbon generated from vegetable tanned leather waste with aluminium oxide. Results Surf. Interfaces **14**, 100–197 (2024)
11. D.O. Cooney, *Adsorption Design for Wastewater Treatment* (Lewis Publishers, Boca Raton, 1999)
12. A.M. Youssef, T. El-Nabarawy, S.E. Samra, "Sorption properties of chemically activated carbons," 1. Sorption of cadmium (II) ions. Colloids Surf. A **235**, 153–163 (2004)
13. D. Mohan, K.P. Singh, V.K. Singh, Removal of hexavalent chromium from aqueous solution using low-cost activated carbons derived from agricultural waste materials and activated carbon fabric cloth. Ind. Eng. Chem. Res. **44**(4), 1027–1042 (2005)
14. L.D. Benefield, C.W. Randall, *Biological Process Design for Wastewater Treatment*, 1st edn. (Prentice-Hall, Inc., Englewood Cliffs, 1980)
15. F.G. Helfferich, J.S. Dranoff, *Ion Exchange* (McGraw Hill, New York, 1962), p. 299
16. S. Kundu, A.K. Gupta, Adsorption characteristics of As(III) from aqueous solution on iron oxide coated cement (IOCC). J. Hazard. Mater. **142**, 97–104 (2007). https://doi.org/10.1016/j.jhazmat.2006.07.059
17. J.S. Noh, J.A. Schwarz, Estimation of the point of zero charge of simple oxides by mass titration. J. Colloid Interface Sci. **130**, 157–164 (1989)
18. Z. Ren, H. Tang, H. Li, Q. Jing, Column experimental study on the removal of hexavalent chromium from water by modified cellulose filter paper loaded with nano zero-valent iron. J. Water Process Eng. **59**, 104920 (2024)
19. P. Das, V. Chakraborty, Experimental study on removal of chromium using nano-silica coated pyrolysed biochar in batch, fixed bed and fluidized bed reactor: Synthesis, characterization, experimentation and mechanism. Waste Manag. Bull. **1**(4), 60–66 (2024)
20. U. Mahapatra, A.K. Manna, A. Chatterjee, Modeling of breakthrough curves in removal of Cr (VI) and methylene blue using immobilized activated carbon from natural rubber waste. Bioresour. Technol. Rep. **24**, 101–600 (2023)
21. S.K. Maji, Y.-H. Kao, C.-J. Wang, G.-S. Lu, J.-J. Wu, C.-W. Liu, Fixed bed adsorption of As (III) on iron-oxide-coated natural rock (IOCNR) and application to real arsenic-bearing groundwater. J. Chem. Eng. **203**, 285–293 (2012). https://doi.org/10.1016/j.cej.2012.07.033
22. A.S. Michaels, Simplified method of interpreting kinetic data in fix-bed ion exchange. Ind. Eng. Chem. **44**(8), 1922–1930 (1952)
23. C.S. Oulman, The logistic curve as a model for carbon bed design. J. Am. Water Works Assoc. **72**, 50–53 (1980)

Arsenic Removal from Drinking Water by Iron–Copper Mixed Hydroxide Precipitate

Saraswat Dey and Chanchal Majumder

Abbreviations

DD	Doehlert's design
IC-HP	Iron–copper mixed hydroxide precipitate
PZC	Point of zero charge
RSM	Response surface methodology
SEM	Scanning electron microscope

1 Introduction

Arsenic is widely recognized as a potent carcinogen, making it one of the most hazardous chemicals in the world. It is primarily found in the form of sulfides and sulfosalts such as arsenopyrites (FeAsS), arsenic trisulfide (As_2S_3), realgar (As_4S_4), and enargite (Cu_3AsS_4) [1–3]. It exists in two main oxidation states: As (+3) in anaerobic conditions and As (+5) in aerobic environments [4]. In India, the Ganga–Brahmaputra (GB) and Padma–Meghna (PM) river plains aquatic system is greatly contaminated by arsenic [5].

Various sources, including weathering of rocks, volcanic emissions, mining, burning pyritic fuels, arsenic-based herbicides/pesticides, and livestock feed additives, lead to arsenic accumulation in the food chain, impacting humans and animals [6–9]. The World Health Organization (WHO) has suggested a permissible limit of 10 ppb for As in drinking water, which is widely adopted by many countries

S. Dey (✉) · C. Majumder
Civil Engineering Department, Indian Institute of Engineering Science and Technology, Shibpur, Howrah, West Bengal, India
e-mail: chanchal@civil.iiests.ac.in

D. Mazumder (ed.), *Sustainable Advanced Technologies for Environmental Management*, Springer Proceedings in Earth and Environmental Sciences,
https://doi.org/10.1007/978-3-031-64006-3_11

(including India and China). Arsenic is transported and accumulated in plants in the order of roots to shoots and shoots to grains [10]. Directly or indirectly consumed arsenic builds up in the human body in the following order: kidney > liver > muscle [11]. Prolonged arsenic exposure or consumption leads to serious health issues including lung and bladder cancer, heart and pulmonary diseases, skin disorders such as keratosis, melanosis, black foot disease, and kidney damage [12–14].

Various technologies, including coagulation–flocculation, electrokinetics, membrane filtration (e.g., nanofiltration, reverse osmosis), phytoremediation, ion exchange, and adsorption, demonstrate varying efficacy in arsenic removal from water [15–18]. Adsorption is a reliable technique for effectively removing contaminants from water environments [19, 20]. Notably, the adsorption method stands out for its flexibility, simplicity, and high removal efficiency. Moreover, it supports compound recovery by capturing pollutants on the adsorbent surface in their native form [21]. Iron copper hydroxide coprecipitate (IC-HP) nano-adsorbent has a large potential for arsenic removal due to its high surface area, higher PZC value, and multilayer adsorption sites [22, 23].

Although efficient, arsenic removal through adsorption demands careful optimization of the adsorbent's usages. Understanding factor interactions is vital, and this is achieved through mathematical modeling and optimization. The adsorption performance was modeled by response surface methodology (RSM) using a statistical tool such as Doehlert's design (DD) matrix. The DD matrix exhibited notable advantages, characterized by accurate output, and a reduced number of experiments required for the optimization process than other methods. It provides flexibility in selecting variable levels and establishes a spherical experimental domain, ensuring uniformity in space-filling [21, 24]. In this study, the uniform distribution of experimental points by implementing a DD matrix with four factors enabled the development of a second-order mathematical model using only 21 experimental runs.

In this research, arsenic removal was investigated by IC-HP adsorbent and was accomplished by the use of Doehlert's design matrix formulation. The DD matrix, comprising four factors (pH, contact time, adsorbent dose, and initial arsenic concentration), facilitated the exploration of uniformly distributed experimental points. A mathematical model was developed for predicting arsenic removal using main factors, quadratic terms, and two-factor interactions only. The study further involved the development of response surfaces and contour plots to comprehensively analyze the combined effects of multiple factors on arsenic removal.

2 Materials and Methods

2.1 Materials

Analytical-grade chemical $CuSO_4{\cdot}5H_2O$, NaOH, $AsHNa_2O_4{\cdot}7H_2O$, HCl (Merck), and $FeCl_3{\cdot}6H_2O$, $(CH_3)_2CO$ (Loba Chemie) were used without additional purification. Glassware and plasticware were cleaned with 1% HCl and NaOH or Acetone, followed by thoroughly rinsing with deionized (DI) water. As(V) working solution was made with DI water using disodium hydrogen arsenate ($AsHNa_2O_4{\cdot}7H_2O$). Arsenic calibrating standard solutions for atomic absorption spectrophotometer (AAS) were freshly prepared by diluting (1005 ± 6 ppm arsenic with 2% (v/v) HNO_3) arsenic stock solutions (Inorganic Ventures) with DI water. The precursor salts are blended using a magnetic stirrer (Skybound). The solution's pH was determined using a pH meter (Orion pH meter model 420 A+, Thermo, USA), and arsenic content was quantified using an atomic absorption spectrophotometer equipped with a hydride vapor generator (Thermo Fisher Scientific).

2.2 Synthesized IC-HP Adsorbent

The synthesis of IC-HP was following through the coprecipitation process [22, 25, 26]. In a typical batch synthesis, 250 mL each of 0.049 moles of $FeCl_3{\cdot}6H_2O$ and 0.150 moles of $CuSO_4{\cdot}5H_2O$ salt were separately dissolved. After completely dissolved, two 250 mL solutions were then combined in a 1-L beaker, and the initial pH was recorded as 2.35 ± 0.3. 5 M NaOH solution was added drop by drop during stirring (100 rpm) until a pH of 9.00 ± 0.3 was achieved. The obtained material leaves for an aging period of 18 h at room temperature (31 ± 2 °C). The resulting precipitate was filtered using commercial filter paper and washed with DI water until material charge neutrality was obtained. Then it was dried at 105 °C for 24 h and stored in a desiccator for further use.

2.3 Characterization of IC-HP Adsorbent

The scanning electron microscope (SEM) was used to understand the surface morphology of the IC-HP adsorbent (JEOL USA, S-3400). The investigation of the point of zero charge (PZC) for the synthesized material was undertaken to elucidate their surface properties [27, 28].

2.4 Four-Factor Doehlert's Design Matrix

A four-factor Doehlert's design (DD) with a uniform shell design was used for model development. Four factors used in this study are listed in Table 1 with their lower and higher levels. The significant factors should be studied at more number of levels. In this study, for four factors, namely pH [x_1], contact time [x_2], adsorbent dose [x_3], and initial As concentration [x_4], the levels studied were 5, 7, 7, and 3 respectively. The responses of the factors were first assessed through batch studies, and corresponding ranges were selected to develop the mathematical model [29].

Any simplex of dimension D can be formulated from its immediate lower dimensional simplex. The generalized expression to derive any simplex of dimension D is given in the following expression (1) by the addition of $(D + 1)$th run [21]:

Table 1 Lower- and upper-level factors studied

Name of factor	Lower level	Upper level
pH	3	12
Contact time (min)	10	120
Adsorbent dose (g/L)	0.1	1.0

Table 2 Initial simplex of four-factors DD matrix

Factors with coded value				
Run number	pH $[x_1]$	Reaction time (min) $[x_2]$	Adsorbent dose (gm/L) $[x_3]$	Initial arsenic conc. (ppb) $[x_4]$
1	0	0	0	0
2	1	0	0	0
3	0.5	0.866	0	0
4	0.5	0.287	0.816	0
5	0.5	0.287	0.204	0.791

$$\left(\frac{1}{2}\right), \left(\frac{1}{2\sqrt{3}}\right), \left(\frac{1}{2\sqrt{6}}\right), \ldots, \left(\frac{1}{\sqrt{2(D-1)(D-2)}}\right), \left(\frac{1}{\sqrt{2D(D-1)}}\right), \left(\frac{\sqrt{(D+1)}}{\sqrt{2D}}\right) \tag{1}$$

The number of experimental runs (N) in a Doehlert's design (DD) required is given by the formula, $N = D^2 + D + 1$ [29]. For the four factors Doehlert's design experiment, we required a total of 21 runs to develop the model. Within a DD model, both the linear and squared terms of the factors are incorporated, along with linear interaction effects. The initial simplex for four factors DD matrix is presented in Table 2.

From natural value, the coded value can be obtained by Eq. (2), where C_v implies the coded value, C_n stands for the natural value, C_h refers to the higher level, C_l indicates the lower level, and γ represents the factor's range studied [30]:

$$C_v = \frac{C_n - (C_h + C_l)/2}{(C_h - C_l)/2 * \gamma} \tag{2}$$

A polynomial model was created in MATLAB R2019b to predict the arsenic removal performance (%) using IC-HP adsorbent that can be represented as per the following Eq. (3):

$$Y = A_0 + \sum\nolimits_{i=1}^{4} B_i x_i + \sum\nolimits_{i=1}^{4} C_i x_i^2 + \sum\nolimits_{i \neq j=1}^{4} D_{ij} x_i x_j + \varepsilon_{ij} \tag{3}$$

where A_0, B_i, C_i, and D_{ij} are model regression coefficients, ε_{ij} is the error in prediction, and Y is the model predicted removal (%) by IC-HP adsorbent.

2.5 Experimental Study

A 100 mg/L arsenic stock was prepared in DI water and diluted as needed. Arsenic removal was measured using AAS with a hydride vapor generator. Batch adsorption experiments were conducted at a temperature of 31 (± 2) °C with an orbital shaker of 100 (± 5) rpm. Sample pH was monitored and adjusted slowly using 0.1 N NaOH or 0.1 N H_2SO_4.

3 Results and Discussion

3.1 Characterization of Adsorbent Materials

SEM analysis (Fig. 1) revealed agglomerated and cluster-like structures, contributing to a rough and porous surface of the adsorbent. The IC-HP adsorbent exhibited a point of zero charge (PZC) at pH 7.8 (Fig. 2), influencing a broader pH range for effective arsenic removal. At pH values below the PZC, a positive surface charge generates and enhance attraction of negatively charge anions. Similar PZC value of IC-HP are reported by other researchers [22].

3.2 Adsorption Experiment Using DD Matrix

The prevailing studies focus on the arsenic treatment process, predominantly considering single factor batch studies. However, real-world scenarios involve multiple factors influencing treatment efficiency, introducing interactive effects, and yielding varying output results. Table 3 provides the factors studied, their levels in coded form, experimental removal (%), predicted removal (%), and percentage error in

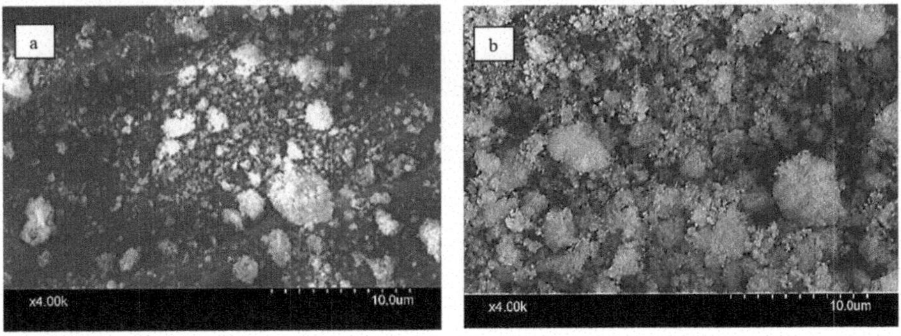

Fig. 1 (**a**, **b**) SEM image of IC-HP adsorbent

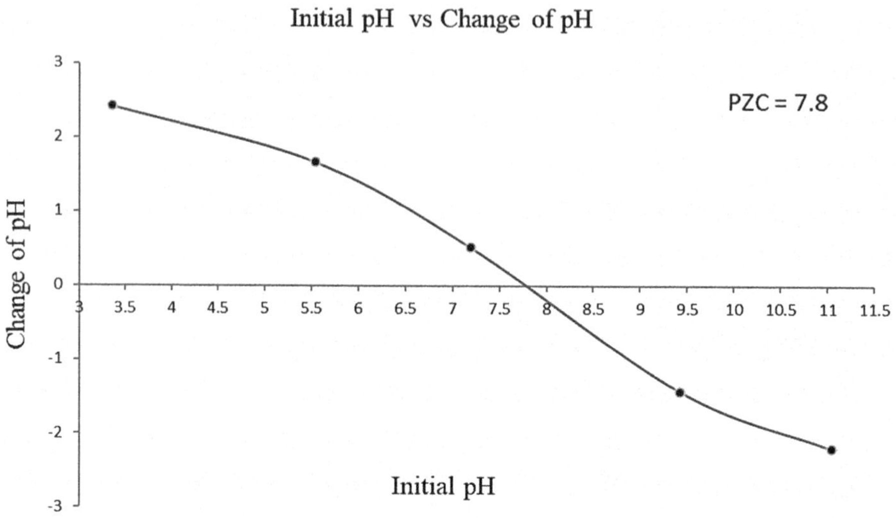

Fig. 2 Determination of PZC of IC-HP adsorbent

prediction. It is clear from the table that the maximum error in As(V) removal remains within (±3.17%).

The necessary analysis was carried out using MATLAB (R2019b) software. Variance analysis was conducted, considering up to two-factor interactions. The formulated model equation for arsenic removal by the IC-HP adsorbent was expressed in polynomial form as given in Eq. (4). The coefficient of determination (R^2) of the developed model was determined to be 0.9843.

$$\begin{aligned} Y\,(\%) = {} & 90.7 - 21.31\,x_1 + 6.4375\,x_2 + 2.2984\,x_3 + 2.1345\,x_4 - 25.2\,x_1^2 \\ & - 12.2\,x_2^2 - 12.325\,x_3^2 - 7.115\,x_4^2 + 1.1547\,x_1x_2 - 14.0642\,x_1x_3 \\ & + 4.0951\,x_1x_4 - 9.0863\,x_2x_3 - 7.1478\,x_2x_4 - 8.4044\,x_3x_4 \end{aligned} \tag{4}$$

The Hessian matrix of the above function was obtained below using Eq. (4) and presented in Eq. (5). To assess the extrema (local maxima or minima), the determinant (H) was computed and was found positive (1.9174×10^5). A determinant less than zero indicates the presence of a saddle point, while a determinant greater than zero suggests either a maximum or minimum. To check the minimum or maximum, the second-order partial derivative was checked and found to be negative, which implies the existence of maxima.

Table 3 Experimental design along with the corresponding arsenic removal, predicted removal, and model error

Run number	Variables in coded value: pH [x_1]	Contact time (min) [x_2]	Adsorbent dose (gm/L) [x_3]	Initial arsenic conc. (ppb) [x_4]	Experimental removal (%)	Predicted removal (%)	Error (%)
1	0.000	0.000	0.000	0.000	90.7	90.70	0.00
2	1.000	0.000	0.000	0.000	42.1	44.19	−2.09
3	0.500	0.866	0.000	0.000	71.4	70.67	0.73
4	0.500	0.289	0.816	0.000	63.7	60.53	3.17
5	0.500	0.289	0.204	0.791	66.8	68.61	−1.81
6	−1.000	0.000	0.000	0.000	88.9	86.81	2.09
7	−0.500	−0.866	0.000	0.000	80.1	80.83	−0.73
8	−0.500	−0.289	−0.816	0.000	71.2	74.37	−3.17
9	−0.500	−0.289	−0.204	−0.791	83.7	81.89	1.81
10	0.500	−0.866	0.000	0.000	59.4	58.52	0.88
11	0.500	−0.289	−0.816	0.000	66.1	64.21	1.89
12	0.500	−0.289	−0.204	−0.791	59.2	59.88	−0.68
13	0.000	0.577	−0.816	0.000	86.2	84.54	1.66
14	0.000	0.577	−0.204	−0.791	84.7	86.21	−1.51
15	0.000	0.000	0.612	−0.791	85.8	85.42	0.38
16	−0.500	0.866	0.000	0.000	90.1	90.98	−0.88
17	−0.500	0.289	0.816	0.000	91.1	92.99	−1.89
18	−0.500	0.289	0.204	0.791	89.9	89.22	0.68
19	0.000	−0.577	0.816	0.000	79.2	80.86	−1.66
20	0.000	−0.577	0.204	0.791	84.6	83.09	1.51
21	0.000	0.000	−0.612	0.791	85.6	85.98	−0.38

$$H=\begin{vmatrix} \frac{\partial^2 Y}{\partial x_1^2} & \frac{\partial^2 Y}{\partial x_1 \partial x_2} & \frac{\partial^2 Y}{\partial x_1 \partial x_3} & \frac{\partial^2 Y}{\partial x_1 \partial x_4} \\ \frac{\partial^2 Y}{\partial x_2 \partial x_1} & \frac{\partial^2 Y}{\partial x_2^2} & \frac{\partial^2 Y}{\partial x_2 \partial x_3} & \frac{\partial^2 Y}{\partial x_2 \partial x_4} \\ \frac{\partial^2 Y}{\partial x_3 \partial x_1} & \frac{\partial^2 Y}{\partial x_3 \partial x_2} & \frac{\partial^2 Y}{\partial x_3^2} & \frac{\partial^2 Y}{\partial x_3 \partial x_4} \\ \frac{\partial^2 Y}{\partial x_4 \partial x_1} & \frac{\partial^2 Y}{\partial x_4 \partial x_2} & \frac{\partial^2 Y}{\partial x_4 \partial x_3} & \frac{\partial^2 Y}{\partial x_4^2} \end{vmatrix}$$

$$=\begin{vmatrix} -50.4 & 1.1547 & -14.0642 & 4.0951 \\ 1.1547 & -24.4 & -9.0863 & -7.1478 \\ -14.0642 & -9.0863 & -24.65 & -8.4044 \\ 4.0951 & -7.1478 & -8.4044 & -14.23 \end{vmatrix} = 1.9174\times 10^5 \quad (5)$$

The experimental condition where maximum removal occurred could be obtain by solving the gradient vectors as given in Eq. (6).

$$\delta Y/\delta x_i = 0; \qquad \text{where } i = 1\text{–}4 \tag{6}$$

As percent removal function Y was expressed in coded value, the optimum condition after solving the set of equations obtained from Eq. (6) was found to be (−0.5941), (0.1645), (0.5095), and (−0.4045), respectively. The corresponding optimum real values of the factors using Eq. (2) are found to be x_1 (pH) = 4.83, x_2 (contact time, in min) = 75.44 min, x_3 (adsorbent dose, in gm/L) = 0.83 gm/L, and x_4 (initial arsenic concentration, in ppb) = 98.83 ppb. The maximum removal obtained at optimal condition was found to be 97.71%. Similar adsorptive capacity are reported by another researcher [23].

3.3 Effect of the Factors

The regression coefficient indicates the direction of the effect of each factor on removal performance. A positive coefficient suggests that an increase in the factor enhances removal, while a negative coefficient implies that an increase diminishes removal efficiency. The positive effects on arsenic removal were observed for contact time (x_2) and adsorbent dose (x_3). Arsenic removal increase with increased of contact time and adsorbent dose. For initial arsenic concentration, Eq. (4) showed positive effect and removal increase with increase of initial arsenic concentration. However, effluent arsenic concentration also increased and it exceed the permissible limit (10 ppb). Effect of pH has been found to be most influencing for arsenic removal. There is an optimum pH for maximum removal occurred. Similar trends were obtained during single factor batch experimental conditions for arsenic removal [22, 23].

In Fig. 3, the Pareto chart utilizes the absolute magnitudes of regression coefficients. Notably, quadratic effects from solution pH, contact time, and adsorbent dose exerted the most substantial influence on yield variability and prediction, constituting approximately 26%, 12%, and 12%, respectively. The factor effects depicted in the plot collectively accounted for 97.71% of the predicted arsenic removal.

3.4 Checking Model Adequacy

The model assumes that errors follow a normal distribution. To verify this, a plot of predicted removals (%) against residuals (%) (experimental minus predicted percentage values) was tested and presented in Fig. 4. Result indicates a random distribution of residuals, affirming the normal distribution assumption. Figure 5 depicts the experimental removal (%) against model-predicted removal (%), revealing a high R^2 value of 0.9843. This signifies a precise and accurate method for designing.

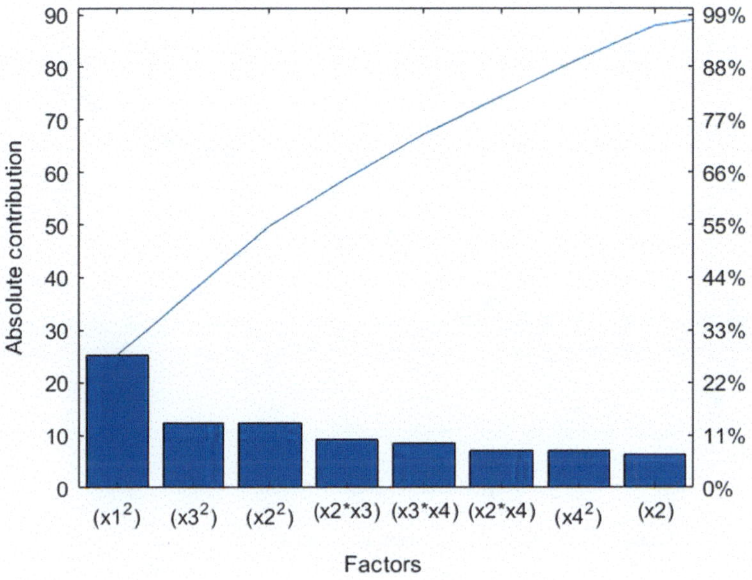

Fig. 3 Pareto plot of the factor affected in arsenic removal

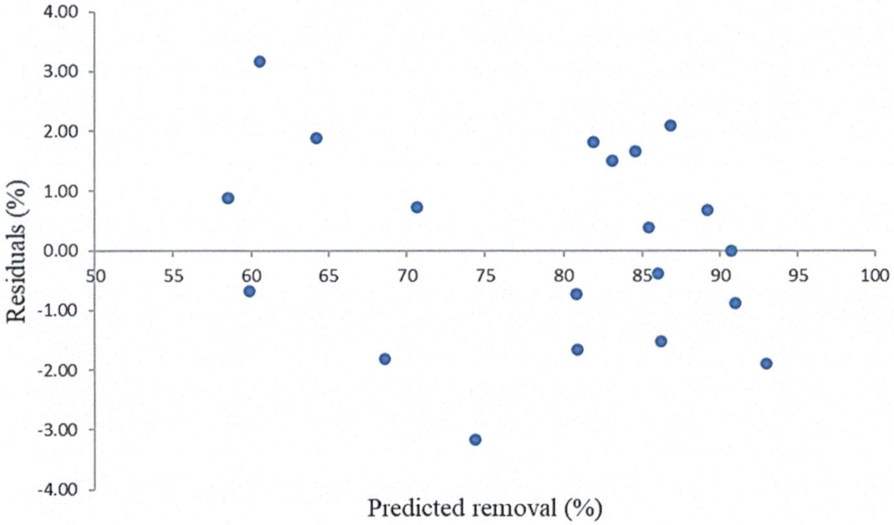

Fig. 4 Predicted removal (%) versus residuals (%)

3.5 *Interaction Effects*

Interaction effects (up to two factors) were presented by response surface and contour plots. The interpretation of Fig. 6a–c suggests that there is an optimal pH

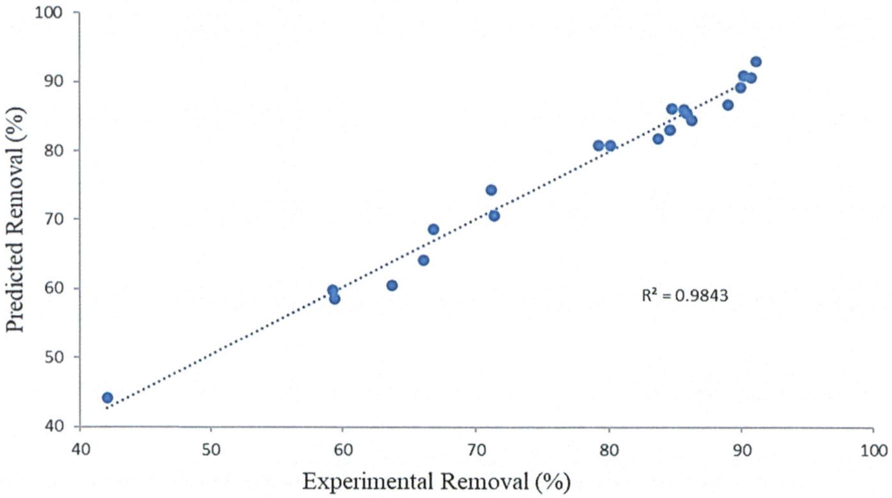

Fig. 5 Predicted removal (%) versus experimental removal (%)

for maximum arsenic removal. Similar result was reported by other researchers also [22, 23]. Figure 6a demonstrates that at a particular pH, if adsorbent dose was increased then the removal also increased. At optimal pH, arsenic removal increased with increasing the adsorbent dose [23]. From Fig. 6b, it was clear that at a particular pH, if contact time was increased, the arsenic removal also increased and there after remains constant as saturation was reached [22]. Figure 6c indicates that at a particular initial arsenic concentration, there is an optimal pH for maximum arsenic removal and conversely, for any particular pH, with increasing initial arsenic concentration, the removal did not improve significantly. This may be due the fact that at a certain pH value, only a particular type of ionic species of As(V) exists [31].

It was clear from the Fig. 6d that minimum arsenic removal occurred at the lowest contact time and lowest adsorbent dose. This may be due to the fact that at lowest point, limited adsorption sites and insufficient interaction time occurred. Increasing the adsorbent dose and contact time enhanced the arsenic removal until equilibrium was reached, which was consistent with other studies [22, 32, 33]. In Fig. 6e, the percentage improvement in arsenic removal is not clearly noticeable with increasing initial arsenic concentration at any given contact time. However, a decrease in contact time at a specific initial arsenic concentration leads to reduced removal efficiency, attributed to insufficient interaction time [22]. Figure 6f shows that at an optimum adsorbent dose, percent increase of arsenic removal is not significant with increasing initial concentration. However, it is noticeable that at any initial arsenic concentration, as the adsorbent dose decreases, the removal efficiency also decreases and vice versa. This trend is attributed to increased adsorption sites with

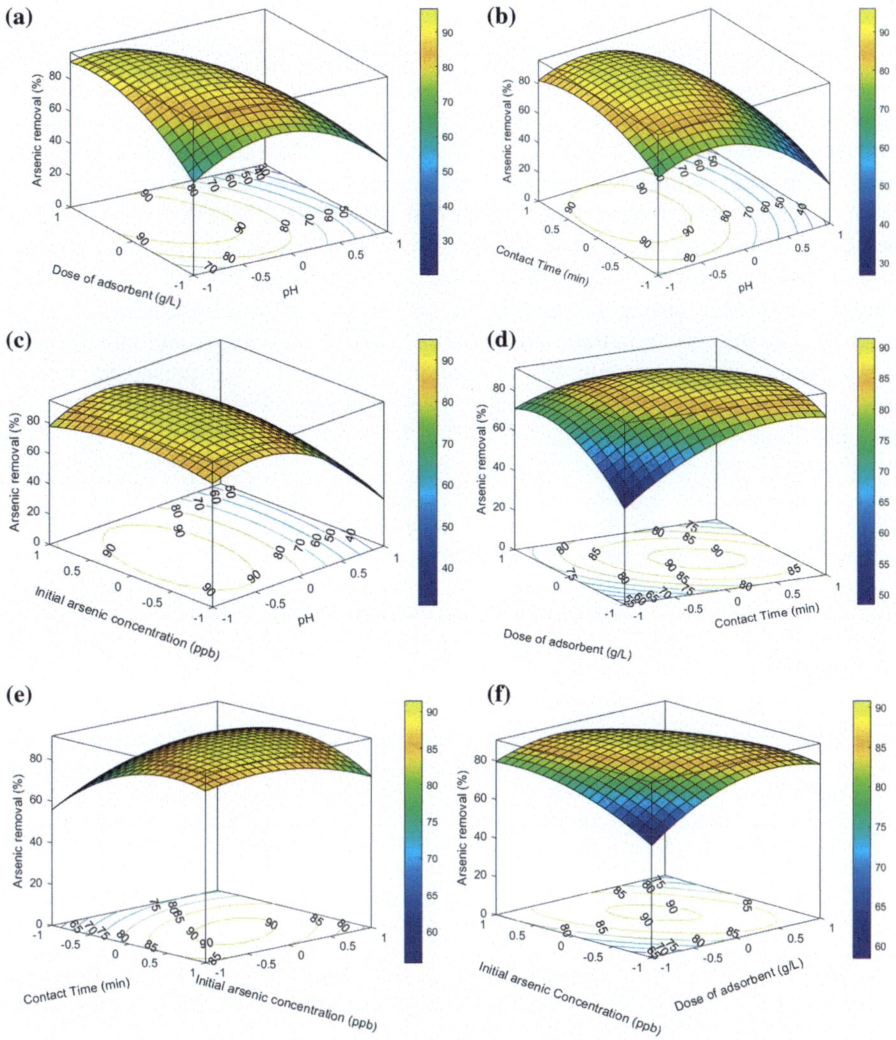

Fig. 6 (**a**) Interaction of pH vs. adsorbent dose (gm/L). (**b**) Interaction of pH vs. contact time (min). (**c**) Interaction of pH vs. initial arsenic concentration (ppb). (**d**) Interaction of contact time vs. adsorbent dose (gm/L). (**e**) Interaction of contact time (min) vs. initial arsenic concentration (ppb). (**f**) Interaction of adsorbent dose (gm/L) vs. initial arsenic concentration (ppb)

introducing higher adsorbent doses, leading to enhanced arsenic removal. Similar findings have been reported by previous studies [22, 33].

4 Conclusion

The model effectively simulated the removal of arsenic (As(V)) using the IC-HP adsorbent. In the four-factor DD matrix, the considered factors were pH, contact time, adsorbent dose, and initial As(V) concentration. DD model identified pH as the most influential factor, followed by adsorbent dose and contact time in arsenic removal. From optimization process, optimal conditions of factors were found to be pH 4.83, contact time of 75.44 minutes, adsorbent dose of 0.83 gm/L, and initial arsenic concentration of 98.83 ppb, resulting in a predicted 97.71% removal. The model exhibited a strong correlation ($R^2 = 0.9843$) with a maximum error of $\pm 3.17\%$. Notably, the quadratic effects of pH, adsorbent dosage, and contact time were significant, and meaningful two-factor interactions were presented. Future research should focus on more intricate models that consider interfering solutes for potential large-scale implementation.

Acknowledgments The author would like to express his gratitude to the Environmental Engineering section at IIEST, Shibpur, India, for granting permission to utilize various instruments and chemical facilities during the course of this research.

References

1. F. Habashi, Arsenic, antimony, and bismuth production, in *Encyclopedia of Materials: Science and Technology*, (Elsevier, 2001), pp. 332–336. https://doi.org/10.1016/B0-08-043152-6/00069-3
2. B.K. Santra, Arsenic contamination of groundwater in West Bengal: Awareness for health and social problems. Int. J. Appl. Sci. Eng. **5**, 43 (2017). https://doi.org/10.5958/2322-0465.2017.00005.3
3. M. Seidl, G. Balázs, M. Scheer, The chemistry of yellow arsenic. Chem. Rev. **119**, 8406–8434 (2019). https://doi.org/10.1021/acs.chemrev.8b00713
4. G. Durante-Rodríguez, H. Fernández-Llamosas, E. Alonso-Fernandes, M.N. Fernández-Muñiz, R. Muñoz-Olivas, E. Díaz, M. Carmona, ArxA from Azoarcus sp. CIB, an anaerobic arsenite oxidase from an obligate heterotrophic and mesophilic bacterium. Front. Microbiol. **10** (2019). https://doi.org/10.3389/fmicb.2019.01699
5. A. Ranjan, K.K. Mandal, M.K. Meena, M.L. Meena, Hydro-social interactions in the arsenic hotspot regions of India. Natl. Geogr. J. India **66**, 356–370 (2020). https://doi.org/10.48008/ngji.1754
6. S. Shankar, U. Shanker, Shikha, Arsenic contamination of groundwater: A review of sources, prevalence, health risks, and strategies for mitigation. Sci. World J. **2014** (2014). https://doi.org/10.1155/2014/304524
7. D. Mohan, C.U. Pittman, Arsenic removal from water/wastewater using adsorbents-A critical review. J. Hazard. Mater. **142**, 1–53 (2007). https://doi.org/10.1016/j.jhazmat.2007.01.006
8. V. Hare, P. Chowdhary, B. Kumar, D.C. Sharma, V.S. Baghel, Arsenic toxicity and its remediation strategies for fighting the environmental threat, in *Emerging and Eco-Friendly Approaches for Waste Management*, (Springer Singapore, 2018), pp. 143–170. https://doi.org/10.1007/978-981-10-8669-4_8
9. W. Ali, A. Rasool, M. Junaid, H. Zhang, A comprehensive review on current status, mechanism, and possible sources of arsenic contamination in groundwater: A global perspective with

prominence of Pakistan scenario. Environ. Geochem. Health **41**, 737–760 (2019). https://doi.org/10.1007/s10653-018-0169-x
10. W. Alam, Y. Gyanendra, M.T. Neihsial, N. Nesa, J. Earth Sci. India, 38–52 (2019). https://doi.org/10.31870/ESI.12.1.2019.03
11. P.C.T. Perera, T.V. Sundarabarathy, T. Sivananthawerl, S.P. Kodithuwakku, U. Edirisinghe, Arsenic and cadmium contamination in water, sediments and fish is a consequence of paddy cultivation: Evidence of river pollution in Sri Lanka. Achiev. Life Sci. **10**, 144–160 (2016). https://doi.org/10.1016/j.als.2016.11.002
12. J.O. Fatoki, J.A. Badmus, Arsenic as an environmental and human health antagonist: A review of its toxicity and disease initiation. J. Hazard. Mater. Adv. **5** (2022). https://doi.org/10.1016/j.hazadv.2022.100052
13. A.P. Sage, B.C. Minatel, K.W. Ng, G.L. Stewart, T.J.B. Dummer, W.L. Lam, V.D. Martinez, Oncogenomic disruptions in arsenic-induced carcinogenesis. Oncotarget **8**, 25736–25755 (2017). https://doi.org/10.18632/oncotarget.15106
14. S. Muzaffar, J. Khan, R. Srivastava, M.S. Gorbatyuk, M. Athar, Mechanistic understanding of the toxic effects of arsenic and warfare arsenicals on human health and environment. Cell Biol. Toxicol. **39**, 85–110 (2023). https://doi.org/10.1007/s10565-022-09710-8
15. S. Alka, S. Shahir, N. Ibrahim, M.J. Ndejiko, D.V.N. Vo, F.A. Manan, Arsenic removal technologies and future trends: A mini review. J. Clean. Prod. **278** (2021). https://doi.org/10.1016/j.jclepro.2020.123805
16. R. Zakhar, J. Derco, F. Čacho, An overview of main arsenic removal technologies. Acta Chim. Slov. **11**, 107–113 (2018). https://doi.org/10.2478/acs-2018-0016
17. A. Shukla, S. Srivastava, A review of phytoremediation prospects for arsenic contaminated water and soil, in *Phytomanagement of Polluted Sites*, (Elsevier, 2019), pp. 243–254. https://doi.org/10.1016/B978-0-12-813912-7.00008-9
18. P.S. Ghosal, K.V. Kattil, M.K. Yadav, A.K. Gupta, Adsorptive removal of arsenic by novel iron/olivine composite: Insights into preparation and adsorption process by response surface methodology and artificial neural network. J. Environ. Manag. **209**, 176–187 (2018). https://doi.org/10.1016/j.jenvman.2017.12.040
19. J.-B. Huo, K. Gupta, C. Lu, H. Chr, M.-L.F. Bruun Hansen, Recyclable high-affinity arsenate sorbents based on porous Fe2O3/La2O2CO3 composites derived from Fe-La-C frameworks. Colloids Surf. A Physicochem. Eng. Asp. **585**, 124018 (2020). https://doi.org/10.1016/j.colsurfa.2019.124018
20. H. Sadegh, G.A.M. Ali, S. Agarwal, V.K. Gupta, Surface modification of MWCNTs with carboxylic-to-amine and their superb adsorption performance. Int. J. Environ. Res. **13**, 523–531 (2019). https://doi.org/10.1007/s41742-019-00193-w
21. R. Chatterjee, C. Majumder, Process optimization of 2-methylpyridine removal by graphene nanoparticles. J. Indian Chem. Soc. **97**, 501–506 (2020)
22. G. Zhang, Z. Ren, X. Zhang, J. Chen, Nanostructured iron(III)-copper(II) binary oxide: A novel adsorbent for enhanced arsenic removal from aqueous solutions. Water Res. **47**, 4022–4031 (2013). https://doi.org/10.1016/j.watres.2012.11.059
23. T. Wang, Y. Jiao, M. He, W. Ouyang, C. Lin, X. Liu, Facile co-removal of As(V) and Sb(V) from aqueous solution using Fe-Cu binary oxides: Structural modification and self-driven force field of copper oxides. Sci. Total Environ. **803** (2022). https://doi.org/10.1016/j.scitotenv.2021.150084
24. A.R. de Sena, G.L. Valasques, I.K.S.P. Barretto, S.A. Assis, Application of Doehlert experimental design in the optimization of experimental variables for the Pseudozyma sp. (CCMB 306) and Pseudozyma sp. (CCMB 300) cell lysis. Cienc. Tecnol. Aliment. **32**, 762–767 (2012). https://doi.org/10.1590/S0101-20612012005000118
25. S.U. Khan, R. Zaidi, S.Z. Hassan, I.H. Farooqi, A. Azam, Application of Fe-Cu binary oxide nanoparticles for the removal of hexavalent chromium from aqueous solution. Water Sci. Technol. **74**, 165–175 (2016). https://doi.org/10.2166/wst.2016.172

26. Y. Li, X. Hu, B. Ren, J. Yue, W. Yang, Preparation of iron-copper binary oxide and its effective removal on antimony(V) from water. Desalin. Water Treat. **57**, 26461–26471 (2016). https://doi.org/10.1080/19443994.2016.1175971
27. T. Mahmood, M.T. Saddique, A. Naeem, P. Westerhoff, S. Mustafa, A. Alum, Comparison of different methods for the point of zero charge determination of NiO. Ind. Eng. Chem. Res. **50**, 10017–10023 (2011). https://doi.org/10.1021/ie200271d
28. E.A. Al-Maliky, H.A. Gzar, M.G. Al-Azawy, Determination of point of zero charge (PZC) of concrete particles adsorbents. IOP Conf. Ser. Mater. Sci. Eng. **1184**, 012004 (2021). https://doi.org/10.1088/1757-899x/1184/1/012004
29. R. Chatterjee, C. Majumder, Uptake of azaarenic 2-Methylpyridine by pre-cooled carboxyl functionalized graphene nanocomposite: Detection, sorption and optimization. J. Water Process Eng. **39** (2021). https://doi.org/10.1016/j.jwpe.2020.101686
30. C. Majumder, Arsenic(V) removal using activated alumina: Kinetics and modeling by response surface. J. Environ. Eng. **144** (2018). https://doi.org/10.1061/(asce)ee.1943-7870.0001337
31. A. Shakya, M. Dodson, J.F. Artiola, M. Ramirez-Andreotta, R.A. Root, X. Ding, J. Chorover, R.M. Maier, Arsenic in drinking water and diabetes. Water (Switzerland) **15** (2023). https://doi.org/10.3390/w15091751
32. Y. Babaee, C.N. Mulligan, M.S. Rahaman, Removal of arsenic (III) and arsenic (V) from aqueous solutions through adsorption by Fe/Cu nanoparticles. J. Chem. Technol. Biotechnol. **93**, 63–71 (2018). https://doi.org/10.1002/jctb.5320
33. P. Nakseedee, V. Tanboonchuy, P. Khemthong, N. Grisdanurak, C.H. Liao, Role of Cu on zero valent bimetallic Cu—Fe in arsenic removal with gas bubbling. Environ. Prog. Sustain. Energy **36**, 1449–1457 (2017). https://doi.org/10.1002/ep.12603

A Study on Adsorption Kinetics and Isotherms of Heavy Metals on Sawdust and Optimization

Tanmoy Bir, Sumit Baran Chowdhury, and Dulal Ch. Mukherjee

Abbreviations

AAS	Atomic adsorption spectroscopy
CCD	Central composite design
FTIR	Fourier transform infrared spectroscopy
RSM	Response surface method

1 Introduction

A wide range of pollutants, including both organic and inorganic substances, present a substantial risk to the environment. Currently, there is significant concern regarding heavy metal ions because of their extremely toxic and cancer-causing properties, which pose risks to the worldwide ecosystem and the food chain. Human activities, spanning industries such as chemicals, textiles, mining, plastics, and others, contribute to the release of heavy metals into the environment [1–3].

Aquatic environments are contaminated with hazardous pollutants known as heavy metals, which are introduced through human activities such as industrial

T. Bir (✉)
Department of Civil Engineering, Indian Institute of Engineering and Science Technology, Shibpur, Howrah, West Bengal, India

S. B. Chowdhury
Envirocheck, Kolkata, India

D. Ch. Mukherjee
Department of Chemical Engineering Heritage Engineering College, Anandapur, Mundapara, Kolkata, West Bengal, India

D. Mazumder (ed.), *Sustainable Advanced Technologies for Environmental Management*, Springer Proceedings in Earth and Environmental Sciences,
https://doi.org/10.1007/978-3-031-64006-3_12

processes, mining, and agriculture. While methods such as chemical precipitation, complexation, solvent extraction, and membrane processes exist for eliminating metal ions from solutions, many are either expensive or overly intricate for metallic materials. As a reaction to this, a range of technologies such as electrochemical and chemical treatments, biological processes, membrane techniques, and adsorption have been devised. Among these, adsorption stands out as the most convenient and widely adopted approach for removing heavy metals from industrial effluents [4–6]. In comparison to physicochemical methods, as highlighted by Jang et al. [7], adsorption procedures offer numerous advantages, exhibiting high selectivity and efficiency in removing various concentrations of soluble heavy metals from solutions.

Sawdust, a residual material produced by the wood industry, has garnered recognition as an ecologically sustainable and economically advantageous absorbent for extracting heavy metals from water-based solutions. Current research has primarily concentrated on investigating the effectiveness of sawdust, a waste material that offers notable environmental benefits in terms of its ability to be reused, for the purpose of removing heavy metal pollutants [7–9]. It is important to note that the behavior of sawdust depends on the chemical com position. The major chemical constituents of sawdust are hemicellulose, cellulose, and lignin. On average, the quantitative percentages of hemicellulose, cellulose, and lignin in sawdust are observed in the range of 15–35, 35–60, and 15–30%, respectively [10]. This study aims to assess the potential of sawdust in adsorbing heavy metals from water solutions, considering factors such as adsorption rate and capacity per unit mass, which impact the replacement frequency of the sorbent and the economic feasibility of the system. Prior to implementing an adsorption system, it is essential to comprehend the characteristics and mechanisms that impact the adsorption of metals by sawdust.

The study examines the rates and patterns of metal adsorption on sawdust by employing various models, while analyzing the impact of factors such as initial concentration and pH. Column experiments were performed to assess the effectiveness of the sawdust adsorption system in treating heavy metals in aquatic settings. The findings make substantial advancements in the field, providing opportunities for implementing effective and eco-friendly approaches to address the problem of heavy metal contamination.

2 Materials and Methods

2.1 Material

The sawdust originated from a sawmill engaged in the milling of sal (*Shorea robusta*) trees. The sawdust was gathered and underwent a procedure of rinsing with distilled water, thereafter, being exposed to sunlight for drying purposes, with the intention of decreasing the moisture content to a level lower than 1%. The following presents the physical characterization of sawdust:

1. Mesh size: 20–40 mesh
2. The apparent specific gravity is 0.18
3. Porosity: 0.80%
4. Water retention rate: 0.77%
5. The ash content is 0.65% by weight
6. Water retention is at a level of 50%

Sawdust consists of 40% cellulose, 30% lignin, and the remaining portion is composed of hemicellulose and resins.

An investigation was conducted to analyze the alterations in the functional groups and surface morphology of the sawdust using multiple analytical methods. The surface functional groups of sawdust were analyzed before and after adsorption using attenuated total reflectance. The spectroscopy technique used is Fourier transform infrared spectroscopy, specifically attenuated total reflection Fourier transform infrared spectroscopy (ATR-FTIR).

2.2 *Methods*

The sawdust obtained from *Shorea robusta* was treated by washing with deionized water and subsequent sun-drying to reduce its moisture content to less than 1%. A specific amount of the dried sawdust was introduced into 1 L of simulated wastewater containing target heavy metals at different concentrations. After adding the sawdust, the entire solution was stirred in a mixer at a constant rotational speed of 100 rpm, and a sample was collected from the outlet with a stopcock. The 10 mL aliquot obtained was filtered and subjected to analysis using atomic absorption spectroscopy (AAS) with a graphite furnace. Standard solutions of 10 ppm for each heavy metal (Cu, Ni, Cr) were prepared from a 1000 ppm stock solution. Subsequently, varying amounts of the adsorbent (sawdust) ranging from 1 to 5 g per 1000 mL were mixed with the solution. The solution's pH was subsequently modified within the range of 1–12, and mechanical agitation was utilized. Figure 1 illustrates the experimental arrangement for the absorption of heavy metals utilizing sawdust.

3 Results and Discussion

Figure 2 shows the predominant factors affecting the performance of adsorption of sawdust.

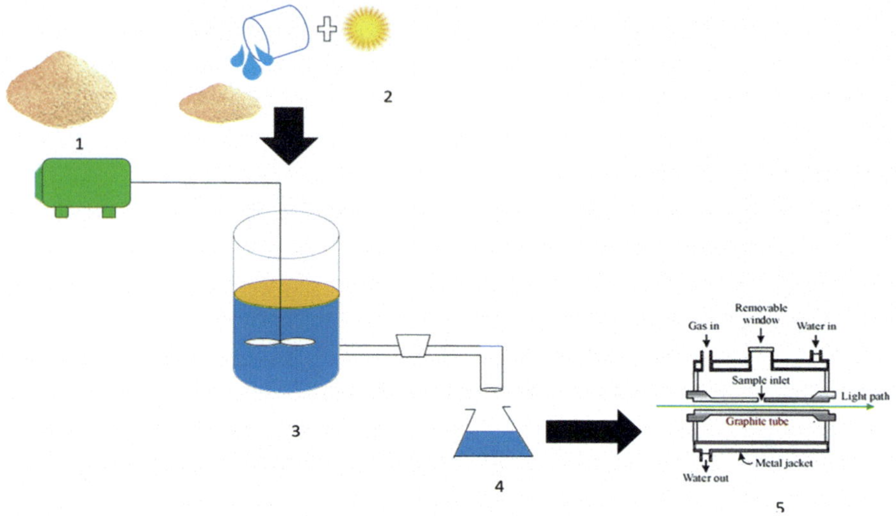

Fig. 1 Flow chart of experimental plan: (1) sawdust, (2) washing of sawdust and sun-drying, (3) beaker with stop cock with rotor for uniform mixing (10 ppm of Cu, Ni, Cr(VI)), (4) collection beaker after adsorption, (5) graphite plate AAS

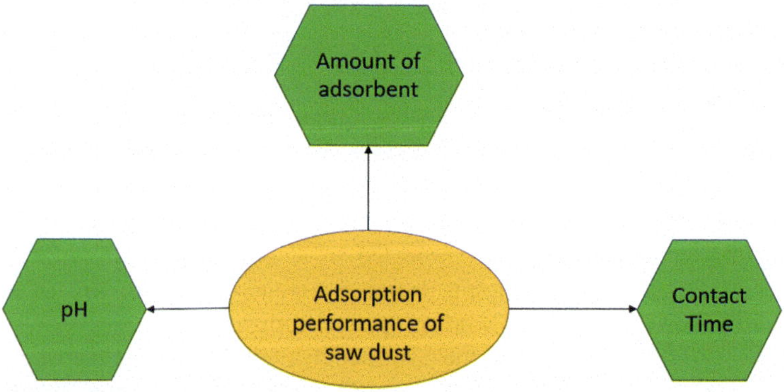

Fig. 2 Factors affecting adsorbent capacity of sawdust

3.1 The Impact of pH, Amount of Adsorbent, and Contact Time

To investigate the impact of different quantities of the adsorbent, specifically sawdust, on the process of adsorption, three distinct amounts of the adsorbent were utilized. It is imperative to highlight that the starting concentration of the metal ion solution remained consistent throughout the experiment. The results unambiguously demonstrate a direct correlation between the adsorption capacity and the amount of sawdust utilized. This relationship can be explained by the observation that increasing the amount of adsorbent also increases the active surface area available for adsorption. Similarly, both pH and contact time exert a comparable influence on the adsorption capacity of sawdust. Regarding contact time, an extended duration enhances performance, with equilibrium reached at 60 min. The maximum adsorption capacity is attained at a pH of 5, falling within the pH range of 2–8. However, beyond this pH level, the adsorption capacity diminishes due to the formation of soluble hydroxide. This study emphasizes the complex relationship between the amount of adsorbent, pH, and contact time in affecting the effectiveness of the adsorption process. Figures 3, 4, and 5 describe the influence of pH, weight of adsorbent, and contact time on the performance of adsorption of different heavy metals.

3.2 Development of Isotherm

The correlation between the equilibrium adsorption of heavy metals, represented as q_e (mg/g), and the corresponding optimum concentration, C_e (mg/L), can be determined by employing the linearized form of the Langmuir isotherm model:

$$\frac{C_e}{q_e} = \frac{1}{K_L S_m} + \frac{C_e}{S_m} \tag{1}$$

or the isotherm model developed by Freundlich:

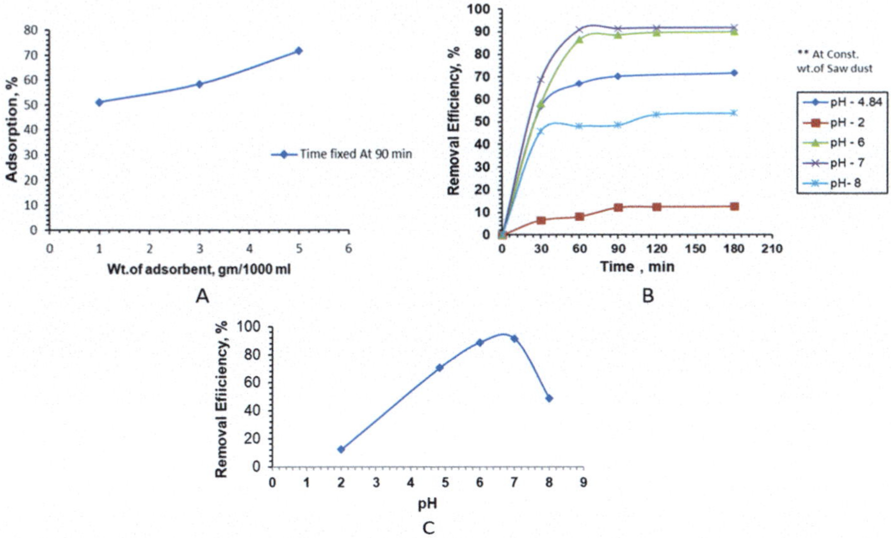

Fig. 3 The impact of (**a**) adsorbent weight, (**b**) contact duration, and (**c**) pH level on the elimination of copper

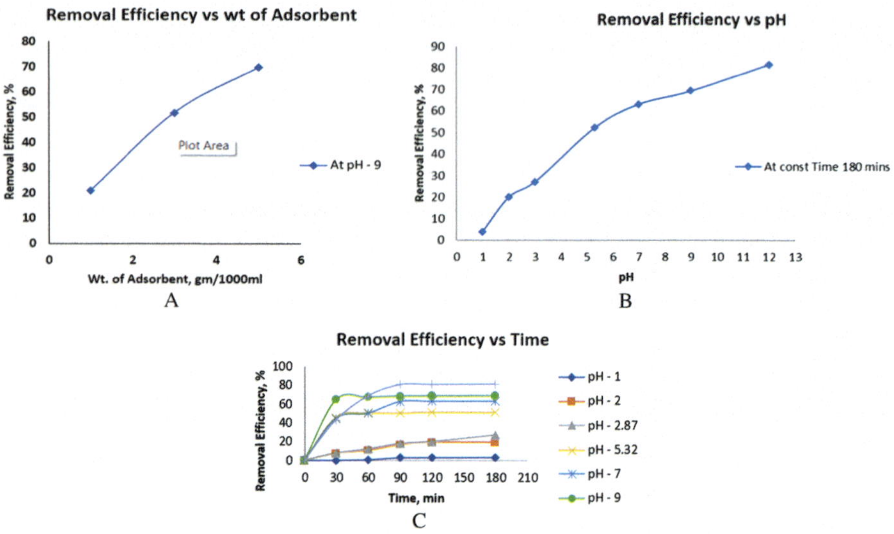

Fig. 4 The impact of (**a**) adsorbent weight, (**b**) contact duration, and (**c**) pH level on the elimination of nickel

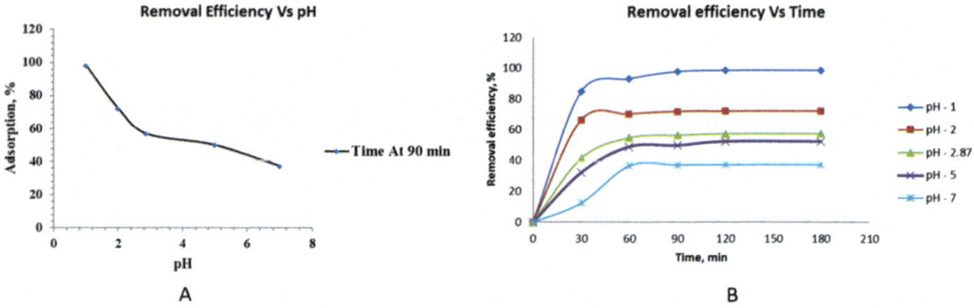

Fig. 5 The influence of (**a**) pH and (**b**) contact time on the elimination of chromium (Cr)

$$\log q_e = \log\ K_f + \frac{1}{n} \log\ C_e \tag{2}$$

A maximum adsorption capacity (S_m) and an energy constant (K_L) associated with the heat of adsorption are introduced by the Langmuir model. The Freundlich model also includes the adsorption capacity constant K_f, where $1/n$ represents the adsorption intensity. For adsorption on a single layer, we have the Langmuir isotherm equation; for adsorption on multiple layers, we have the Freundlich isotherm equation. This is the expression of the Dubinin–Radushkevich isotherm equation according to [11–13]:

$$q_e = q_{\max} \exp\left(-\beta_{\mathrm{DR}} \epsilon^2\right) \tag{3}$$

The activity coefficient (β_{DR}) is equivalent to $\mathrm{mol}^2\ \mathrm{KJ}^{-2}$ and the Polanyi potential (ϵ) is equivalent to $\mathrm{kJ\ mol}^{-1}$, and they can be expressed as

$$\epsilon = RT \ln\left(1 + \frac{1}{C_e}\right) \tag{4}$$

where T is the absolute temperature (298 K), R is the gas constant (0.00831447 kJ $\mathrm{K}^{-1}\ \mathrm{mol}^{-1}$), and C_e is the equilibrium metal concentration (g g^{-1}). There is a linear form for the Dubinin–Radushkevich equation:

$$\ln\ q_e = \ln\ q_{\max} - \beta\mathrm{DR}\ \epsilon^2 \tag{5}$$

Table 1 shows the isotherm constants of heavy metal for heavy metals such as Cu, Ni, and Cr.

Figures 6, 7, 8, and 9 show the best-fit isotherm of Cu, Ni, and Cr after adsorption using *Shorea robusta* sawdust.

Table 1 Isotherm constants of adsorption of heavy metal

	Freundlich			Dubinin–Radushkevich			
Metal	n	k	R^2	qm	k	E (kJ/mol)	R^2
Cu	1	0.19	0.988	1.21	8.00E−08	2.5	0.989
Ni	1.8	2.07	0.96				
Cr	2	1.52	0.95				

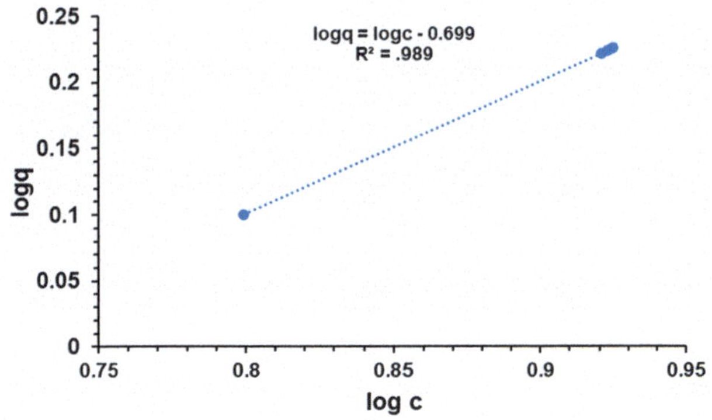

Fig. 6 Freundlich isotherm plot of adsorption of Cu

Fig. 7 Dubinin–Radushkevich plot for Cu adsorption

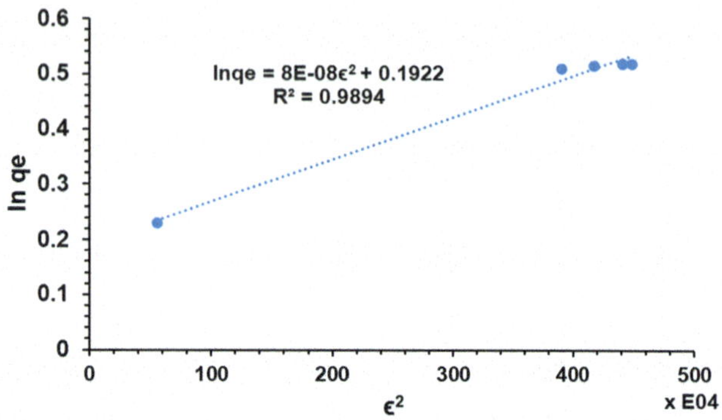

Fig. 8 Freundlich isotherm for Ni adsorption

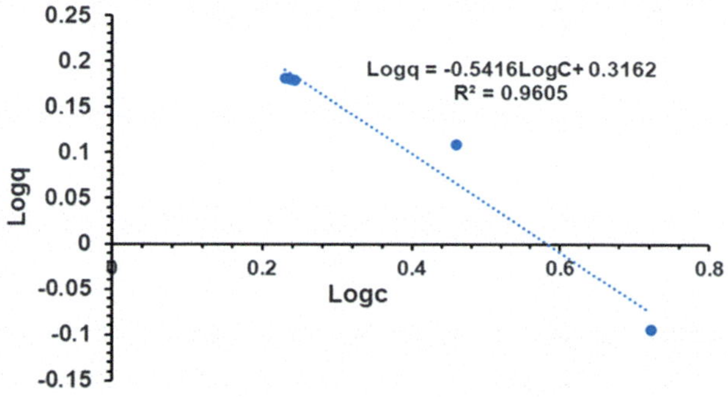

Fig. 9 Freundlich isotherm for Cr adsorption

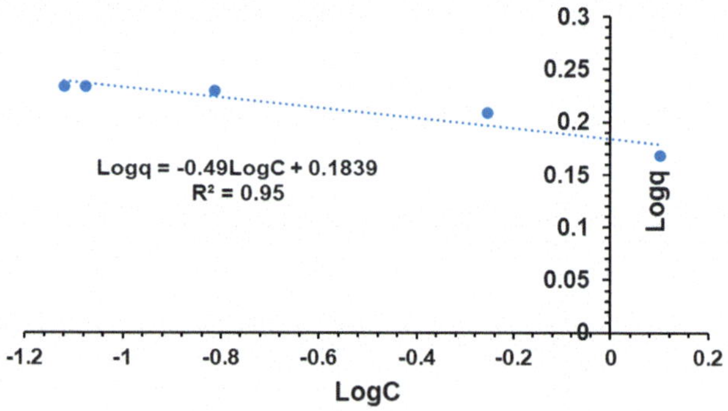

3.3 *Development of Kinetics of Adsorption*

Important information for process design is provided by the kinetic parameter, which aids in predicting the adsorption rate. Various kinetic models, including pseudo-first-order and pseudo-second-order kinetic models, are used to analyze the adsorption data kinetics [14]. When developing an adsorption process, it is crucial to anticipate the rate-limiting step. Solute transfer in solid–liquid sorption processes is typically

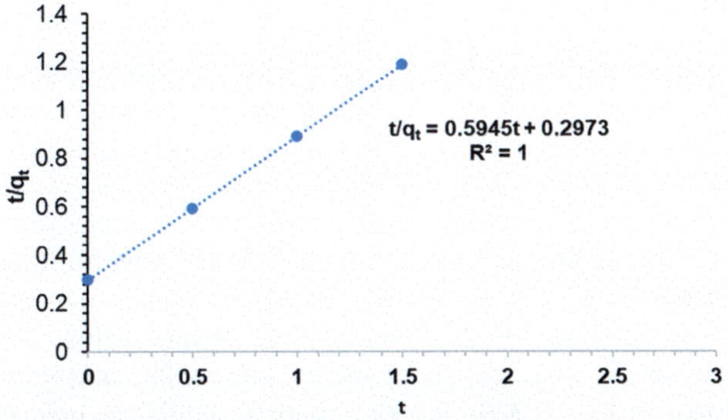

Fig. 10 Pseudo-second-order kinetics plot (Cu)

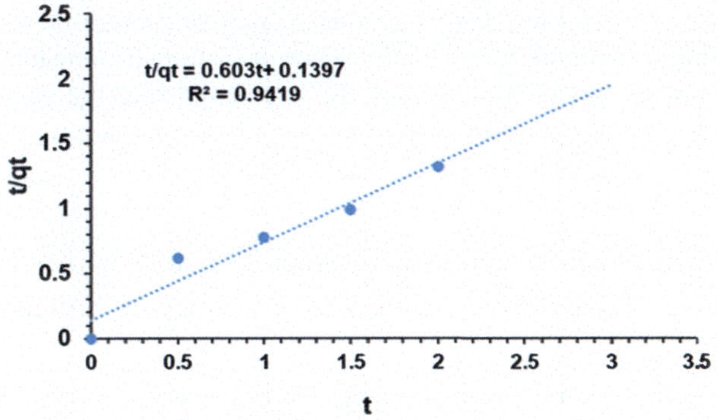

Fig. 11 Pseudo-second-order kinetics plot (Ni)

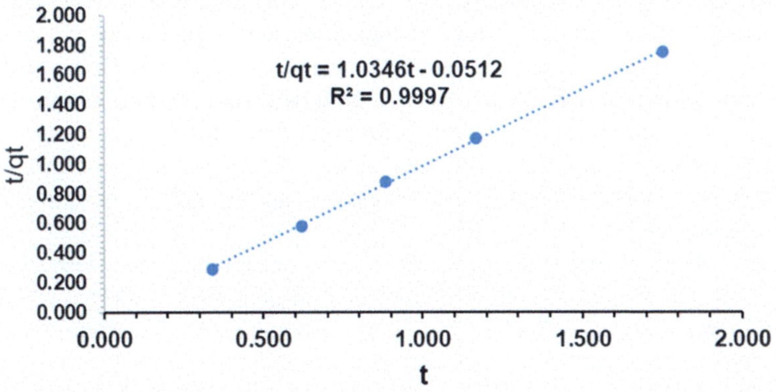

Fig. 12 Pseudo-second-order kinetics plot (Cr)

defined by either intraparticle diffusion or external mass transfer (boundary layer diffusion) [15]. Figures 10, 11, and 12 show the best-fit kinetics plot for the adsorption of Cu, Ni, and Cr.

Table 2 Pseudo-second-order kinetic constants for heavy metal (Cu, Ni, Cr)

	Kinetic constants (pseudo-second order)		
Metal	q_e (mg/g)	k_2 (h^{-1})	R^2
Cu	1.69	0.104	1
Ni	1.65	0.71	0.9419
Cr	1.07	0.0544	0.99

Table 2 displays the constants associated with the pseudo-second-order kinetic model.

Table 2 shows that of the three metals tested, copper (Cu) has the highest adsorption capacity, followed by nickel (Ni) and chromium (Cr). According to Meena et al. (2008), the *RL* parameter indicates how favorable the adsorption is, with values between 0 and 1. Both isotherm models correctly depict the equilibrium data for all three heavy metals, which suggests that the sawdust surface forms a single layer of metal ions. When it comes to sawdust, the pseudo-second-order kinetics model fits the data the best, capturing the chemisorption preference over the physisorption one. Ion exchange is the most probable process for heavy metal removal, including Ni, Cu, and Cr [9].

3.4 Characterization and Analysis of Sawdust

Figures 13 and 14 illustrate the FTIR spectra of *Shorea robusta* sawdust before and after the adsorption of Cu, Ni, and Cr ions from an aqueous solution. The primary components present in agricultural wastes that are accountable for the adsorption of heavy metals consist of hemicellulose, monosaccharides, lignin, proteins, lipids, and starch, which possess a diverse range of functional groups [16]. The constituents contain various functional groups such as carbonyl, acetamido, phenolic, amido, carboxyl, amino, alcohols, esters, and sulfhydryl. These functional groups have an affinity for forming complexes with metals [17]. The hydroxyl (OH), amine (NH), carbonyl (C=O), and carboxylic (COOH) groups are the primary sorption sites of utmost significance [18].

Figure 13 displays the FTIR spectrum analysis of untreated sal (*Shorea robusta*) sawdust. The analysis shows a wide spectrum ranging from 3500 to 3200 cm^{-1}, which indicates a strong stretching of the O–H bond before the adsorption of the heavy metals under consideration, namely Cu, Ni, and Cr. Figure 14 shows that the FTIR spectra of the sawdust after adsorption demonstrate a reduction in the width of peaks within the 3500–3200 cm^{-1} range for all three cases (Cu, Ni, and Cr). The expansion of peaks in the 1700–1000 cm^{-1} range confirms that C=O and C–O groups are involved in the adsorption of the three heavy metals (Cu, Ni, and Cr).

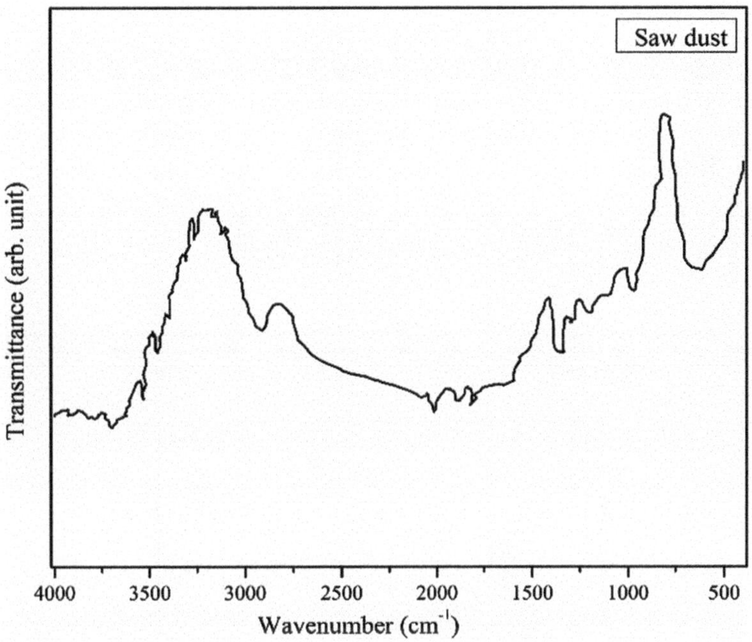

Fig. 13 Pre-adsorption FTIR spectra of sawdust (*Shorea robusta*)

3.5 *Optimization of Parameters of Adsorption*

Using mathematical programming, optimization seeks to maximize or minimize a function or functions within a given set of constraints. It is a different way to find the optimal performance within the constraints you have set, by maximizing the desirable variables and minimizing the undesirable ones. The fundamental goal of applying optimization methods is to enhance performance while decreasing the total cost of experimental work by conducting fewer experiments [19, 20].

3.5.1 Response Surface Methodology (RSM)

Response surface methodology (RSM) employs a predetermined number of carefully planned experiments to achieve a desired outcome by conducting a limited number of trials. This statistical technique is suitable for multifactor experiments and provides the relationship between different parameters for optimal operational conditions [21, 22].

The number of trial runs (N) in response surface method is given by following equation:

$$N = 2\,K(K-1) + C_o \tag{6}$$

where K is variable number and C_o is central point.

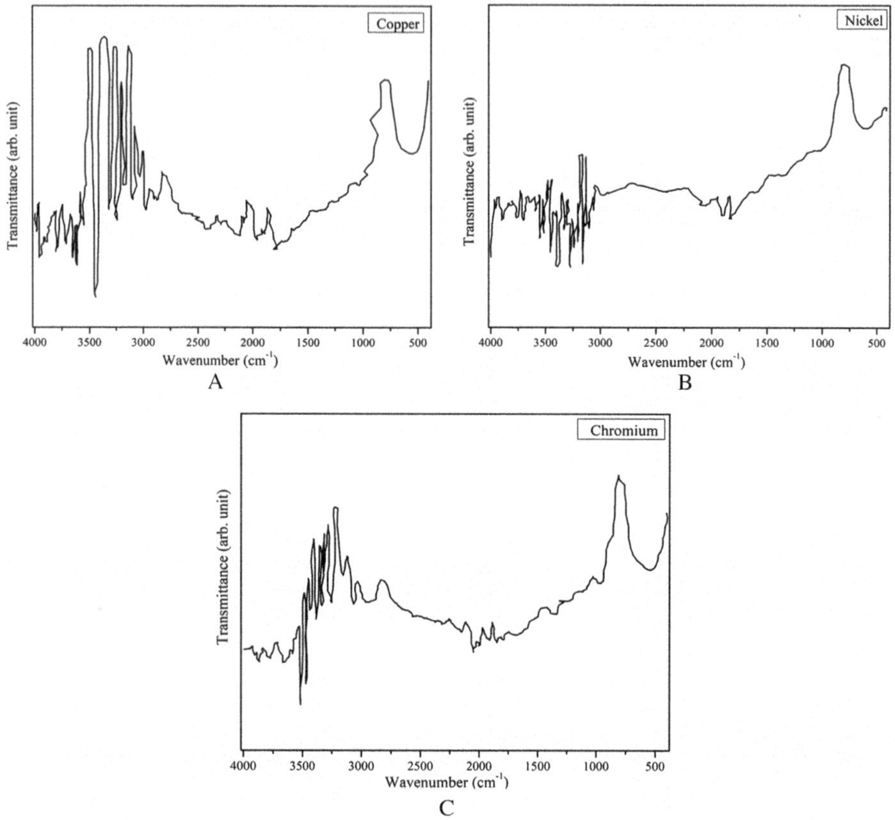

Fig. 14 FTIR spectra of sawdust following the adsorption of (**a**) copper, (**b**) nickel, and (**c**) chromium

The behavior of the system is explained by the following quadratic equation:

$$Y = \beta_0 + \sum\nolimits_{m=1}^{K} \beta_m X_m + \sum\nolimits_{m=1}^{k} \beta_{mn} X_m^2 + \sum\nolimits_{m<n}^{k} \beta_{mn} X_m X_n + \varepsilon \quad (7)$$

where Y is predicted response, β_0 is constant coefficient, β_m is linearity coefficient, β_{mn} is quadratic coefficient, β_{mn} is interaction coefficients, and ε is residual of the model.

Since central composite design (CCD) is a well-known and widely used statistical modeling technique for multivariate nonlinear models, it is selected as the best approach for optimizing biosorption process variables. This approach, along with other approaches such as the Box–Behnken and Plackett–Burman designs, is unique in that it uses a series of well-planned experiments to identify regression model equations and ideal operating conditions. As shown by Chattoraj et al.'s work [23], the CCD is especially useful for examining the interactions between different parameters influencing the biosorption process. Additionally, demonstrating the effectiveness of CCD in fitting a second-order model and the low number of experiments needed for accurate modeling is Kumar et al. [24]. The CCD consists of $2n$ factorial runs (coded using standard ± notation) with $2n$ accentor runs ((six replicates, 0, 0, 0, 0), $\pm a$, 0, 0, . . ., 0), (0, $\pm a$, 0, 0, . . ., 0), . . ., (0, 0, $\pm a$). The number of runs for a full replicate of the design, as provided by Eq. (7), increases with the number of factors n.

$$N = 2^n + +2n + n_c \tag{8}$$

Surface and contour plots are used to graphically depict the response, illustrating the interplay between the primary factors and the system's ideal conditions. The methodical framing of experimental parameters, such as pH, contact time, and adsorbent weight, was a part of the optimization process. These parameters were meticulously changed and examined using the central composite design (CCD) to determine how they affected the biosorption process. Finding the ideal values for pH, contact time, and adsorbent weight was the goal, and their interactions were investigated using the CCD methodology. The results of this systematic study provide a better understanding of the landscape of biosorption optimization and make it possible to establish an effective process operating under the determined ideal parameters.

3.5.2 Results of Optimization Study

In the present study, CCD was run under three-factor optimization process where pH, weight of adsorbent, and contact time was selected as variable, whereas removal in % is selected as response to be optimized. In each three cases, i.e., Cu, Ni, and Cr, the optimization process has been run for 20 number of trials. After final optimization, the final optimized Eqs. (9, 10, and 11) has been considered:

$$\begin{aligned}\text{Removal Rate }(\%) &= 50.31\text{–}1.71\text{ pH} + 29.67^{*}\text{ time}\\ &+ 4.34^{*}\text{ wt.of adsorbent} - 0.645^{*}\text{ time}^{*}\text{ pH}\\ &- 1.24^{*}\text{ pH}^{*}\text{ wt.of adsorbent} - 1.25^{*}\text{ time}^{*}\text{ wt.of adsorbent}\\ &+ 1.40^{*}\text{ pH}^2\text{–}6.20^{*}\text{ time}^2 + .56^{*}\text{ wt.of adsorbent}^2\end{aligned} \tag{9}$$

for Cu.

$$\begin{aligned}\text{Removal Rate} &= +131.80511\text{–}14.80241^{*}\text{ wt of Adsorbent}\\ &- 0.091299^{*}\text{ Time} - 0.771448^{*}\text{ pH} - 0.015326^{*}\text{ wt of Adsorbent}^{*}\text{ Time}\\ &+ 0.103375^{*}\text{ wt.of Adsorbent}^{*}\text{ pH} + 0.018297\text{ Time}^{*}\text{ pH}\\ &+ 2.6380^{*}\text{ wt of Adsorbent}^2 + 7.22891\text{E} - 06^{*}\text{ Time}^2 + 0.013939^{*}\text{ pH}^2\end{aligned} \tag{10}$$

for Ni

$$\begin{aligned}\text{Removal Rate} = {} & 102.09229–1.24441^{*}\text{ wt.of Adsorbent} - 0.035512^{*}\text{ time} \\ & - 0.347621^{*}\text{ pH} + 0.010177^{*}\text{ wt.of Adsorbent}^{*}\text{ time} \\ & + 0.124660^{*}\text{ wt.of Adsorbent}^{*}\text{ pH} + 0.002596^{*}\text{ time}^{*}\text{ pH} \\ & - 0.584259^{*}\text{ wt.of Adsorbent}^{2}–6.66616\text{E} - 08^{*}\text{ time}^{2} + 0.050867^{*}\text{ pH}^{2} \end{aligned} \tag{11}$$

for Cr.

The aforementioned three equations show that the three parameters—pH, adsorbent weight, and contact time—have a strong intercorrelation. Positive coefficients in the equations indicate that the corresponding factor has increased within the tested range, whereas negative coefficients show that the interactions of individual or combined factors have a negative effect on adsorption, leading to a decrease in the percentage of adsorption. The adequacy of the model is confirmed by the p-values for Cu, Ni, and Cr being less than 0.0001. This is important because an ANOVA test result with a p-value of less than 0.05 indicates a high degree of significance. The low p-values provide more evidence for the model's significance. The regression coefficient (R^2) values, which fall between 0.90 and 0.99 for Cu, Ni, and Cr in particular, show a strong correlation between the parameters in question for each case. This high R^2 value highlights the regression model's ability to reliably predict how the biosorption process for copper (Cu), nickel (Ni), and chromium (Cr) will behave by demonstrating how well it captures and explains the variation in the data.

The tabular representation of Tables 3, 4, and 5 serves as evidence of the model's adequacy, which was generated through the central composite design (CCD). The previously mentioned p-values and F-values are provided for each of the three cases, corresponding to the adsorption of Cu, Ni, and Cr. This presentation in tabular form enhances the clarity and accessibility of the statistical results, affirming the reliability and effectiveness of the CCD-generated model in explaining and predicting the adsorption behavior for copper (Cu), nickel (Ni), and chromium (Cr).

Furthermore, graphical representations for each adsorption study will depict the interplay of three factors—pH, time, and adsorbent weight—on the removal rate (%). These visual plots serve to elucidate how alterations in these parameters influence the efficiency of the adsorption process. Examining the trends and patterns in these graphs enables a detailed comprehension of how time, pH, and weight of adsorbent either cooperate or counteract each other, affecting the overall rate of removal. This enhances the overall understanding of the experimental results.

Figures 15, 16, and 17 depict a 3D surface plot illustrating the mutual interaction of three key factors–adsorbent weight, pH, and contact time–that influence the adsorption of heavy metals. The rate of adsorption in the RSM plot of Cu initially increased with longer contact time and higher pH. However, over time, the removal effectiveness dropped. The removal effectiveness of Ni, as shown in Fig. 16, is

Table 3 ANOVA result for optimization of Cu adsorption

Source	Sum of square	*df*	Mean square	*F*-value	*p*-value	
Model	1665.02	9	185.00	7.07	0.0026	Significant
A—pH	162.65	1	162.65	6.22	0.0318	
B—time	1.02	1	1.02	0.0391	0.8472	
C—weight of adsorbent	138.09	1	138.09	5.28	0.0444	
AB	6.39	1	6.39	0.2443	0.6318	
AC	182.12	1	182.12	6.96	0.0248	
BC	49.95	1	49.95	1.91	0.1971	
A^2	386.15	1	386.15	14.76	0.0033	
NB^2	555.73	1	555.73	21.25	0.0010	
C^2	73.00	1	73.00	2.79	0.1257	
Residual	261.54	10	26.15			
Inadequate Conformity	239.62	5	47.92	10.93	0.0101	Significant
Error	21.92	5	4.38			
Total	1926.56	19				

Table 4 ANOVA result for optimization of Ni adsorption

Source	Sum of square	*df*	Mean square	*F*-value	*p*-value	
Block	897.52	2	448.76			
Model	2900.29	9	322.25	21.37	0.0001	Significant
A—weight of adsorbent	7.47	1	7.47	0.4955	0.5015	
B—time	6.90	1	6.90	0.4573	0.5179	
C—pH	650.74	1	650.74	43.15	0.0002	
AB	60.89	1	60.89	4.04	0.0794	
AC	8.55	1	8.55	0.5668	0.4731	
BC	542.36	1	542.36	35.96	0.0003	
A^2	1603.37	1	1603.37	106.31	<0.0001	
B^2	0.0494	1	0.0494	0.0033	0.9558	
C^2	1.75	1	1.75	0.1159	0.7422	
Residual	120.65	8	15.08			
Inadequate Conformity	102.86	5	20.57	3.47	0.1675	Not significant
Error	17.80	3	5.93			
Total	3918.47	19				

Table 5 ANOVA result for optimization of Cr adsorption

Source	Sum of square	*df*	Mean square	*F*-value	*p*-value	
Model	801.97	9	89.11	28.34	<0.0001	Significant
A—weight of adsorbent	641.58	1	641.58	204.06	<0.0001	
B—time	2.78	1	2.78	0.8836	0.3694	
C—pH	50.20	1	50.20	15.97	0.0025	
AB	25.30	1	25.30	8.05	0.0176	
AC	4.45	1	4.45	1.42	0.2615	
BC	3.71	1	3.71	1.18	0.3031	
A^2	79.18	1	79.18	25.18	0.0005	
B^2	2.900E−06	1	2.900E−06	9.222E−07	0.9993	
C^2	3.04	1	3.04	0.9664	0.3488	
Residual	31.44	10	3.14			
Inadequate Conformity	5.50	5	1.10	0.2119	0.9431	Not significant
Error	25.94	5	5.19			
Total	833.41	19				

significantly affected by time and pH due to a decrease in adsorption capacity as these parameters increase. The adsorption capacity initially dropped as the weight of the adsorbent increased, then reached a maximum limit at 5 g. Figure 17 illustrates that as the weight of the adsorbent increases, the adsorption capacity of Cr decreases. However, at low pH levels within the acidic range and with longer contact times, the adsorption capacity increases.

Following the development of the model, a comparative analysis was conducted between the predicted outcomes and the actual results obtained from the set of experiments. This comparison is conducted to evaluate the precision and dependability of the model in forecasting the observed responses. It offers valuable insights into the model's performance and its capacity to comprehend the complexities of the studied phenomena. The graph depicts the relationship between the observed and projected values.

The graphs presented in Figs. 18, 19, and 20 provide definitive evidence for the validity of the model. They demonstrate a robust correlation between the observed removal rates in the actual experiments and the rates predicted by the model. This graph provides definitive evidence that the model effectively forecasts the results of the adsorption experiments for the three designated heavy metals—Cu, Ni, and Cr.

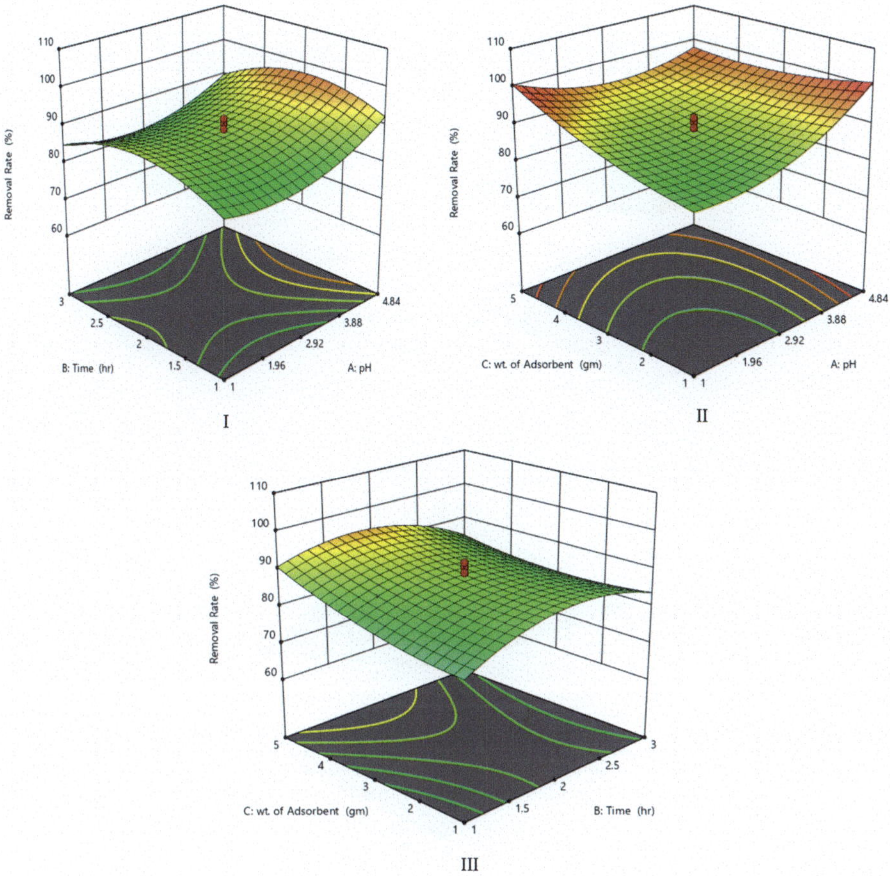

Fig. 15 RSM plot for (**a**) adsorbent weight (**b**) time, and (**c**) pH for adsorption of Cu

4 Conclusion

The use of sawdust as an adsorbent for heavy metal removal from water-based systems demonstrates a clear relationship between the concentration of the substances to be removed and the pH levels. The strong agreement between the Freundlich and Dubinin–Radushkevich models in the equilibrium studies demonstrates that sawdust is a reliable adsorbent. Kinetic studies also illuminate the adsorption dynamics, showing that a pseudo-second-order adsorption model is prevalent, which in turn suggests that chemical interaction plays a substantial role in the adsorption process. For the adsorption process to be optimized, statistical

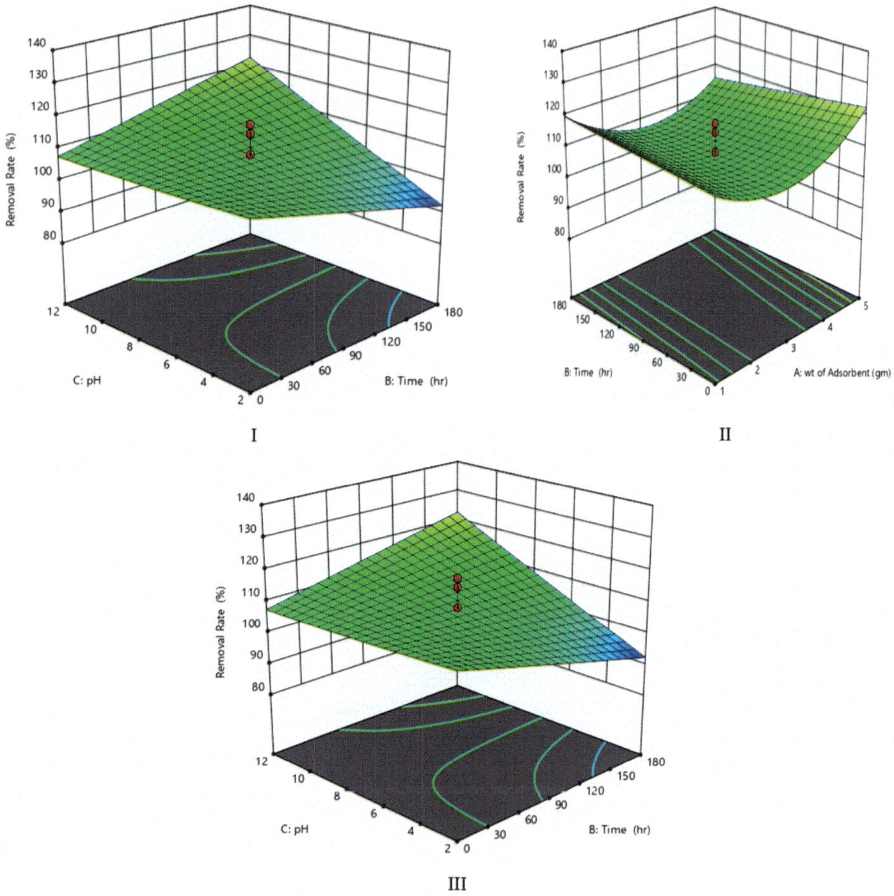

Fig. 16 RSM plot for (**a**) adsorbent weight (**b**) pH, and (**c**) time for adsorption of Ni

analysis must be applied. An optimization study is carried out to develop a refined model using central composite design (CCD). This study elucidates the intercorrelation between key factors, which include pH, adsorbent weight, and contact time. By taking a methodical approach, we can better comprehend how these factors work together, which in turn helps us determine the ideal conditions for achieving maximum adsorption efficiency. Collectively, the *p*-value, *F*-value, and R^2—statistical parameters derived from the optimization study—confirm the developed model's robustness and appropriateness. The model's predictive accuracy in capturing the complexities of the adsorption process is underscored by these parameters, which also confirm its statistical significance. Using sawdust in an adsorption

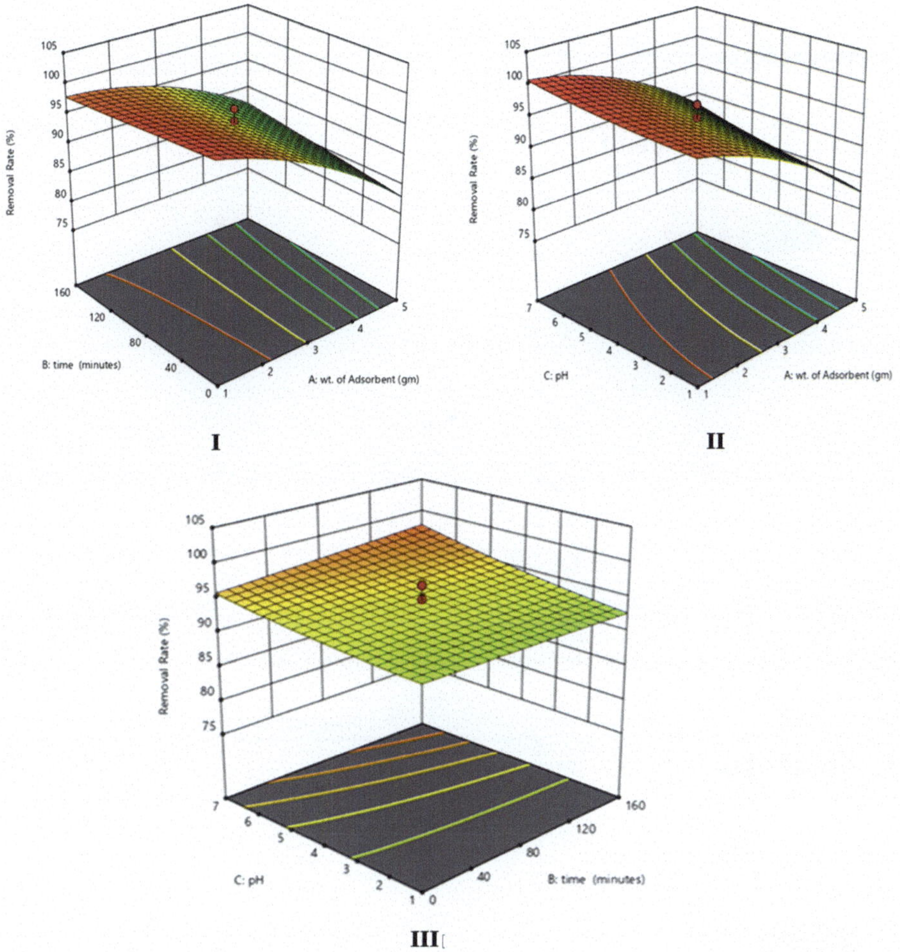

Fig. 17 RSM plot for (**a**) adsorbent weight (**b**) time, and (**c**) pH for adsorption of Cr

system could be a sustainable and economically viable way to treat heavy metals in water, according to this study's implications, which go beyond the lab. In addition, there is a promising avenue for further optimization and improvement of the effectiveness of this eco-friendly remediation method for heavy metal-contaminated water: increasing the adsorption capacity of sawdust by adding organic acids.

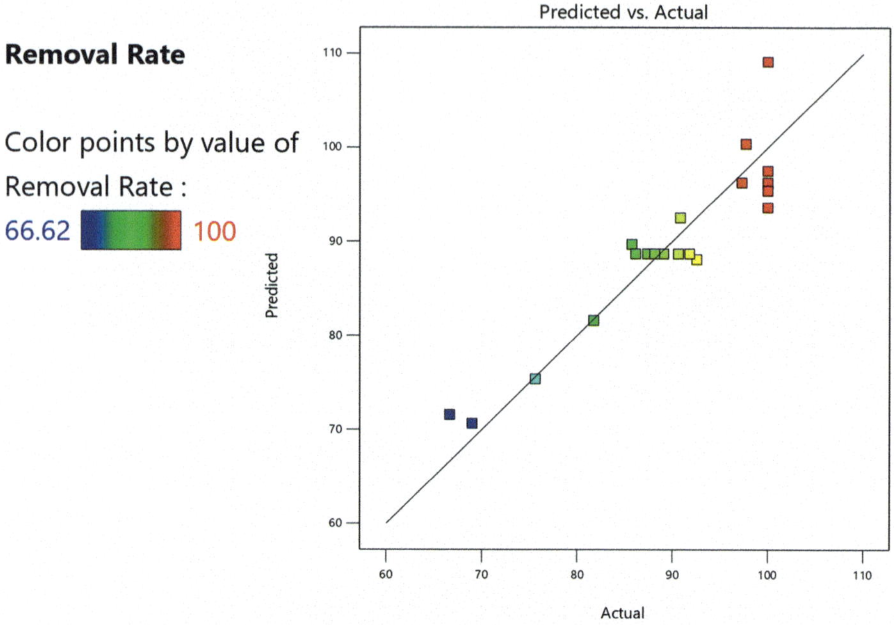

Fig. 18 Correlation between the observed and anticipated reduction in Cu

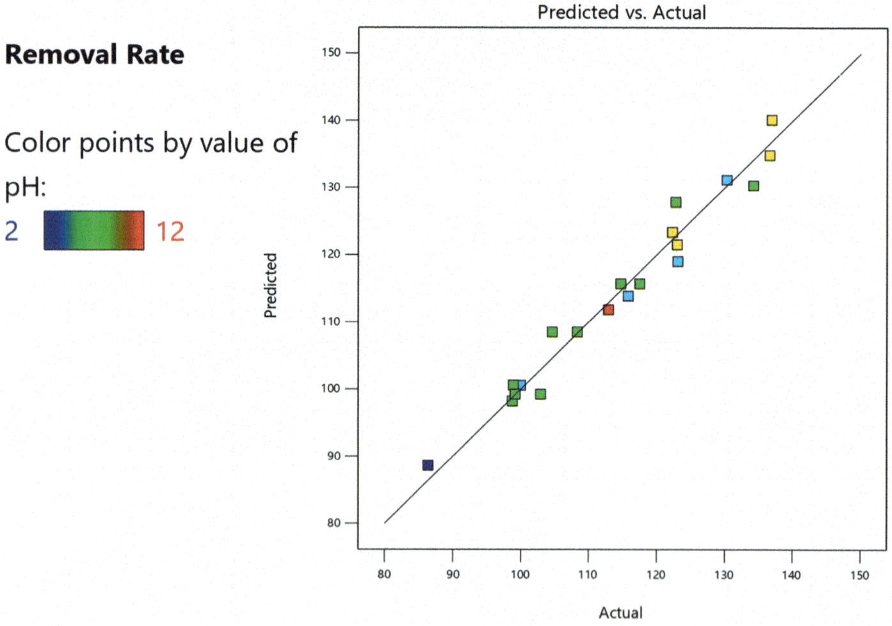

Fig. 19 Correlation between the observed and anticipated reduction in Ni

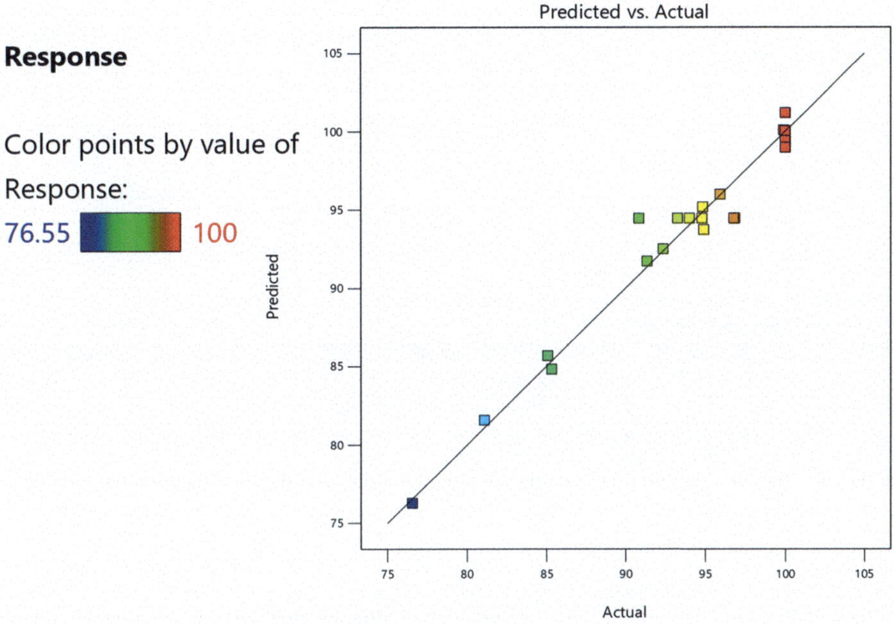

Fig. 20 Correlation between the observed and anticipated reduction in Cr

Acknowledgments This research work is supported by the research fellowship program of Ministry of Human Resource Development, Government of India.

References

1. S. Babel, T.A. Kurniawan, Cr (VI) removal from synthetic wastewater using coconut shell charcoal and commercial activated carbon modified with oxidizing agents and/or chitosan. Chemosphere **54**(7), 951–967 (2004)
2. S.P. Dubey, K. Gopal, Adsorption of chromium (VI) on low cost adsorbents derived from agricultural waste material: A comparative study. J. Hazard. Mater. **145**(3), 465–470 (2007)
3. B. Geng, Z. Jin, T. Li, X. Qi, Preparation of chitosan stabilized Fe0 nanoparticles for removal of hexavalent chromium in water. Sci. Total Environ. **407**(18), 4994–5000 (2009)
4. R.M. Nthumbi, J.C. Ngila, B. Moodley, A. Kindness, L. Petrik, Application of chitosan/polyacrylamide nanofibres for removal of chromate and phosphate in water. Phys. Chem. Earth, Parts A/B/C **50**, 243–251 (2012)
5. A. Demirbas, Heavy metal adsorption onto agro-based waste materials: A review. J. Hazard. Mater. **157**(2–3), 220–229 (2008)
6. M.S. Rahman, M.R. Islam, Effects of pH on isotherms modeling for Cu (II) ions adsorption using maple wood sawdust. Chem. Eng. J. **149**(1–3), 273–280 (2009)
7. H.D. Choi, J.M. Cho, K. Baek, J.S. Yang, J.Y. Lee, Influence of cationic surfactant on adsorption of Cr (VI) onto activated carbon. J. Hazard. Mater. **161**(2–3), 1565–1568 (2009)
8. A. Jang, Y. Seo, P.L. Bishop, The removal of heavy metals in urban runoff by sorption on mulch. Environ. Pollut. **133**(1), 117–127 (2005)

9. F.V. Pereira, L.V. Gurgel, S.F. de Aquino, L.F. Gil, Removal of Zn 2+ from electroplating wastewater using modified wood sawdust and sugarcane bagasse. J. Environ. Eng. **135**(5), 341–350 (2009)
10. R. Chikri, N. Elhadiri, M. Benchanaa, Y. Maguana, Efficiency of sawdust as low-cost adsorbent for dyes removal. J. Chem. **2020**, 1–17 (2020)
11. F. Gode, E.D. Atalay, E. Pehlivan, Removal of Cr (VI) from aqueous solutions using modified red pine sawdust. J. Hazard. Mater. **152**(3), 1201–1207 (2008)
12. Y. Bulut, Removal of heavy metals from aqueous solution by sawdust adsorption. J. Environ. Sci. **19**(2), 160–166 (2007)
13. B.P. Bering, M.M. Dubinin, V.V. Serpinsky, On thermodynamics of adsorption in micropores. J. Colloid Interface Sci. **38**(1), 185–194 (1972)
14. M.M. Dubinin, The equation of the characteristic curve of activated charcoal. Dokl. Akad. Nauk. SSSR **55**, 327–329 (1947)
15. M. Polanyi, Section III.—Theories of the adsorption of gases. A general survey and some additional remarks. Introductory paper to section III. Trans. Faraday Soc. **28**, 316–333 (1932)
16. Y.S. Al-Degs, M.I. El-Barghouthi, A.A. Issa, M.A. Khraisheh, G.M. Walker, Sorption of Zn (II), Pb (II), and Co (II) using natural sorbents: Equilibrium and kinetic studies. Water Res. **40**(14), 2645–2658 (2006)
17. M.H. Kalavathy, T. Karthikeyan, S. Rajgopal, L.R. Miranda, Kinetic and isotherm studies of Cu (II) adsorption onto H3PO4-activated rubber wood sawdust. J. Colloid Interface Sci. **292**(2), 354–362 (2005)
18. APHA, *APHA- Standard Method for the Examination of Water and Waste Water* (American Water Works Association and Water Pollution; Control Federation, Washington, DC, 1989)
19. A.E. Ofomaja, Y.S. Ho, Effect of pH on cadmium biosorption by coconut copra meal. J. Hazard. Mater. **139**(2), 356–362 (2007)
20. F. Chigondo, B.C. Nyamunda, S.C. Sithole, L. Gwatidzo, Removal of lead (II) and copper (II) ions from aqueous solution by baobab (Adononsia digitata) fruit shells biomass. IOSR J. Appl. Chem. **5**(1), 43–50 (2013)
21. R. Zare-Dorabei, S.M. Ferdowsi, A. Barzin, A. Tadjarodi, Highly efficient simultaneous ultrasonic-assisted adsorption of Pb (II), Cd (II), Ni (II) and Cu (II) ions from aqueous solutions by graphene oxide modified with 2, 2′-dipyridylamine: Central composite design optimization. Ultrason. Sonochem. **32**, 265–276 (2016)
22. A. Keramat, R. Zare-Dorabei, Ultrasound-assisted dispersive magnetic solid phase extraction for preconcentration and determination of trace amount of Hg (II) ions from food samples and aqueous solution by magnetic graphene oxide (Fe3O4@ GO/2-PTSC): Central composite design optimization. Ultrason. Sonochem. **38**, 421–429 (2017)
23. S. Chattoraj, B. Sadhukhan, N.K. Mondal, Predictability by Box-Behnken model for carbaryl adsorption by soils of Indian origin. J. Environ. Sci. Health B **48**(8), 626–636 (2013)
24. R. Kumar, R. Singh, N. Kumar, K. Bishnoi, N.R. Bishnoi, Response surface methodology approach for optimization of biosorption process for removal of Cr (VI), Ni (II) and Zn (II) ions by immobilized bacterial biomass sp. Bacillus brevis. Chem. Eng. J. **146**(3), 401–407 (2009)

Arsenic Organic Interaction: Removal by Synthesized Iron Functional Carbon Nanocomposite Through Batch Study

Soumya Kanta Ray and Chanchal Majumder

Abbreviations

DI	Deionized water
DMF	Dimethylformamide
DSTBT	Department of Science and Technology and Biotechnology
FA	Fulvic acid
fRGO	Iron-functionalized reduced graphene oxide
GO	Graphene oxide
HA	Humic acids
HAA	Haloacetic acids
JCPDS	Joint Committee on Powder Diffraction Standards
NOM	Natural organic material
ORP	Oxidation–reduction potential
PZC	Point of zero charge
RGO	Reduced graphene oxide
THM	Trihalomethanes
TOC	Total organic carbon
US EPA	United States Environmental Protection Agency
WHO	World Health Organization
XRD	X-ray diffraction

S. K. Ray (✉) · C. Majumder
Civil Engineering Department, Indian Institute of Engineering Science and Technology, Howrah, India
e-mail: chanchal@civil.iiests.ac.in

D. Mazumder (ed.), *Sustainable Advanced Technologies for Environmental Management*, Springer Proceedings in Earth and Environmental Sciences,
https://doi.org/10.1007/978-3-031-64006-3_13

1 Introduction

Arsenic, a highly toxic and carcinogenic substance, has contaminated groundwater sources worldwide, posing serious health risks. Alarmingly, over 60% of wells in Southeast Asia, especially in Bangladesh and West Bengal, India, exceed the safe drinking water limit of 10 (ppb) as per the World Health Organization (WHO) [1]. It has been classified as a Group 1 carcinogen [2]. Long-term consumption of arsenic-contaminated water can lead to various cancers, including skin, lung, bladder, kidney, and even infant mortality [3]. After 5–10 years of exposure, individuals may develop arsenicosis, a condition characterized by initial skin darkening on palms, chest, back, and limbs, progressing to spotted melanosis [4, 5]. Subsequently, the skin hardens and becomes fibrous (keratosis), signifying severe toxicity. Ultimately, this can contribute to skin cancer development, such as basal cell carcinoma, squamous cell carcinoma, and Bowen's disease [6].

Natural organic matter (NOM), a very common aquatic component, comprises of complex organic substances, impacting water flavor, appearance, and generating coordinated compounds with heavy metals and organic pollutants [7]. Derived from terrestrial vegetation breakdown and bacterial, algal, and aquatic plant processes, NOM mixes contain functional groups such as esters, phenols, quinones, carboxylics, hydroxyls, aminos, and nitrosos, often negatively charged at neutral pH. NOM has hydrophobic and hydrophilic constituents, with hydrophobic acids predominating over 50% of the total organic carbon [8]. These hydrophobic acids, called humic compounds, include fulvic acids (FA) and humic acids (HA) with common functional groups such as carbonyls, alcohols, phenols, and carbonyls [9]. However, FA has more carboxylic and oxygen groups, resulting in lower carbon content per mass than HA.

The ternary complexation process involves the formation of strong bonds between cationic metals, functional groups in natural organic matter (NOM), and arsenic oxyanions [10]. To assess its significance, researchers examined how well each NOM-containing water sample could facilitate the complexation of arsenate with metal oxides. Metal bridging provides a plausible explanation for the binding of NOM and arsenic [11]. Additionally, NOM samples influenced the redox speciation of arsenic, with As(III) oxidation patterns mirroring those seen in aqueous complexation studies. The presence of hematite notably enhanced the ability of NOM to reduce As(V) to As(III) [12]. This enhanced reduction likely occurred via hematite acting as either a surface catalyst or an electron transfer intermediate between reduced NOM moieties and As(V), consistent with the observed variation among NOM samples [13–15].

Different nanomaterials, including iron oxides/hydroxides, copper oxide, zinc oxide, aluminum oxides, titanium oxide, carbon nanotubes, and mixed metal oxides, are employed for arsenic removal [16]. Graphene and functionalized graphene oxide exhibit high efficacy in eliminating both inorganic and organic pollutants during water treatment [17, 18]. Porous graphene oxide and inverse spinel nickel ferrite nanocomposites can effectively remove As(III) and As(V) with a 99% success rate [19]. The combination of reduced graphene oxide and Fe_3O_4 removes organic dyes and inorganic As(V) from polluted water. Iron–manganese binary oxide on modified reduced graphene oxide eliminates As(III), As(V), and phosphate from synthetic contaminated water [20, 21]. Arsenic-contaminated water can be treated using RGO–copper–zeolite composites. Reduced graphene oxide functionalized with Mg $(OH)_2$–MgO reported to have high As(III) removal [22].

The main objective of this study was to assess the effectiveness of iron-functionalized carbon nanomaterial in removal of arsenic from water having high organic content.

2 Materials and Methods

FA was used as model organic source and was obtained from SNEH Corporation, Ahmadabad, India. Graphite powder, concentrated sulfuric acid (H_2SO_4), sodium nitrate ($NaNO_3$), potassium permanganate ($KMnO_4$), hydrogen peroxide (H_2O_2), hydrochloric acid (HCl), ammonium hydroxide (NH_4OH), and ferrous sulfate ($FeSO_4$) were all procured from Merck, Dermasdat, Germany. Hydrazine hydrate (N_2H_4) (50–60%) and *N*,*N*-dimethylformamide (DMF) were provided by Sigma-Aldrich, Taufkirchen, Germany.

2.1 Synthesis of Iron Functionalized Graphene Oxide (fRGO)

The Hummers' method (Scheme 1) was utilized to synthesize graphene oxide (GO) from powdered graphite. A typical batch synthesis was performed by mixing 5.0 g of graphite powder with 2.5 g of $NaNO_3$. Then, 120 mL of 95% H_2SO_4 was introduced into a 1.2 L borosilicate glass beaker. Afterward, a magnetic stirrer was used to thoroughly blend the mixture for an extra 30 min in a container packed with ice. While agitating vigorously, 15.0 g of $KMnO_4$ were introduced into the

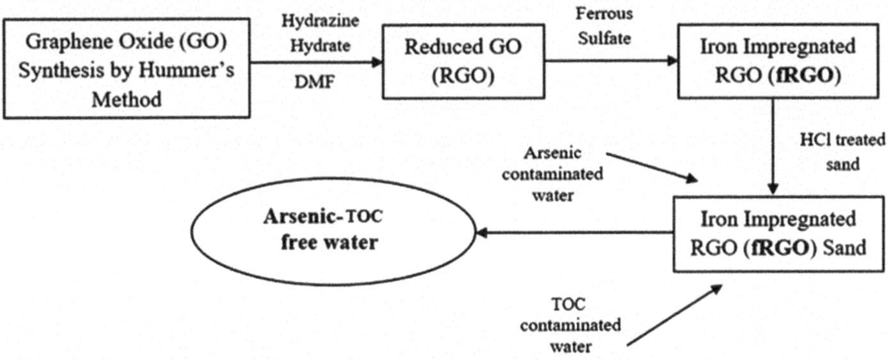

Scheme 1 Flow diagram of TOC removal using iron-functionalized reduced graphene oxide

suspension. To maintain the temperature of the mixture below 20 °C, the rate of $KMnO_4$ mixing was meticulously controlled. Subsequently, the ice bath was eliminated, and the concoction was heated to a precise temperature of 35 °C and vigorously stirred for a duration of 30 min. Throughout the operation, the chemical gradually transformed into a viscous consistency and acquired a light brown color. Following the completion of the reaction, a total of 150 mL of water was slowly incorporated into the paste with vigorous stirring. The temperature of the combination was rapidly elevated to 98 °C and maintained at that level for 15 min with the use of external heating. Subsequently, effervescence was detected. Further oxidation was restricted by adding 450 mL of water after the fumes had been stopped from exfoliated paste for 20 min in a water bath. The content was then mixed with 15 mL aliquot of 30% H_2O_2, and the color became yellow. The mixture was refined by centrifuging and washing it separately with 10% HCl and deionized (DI) water multiple times until the pH level reached 7. By using 250 mL H_2O:100 mg GO sonication to disperse the gray-brownish GO powder in DI water, particle debris in the solution was eliminated (Scheme 1). The mixture was heated for 2 h at 100 °C in an oil bath of commercial sunflower oil with magnetic stirring to maintain a homogeneous suspension after adding 100 mL of N_2H_4. The solution was then transported to a condenser that uses water for cooling. Black solids that obtained were reduced graphene oxide (RGO) precipitate. For uniform RGO dispersion in the solution, DMF was sonicated into the previously described solution in a 9:1 volume ratio of DMF:H_2O. To prepare the RGO solution for additional studies, it was concentrated to 20 mg/mL.

2.2 Functionalization of RGO

RGO of 3 mL solution was ultrasonically mixed for 30 min with 147 mL of DI water. Add 15 mg $FeSO_4 \cdot 7H_2O$ to this solution. The solution was stirred for 24 h at 27 °C. A total of 5 mL of 0.1 N NH_4OH was combined after 24 h until the pH was 8. Ammonium hydroxide was added, and the color of the compound changed to a brownish-red hue. The dark brownish-red color mixture indicated the presence of ferric hydroxide. The water of the compound was evaporated using a hot plate. A solid brownish mass was formed when the water from the combination evaporated. This material was then suspended and repeatedly cleaned with warm water. Afterward, the brownish mass was subjected to a 12 h drying procedure at 85 °C in a vacuum oven to fully functionalize the RGO nanomaterial.

2.3 fRGO Coating on Sand

To remove various pollutants prior to use, natural sand with particle sizes that pass through 300 mm and retentive on 150 mm was sieved, gathered, and washed with a 10% HCl solution. To eliminate any organic contaminants, the acid-washed sand sample was then burned to 550 °C for 1 h. After cooling to ambient temperature, it was kept for future use in a vacuum desiccator. An amount of 50 g of this cleaned sand was added to 3 mL of 20 mg/mL RGO in a 250 mL beaker to cover the sand particles with RGO. Subsequently, the combination was made uniform by adding 15 mg of $FeSO_4$ dissolved in 150 mL of deionized water. The combination was subsequently exposed to a temperature of 105 °C for a period of 3 h in a vacuum oven, until the nanocomposite hardly adhered to the sand surface. The sand coated with fRGO nanocomposite was subjected to annealing in a muffle furnace at a temperature of 450 °C to complete the strengthening process [21]. The flow diagram is shown in Scheme 1.

2.4 *Characterization of Functionalized Reduced Graphene (fRGO)*

The produced material was characterized for physicochemical properties. Analyses were done using a Bruker D8 Advance X-ray diffractometer. CuKα1 irradiation (=1.5405980) was conducted at room temperature using a 40 kV, 30 mA power supply. At a scan rate of 0.02°/s, the analysis covered the 5–80° (2θ) range. Use Joint Committee on Powder Diffraction Standards (JCPDS) to analyze results. The average size of fRGO was calculated using the Scherrer eq. ($D = K/d \cos \theta$), where D is the crystallite size, K is a constant of 0.9, d is the FWHM of the diffraction peak, and θ is the angle in radians. A pH meter (WTW, Germany) was used to determine the point of zero charge (PZC) for fRGO and fulvic acid using the pH drift method.

2.5 *Arsenic Removal Study*

The first and ultimate arsenic concentrations were determined using an atomic absorption spectrometer (Thermo Scientific, USA, iCE 3000 AA05182611 v1.30) in flame mode with an air/acetylene flame type at 193.7 nm. Batch experiments were used to determine how operational parameters affected solution arsenic removal. Studies were grouped by situation. Arsenic is removed by fRGO first without dissolved organic carbon and then with it.

2.5.1 Arsenic Removal by fRGO in the Absence of Organic Matter

Batch experiments were performed at different concentrations of total arsenic removal by fRGO. The six different solution concentrations of total arsenic having values 0.05, 0.075, 0.1, 0.15, 0.2, and 0.25 mg/L were prepared, respectively. Every single total arsenic sample was then mixed with 0.1 g of fRGO in 100 mL of deionized (DI) water.

2.5.2 Arsenic Removal by fRGO in the Presence of Organic Matter

The study maintained a constant concentration of total arsenic at 0.2 mg/L. The effects of total organic carbon (TOC) were examined at six distinct levels of TOC, namely 0.443, 0.886, 1.772, 2.659, 3.545, and 4.431 mg/L, respectively. Each sample was combined with 0.1 g of fRGO and was diluted with 100 mL of DI water.

2.5.3 Effect of Time

Batch kinetics tests were conducted by taking 0.1 g of fRGO in a 100 mL solution with 0.2 mg/L total arsenic. In the absence and presence (4.431 mg/L) of TOC, synthetic samples were tested for arsenic removal. All samples were adjusted to pH 7, the equilibrium pH. The samples were agitated for 10–12 h. Then, The samples were then taken at various times and filtered using a 0.45 μm filter. The arsenic level was measured later.

2.5.4 Effect of Dose of Adsorbent

Batch adsorption tests were carried out utilizing different quantities of fRGO, both in the presence and absence of TOC. To assess the influence of varying quantities of adsorbent on the elimination of total arsenic, we employed six different dosages: 2.0, 4.0, 6.0, 8.0, 10.0, and 12.0 g of fRGO-sand. The given doses correspond to 0.013, 0.026, 0.039, 0.052, 0.066, and 0.079 mg of fRGO. The adsorbent was added to a 100 mL solution at a concentration of 0.2 mg/L of total arsenic. The pH of the solution was modified to 7. Subsequently, the samples were agitated at a speed of 150 rpm for duration of 5 h using a shaker. Subsequently, the samples underwent filtration using a 0.45 μm filter, and the quantities of arsenic were quantified.

2.5.5 Effect of pH

In order to investigate the effects of pH on the removal of arsenic from TOC-bearing synthetic solution, a number of batch experiments were carried out. Six distinct solutions with pH values ranging from 2 to 12 were created. Next, 0.2 mg/L of total arsenic was added to a 100 mL solution containing 0.1 g of the adsorbent and 4.431 mg/L of TOC. A mechanical shaker was then used to agitate the samples for 5 h at a speed of 150 rpm. The samples were agitated, then filtered, and the amount of arsenic in the effluent was measured. After the adsorption procedure, the pH values of the solutions were evaluated to determine the pH value that was closest to equilibrium.

2.5.6 Effect of Concentration of Arsenic

In a series of batch experiments, the influence of pH on arsenic removal, with and without TOC, was investigated across a pH range of 2–12. Six solutions were prepared, each containing 0.1 g of the adsorbent in a 100 mL solution with 0.2 mg/L total arsenic. The mixtures were agitated for 5 h at 150 rpm. After agitation, samples were filtered, and the arsenic content in the effluent was quantified. To ascertain the closest equilibrium pH, the pH values of the solutions were assessed after adsorption. This process aimed to evaluate the role of pH in achieving complete arsenic removal under various conditions.

2.5.7 Effect of Temperature

A temperature-based batch experiment investigated effects at 290, 300, and 310 K, considering the presence of TOC. In a simulated sample, 0.1 g of fRGO was added to a 100 mL solution containing 0.2 mg/L total arsenic at pH 7. After filtration, the quantification of arsenic concentration in the effluent was conducted, examining the impact of temperature on arsenic removal in the presence and absence of TOC.

2.5.8 Effect of Agitation Speed

Batch studies were undertaken to analyze the impact of agitation speed on the system, considering both the presence and absence of TOC. Agitation speed varied from 100 to 200 rpm. In a simulated sample, 0.1 g of fRGO was added to a 100 mL solution containing 0.2 mg/L arsenic with a pH adjustment to 7. After filtration, the quantification of arsenic levels in the effluents was conducted, elucidating the influence of different agitation speeds on the system under varied TOC conditions.

3 Results and Discussion

3.1 Characterization of Functionalized Reduced Graphene (fRGO)

The different physicochemical behaviors were explored by X-ray diffraction (XRD) and determination of point of zero –charge (PZC). The characterization techniques are detailed in the following subsections.

3.1.1 X-Ray Diffraction (XRD)

The examination of a powder sample, encompassing phase identification, sample purity, crystallite size, and occasionally, shape, provides essential details through X-ray diffraction (XRD) analysis, complementing various microscopic and spectroscopic techniques. As a bulk method, XRD results can be cross-referenced with microscopic data to ascertain the representativeness of microscopic findings on a small particle sample for the entire specimen. Figure 1 illustrates X-ray diffraction patterns for GO, RGO, and fRGO. In the XRD pattern of GO, a prominent peak at a 2θ angle of 10.82° (001) indicates the transformation of graphite powder to GO through oxidation. The carbon peak (001) in GO corresponds to a 0.80 nm d-spacing, as determined by Powder X software [22]. The Hummer's oxidation and exfoliation process significantly increased the interlayer spacing of GO compared to original graphite, attributed to the presence of carboxyl, hydroxyl, and epoxide groups on the graphene plane.

The reduction of oxygen-containing functional groups in GO to RGO is marked by the absence of the (001) peak during the process involving hydrazine hydrate (N_2H_4). Instead, a new peak, (002), emerges at $2\theta = 24.5°$, suggesting a highly consistent interlayer distance of approximately 0.84 nm. The interlayer spacing of RGO (0.38 nm), while smaller than that of GO, remains larger than pure graphite (0.32 nm), indicating structural irregularities resulting from the oxidation of graphite

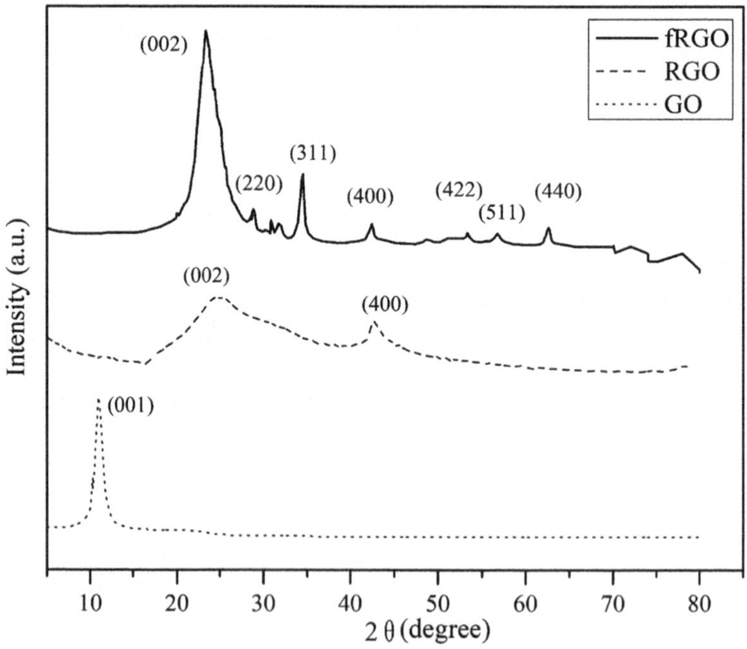

Fig. 1 XRD pattern of GO, RGO, and fRGO nanocomposite

powder. The presence of defects is corroborated by the (002) peak of RGO, revealing highly disordered and amorphous stacking of RGO sheets, as indicated by the weak and broad diffraction peak (400) at $2\theta = 42.67°$.

After the formation of RGO, $FeSO_4$ was added carefully and annealed at 450 °C in a muffle furnace for functionalization (fRGO) by iron moieties. The XRD pattern of the fRGO structure reveals prominent diffraction peaks at 2θ values of 28.85°, 34.51°, 42.28°, 53.34°, 56.75°, and 62.55°, corresponding to the (220), (311), (400), (422), (511), and (440) planes, respectively. These peaks align with the cubic crystal planes of Fe_3O_4 (JCPDS 75-0033), signifying the incorporation and dispersion of thermally treated iron precursors across the RGO surface. Following annealing, the fRGO nanocomposite solution is ultrasonically processed to achieve a homogenous dispersion with a concentration of 20 mg/mL, and the Scherrer formula calculates an average fRGO size of about 52 nm [23–28].

3.1.2 Determination of Point of Zero Charge (PZC)

The point of zero charge (PZC) was determined to ascertain the pH at which the surface of the particles remains neutral charged condition. Figure 2 demonstrates that the line intersects the abscissa at a value of 8.01 for fRGO, implying that it exhibited cationic character below pH 8.01, whereas it becomes negatively charged at pH levels beyond 8.01. In this study it is seen that the percentage removal of FA is higher at pH 4.03. The pK value of FA is in the range of 3.0–5.0, therefore within

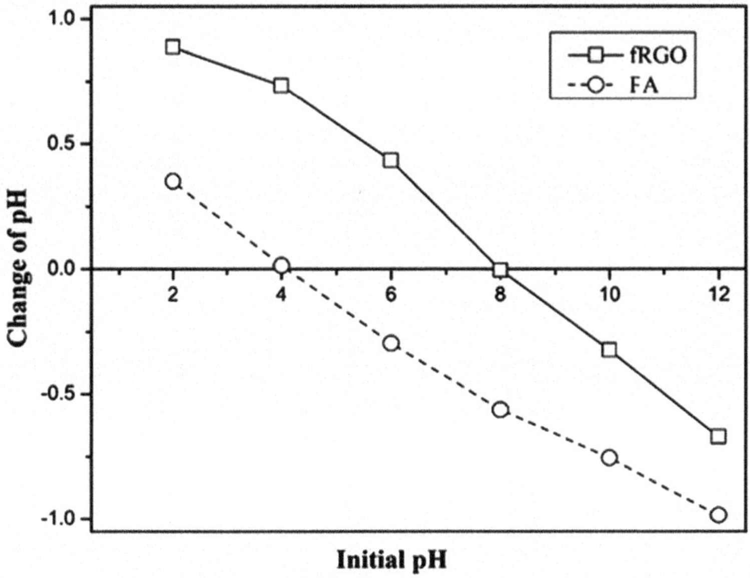

Fig. 2 Point of zero charge of fRGO and FA

this pH range, FA species predominate, which is adsorbed by positively charged fRGO nanocomposite coated on sand [29–33].

The regions of fRGO in contact with water demonstrate incomplete coordination, generating electric charge. This phenomenon is ascribed to the Brönsted acid–base properties of surface oxides, typical in oxide materials exposed to aqueous environments. The charged sites on the nanostructure's surface arise from amphoteric acid–base dissociation reactions of hydrolyzed groups.

$$MOH \Leftrightarrow MO^- + H^+$$

$$MOH^+{}_2 \Leftrightarrow MOH + H^+$$

This is impacted by the pH_{pzc} phenomenon, where extracting hydrogen ions from surface hydroxyl groups elevates H^+ ion concentration in the adjacent water and O^- ions on the fRGO surface. This results in a negatively charged surface, causing a shift in media pH toward approximate neutrality. The presence of surface hydroxyl groups significantly affects pH_{pzc} changes due to proton transfer to or from these groups [29, 30]. The preference for pH drift over conventional methods such as titrations is due to its time efficiency, streamlined process, cost-effectiveness, and consistent results. This approach aligns with alternative methods [33–35]. Below the pH threshold, fRGO maintains a positive charge, fostering electrostatic attraction with negatively charged arsenate species and FA. Above pH_{pzc}, both fRGO and arsenate species–FA maintain negative charges, resulting in electrostatic repulsion and reduced arsenic species–FA combination elimination [29, 31, 36].

A higher pH is employed for regenerating fRGO and detaching arsenic species using the FA combination. pH and oxidation–reduction potential (ORP) measurements assess the pe-pH-Eh state of the water system under equilibrium conditions. The transition As(III) to As(V) is influenced by the oxidation–reduction environment, monitoring ORP effectively describes the arsenic removal mechanism. In the presence of FA, fRGO selectively removes As(III) over As(V). The nanocomposite study focuses on total arsenic removal from contaminated water, exploring scenarios with presence and absence of TOC, as demonstrated by fRGO [34, 35, 37–39].

3.2 *Arsenic Removal Study*

The study on arsenic removal aimed to investigate the impact of the following parameters.

3.2.1 Effect of Time

The experiment investigating the impact of time were conducted using 0.1 g of sand coated with fRGO, which corresponds to 0.6552 mL of fRGO. The mixture was combined with a 100 mL solution containing 0.2 mg/L of total arsenic in the

Table 1 The varied experimental settings utilized to examine the impact of contact time on the removal of total arsenic

Sl. no.	Time (h)	Initial arsenic dose (mg/L)	Absence of TOC		Presence of TOC	
			Residual arsenic (mg/L)	Arsenic removal (%)	Residual arsenic (mg/L)	Arsenic removal (%)
1.	0.5	0.2	0.15	25	0.17	15
2.	1	0.2	0.12	40	0.145	27.5
3.	1.5	0.2	0.095	52.5	0.125	37.5
4.	2	0.2	0.07	65	0.107	46.5
5.	3	0.2	0.035	82.5	0.085	57.5
6.	4	0.2	0.015	92.5	0.074	63
7.	5	0.2	0.008	96	0.07	65

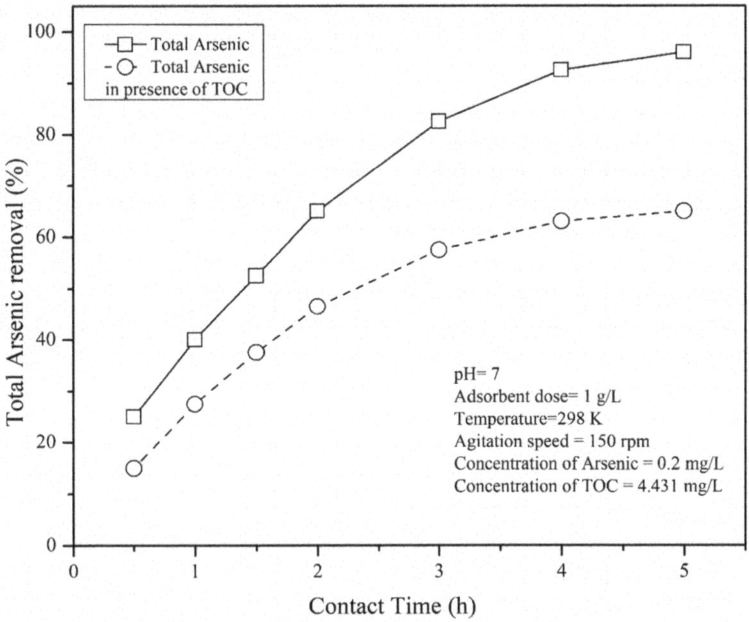

Fig. 3 Effect of contact time on the adsorption of arsenic by the adsorbent fRGO

presence and absence of TOC. Total arsenic was thoroughly mixed for durations ranging from 10 min to 5 h, with samples collected at various time intervals. The experimental scenarios and test results are presented in Table 1.

It was clear from Fig. 3 that due to increasing time of adsorption, the rate of removal of total arsenic was decreased over time and get saturation after 4 h. The total arsenic adsorption was considerably reduced in the presence of TOC in the total arsenic solution [13–15, 34, 35].

3.2.2 Effect of Dose of Adsorbent

The study included performing batch adsorption trials with fRGO-coated sand at six different dosages. The doses of the adsorbent employed were 2 g/L, 4 g/L, 6 g/L, 8 g/L, 10 g/L, and 12 mg/L. The fRGO dosages were 1.31, 2.62, 3.93, 5.24, 6.55, and 7.86 mg/L. The initial arsenic concentration was 0.2 mg/L, pH was set at 7, contact duration for adsorption at 4 h, temperature at 300 K, and agitation speed at 150 rpm. The different experimental conditions to determine the effect of dose of adsorbent (fRGO) for the removal of total arsenic were presented in Table 2.

The percentage arsenic removal and corresponding adsorption capacities were reported against active mass of fRGO. The temperature was maintained at 300 K and the agitation speed at 150 rpm. It was clear from Fig. 4 that the maximum percentage of total arsenic removal was successively reduced in the presence of TOC by fRGO [13–15, 34, 35].

3.2.3 Effect of pH

The study involved conducting batch trials at six distinct pH levels ranging from 2.0 to 12.0 to investigate the impact of pH on the overall removal of arsenic by fRGO. 0.1 g of the adsorbents, namely fRGO-coated sand, were put to a 100 mL sample containing 0.2 mg/L of total arsenic. The samples were thereafter agitated for a duration of 4 h at a speed of 150 rpm in a shaker. Subsequently, the solutions underwent filtration, and the resulting effluent was quantified for its total arsenic content. The removals were reported in relation to the pH at which equilibrium is achieved. The different experimental conditions to determine the effect of pH for the removal of total arsenic were presented in Table 3.

As observed in Fig. 5, arsenic removal diminishes at higher pH levels. The presence of TOC in the arsenic solution significantly reduces total arsenic adsorption, given the pH_{pzc} value of fRGO of approximately 8.1 [13–15, 34, 35, 40, 41]. Arsenic adsorption is notably affected by solution pH, with an increase from

Table 2 The different experimental conditions to determine the effect of dose for the removal of total arsenic

			Absence of TOC		Presence of TOC	
Sl. no.	Dose of adsorbent fRGO (mg/L)	Initial arsenic dose (mg/L)	Residual arsenic (mg/L)	Arsenic removal (%)	Residual arsenic (mg/L)	Arsenic removal (%)
1.	1.310	0.2	0.056	72	0.148	26
2.	2.621	0.2	0.03	85	0.118	41
3.	3.931	0.2	0.018	91	0.095	52.5
4.	5.242	0.2	0.012	94	0.08	60
5.	6.552	0.2	0.008	96	0.072	64
6.	7.862	0.2	0.006	97	0.068	66

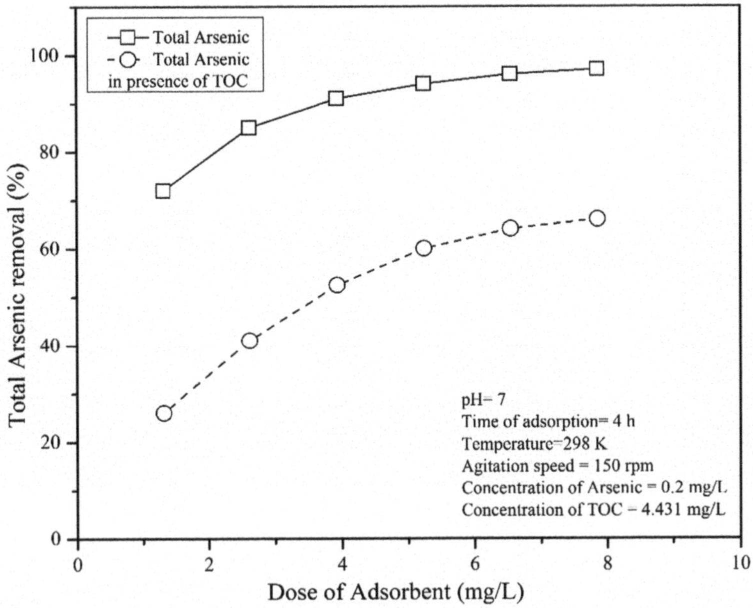

Fig. 4 Plot of percentage of total arsenic removal vs. weight of adsorbent

Table 3 The various experimental settings utilized to assess the impact of pH on the elimination of total arsenic

Sl. no.	Equilibrium pH	Initial arsenic dose (mg/L)	Absence of TOC		Presence of TOC	
			Residual arsenic (mg/L)	Arsenic removal (%)	Residual arsenic (mg/L)	Arsenic removal (%)
1.	2.2	0.2	0.16	20	0.18	10
2.	4.2	0.2	0.06	70	0.12	40
3.	6.5	0.2	0.006	97	0.068	66
4.	6.9	0.2	0.008	96	0.072	64
5.	7.8	0.2	0.06	70	0.11	45
6.	9.8	0.2	0.17	15	0.185	7.5
7.	11.7	0.2	0.19	5	0.195	2.5

20% to 34% when reducing the pH from 12.0 to 2.0. Interactions such as electrostatic attractions and repulsions, hydrophobic effects, ligand exchange, van der Waals forces, and pH play vital roles in arsenic adsorption [39, 40]. The surface charge of fRGO varies with solution pH, influencing its interaction with arsenic, which can adopt diverse microstructures [40]. At lower pH, arsenic resides predominantly in an undissociated or surface protonated state, enhancing adsorbability due to increased hydrophobicity. However, at higher pH, electrostatic repulsive interactions between arsenic molecules intensify, reducing adsorption as arsenic exhibits

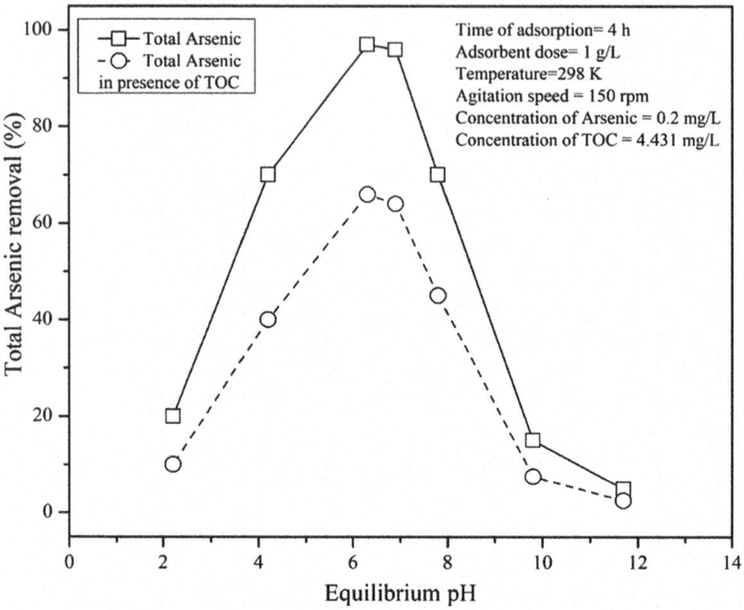

Fig. 5 Percentage (%) removal of total arsenic vs. pH of solutions measured after adsorption by fRGO

lower hydrophobicity and increased solubility. The pH-dependent changes in surface charge of fRGO contribute to the modulation of electrostatic interactions, impacting the adsorption behavior of arsenic onto fRGO. These findings highlight the complex interplay of various factors in the arsenic–fRGO system, elucidating the importance of pH in governing arsenic adsorption dynamics. Consequently, a smaller amount of arsenic is absorbed at elevated pH levels [41].

3.2.4 Effect of Concentration of TOC

The effect of concentration of arsenic removal by fRGO was studied with a sample volume of 100 mL having pH of 7. Within this volume, 1 g of fRGO-coated sand was dosed (equivalent to 0.655 mg of active fRGO). Six different concentration values such as 0.05, 0.1, 0.2, 0.3, 0.4, and 0.5 mg/L of arsenic solution were studied. The different experimental conditions to verify the influence of concentration of TOC for the removal of total arsenic were presented in Table 4.

It was clear from Fig. 6 that the total arsenic adsorption was considerably reduced in the presence of TOC, the negative charged ions increased consistently, and a decreasing rate of total arsenic adsorption occurred in the total arsenic solution [13–15, 34, 35].

Table 4 The distinct experimental parameters employed to investigate the influence of concentration of TOC on the removal of total arsenic

Sl. no.	Concentration of TOC (mg/L)	Initial arsenic dose (mg/L)	Absence of TOC		Presence of TOC	
			Residual arsenic (mg/L)	Arsenic removal (%)	Residual arsenic (mg/L)	Arsenic removal (%)
1.	0.4431	0.2	0.01	95	0.07	65
2.	0.8862	0.2	0.012	94	0.075	62.5
3.	1.7724	0.2	0.018	91	0.08	60
4.	2.6586	0.2	0.026	87	0.085	57.5
5.	3.5448	0.2	0.032	84	0.09	55
6.	4.431	0.2	0.04	80	0.095	52.5
7.	0.4431	0.2	0.01	95	0.07	65

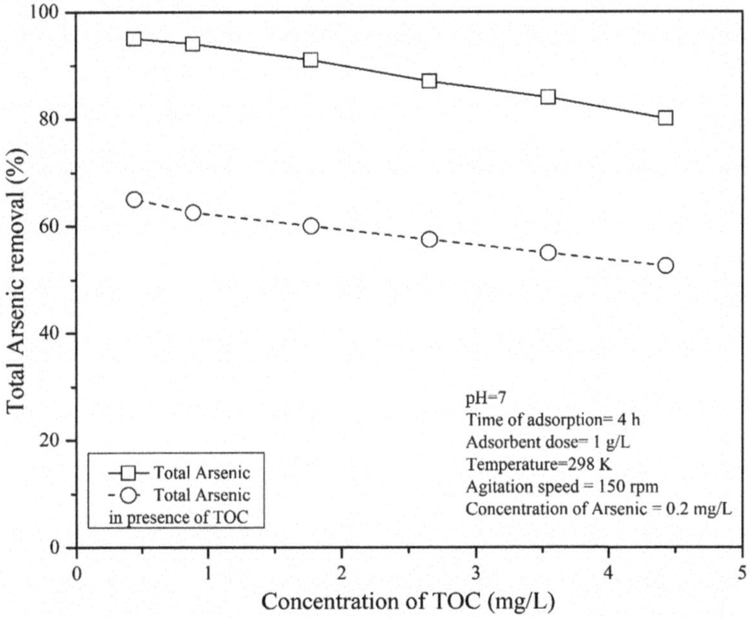

Fig. 6 Effect of concentration of arsenic of solutions measured after adsorption of arsenic by fRGO

3.2.5 Effect of Temperature

The impact of temperature on the removal of arsenic by fRGO was investigated using a sample volume of 100 mL containing arsenic at an intensity of 0.2 mg/L. In this amount, 1 g of sand coated with fRGO was added (equivalent to 0.655 mg of active fRGO). The study examined three distinct temperature values: 290, 300, and 310 K. Table 5 presents the various experimental conditions used to evaluate the impact of temperature on the elimination of total arsenic.

Table 5 The distinct experimental parameters employed to investigate the influence of temperature on the removal of total arsenic

Sl. no.	Temperature (K)	Initial arsenic dose (mg/L)	Absence of TOC		Presence of TOC	
			Residual arsenic (mg/L)	Arsenic removal (%)	Residual arsenic (mg/L)	Arsenic removal (%)
1.	290	0.2	0.018	91	0.075	62.5
2.	300	0.2	0.008	96	0.07	65
3.	310	0.2	0.006	97	0.068	66

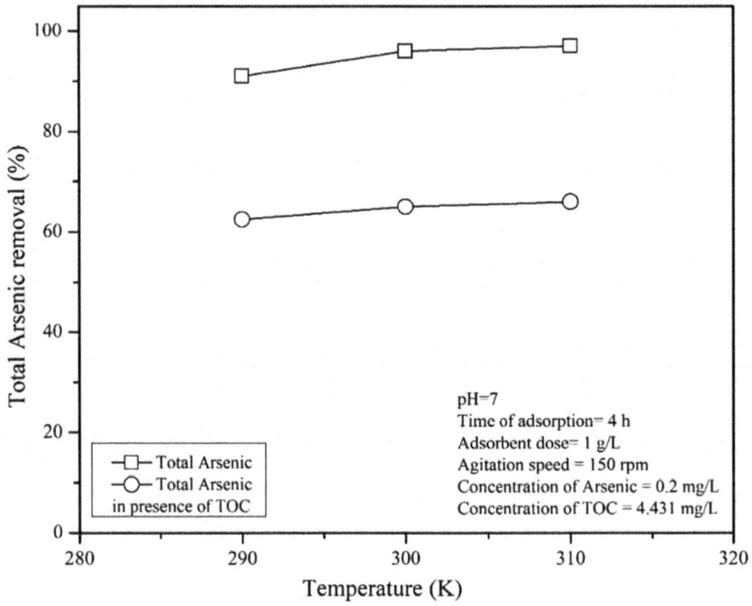

Fig. 7 Plot of temperature vs. arsenic removal (%)

The highest arsenic removal by fRGO was obtained to be 97%. It was clear from Fig. 7 that within the level of fRGO studied, this may be due to the fact that fRGO has different types of function groups such as –COOH, –OH, and –C=O and higher specific surface area [42–46]. It was also clear that beyond temperature 300 K did not explain any significant improvement in arsenic removal within the study range. The total arsenic adsorption was considerably reduced in the presence of TOC in the total arsenic solution [13–15, 34, 35].

3.2.6 Effect of Agitation Speed

The study investigated the influence of agitation speed on the removal of arsenic by fRGO using a 100 mL sample having arsenic concentration of 0.2 mg/L. For this volume, 1 g of fRGO-coated sand was added, equivalent to 0.655 mg of active

Table 6 The distinct experimental parameters employed to investigate the influence of agitation speed on the removal of total arsenic

			Absence of TOC		Presence of TOC	
Sl. no.	Agitation speed (rpm)	Initial arsenic dose (mg/L)	Residual arsenic (mg/L)	Arsenic removal (%)	Residual arsenic (mg/L)	Arsenic removal (%)
1.	100	0.2	0.016	92	0.072	64
2.	150	0.2	0.008	96	0.07	65
3.	200	0.2	0.006	97	0.068	66

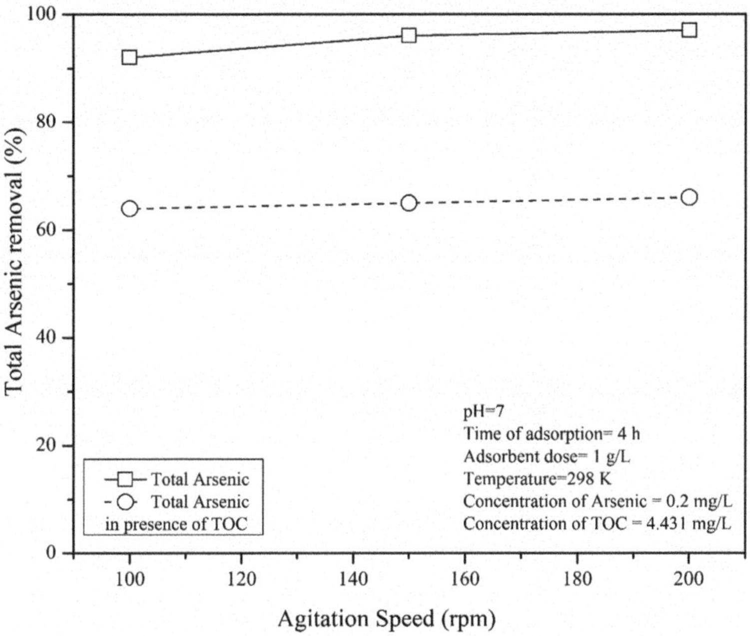

Fig. 8 Plot of agitation speed vs. arsenic removal (%)

fRGO. The investigation considered three different agitation levels: 100, 150, and 200 rpm. Table 6 details the diverse experimental conditions employed to evaluate the result of agitation speed on the removal of total arsenic.

The data presented in Fig. 8 demonstrates that increasing the agitation speed did not result in any notable improvement in arsenic removal within the range examined. The total arsenic adsorption was considerably reduced in the presence of TOC in the total arsenic solution [13–15, 34, 35].

4 Conclusion

The research demonstrated the improved efficiency of iron-functionalized reduced graphene oxide (fRGO) for enhanced arsenic removal. The use of natural sand demonstrated it to be used as a good supportive base material to hold fRGO. The fRGO nano adsorbent was characterized by X-ray diffraction (XRD) and point of zero charge (PZC) analysis to confirm the effectiveness of the functionalization. The study demonstrated that fRGO had improved the arsenic removal performance when there is no organic matter. The presence of dissolved total organic carbon has a detrimental effect on the adsorption process as it competes with the fRGO molecule for the active adsorption sites. This comparison study holds significant utility in the analysis of adsorbent efficiency over extended periods of time under real-life situations. More study is required with real water system having other competing co-occurring ions.

Acknowledgments This research is funded by the Department of Science and Technology and Biotechnology (DSTBT), Government of West Bengal, India, through their Research Fellowship Program vide grant Memo No. 1692(Sanc.)/ST/P/S&T/6G-7/2019, date: 06/02/2020.

References

1. US-EPA, Technologies and costs for removal of arsenic from drinking water, US Environmental Protection Agency, EPA/815/R-00/028 (2000)
2. IARC Working Group on the Evaluation of Carcinogenic Risks to Humans, World Health Organization and International Agency for Research on Cancer, some drinking-water disinfectants and contaminants, including arsenic. IARC **84** (2004)
3. M.I. Fernández, J.F. López, B. Vivaldi, F. Coz, Long-term impact of arsenic in drinking water on bladder cancer health care and mortality rates 20 years after end of exposure. J. Urol. **187**, 856–861 (2012)
4. World Health Organization, *Drinking Water Quality in the South-East Asia Region (No. SEA-EH-567)* (WHO Regional Office for South-East Asia, 2010)
5. M.M. Rahman, M.K. Sengupta, S. Ahamed, U.K. Chowdhury, M.A. Hossain, B. Das, D. Lodh, K.C. Saha, S. Pati, I. Kaies, A.K. Barua, The magnitude of arsenic contamination in groundwater and its health effects to the inhabitants of the Jalangi – One of the 85 arsenic affected blocks in West Bengal, India. Sci. Total Environ. **338**, 189–200 (2005)
6. M.R. Karagas, A. Gossai, B. Pierce, H. Ahsan, Drinking water arsenic contamination, skin lesions, and malignancies: A systematic review of the global evidence. Curr. Environ. Health Rep. **2**, 52–68 (2015)
7. S.A. Baghoth, S.K. Sharma, G.L. Amy, Tracking natural organic matter (NOM) in a drinking water treatment plant using fluorescence excitation-emission matrices and PARAFAC. Water Res. **45**, 797–809 (2011)
8. K.M. Elkins, D.J. Nelson, Fluorescence and FT-IR spectroscopic studies of Suwannee River fulvic acid complexation with aluminum, terbium and calcium. J. Inorg. Biochem. **87**, 81–96 (2001)
9. E.L. Sharp, P. Jarvis, S.A. Parsons, B. Jefferson, Impact of fractional character on the coagulation of NOM. Colloids Surf. A Physicochem. Eng. Asp. **286**, 104–111 (2006)

10. S.W. Krasner, H.S. Weinberg, S.D. Richardson, S.J. Pastor, R. Chinn, M.J. Sclimenti, G.D. Onstad Jr., A.D. Thruston, Occurrence of a new generation of disinfection byproducts. Environ. Sci. Technol. **40**, 7175–7185 (2006)
11. P.C. Singer, DPBs in drinking water. J. Am. Water Works Assoc. **98**, 73–79 (2006)
12. A. Matilainen, M. Vepsalainen, M. Sillanpaa, Removal of natural organic matter from drinking water by advanced oxidation. Adv. Colloid Interf. Sci. **159**, 189–197 (2010)
13. A.D. Redman, D.L. Macalady, D. Ahmann, Natural organic matter affects arsenic speciation and sorption onto hematite. Environ. Sci. Technol. **36**, 2889–2896 (2002)
14. A.B. Giasuddin, S.R. Kanel, H. Choi, Adsorption of humic acid onto nanoscale zerovalent iron and its effect on arsenic removal. Environ. Sci. Technol. **41**, 2022–2027 (2007)
15. H. Xu, B. Allard, A. Grimvall, Effects of acidification and natural organic materials on the mobility of arsenic in the environment. Water Air Soil Pollut. **57**, 269–278 (1991)
16. T. Kuila, S. Bose, A.K. Mishra, P. Khanra, N.H. Kim, J.H. Lee, Chemical functionalization of graphene and its applications. Prog. Mater. Sci. **57**, 1061–1105 (2012)
17. H. Su, Z. Ye, N. Hmidi, R. Subramanian, Carbon nanosphere–iron oxide nanocomposites as high-capacity adsorbents for arsenic removal. RSC Adv. **7**, 36138–36148 (2017)
18. B. Paul, V. Parashar, A. Mishra, Graphene in the Fe_3O_4 nano-composite switching the negative influence of humic acid coating into an enhancing effect in the removal of arsenic from water. Environ. Sci. Water Res. Technol. **1**, 77–83 (2015)
19. J. Zhu, Z. Lou, Y. Liu, R. Fu, S.A. Baig, X. Xu, Adsorption behavior and removal mechanism of arsenic on graphene modified by iron–manganese binary oxide (FeMnOx/RGO) from aqueous solutions. RSC Adv. **5**, 67951–67961 (2015)
20. M. Khatamian, N. Khodakarampoor, M.S. Oskoui, N. Kazemian, Synthesis and characterization of RGO/zeolite composites for the removal of arsenic from contaminated water. RSC Adv. **5**, 35352–35360 (2015)
21. S.K. Ray, C. Majumder, P. Saha, Functionalized reduced graphene oxide (fRGO) for removal of fulvic acid contaminant. RSC Adv. **7**, 21768–21779 (2017)
22. C. Dong, PowderX: Windows-95-based program for powder X-ray diffraction data processing. J. Appl. Crystallogr. **32**, 838–838 (1999)
23. S. Mahpishanian, H. Sereshti, One-step green synthesis of β-cyclodextrin/iron oxide-reduced graphene oxide nanocomposite with high supramolecular recognition capability: Application for vortex-assisted magnetic solid phase extraction of organochlorine pesticides residue from honey samples. J. Chromatogr. A **1485**, 32–43 (2017)
24. D. Wang, H. Duan, J. Lü, C. Lü, Fabrication of thermo-responsive polymer functionalized reduced graphene oxide@ Fe_3O_4@ Au magnetic nanocomposites for enhanced catalytic applications. J. Mater. Chem. A **5**, 5088–5097 (2017)
25. S. Zhu, J. Guo, J. Dong, Z. Cui, T. Lu, C. Zhu, D. Zhang, J. Ma, Sonochemical fabrication of Fe_3O_4 nanoparticles on reduced graphene oxide for biosensors. Ultrason. Sonochem. **20**, 872–880 (2013)
26. A. Prakash, S. Chandra, D. Bahadur, Structural, magnetic, and textural properties of iron oxide-reduced graphene oxide hybrids and their use for the electrochemical detection of chromium. Carbon **50**, 4209–4219 (2012)
27. J. Li, S. Zhang, C. Chen, G. Zhao, X. Yang, J. Li, X. Wang, Removal of Cu (II) and fulvic acid by graphene oxide nanosheets decorated with Fe_3O_4 nanoparticles. ACS Appl. Mater. Interfaces **4**, 4991–5000 (2012)
28. X. Luo, C. Wang, S. Luo, R. Dong, X. Tu, G. Zeng, Adsorption of As (III) and As (V) from water using magnetite Fe_3O_4-reduced graphite oxide–MnO_2 nanocomposites. Chem. Eng. J. **187**, 45–52 (2012)
29. D.L. De Faria, S. Venâncio Silva, M.T. De Oliveira, Raman microspectroscopy of some iron oxides and oxyhydroxides. J. Raman Spectrosc **28**, 873–878 (1997)
30. M.P. Gatabi, H.M. Moghaddam, M. Ghorbani, Point of zero charge of maghemite decorated multiwalled carbon nanotubes fabricated by chemical precipitation method. J. Mol. Liquids **216**, 117–125 (2016)

31. G. Zelmanov, R. Semiat, Selenium removal from water and its recovery using iron (Fe^{3+}) oxide/hydroxide-based nanoparticles sol (NanoFe) as an adsorbent. Sep. Purif. Techmol. **103**, 167–172 (2013)
32. H. Chen, B. Gao, H. Li, Functionalization, pH, and ionic strength influenced sorption of sulfamethoxazole on graphene. J. Environ. Chem. Eng. **2**, 310–315 (2014)
33. W. Stumm, J.J. Morgan, *Aquatic Chemistry: An Introduction Emphasizing Chemical Equilibria in Natural Waters* (Wiley, New York, 1981)
34. P. Warwick, E. Inam, N. Evans, Arsenic's interaction with humic acid. Environ. Chem. **2**, 119–124 (2005)
35. Z. Li, J.S. Jean, W.T. Jiang, P.H. Chang, C.J. Chen, L. Liao, Removal of arsenic from water using Fe-exchanged natural zeolite. J. Hazard. Mater. **187**, 318–323 (2011)
36. S.T. Yang, Y. Chang, H. Wang, G. Liu, S. Chen, Y. Wang, Y. Liu, A. Cao, Folding/aggregation of graphene oxide and its application in Cu^{2+} removal. J. Colloid Interface Sci. **351**, 122–127 (2010)
37. D. Ghosh, A. Gupta, Economic justification and eco-friendly approach for regeneration of spent activated alumina for arsenic contaminated groundwater treatment. Resour. Conserv. Recycl. **61**, 118–124 (2012)
38. J. Zhang, M.S. Azam, C. Shi, J. Huang, B. Yan, Q. Liu, H. Zeng, Poly (acrylic acid) functionalized magnetic graphene oxide nanocomposite for removal of methylene blue. RSC Adv. **5**, 32272–32282 (2015)
39. X. Fang, X. Xu, S. Wang, D. Wang, Adsorption kinetics and equilibrium of Cu (II) from aqueous solution by polyaniline/coconut shell–activated carbon composites. J. Environ. Eng. **139**, 1279–1284 (2013)
40. S.G. Wang, X.W. Liu, W.X. Gong, W. Nie, B.Y. Gao, Q.Y. Yue, Adsorption of fulvic acids from aqueous solutions by carbon nanotubes. J. Chem. Technol. Biotechnol. Int. Res. Proc. Environ. Clean Technol. **82**, 698–704 (2007)
41. M.A. Islam, D.W. Morton, B.B. Johnson, M.J. Angove, Adsorption of humic and fulvic acids onto a range of adsorbents in aqueous systems, and their effect on the adsorption of other species: A review. Sep. Purif. Technol. **247**, 116949 (2020)
42. M. Joudi, H. Nasserlah, H. Hafdi, J. Mouldar, B. Hatimi, M.A. El Mhammedi, M. Bakasse, Synthesis of an efficient hydroxyapatite–chitosan–montmorillonite thin film for the adsorption of anionic and cationic dyes: Adsorption isotherm, kinetic and thermodynamic study. SN Appl. Sci. **2**(6), 1–13 (2020)
43. R. Gusain, N. Kumar, E. Fosso-Kankeu, S.S. Ray, Efficient removal of Pb (II) and Cd (II) from industrial mine water by a hierarchical MoS2/SH-MWCNT nanocomposite. ACS Omega **4**, 13922–13935 (2019)
44. X.H. Cai, R.H. Lin, J. Wu, J.B. He, Y.C. Wu, X.Y. Wang, Adsorption of ethylenediaminetetraacetic dianhydride modified oxalate decarboxylase on calcium oxalate. Biotech. Histochem. **93**, 220–229 (2018)
45. Z. Wu, H. Zhong, X. Yuan, H. Wang, L. Wang, X. Chen, G. Zeng, Y. Wu, Adsorptive removal of methylene blue by rhamnolipid-functionalized graphene oxide from wastewater. Water Res. **67**, 330–344 (2014)
46. D. Zhao, W. Zhang, C. Chen, X. Wang, Adsorption of methyl orange dye onto multiwalled carbon nanotubes. Procedia Environ. Sci. **18**, 890–895 (2013)

Part III
Innovative Techniques for Measuring Environmental Pollutants

Life Cycle Assessment (LCA) of a Low-Cost Conventional Building in Rishra, West Bengal: A Case Study

Aritra Majumder, Rohan Das, Anupam Debsarkar, and Kaushik Dutta Roy

Abbreviations

VA	Vernacular architecture
LCA	Life cycle assessment
PMAY	Pradhan Mantri Awas Yojana
GWP	Global warming potential
ADP	Abiotic depletion potential
PCC	Plain cement concrete
RCC	Reinforced cement concrete
HYSD	High yielding strength deformed steel
BIM	Building information modeling
BREEAM	Building Research Establishment Environmental Assessment Methodology
LEED	Leadership in Energy and Environmental Design
EN	European Standards (European Norm)
ISO	International Organization for Standardization
LCI	Life cycle inventory
LCIA	Life cycle impact assessment

A. Majumder (✉) · R. Das · A. Debsarkar
Department of Civil Engineering, Jadavpur University, Kolkata, West Bengal, India
e-mail: aritra.civil.rs@jadavpuruniversity.in; rohand.civil.pg@jadavpuruniversity.in; anupamdeb.sarkar@jadavpuruniversity.in

K. D. Roy
Kolkata Metropolitan Development Authority, Kolkata, West Bengal, India

D. Mazumder (ed.), *Sustainable Advanced Technologies for Environmental Management*, Springer Proceedings in Earth and Environmental Sciences,
https://doi.org/10.1007/978-3-031-64006-3_14

1 Introduction

1.1 Low-Cost Building: Significance

The majority of the population of India belongs to a low-income demographic and faces various economic challenges. Targeted efforts are thus needed to alleviate poverty and improve living conditions. Low-cost housing is crucial for providing affordable shelter and a decent standard of living for those struggling to afford housing in urban areas. These initiatives can also help address homelessness among the urban poor and improve overall social welfare in the country.

1.2 Pradhan Mantri Awas Yojana (PMAY) and Its Objective

The Pradhan Mantri Awas Yojana (PMAY) is a government scheme aimed at addressing the housing crisis among Indian slum-dwellers and the urban poor. It involves the rehabilitation of slums, the promotion of affordable housing through credit subsidies, and the offering of subsidies to the beneficiaries to construct houses. The project aims to improve basic amenities such as water, sewage, drainage, street lighting, and waste management in slum areas. It also aims to create affordable housing with civic amenities for ownership by these groups. Additionally, it focuses on cleanliness and awareness through the construction of pay-and-use toilets and community education.

1.3 Features of PMAY Buildings

A typical building plan for the PMAY scheme includes 32 m^2 of dwellings with two bedrooms, a kitchen, and a sanitation facility. The design must comply with municipal regulations and consider factors such as local climatic conditions, beneficiaries' preferences, and community traditions. Materials include PCC (1:3:6) for the foundation, RCC (M-20 grade) for the substructure and superstructure, HYSD steel, first-class brick masonry, 1:6 plaster, and IPS flooring. Structural design includes M-20 grade concrete and Fe-415 grade steel, with plinth beams as tie beams and pedestals extending to the ground level. Wall thickness varies with external walls being 250 mm thick and internal walls being 125 mm. Seismic loads are also considered at the time of design.

1.4 Literature Review

Numerous studies have employed LCA to evaluate different aspects of sustainable building practices across various regions. In Catamarca, Argentina, a study compared traditional and contemporary social houses to assess their energy use and CO_2 emissions [1]. A mixed building system has been proposed that utilizes natural, reused, and local materials to reduce transport-related emissions and embodied carbon, emphasizing the importance of balancing environmental impact and economic considerations in sustainable building design [2]. Another study examined the environmental impact of a modern villa in Dhahran, Saudi Arabia, using a "cradle-to-gate" approach [3]. India has also been a focus, with research discussing the energy implications of building materials and construction schemes in vernacular architecture [4] and exploring the potential of building information modeling (BIM) for sustainable design in southern India [5]. The need for new tools and methods to measure and compare environmental impacts, including degradation and resource consumption emissions, for achieving sustainability goals has also been emphasized [6]. Furthermore, these studies delve into various aspects of the sustainability of building materials using LCA. For instance, an examination of earthen materials in Portugal found that they can reduce environmental impacts by 50% compared to conventional materials [7]. Similarly, another study conducted in Egypt's Dakhla Oasis assessed the embodied energy and life cycle carbon emissions of fired clay and sun-dried bricks, highlighting the significant potential for emissions reduction [8]. A holistic life-cycle perspective was adopted to understand the sustainability of tall buildings, emphasizing their role in energy conservation and sustainable design [9]. Lastly, some studies demonstrate the advantages of applying vernacular architecture principles to contemporary construction practices. An investigation in Iran's building sector compared a passive building involving traditional materials and techniques to a conventional one, revealing significant reductions in carbon emissions and embodied energy, showcasing the potential of traditional practices for sustainability in the Iranian construction industry [10]. Overall, these studies contribute enormously toward promoting a more sustainable construction sector and offer insights into the benefits of integrating traditional building methods into modern construction.

1.5 Objective of the Study

To ensure that low-cost houses constructed under the PMAY scheme are not only affordable but also sustainable and environmentally friendly, the present study evaluated low-cost houses under the scheme in Rishra, West Bengal, using life cycle assessment (LCA) methodology. It evaluated energy efficiency and environmental impact, analyzing factors such as construction materials, design elements, and energy consumption patterns. The findings aim to improve sustainability in future projects.

Additionally, this research aimed to gauge the extent to which climate change potential was factored into the construction of these houses. Given the substantial number of housing units constructed under the PMAY scheme to fulfill the housing needs of India's large underprivileged population, including a significant number of 20.76 lakh houses in West Bengal and 1.75 crore houses nationwide, the study sought to quantify the CO_2 emissions and global warming potential associated with the widespread presence of such housing developments throughout India. In essence, it aimed to understand the environmental implications and carbon footprint of these housing projects in the context of climate change and sustainability.

2 Materials and Methods

Low-cost buildings are constructed from the perspective of providing affordable housing to low-income group members. Such buildings have been constructed at various places in West Bengal, India. A low-cost building constructed under the PMAY scheme at Rishra in the Hooghly district of West Bengal was surveyed in this study. Primary data regarding its structural features, electrical features, architectural features, plumbing, and so on was observed and collected. The carbon footprint and energy usage have been assessed through the life cycle assessment (LCA) methodology.

LCA is a newly emerged technique adopted to assess the environmental footprint of a product's life from cradle to grave. Also referred to as cradle-to-grave analysis, eco-balance, or life cycle analysis, it entails the examination of all stages of a product's entire lifespan, spanning from the extraction of raw material to manufacturing, distribution, and end-use. Adhering to international standards such as International Organization for Standardization (ISO) 14040, the LCA method comprises a fixed structure, reflecting the distinct stages of the process. Involving several phases including the life cycle inventory (LCI) preparation, life cycle impact assessment (LCIA), and life cycle interpretation, LCA embodies a systematic approach to comprehending the environmental impacts of a product system's life cycle. During the LCI phase, requisite data is gathered to fulfill the aims of the study, while the LCIA provides supplementary insights to evaluate the environmental impacts of the system. The conclusive phase, i.e., life cycle interpretation summarizes and deliberates upon the findings to derive conclusions, recommendations, and decisions aligned with the objectives and scopes of the study.

The international standards encompasses two types of studies: LCI and LCA. LCI studies, such as LCA studies but lacking the LCIA phase, share similarities yet should not be confused with the LCI phase of an LCA study. While LCA represents one of several environmental management techniques, it may not apply to all scenarios. Typically, LCA does not encompass economic or social aspects of a product; however, the life cycle approach and methodologies described in the international standards can be adopted to address these aspects.

2.1 *Tools of Life-Cycle Methodology*

The model of the low-cost housing structure in Rishra was developed using Auto-Desk Revit. Subsequently, life cycle analysis was conducted using the One Click LCA software, which serves as a comprehensive tool for life cycle assessment and costing, ensuring compliance with more than 30 standards. It facilitates life cycle assessments for both civil engineering and building projects, aligning with international standards such as ISO 14040, ISO 14044, and ISO 21930, as well as European Standards such as EN 15978, EN 15804+A1, and EN 15804+A2. These standards establish a unique life-cycle model for construction and building work, encompassing product and end-of-life stages for building-level assessments, allowing for optional omissions or replacements. The analysis aimed to assess the life cycle of the project with a focus on cradle-to-cradle sustainability using the specified service life and replacement building use indicators.

The life cycle information of the building concerning the various stages (Fig. 1) of its lifespan can be classified into the Product Stage (A1–A3), Construction Process Stage (A4–A5), Use Stage (B1–B7), and End of Life Stage (C1–C4). Furthermore, there is scope for the assessment of additional information regarding the building beyond its life cycle with respect to its Reuse–Recovery–Recycling–Potential (D). Subsequently, the study assessed each stage of the building's life cycle keeping in mind specific impact categories (Table 1). These stages, beginning from the extraction of raw materials to disposal and beyond, are evaluated with regard to their global

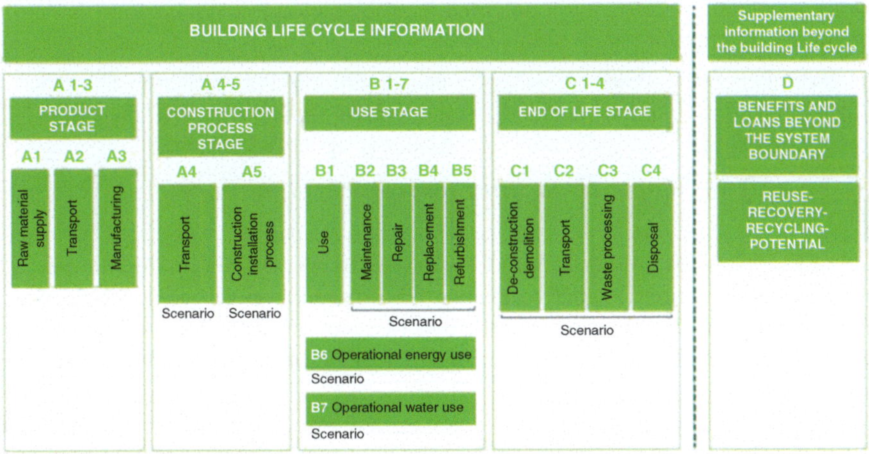

Fig. 1 Life cycle information and life cycle stages of a building

Table 1 Description of different impact categories

Impact category	Unit(s)	Description
Global warming potential	kg CO_2eq	The changes in surface temperatures result from increased levels of greenhouse gases in the atmosphere. Greenhouse gas emissions from burning fossil fuels are linked to acidification and smog formation termed "carbon footprint"
Acidification potential	kg SO_2eq	Acidification process occurring in the environment. Substances such as carbon dioxide easily dissolve in water, leading to enhanced acidity levels and resulting in adverse impacts on water ecosystems
Eutrophication potential	CML: kg PO_4^{3-} eq TRACI: kg Neq	The impact of supplementing soil or water with mineral nutrients, leads to the dominance of species within an ecosystem, jeopardizing the survival of other organisms and leading to the extinction of entire populations
Ozone depletion potential	kg CFC-11eq	Role of certain substances present in the atmosphere in degrading the ozone layer, absorbing and shielding Earth's surface from harmful solar UV rays
Formation of ozone of lower atmosphere ("tropospheric")	CML: kg C_2H_4eq TRACI: kg O_3eq	The role of atmospheric substances in the formation of photochemical smog. They are also known as summer smog
Abiotic depletion potential (ADP) for fossil resources	MJ	The extent of abiotic resources consumed that contribute toward the depletion of fossil energy resources, such as oil and coal
Water use	m^3 deprived	Water (user) deprivation, and deprivation-weighted water consumption (WDP), characterized by available WAter REmaining (AWARE)

warming potential, acidification potential, eutrophication potential, ozone depletion potential, formation of ozone of lower atmosphere, abiotic depletion potential-fossil resources, and water usage.

2.2 Analysis Scope, Data Sources, and Assumptions

The LEED v4 Whole Building Life Cycle Assessment and BREEAM Mat 01 Life Cycle Assessment are designed to evaluate the structure, envelope materials, operational energy, and construction site impacts of a building for 50 years. The analysis has been conducted using the data sources (Table 2) for building information.

Table 2 Sources of different categories of data

Data type	Data source
Material quantities (A1–A3)	The client and their designated designers provide construction drawings, BIM models, and bills of quantities
Material transport distances (A4)	One Click LCA offers transportation scenarios tailored to specific regions, reflecting typical distances and methods for various product types, particularly relevant when supplier decisions remain undecided
Construction and installation (A5)	This analysis utilizes conservative default values derived from One Click LCA and estimates site utility consumption excluding impacts
Material impacts in use (B1–B5)	The service lives of materials are determined using typical values assessed for project suitability and modified as needed. Maintenance and repair jobs are not considered; it is assumed materials will be replaced upon reaching the end of their service life
Use phase energy consumption (B6)	The analysis prioritizes the detailed energy analysis specific to this project, excluding assessment of impacts
Use phase water consumption (B7)	Project-specific impact values are excluded from the analysis in this project
End-of-life impacts (C1–C4)	End-of-life impacts are assessed based on One Click LCA's scenarios, which are aligned with the standard end-of-life processing methods for material types as per the requirements of the EN 15804 + A1

3 Results and Discussion

The study was undertaken to evaluate the environmental impacts of a low-cost conventional building in Rishra, West Bengal, over its lifespan of 50 years. The global warming potential (GWP), water usage, and energy consumption of the building could potentially help identify areas for improvement and guide sustainable design choices. The study found valuable insights into the major contributors to these environmental impacts and their associated effects on the building.

Global Warming Potential (GWP) The total GWP associated with the analyzed low-cost building is estimated at 88,208 kg CO_2 eq. When considering life-cycle stages, the major contributors to GWP are identified as B6 Energy, accounting for 79.8%, and A1–A3 Materials, contributing 14.7% (Fig. 2). It is further evident that electricity consumption is the major driver behind the GWP, accounting for 79.8% of the total kg CO_2 eq. Additionally, components such as the foundation of the building, frame (encompassing beam, columns, and slabs), and external walls make up another 10.9% of the GWP.

Regarding the specific types of resources that significantly impact the GWP, electricity, constituting 79.8%, takes the lead. In addition, ready-mix concrete, used for internal walls, foundations, floors, and external walls contributes 8.9% to the GWP. Structural steel and steel profiles make up 3.9% of this total. While considering these contributions in terms of mass (in kg), the frame, composed of beams, columns, and slabs, is the most substantial element, representing a significant 54%.

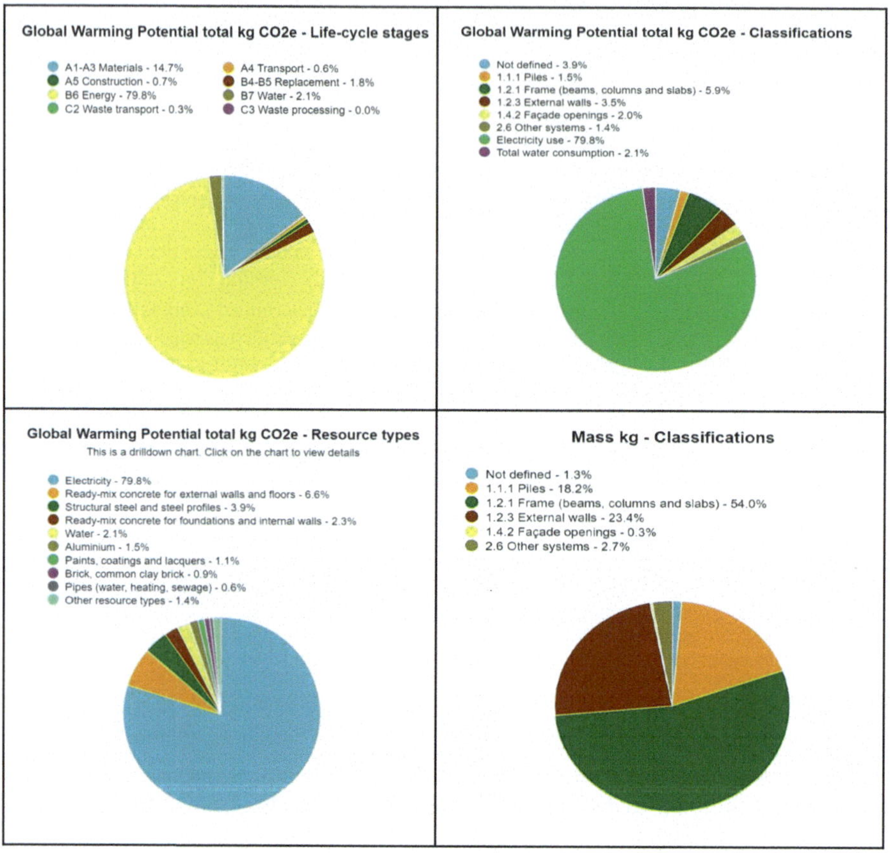

Fig. 2 Percentage share of global warming potential and mass for different resources

External walls account for 23.4% and piles constitute 18.2% of the total mass. These elements significantly contribute to the overall environmental impact of a building, particularly in terms of carbon emissions.

Abiotic Depletion Potential (ADP) from Fossil Resources The total embodied energy in the low-cost building, concerning ADP related to fossil resources, is estimated at 891,000 MJ. In the context of life-cycle stages, the primary contributors to ADP-fossil fuels are B6 Energy, making up 83.6%, and A1–A3 Materials, with a 9.7% share (Fig. 3).

Analyzing the classification of abiotic depletion potential for fossil resources (+A2) MJ, we find that electricity use constitutes a significant 83.6% of the total. The frame, which comprises beams, columns, and slabs, contributes 3.7%, and total water consumption represents 3% of the overall figure. Examining the specific types of resources that significantly contribute to ADP from fossil resources, electricity emerges as the dominant factor, making up a substantial 83.6%. In the same context,

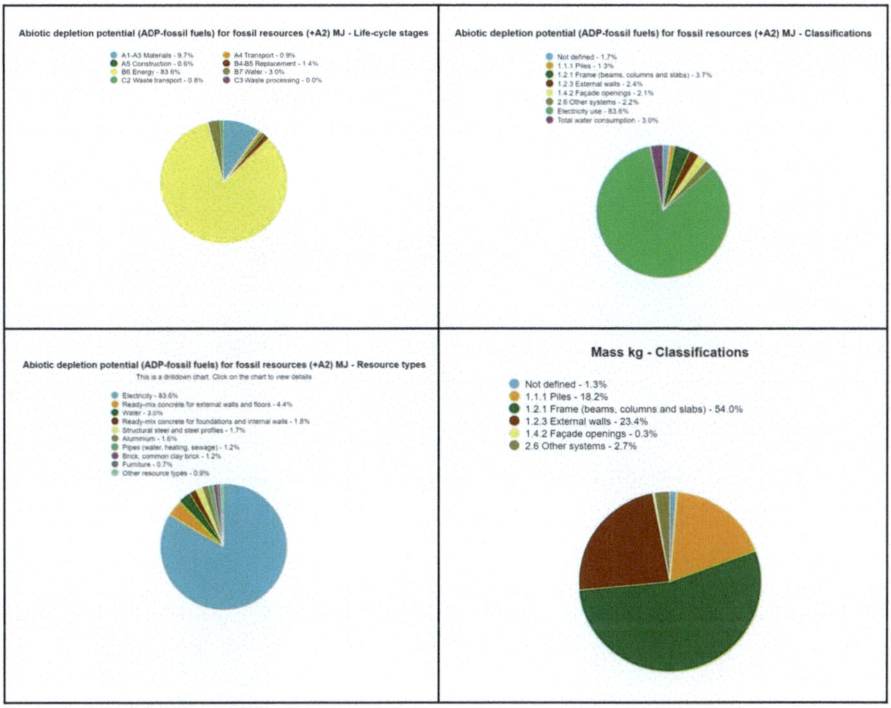

Fig. 3 Abiotic depletion potential (ADP) and mass of fossil resources

ready-mix concrete used for floors and external walls contributes 4.4%. While considering the ADP contributions in terms of mass (in kg), the frame, consisting of beams, columns, and slabs, accounts for a substantial 54% of the contributions, with external walls following closely at 23.4%.

These findings underscore the building's substantial reliance on fossil fuels to meet its energy requirements, thereby carrying significant environmental implications.

Water Consumption The total water usage for the low-cost building over its entire lifespan amounts to 56,300 m^3. While examining the major contributors, A1–A3 Materials constitute 55.1% of water consumption, while B6 Energy contributes 38.8% (Fig. 4). The breakdown of water usage in cubic meters (m^3) shows that the foundation makes up the majority, representing 51.6%, while electricity use follows at 38.8%.

Analyzing the specific types of resources that play a substantial role in water usage in m^3, more than half (52%) is attributed to ready-mix concrete used in foundations and internal walls, while electricity contributes an additional 38.8%. While assessing the water use depletion in terms of mass (in kg), the foundation,

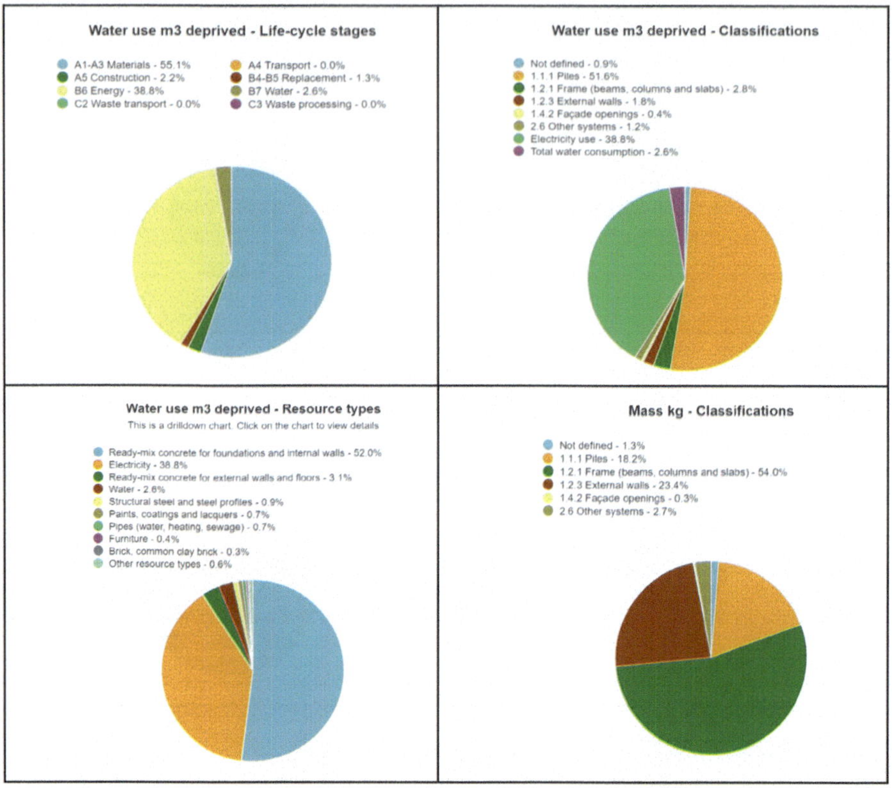

Fig. 4 Water usage and mass for different resources

consisting of beams, columns, and slabs, accounts for a significant 54%, with external walls also emerging as a major contributor at 23.4%.

These findings emphasize the quantity of water requirements associated with both material usage and energy consumption in the building. The findings emphasize the need for sustainable design and construction practices, shifting to renewable energy sources, integrating energy-efficient technologies, and selecting materials that are capable of curtailing the adverse environmental impacts and promoting a more sustainable built environment. Addressing these factors can enhance the building's sustainability and reduce its ecological footprint.

The chart (Fig. 5) displays the results derived for the building life cycle across all reported impact categories and stages, aiding in identifying key drivers for various environmental impacts. The chart also displays the global warming potential, i.e., carbon footprint of a building's entire lifespan, highlighting impacts before use, deconstruction, and outside the life cycle. It helps identify key drivers for environmental impacts across all impact categories and stages.

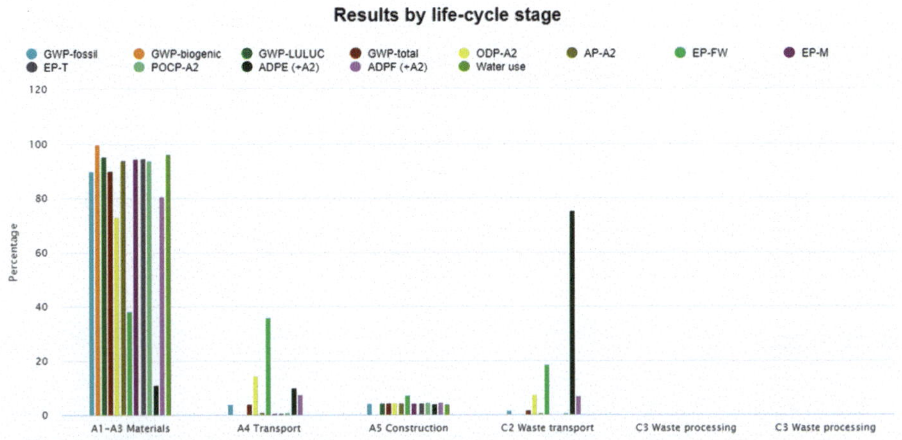

Fig. 5 Stagewise results of the LCA

4 Conclusion

LCA is a multiutility methodology developed for optimizing and assessing the environmental effects of construction materials and activities. An in-depth review of the literature reveals the widespread application of LCA to reduce the environmental impact caused by construction activities involving diverse materials across different bioclimatic regions. This study evaluated the energy efficiency and environmental impact of low-cost housing using LCA methodology, focusing on GWP to determine their carbon footprint, energy consumption, and water consumption in the context of climate change. The analysis revealed that the major contributions to the environmental impacts of low-cost housing were from B6 Energy and A1–A3 Materials. B6 Energy refers to the energy consumption associated with the operation of the housing, such as electricity usage for lighting, heating, and cooling. A1–A3 Materials encompass the environmental consequences associated with the production, extraction, and transportation of construction materials. These impacts can include carbon emissions from manufacturing processes, energy consumption during extraction, and environmental degradation. On identification and quantification of these impacts, it becomes clear that there is a need to address them to reduce the overall environmental footprint of low-cost housing.

Possible solutions to help reduce energy consumption and promote greater sustainability in the construction of such low-cost housing may include the substitution of materials with those used in traditional architecture that are climatically compatible, environmentally friendly, and energy efficient. For example, in regions with hot climates, materials such as clay and terra-cotta can be used in low-cost housing construction. These materials have excellent thermal properties, providing natural insulation that reduces the need for energy-intensive cooling systems. Additionally, they are locally available and do not require significant energy inputs for their production. In regions with cold climates, materials such as wood can be used, as they have good insulation properties and can help reduce heating energy requirements. By incorporating these materials into low-cost housing construction, energy consumption can be reduced and thus overall sustainability can be promoted.

Acknowledgments The authors gratefully acknowledge the All India Council for Technical Education (AICTE), India, for providing the communicating author with the ADF fellowship as financial support. The authors would also like to express our gratitude to the anonymous reviewers who provided insightful suggestions and advice.

References

1. L.A.M. Gazquez, F.F. Hernández, J.M.C. López, A comparison of traditional and contemporary social houses in Catarmarca (Argentina). Comfort conditions and life cycle assessment. Sustain. Cities Soc. **82**(April) (2022). https://doi.org/10.1016/j.scs.2022.103891
2. P. Mendonca, C. Vieira, Embodied carbon and economic cost analysis of a contemporary house design using local and reused materials. Sustain. Futures **4**(November) (2022). https://doi.org/10.1016/j.sftr.2022.100100
3. M. Asif, A.H.A. Dehwah, F. Ashraf, H.S. Khan, M.M. Shaukat, M.T. Hassan, Life cycle assessment of a three- bedroom house in Saudi Arabia. Environments **4**(November) (2017). https://doi.org/10.3390/environments4030052
4. V. Sampathkumar, S. Loganathan, S. Manoj, N. Jothi, Life cycle energy analysis of a traditional building in India (a case study). J. Eng. Res. Appl. **6**(May), 47 (2016) [Online]. Available: www.ijera.com
5. G.T.N. Veerendra, S. Dey, A.V.P. Manoj, B. Kumaravel, Life cycle assessment for a suburban building located within the vicinity using Revit architecture. J. Build. Pathol. Rehabil. **7**(1), 1–11 (2022). https://doi.org/10.1007/s41024-022-00199-6
6. A. Yadav, A. Jamwal, R. Agrawal, A. Yadav, J.K. Jain, *Life Cycle Assessment in Buildings: Indian Perspective*, [Online]. Available: https://ssrn.com/abstract=3761797
7. J. Fernandes, M. Peixoto, R. Mateus, H. Gervásio, Life cycle analysis of environmental impacts of earthen materials in the Portuguese context: Rammed earth and compressed earth blocks. J. Clean. Prod. **241** (2019). https://doi.org/10.1016/j.jclepro.2019.118286
8. M. Dabaieh, D. Maguid, D. El-Mahdy, Circularity in the new gravity – Re-thinking vernacular architecture and circularity. Sustain. For. **14** (2022). https://doi.org/10.3390/su14010328
9. D. Trabucco, A. Wood, LCA of tall buildings: Still a long way to go. J. Build. Eng. **7**(July) (2016). https://doi.org/10.1016/j.jobe.2016.07.009
10. A. Pakdel, H. Ayatollahi, S. Sattary, Embodied energy and CO2 emissions of life cycle assessment (LCA) in the traditional and contemporary Iranian construction systems. J. Build. Eng. **39**(June), 2021 (2020). https://doi.org/10.1016/j.jobe.2021.102310

An Eco-benign Seed Treatment Strategy for Rosehips Using Deep Eutectic Solvents

Debshikha Dutta Roy, Samuel Jacob, and S. P. Jeevan Kumar

Abbreviations

DES	Deep eutectic solvent
HBA	Hydrogen bond donor
HBD	Hydrogen bond acceptor
FE-SEM	Field emission scanning electron microscopy
XRD	X-ray diffractometer
FTIR	Fourier transform infrared spectroscopy
GC–MS	Gas chromatography–mass spectrometry
TTC	Triphenyl tetrazolium chloride
ChCl	Choline chloride

1 Introduction

The usage of conventional seed treatments for germination adversely affects our environment due to the presence of toxic chemicals. The conventional treatments for seed germination including thermal treatment (10–100 °C) and sulphuric acid treatment (5–30 min) were done on *Psoralea corylifolia* [1]. Sulphuric acid treatment of 30 min produced 30% seed germination, while heat treatment at 70 °C

D. D. Roy · S. Jacob (✉)
Department of Biotechnology, School of Bioengineering, College of Engineering and Technology, Faculty of Engineering and Technology, SRM Institute of Science & Technology, Kattankulathur, Tamil Nadu, India
e-mail: dd1315@srmist.edu.in; samueljb@srmist.edu.in

S. P. J. Kumar
ICAR-Directorate of Floricultural Research, Pune, Maharashtra, India

D. Mazumder (ed.), *Sustainable Advanced Technologies for Environmental Management*, Springer Proceedings in Earth and Environmental Sciences,
https://doi.org/10.1007/978-3-031-64006-3_15

produced 70% seed germination. However, higher concentrations of acid used for germination treatment is corrosive in nature and it damages the seed. Conventional acid treatments also lead to acid leaching of soils, which leads to draining off of nutrients present in soil and alteration in soil pH, leading to decrease in the level of soil fertility [2]. In order to prevent such concerns, there has been a need for an eco-friendly and sustainable seed treatment strategy. Seeds contain 40–60% lignin in their pericarp, which may lead to improved germination. Deep eutectic solvent (DES) is a recent technological advancement intended for the pretreatment of lignocellulosic materials. DES consists of a hydrogen bond acceptor (HBA) and hydrogen bond donor (HBD) with a melting point lesser than its constituents [3, 4]. Due to the hydrogen bond formation between HBD and HBA, charge delocalization occurs, which is accountable for lowering the melting point of deep eutectic solvent [5]. The combination of HBD and HBA causes an acid–base catalysis, which breaks the ether linkages present in the lignin phenyl propane units during the delignification process by DES [6]. A diluted form of choline chloride and urea, known as DES, was utilized to break down the lignocellulosic structure present in chestnut burrs [7]. Achieving a delignification rate of around 40%, this DES pretreatment effectively treated untreated chestnut burrs. Additionally, the efficacy of green solvents such as choline chloride–oxalic acid (CO) and choline chloride–urea (CU) in delignifying *Bambusa bambos* stem (BS) has been investigated, resulting in a delignification rate of 44.3% after treatment [8]. In another study, oil palm empty fruit bunches (OPEFB) were pretreated using ultrasound-assisted DESs [9]. Three types of DESs were examined: choline chloride–lactic acid (ChCl–LA), choline chloride–glycerol (ChCl–G) and choline chloride–urea (ChCl–U), with ChCl–LA demonstrating the lowest total lignin content at 18.8%, followed by ChCl–U (19.4%) and ChCl–G (21.2%).

Compared to conventional treatments, DES offers several advantages, including ease of synthesis, biodegradability, non-flammability and mild treatment conditions, making them environmentally friendly. This study specifically focuses on the use of DES (choline chloride:urea) for delignifying rosehip seeds, with the potential to enhance germination and serve as an alternative to traditional seed treatment methods.

2 Materials and Methods

2.1 Seed Acquisition and Preparation

Rosehip achenes of *Rosa hybrida* were procured from Indian Council of Agricultural Research (ICAR), Directorate of Floriculture, Pune, Maharashtra, India. A total of 600 seeds were collected, dried and kept in cold storage for 1 year before the seed was sent to ensure the lignification of seeds upon storage. After the seeds were procured, they were kept in zip lock cover and stored at −80 °C (Fig. 1).

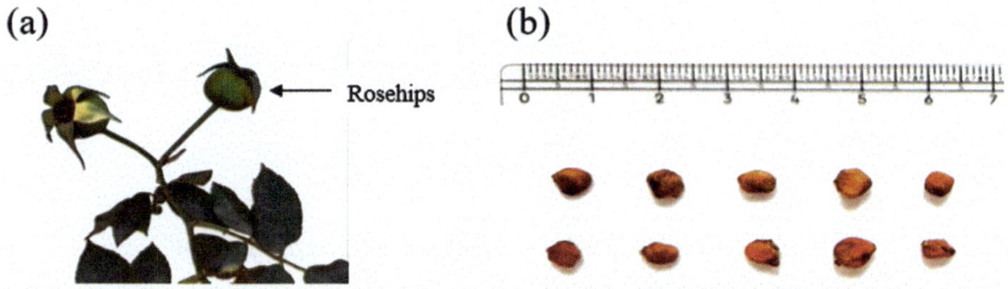

Fig. 1 (**a**) Rosehips of *Rosa hybrida* plant, (**b**) isolated seeds

2.2 DES (Reline) Synthesis and Seed Treatment

DES was prepared by combining urea as the hydrogen bond donor (HBD) and choline chloride (ChCl) as the hydrogen bond acceptor. The DES constituents were mixed in a 1:2 molar ratio of choline chloride to urea. This mixture was placed in a water bath at 60 °C and stirred at 200 rpm until a homogeneous transparent solution (Reln) was formed. Subsequently, the DES was cooled and stored at room temperature (28 ± 2 °C) for future use in seed treatment.

The DES-based seed treatment followed pre-optimized conditions as reported by Umai et al. [10]. Seeds were mixed with the DES solution (ChCl:urea) at a ratio of 1: 12.5 (g/g). The DES-based seed treatment was conducted at 87 °C for 4.63 h in a water bath.

2.3 Estimation of Seed Lignin Content

The iodometric technique was employed to assess the lignin content within the biomass [11]. Initially, 0.01 g of seed was introduced into 63 mL of distilled water. Subsequently, 7.5 mL of 0.1 N potassium permanganate and 7.5 mL of 4 N sulphuric acid were added, followed by an incubation period at room temperature in darkness for 10 min. Upon the addition of 1.5 mL of 1 N potassium iodide and 1 mL of 0.2% starch, the mixture underwent titration against a 0.1 N sodium thiosulphate solution until it transitioned to a colourless solution. The lignin content was determined utilizing the following equation (Eq. (1)), and the percentage of delignification was calculated using Eq. (2):

$$\%\,\text{Lignin content} = \left(\frac{\text{Volume of sodium thiosulphateconsumed}}{\text{Weight of biomass}}\right) \times 0.15 \quad (1)$$

$$\text{Delignification\%} = \left(\frac{\text{Lignin content of raw seed} - \text{Lignin content of treated seed}}{\text{Lignin content of raw seed}}\right) \times 100 \tag{2}$$

2.4 *Structural Characterization of the Pericarp of Raw and Treated rosehip Seeds*

2.4.1 Pericarp Thickness Analysis

The field emission scanning electron microscope (FE-SEM) was utilized to measure the pericarp thickness of both untreated and deep eutectic solvent (DES) treated rosehip seeds. The seeds were dried and affixed onto a specimen stub before being examined using the FE-SEM (Model-FEI, QUANTA 200, USA) at 2000 kV.

2.4.2 X-Ray Diffractometer (XRD)

The crystallinity of the rosehip seeds before and after DES-based seed treatment was analysed by X-ray diffractometer (Model-Malvern Panalytical, AERIS—High resolution bench top XRD, UK) using CuKα irradiation at 40 kV having a range from 10 to 75° (2*θ*) along with a scanning speed of 3°/min. To calculate the crystallinity index (CrI) of DES treated and raw rosehip seeds, an empirical method (Eq. (3)) was used [12]:

$$\text{Crystallinity Index (CrI)\%} = \left(\frac{I002 - I\text{am}}{I002}\right) \times 100 \tag{3}$$

where 'I_{002}' is the peak having highest intensity for cellulose at 2*θ* having range between 22° and 23° and 'I_{am}' is the peak having lowest intensity for cellulose at 2*θ* having range between 18° and 19°.

2.4.3 Seed Functional Group Analysis using FTIR

FTIR analysis was conducted to examine alterations in the functional groups of seed constituents potentially induced by DES-based seed treatment. Both untreated and treated seeds were ground using phosphate buffered saline (PBS), followed by centrifugation for 5 min at 4000 rpm. The resulting pellet was subjected to a temperature of 60 °C and subsequently mounted into thin sections with KBr. Spectra were recorded within the range of 4000–400 cm^{-1} at a spectral resolution of 0.5 cm^{-1} using FTIR (Model-Shimadzu, IR Tracer 100) [13].

2.5 *Lignin Degraded Compound Analysis*

GC–MS analysis was conducted to examine the presence of intermediate degradation products in the liquid hydrolysate following DES-based seed treatment [14]. Hexane was added to the liquid hydrolysate obtained after DES-based seed treatment at a ratio of 2:1 and allowed to undergo phase separation overnight. The aqueous layer was then extracted, and contaminants were removed using a 0.45 μm syringe filter. Subsequently, the filtered solution was subjected to GC–MS (Agilent Technologies, 7890B GC system combined with 5977A MSD system, USA) equipped with an HP 5MS 5% phenyl methyl siloxane column (30 m × 0.25 mm × 0.25 μm). The GC–MS system was operated with an initial and final temperature of 50 °C and 280 °C, respectively. Helium was employed as the carrier gas with a split ratio of 70:1, which maintains at a continuous flow rate of 1 mL/min.

2.6 *Histochemical Analysis*

2.6.1 Lignin Staining

The presence of lignin in raw and DES-treated seeds were visualized using phloroglucinol staining [15]. The cross sections of the seeds were subjected to 1 mL of phloroglucinol stain (0.3 g of phloroglucinol in 10 mL of absolute ethanol and 37 N concentrated hydrochloric acid). The cross sections of the seed were then mounted on a 1-mm microscopy glass slide and a 150-μm thick glass coverslip and observed under confocal microscope at 100× magnification [16].

2.6.2 Tetrazolium Chloride Testing for Seed Viability

The triphenyl tetrazolium chloride (TTC) assay is utilized to determine seed viability. Respiring tissues have the ability to convert colourless tetrazolium chloride to formazan, which has a carmine red coloration. Formazan was formed due to the hydrogen transfer reaction, which is catalysed by the mitochondrial dehydrogenase enzyme. Tetrazolium chloride permeates through both viable and non-viable cells, but only the viable cells catalyse the formation of formazan [14]. Rosehip seeds were immersed in water for a duration of 12–15 h. After by soaking, longitudinal dissection was performed on the seeds with a medical scalpel. Seeds having intact embryos were kept on PCR tubes of 0.5 mL, which was filled with 1%-TTC in demineralized water. After 24 h of incubation in the darkness, the seeds were analysed for viability. Seeds having carmine red coloration is considered as viable and seeds that did not change their colour is considered as non-viable (ISTA., 1996).

3 Results and Discussion

3.1 Delignification

After the lignin estimation, it can be seen that the lignin% of raw seed is 63, which is very high as compared to the lignin% of germinating seeds, which should be in the range of 4.4–6 as reported by Gudin et al. [17]. After DES treatment, it was observed that for DES (ChCl:urea) for open seed (seed dissected before DES treatment) and DES (ChCl:urea) treatment for 4.6 h, the lignin content of the rosehip seeds after treatment is 6.23%, which is close to the range of lignin content of germinating seeds (as shown in Table 1).

3.2 SEM Analysis and Fold Reduction in Thickness of Pericarp

SEM analysis is performed to find the alteration in the pericarp thickness before and after treatment. The thickness of pericarp of Rose achenes (*Rosa hybrida*) was measured to be 1.4 ± 0.4 mm [18]. From the SEM analysis (Fig. 2), it is observed that the thickness of the pericarp of rosehip seed before DES treatment is 1.22 ± 1.06 mm, which correlates to the 63% lignin content in the pericarp of the seed. After treatment with DES (ChCl:urea) for 4.6 h for whole seed and open seed, we observe 52%- and 56%-fold reduction in thickness of pericarp, respectively (Table 2).

3.3 X-Ray Diffraction Analysis

It is well established that seed structure constitutes of cellulose (crystalline) embedded within hemicellulose and lignin (amorphous) layer. From the XRD spectra, the crystallinity index (CrI%) of raw seed is found to be 7.44. The CrI% of the three types of DES treatments (ChCl:urea for 4.6 h, ChCl:urea for 4.6 h (open seed) and ChCl:urea for 30 min) is 27.64, 26.09 and 21.52, respectively (Fig. 3). The

Table 1 Lignin% and delignification% of raw and treated rosehip seeds

Treatment	Lignin%	Delignification%
Raw seed	63.00 ± 0.10	–
DES (ChCl:urea) for 4.6 h (open seed)	6.20 ± 0.10	90
DES (ChCl:urea) for 30 min	40.30 ± 0.14	35
DES (ChCl:urea) for 4.6 h	9.30 ± 0.09	85
Acid treatment for 30 min	15.00 ± 0.11	75

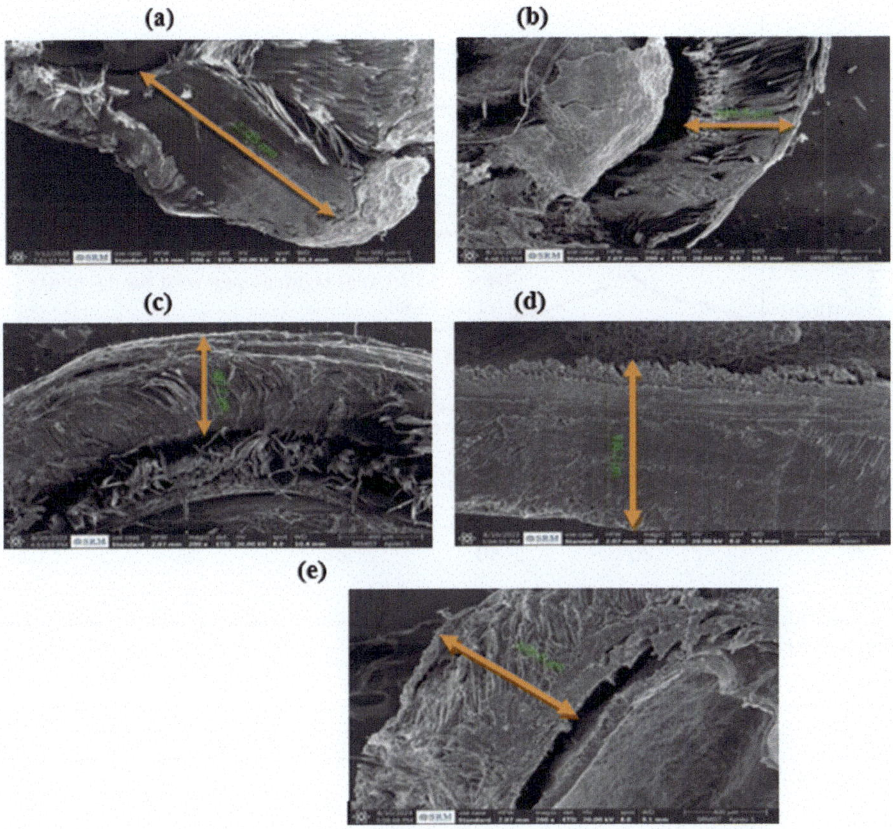

Fig. 2 SEM images of (**a**) raw seed, (**b**) DES (ChCl:urea)-treated seeds for 4.6 h, (**c**) DES (ChCl: urea)-treated seeds for 4.6 h (open seed), (**d**) DES (ChCl:urea)-treated seeds for 30 min, (**e**) acid-treated seed

Table 2 Thickness of pericarp and fold reduction% in thickness of raw and DES-treated seeds

Treatment	Thickness of pericarp (μm)	Fold reduction (%)
Raw seed	1220.00 ± 1.06	–
DES (ChCl:urea) for 30 min	913.95 ± 31.46	25
DES (ChCl:urea) for 4.6 h	585.25 ± 62.57	52
DES (ChCl:urea) for 4.6 h (open seed)	536.90 ± 24.74	56
Acid treatment for 30 min	403.75 ± 118.72	66

increase in crystallinity exhibits the efficient removal of the amorphous regions of the seed contributed by lignin, leading to the exposure of crystalline cellulose [19]. The crystallinity index of DES-treated seed for 4.6 h is almost similar to acid-treated seed (CrI%: 30.47), which denotes that DES treatment provides similar effects as that of acid treatment for seeds, without any corrosive effect.

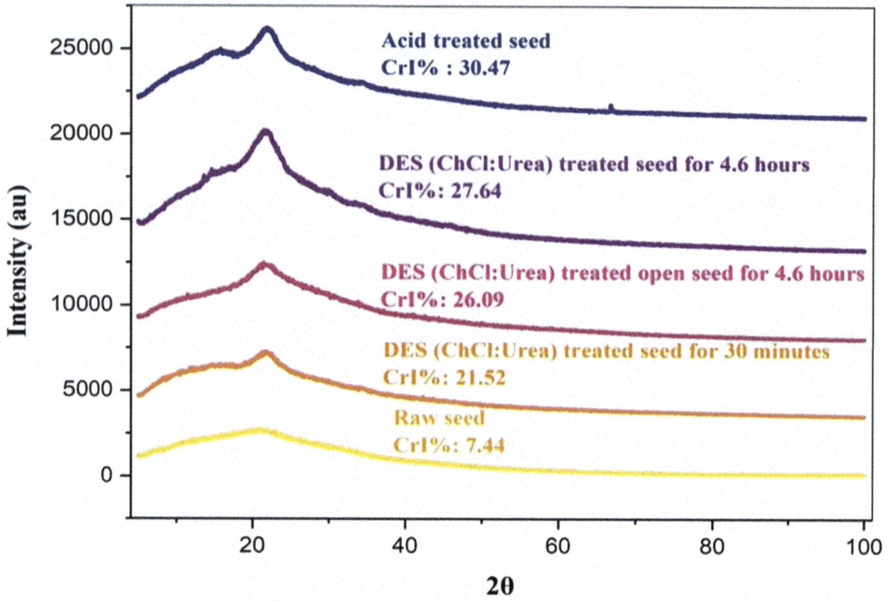

Fig. 3 XRD spectra of raw and treated seeds

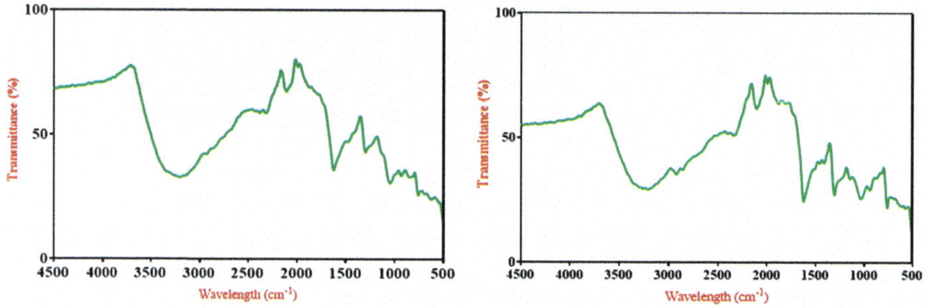

Fig. 4 Comparative analysis of functional groups of (**a**) Raw seed and (**b**) DES-treated seed using FTIR spectra

3.4 *FTIR Analysis*

The structural changes in untreated and treated seeds were analysed using FTIR spectra (Fig. 4), focusing on the absorption band positions indicative of various functional groups (Table 3). For instance, the presence of a peak at 1622.13 cm^{-1} corresponds to the C=C stretching of the lignin aromatic ring in the untreated seed [20]. The absence of this functional group in the DES-treated seed suggests the removal of lignin. Additionally, peaks at 1418 cm^{-1} and 1456.25 cm^{-1} indicate the O–CH_3 vibration of the aryl group of lignin [21], signifying the presence of lignin in the untreated seed. The exposure of cellulose is also evidenced by peaks around

Table 3 FTIR functional group for raw and treated *rosehip* seeds

Band position (cm^{-1})		
Raw seed	DES (ChCl:urea)	Functional group
1418	1456.25	Methyl group
1622.13		Phenyl group
	2854.65	CH_2 of aliphatic hydrocarbon
	2918.30	CH of aliphatic hydrocarbon

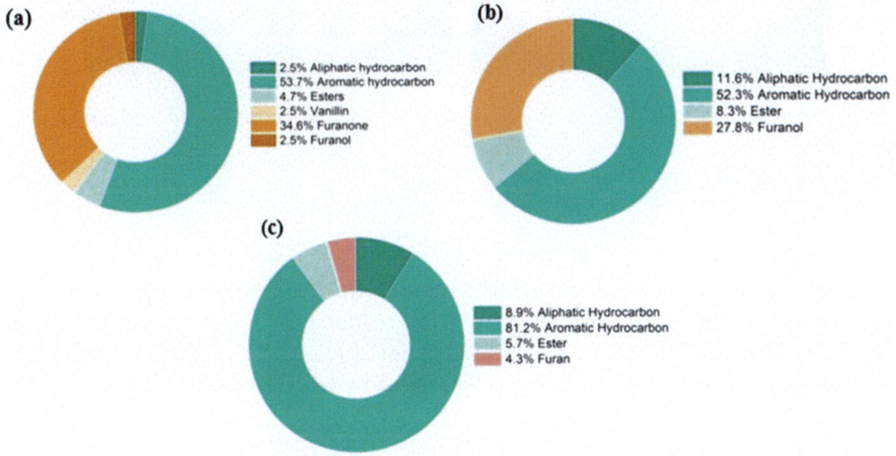

Fig. 5 GC–MS analysis of (**a**) DES (ChCl:urea) for 4.6 h, (**b**) DES (ChCl:urea) for 4.65 h (open seed), (**c**) DES ChCl:urea) for 30 min

2920 cm^{-1} to 2850 cm^{-1}, representing the CH and CH_2 groups of aliphatic hydrocarbons found in cellulose [22, 23]. These variations in functional groups demonstrate delignification following DES treatment.

3.5 Analysis of Lignin Degraded Compounds

DES breaks down β-O-4-aryl ether and ester linkages within the lignin carbohydrate complexes, initiating the formation of hydrogen bonds between lignin monomers and chloride anions. This process yields lignin degradation products such as aliphatic hydrocarbons, aromatic hydrocarbons and esters, which are also identified in the GC–MS analysis of the hydrolysate obtained post-DES treatment (ChCl:urea) [24, 25]. The presence of aromatic hydrocarbons arises from the cleavage of hydroxyl and methoxyl functional groups within the lignin present in the pericarp of rosehip seeds. Furthermore, the production of vanillin following DES (ChCl:urea) treatment for 4.6 h serves as an indicator of delignification. Esters are detected as a result of lignin oxidation (Fig. 5).

(a) (b)

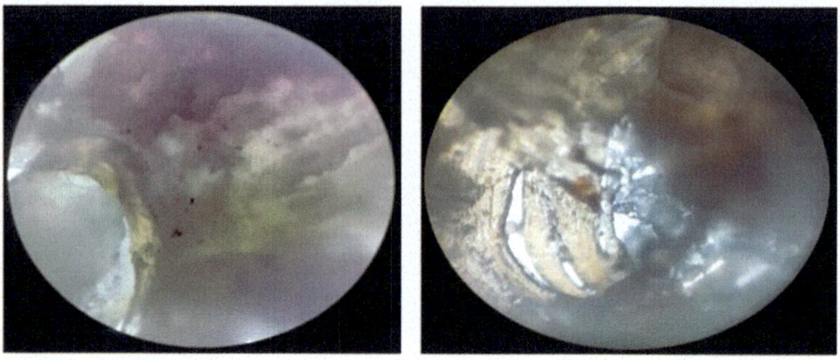

Fig. 6 Phloroglucinol staining of (**a**) raw seed and (**b**) DES (ChCl:urea)-treated seed for 4.6 h

3.6 Phloroglucinol Staining

Phloroglucinol staining is utilized to ascertain the presence of lignin in both untreated and treated seeds. In an acidic hydrochloric medium, phloroglucinol reacts with coniferyl aldehyde and sinapyl aldehyde found in lignin, resulting in the formation of a pink pigment due to the generation of hydroxycinnamoyl aldehyde or a reddish-brown hue due to the production of hydroxybenzaldehyde [26]. Upon phloroglucinol staining, the presence of lignin is evident in the untreated rosehip seed as a pink coloration, which diminishes following DES (ChCl:urea) treatment, indicating the occurrence of delignification. This reduction in coloration aligns with findings from XRD, FTIR and GC–MS analyses, further corroborating the delignification process post-treatment (Fig. 6).

3.7 Tetrazolium Chloride Staining

From the tetrazolium chloride assay, we have observed the formation of carmine red colour after the DES treatment (Fig. 7). This has been caused due to the reduction reaction of tetrazolium chloride in the tissues determining that the respiratory action is occurring in the mitochondria present in the alive seed tissue cells. Therefore, the formation carmine red colour in the seed tissues is a positive indicator of the viability of DES-treated rosehip seeds, which is determined indirectly by detection of the respiratory activity at the cellular level [28].

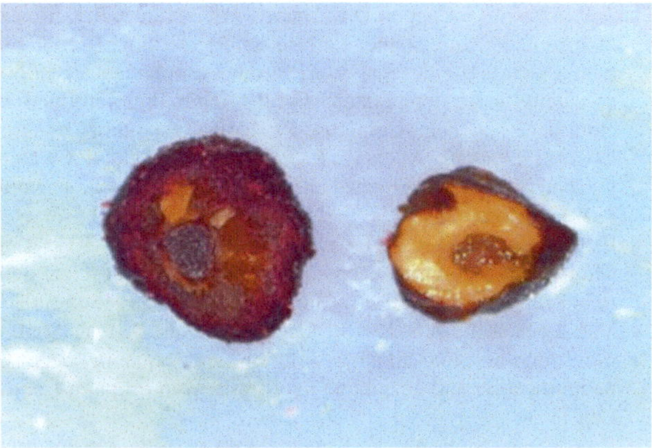

Fig. 7 Tetrazolium chloride testing for DES (ChCl:urea)-treated rosehip seed

4 Conclusion

Based on the experiment, we contrasted the delignification outcomes of rosehip seeds through conventional seed treatment and DES-based seed treatment. Our analysis revealed substantial delignification with both approaches. However, considering the corrosive nature of acid treatment, we concluded that DES-based treatment offers a promising, environmentally friendly alternative to conventional methods, ultimately promoting germination.

Acknowledgement The authors express their gratitude to the Department of Biotechnology, SRM Institute of Science and Technology, Kattankulathur, India for providing financial assistance and infrastructure support to conduct this research.

References

1. G. Siva, S. Sivakumar, G. Premkumar, P. Baskaran, T. Senthilkumar, N. Jayabalan, Enhanced seed germination of *Psoralea corylifolia* L. by heat treatment. World J. Agric. Res. **2**(4), 151–154 (2014)
2. J. Šoch, J. Šonka, J. Ponert, Acid scarification as a potent treatment for an in vitro germination of mature endozoochorous Vanilla planifolia seeds. Bot. Stud. **64**(1), 9 (2023)
3. C.Y. Chu, T.H. Ko, Evaluation of acid leaching on the removal of heavy metals and soil fertility in contaminated soil. J. Chem. (2018)
4. Y.H. Choi, J. van Spronsen, Y. Dai, M. Verberne, F. Hollmann, I.W. Arends, G.J. Witkamp, R. Verpoorte, Are natural deep eutectic solvents the missing link in understanding cellular metabolism and physiology? Plant Physiol. **156**(4), 1701–1705 (2011)
5. Y. Dai, J. van Spronsen, G.J. Witkamp, R. Verpoorte, Y.H. Choi, Natural deep eutectic solvents as new potential media for green technology. Anal. Chim. Acta **766**, 61–68 (2013)

6. A.P. Abbott, G. Capper, D.L. Davies, H.L. Munro, R.K. Rasheed, V. Tambyrajah, Preparation of novel, moisture-stable, Lewis-acidic ionic liquids containing quaternary ammonium salts with functional side chains. Electronic supplementary information (ESI) available: Plot of conductivity vs. temperature for the ionic liquid formed from zinc chloride and choline chloride(2:1). Chem. Commun. **19**, 2010–2011 (2001)
7. I. Costa-Trigo, P. Alicia, O.P. Patricia, O. David, P.G. Nelson, M.D. José, Enhancing the saccharification of pretreated chestnut burrs to produce bacteriocins. J. Biotechnol. **329**, 13–20 (2021)
8. R. Ramesh, A. Nair, A. Jayavel, K. Sathiasivan, M. Rajesh, S. Ramaswamy, K. Tamilarasan, Choline chloride-based deep eutectic solvents for efficient delignification of Bambusa bambos in bio-refinery applications. Chem. Pap. **74**, 4533–4545 (2020)
9. V.Z. Ong, T.Y. Wu, K.K.L. Chu, W.Y. Sun, K.P.Y. Shak, A combined pretreatment with ultrasound-assisted alkaline solution and aqueous deep eutectic solvent for enhancing delignification and enzymatic hydrolysis from oil palm fronds. Ind. Crops Prod. **160**, 112974 (2021)
10. R.D. Umai, S. Jacob, V. Kumar, Deep eutectic solvent pretreatment of water hyacinth for improved holocellulosic saccharification and fermentative co-production of xylitol and lipids using Rhodosporidium toruloides NCIM 3547. Fermentation **8**(11), 591 (2022)
11. M.A. Hussain, M.E. Huq, S.M. Rahman, Z. Ahmed, Estimation of lignin in jute by titration method. Pak. J. Biol. Sci. **5**(1), 521–532 (2002)
12. L.G.J.M.A. Segal, J.J. Creely, A.E. Martin Jr., C.M. Conrad, An empirical method for estimating the degree of crystallinity of native cellulose using the X-ray diffractometer. Text. Res. J. **29**(10), 786–794 (1959)
13. P. Binod, K. Satyanagalakshmi, R. Sindhu, K.U. Janu, R.K. Sukumaran, A. Pandey, Short duration microwave assisted pretreatment enhances the enzymatic saccharification and fermentable sugar yield from sugarcane bagasse. Renew. Energy **37**(1), 109–116 (2012)
14. J. Abati, C. Zucareli, C.R. Brzezinski, I.D.O.N. Lopes, F.C. Krzyzanowski, L.A.C. Moraes, F.A. Henning, Water absorption and storage tolerance of soybean seeds with contrasting seed coat characteristics. Acta Sci. Agron. **44** (2022)
15. L. Blaschek, A. Champagne, C. Dimotakis, R. Nuoendagula, S. Decou, S. Hishiyama, S. Kratzer, S. Kajita, E. Pesquet, Cellular and genetic regulation of conifer aldehyde incorporation in lignin of herbaceous and woody plants by quantitative Wiesner staining. Front. Plant Sci. **11**, 109 (2020)
16. S.H. Moussa, A.A. Tayel, A.A. Al-Hassan, A. Farouk, Tetrazolium/formazan test as an efficient method to determine fungal chitosan antimicrobial activity. J. Mycol., 1–7 (2013)
17. S. Gudin, L. Arene, A. Chavagnat, C. Bulard, Influence of endocarp thickness on rose achene germination: Genetic and environmental factors. HortScience **25**(7), 786–788 (1990)
18. A. Adewuyi, F.V. Pereira, Surface modification of cellulose isolated from Sesamun indicum underutilized seed: A means of enhancing cellulose hydrophobicity. J. Sci. Adv. Mater. Devices **2**(3), 326–332 (2017)
19. L.M. Kline, D.G. Hayes, A.R. Womac, N. Labbé, Simplified determination of lignin content in hard and soft woods via UV-spectrophotometric analysis of biomass dissolved in ionic liquids. Bioresources **5**(3) (2010)
20. H. Yang, X.D. Shen, Dynamic TGA–FTIR studies on the thermal stability of lithium/graphite with electrolyte in lithium-ion cell. J. Power Sources **167**(2), 515–519 (2007)
21. J. Trifol, C. Sillard, D. Plackett, P. Szabo, J. Bras, A.E. Daugaard, Chemically extracted nanocellulose from sisal fibres by a simple and industrially relevant process. Cellulose **24**, 107–118 (2017)
22. T. Kondo, C. Sawatari, A Fourier transform infra-red spectroscopic analysis of the character of hydrogen bonds in amorphous cellulose. Polymer **37**(3), 393–399 (1996)

23. M.A. Hossain, M.S. Rahaman, D. Yelle, H. Shang, Z. Sun, S. Renneckar, J. Dong, S. Tulaphol, N. Sathitsuksanoh, Effects of polyol-based deep eutectic solvents on the efficiency of rice straw enzymatic hydrolysis. Ind. Crop. Prod. **167**, 113480 (2021)
24. C.W. Zhang, S.Q. Xia, P.S. Ma, Facile pretreatment of lignocellulosic biomass using deep eutectic solvents. Bioresour. Technol. **219**, 1–5 (2016)
25. F. Pomar, F. Merino, A.R. Barceló, O-4-Linked coniferyl and sinapyl aldehydes in lignifying cell walls are the main targets of the Wiesner (phloroglucinol-HCl) reaction. Protoplasma **220**, 0017–0028 (2002)
26. J.D.B. França-Neto, F.C. Krzyzanowski, Tetrazolium: An important test for physiological seed quality evaluation. J. Seed Sci. **41**, 359–366 (2019)

A Sustainable Approach for Monitoring of Ambient Air Quality Considering Lichen as a Bioindicator: A Case Study for Two Major Traffic Intersections of Kolkata

Saptaparni Ghosh Majumdar, Bipasha Dinda Chakraborty, Anupam Debsarkar, Mrinal Kanti Biswas, and Rita Saha

Abbreviations

AQI	Air Quality Index
ATP	Adenosine triphosphate
CPCB	Central Pollution Control Board
IAP	Index of Atmospheric Purity
NAAQS	National Ambient Air Quality Standards
N_2	Nitrogen
NO_x	Nitrogen oxides
PM10	Particulate matter 10 micrometres or less in diameter
PM2.5	Particulate matter 2.5 micrometres or less in diameter
QGIS	Quantum Geographic Information System
SO_x	Sulphur oxides
SO_2	Sulphur dioxide

S. G. Majumdar
School of Environmental Studies, Jadavpur University, Kolkata, India

B. D. Chakraborty (✉) · M. K. Biswas · R. Saha
Central Pollution Control Board, Kolkata, Kolkata, India

A. Debsarkar
Environmental Engineering Division, Department of Civil Engineering, Jadavpur University, Kolkata, India

D. Mazumder (ed.), *Sustainable Advanced Technologies for Environmental Management*, Springer Proceedings in Earth and Environmental Sciences,
https://doi.org/10.1007/978-3-031-64006-3_16

1 Introduction

Kolkata, a rapidly evolving metropolis in the Indian subcontinent, has been exposed to significant air pollution levels over the decades. With an increase in overall population, certain major contributors to air pollution in the city have constantly enhanced the overall pollution index. This involves vehicular pollution in major traffic intersections. The enhanced air pollution levels in Kolkata are primarily attributed to escalated vehicular population, resulting due to an average decadal growth rate for two wheelers by 70%, four wheelers by 20% (specifically taxis and cars), nearly 14% for three wheelers and about 10–13% for buses [1]. Based on West Bengal Pollution Control Board Report, the primary reason behind enhanced vehicular emissions is due to inflated average age of vehicles, surface condition of roads, which involves high population density utilizing similar road space, diversified vehicular mode including slow-moving vehicles that are accountable for congested traffic and lesser surface area of roads in the city. According to Das et al. [2], concerns regarding the rapidly deteriorating air quality in these Indian cities, notably due to air pollution from automobiles given the overall tonnage of contaminants in metropolitan areas, have grown in recent years. Road traffic-related air pollution poses a serious threat to both the environment and public health in today's society. In Kolkata, the yearly average pollution level, particularly with reference to $PM_{2.5}$ and PM_{10} and NO_x, did not meet the national limit [3]. In recent years, yearly average NO_x and PM_{10} levels had routinely exceeded the National Ambient Air Quality Standards (NAAQS), while wintertime $PM_{2.5}$ levels have been shown to be far higher than the requirement [4]. Along with those from numerous small-scale enterprises, Kolkata is subjected to the daily emissions of 1.14 million vehicles [5]. The regular occurrence of temperature stratification and the relatively low wind speed during the winter season trap pollutants close to the ground, making the issue of air pollution more severe at this time of year [6].

Monitoring biodiversity changes offers a practical way to assess the condition (quality) of an ecosystem. Loppi [7] states that lichen species that are found in a given area exhibit measurable responses to environmental changes, and lichen biodiversity counts can be used as accurate indicators of the state of the environment, with high values signifying unpolluted or low-polluted conditions and low values signifying polluted ones. Lichens are categorized by their body features or growth forms. In the study area, mostly Crustose and Foliose variety of lichen were observed. Lichen species involves mutualistic symbiosis of an algae or cyanobacteria and fungus. Lichens are extremely sensitive organisms and susceptible to environmental transformations that illustrate them as the most sensitive component in the ecosystem [8]. These organisms are treated as ecological indicators and their sensitivity has been utilized for mapping air pollution levels and also for estimating deposition and decay in SO_2. Organisms such as lichens lack defensive tissues, so they can easily absorb water, nutrients and gases straight from the environment. Lichen species are long living and some of them may even survive for decades or longer. Thus, the cumulative impact of pollution on the lichen community is prominent and can be studied easily. Transformation in physiological parameters and diversity of lichens are utilized as indicators of environmental concerns [9]. These organisms possess the ability to flourish in various forms of substrates and are outlined from different climatic conditions [10].

Green et al. [11] state that these species are poikilohydric and are unable to control their water content, which varies to maintain equilibrium with adjacent environments. The water dependence externally controls physiological activity of lichens as damp environmental conditions activate and hydrate them in arid environmental conditions and it results in inactiveness [11]. Aptroot and Van Herk [12] state that lichens are considered to be slow-growing organisms and their dependence on atmosphere permits them to acknowledge climate change for 5 years in terms of sensitivity to climate change. Morphological performance and lichen morphology involve growth and this has been related to atmospheric sulphur dioxide and the prevalence of acidic precipitation [13]. Growth of lichen virtually ceases in winter season and there might be a substantial difference in growth rate of lichen species in winter and summer in temperate latitudes. Moreover, when lichens are exposed to pollutants, it leads to stress production of ethylene, which generates electrolyte leakage as an outcome of cell membrane disruption, further increasing chlorophyll degradation and diminishing ATP content [14]. Thus, it leads to degradation in photosynthesis, respiration and N_2 fixation. Majumder et al. [9] state that these responses are extremely essential in terms of stress indication and further assist in spotting early indications of changing environmental circumstances.

The present study emphasized on assessment of air pollution level in the study area, which is located in proximity two major traffic intersections using lichen as bioindicator. The identified trees with lichen growth present in the impact area studied for a year for their growth rate and compositional diversity to draw a comprehensive inference about the status of air pollution prevailing in the study area. Assessing air pollution using bioindicator such as lichen can provide a sustainable option for monitoring of air pollution especially for remote areas. Lichen diversity distribution is utilized as it is the cheapest form of air pollution evaluation mechanism. Surveying and mapping lichen diversity is inexpensive and does not require high-end costs such as chemical analysis, permitting cost reduction and complicated sample gridding [15]. The study takes into consideration prior research in terms of lichen adaptability in areas that are exposed to high pollution levels. The study area chosen for this work is also exposed to rising pollution levels due to close proximity to major traffic intersections. The study focussed on assessing potential of thorough evaluation of overall lichen growth in an urban setup with exposure to dominant pollution sources such as vehicular emission to predict air quality trend prevailing at the site, as well as cost-effective tool for air quality monitoring and management.

2 Materials and Methods

The study was designed to compare biomonitoring data of lichen species available in the study area with the air quality data acquired from CPCB's (Central Pollution Control Board) continuous air quality monitoring station, Biomonitoring of lichen in the study area was carried out by grid mapping to assay the coverage and diversity of lichen species. Thereafter, selected lichen species growth continually studied for a period of 1 year and compared with seasonal air quality trends. It was also evaluated whether lichen biomonitoring can be chosen as effective air quality monitoring tool. Index of Atmospheric Purity was also calculated to compare existing air quality scenario.

2.1 Study Area

Lichen samples were gathered from Jadavpur area, Kolkata, India, adjoining two major traffic intersections (22°20′N, 88°22′E) in Eastern segment of India from June 2022 to April 2023. Lichen species were predominantly collected from barks of tree. The investigated site, considering major traffic intersections nearby, was chosen at Jadavpur (22.4955° N, 88.3709° E), Kolkata, India, as shown in Fig. 1 to detect impact of vehicular pollution. In order to confirm continuous vehicular movement, vehicle movement data were acquired from two nearby traffic intersections.

2.2 Meteorological Parameters

During the study period, the ambient temperature of this region ranged between 30 and 35 °C in summer; and between 21 and 27 °C in winter. Average wind speed was observed to be 3.5–5.8 m/s. Wind roses for the study period were constructed to confirm that trees under consideration at the receiving end of the vehicular emission throughout the study period. Windrose analysis was conducted to track correlation among prevalent wind directions and lichen growth. The wind patterns were analysed in terms of seasonal variation, and the wind pattern analysis was comparatively similar every year based on overall wind speed with few variations in wind direction patterns. The predominant wind direction patterns observed during summer months are South, South-West and South-East, and during monsoons, predominant wind direction is observed to be South and South-East. In winters, predominant wind direction is observed to be West, South-West, South-East and East as shown in Fig. 2. Moreover, temperature variations in various seasons also significantly affect wind patterns [16].

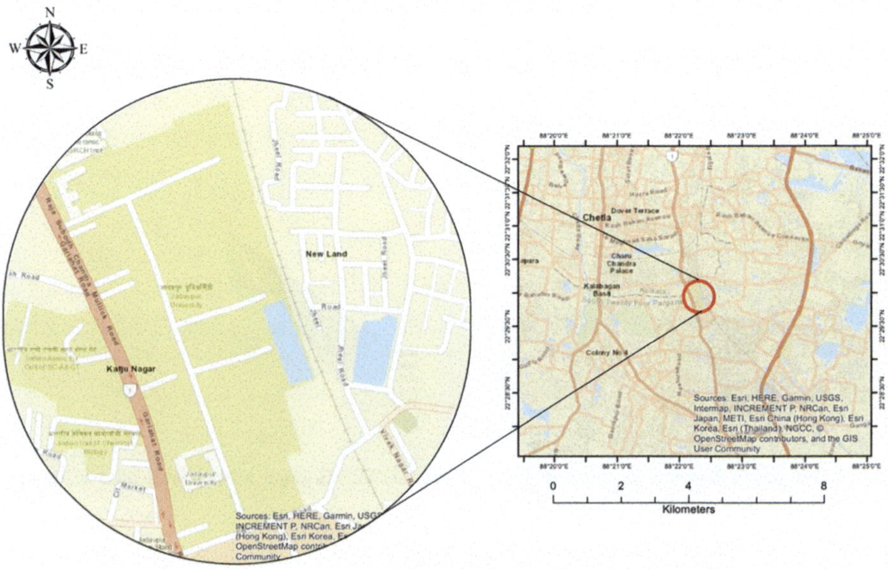

(a)

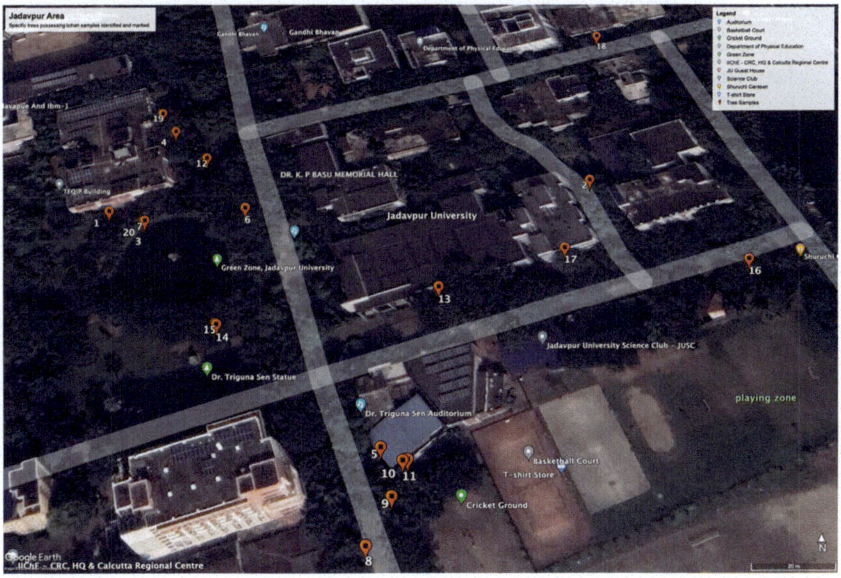

(b)

Fig. 1 Study area with location of trees in which lichen samples were found

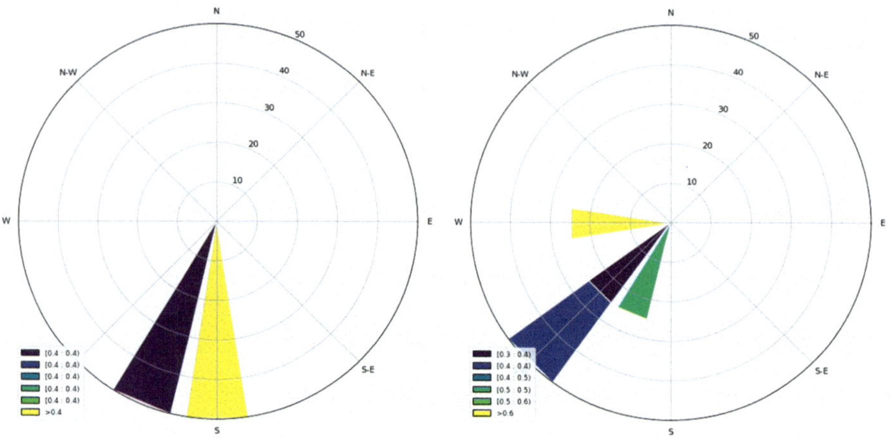

Fig. 2 Windrose diagram for summer '23 and Winter '22. (Source: Model fabricated in Jupyter Notebook using library Windrose Axes)

Table 1 Traffic volume analysed at study sites of Kolkata

Sl. no.	Study site	Traffic volume (vehicles/h)[a]	Number of heavy traffic (bus or trucks/h)[a]
1.	8B Bus Stand traffic intersections	1850	235
2.	Jadavpur Police Station traffic intersections	2165	320

[a]The traffic data provided is subject to approximations

2.3 Collection of Traffic Data to Confirm Impacts of Vehicular Emission

The data was collected by installing cameras at two major traffic intersections of Jadavpur area of Kolkata city. A duration of 1 h was taken into consideration from 11 a.m. to 12 p.m., which was found to have a higher flow of traffic. This data illustrated in Table 1 was acquired for a span of 1 week and the traffic volume was observed to be similar with a difference of 50–60 vehicles depending on the day of week.

2.4 Biomonitoring of Lichen Species

Biomonitoring of lichen species were conducted as they are both sensitive and perennial. Transformation in environmental parameters was effectively analysed as lichens are subjected to elevated levels of pollution. Lichens exist primarily in four forms, which include Fruticose lichens, Foliose lichens, Crustose lichens and

Squamulose lichens. Based on air quality and sulphur dioxide (SO_2) concentration, lichens are classified as Extremely clean air (sensitive species such as Usnea, Lobaria and Ramalina), Clean air (Fruticose and Foliose lichen), Moderate air (Potential and Squamulose lichen), Polluted air (Crustose lichen such as Lecanora) and Heavily polluted air (no lichen, only green algae). The mapping of lichen biodiversity is a common practice in many nations since it provides a gauge of the air quality based on the biological effects of air pollution. For biomonitoring, individual lichens were tracked on a gap of 15 days for 9 months. Tracing of individual lichen species from various trees found in Jadavpur University area was done to analyse the overall growth of lichen in various seasons concerning prevalent air quality. The tracings were further placed in millimetre graphs to assess the surface area in terms of mm^2. Furthermore, each lichen tracing was scanned in 1:1 scale and digitized using QGIS software through geo referencing and converting raster data to vector data. Field method considered for monitoring and recording data of epiphytic lichen species in terms of its stress factors for various trees found in Jadavpur area, predominantly Royal Palm species (*Roystonea regia*). Five substantial trees exposed to more or less comparable environmental circumstances were chosen for each location. These trees' trunks had lichens on them up to a height of 2 m, which were gathered and recorded. The percentage of trees at a given site on which a species of lichen was physically present was used to compute frequency of occurrence (Fr), and the surface area occupied by the species at that location was used to calculate coverage (Co). A translucent sheet was used to trace the outline of each individual lichen patch, and graph paper was used to calculate its surface area. The total surface area of all the individual patches present at that site made up the coverage of any species [17]. By multiplying frequency (Fr) and coverage (Co), the frequency coverage (f) of each species was calculated and expressed as a number on a scale of 1–5 [18].

2.5 Grid Mapping Technique for Assessing Lichen Coverage

Sampling technique implemented in this segment was based on the methods illustrated in a study by Muhammad et al. [19] in which tree specimens were assessed through 'Index of Atmospheric Purity' (IAP) methodology. It is based on frequency, tolerance and number of lichen available in the study area. In this segment, Royal Palm (*Roystonea regia*) species were selected as host trees that possess a circumference higher than 60 cm and are located at a distance of 5 m from vehicular traffic movement. For our study, nearly 20 tree specimens were taken into consideration from Jadavpur University and a grid subdivided into 10 rectangles with size of 30 × 50 cm. Each rectangle comprised lichen species, which were significantly identified and measured as depicted in Fig. 3. Quantity of lichen available in each rectangular segment was analysed. This grid framework was placed over tree bark and was measured in terms of its height from the bottom of the tree. Amount of vehicular movement in adjacent areas were studied based on vehicles in number passing during peak hours. The primary pollutants in air were evaluated through data

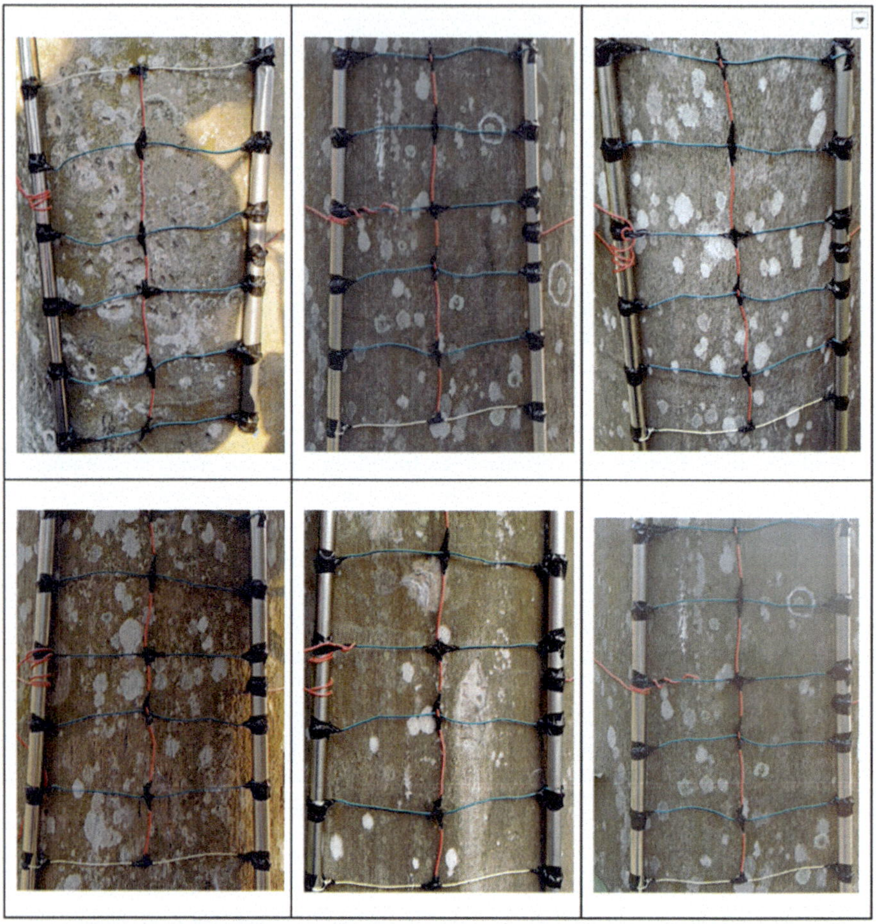

Fig. 3 Grid mapping of tree specimens to identify the concentration of lichen growth in tree bark

acquired from continuous monitoring stations and the overall exposure level was assessed. Moreover, depth of bark crevice (mm) and diameter of tree at breast height (cm) were measured. Based on study of Mežaka et al. [20], the depth of bark crevice was evaluated at a height of 1.20 m and significant tree bark samples were collected for pH measurements.

2.6 Index of Atmospheric Purity

The sampling was done for an area of 2 km, mostly within Jadavpur University campus area, and nearly 20 trees were approached from this site. This site was close to a continuous monitoring station by CPCB to record the air quality in terms of significant parameters. In various compass directions, the abundances of lichen species were calculated and based on these results, lichen indices were calculated in terms of lichen abundance and ecological sensitivity of individual species [21].

$$\mathrm{IAP} = \sum_{1}^{n} (Q \times f)/10$$

where n is the number of species, Q is the factor of accompanying species, f is the cover and frequency of each species and IAP is the Index of Atmospheric Purity.

The Index of Atmospheric Purity enables monitoring of significant air pollutants such as NO_x and SO_x, as well as concentration of heavy metals such as zinc, nickel, copper, manganese, cadmium, lead and chromium. It further assists in zoning and mapping regions adjacent to pollution genesis with a gradient. This method is primarily utilized for bryophytes and lichens that further enables an examination of pollution gradients across a line or belt transect. The abundance and frequency of species can be taken into consideration through the quantity of transects diffusing in various directions from the transect line and with enhanced distance from the pollution source [21]. Thus, it illustrates a significant picture of the long-term effect of pollution levels at a specified site.

3 Results and Discussion

3.1 Selection of Air Quality Features

In this study, the CPCB dataset includes significant parameters such as $PM_{2.5}$, PM_{10}, NO_x, SO_x, ozone, relative humidity, wind direction and benzoyl pyrene. Central and state agencies utilize this parameter to make individuals aware of quality of air. Selection of features is a significant preprocessing tread and it assists in zoning relevant variables in a dataset, further eliminating collinear variables [22]. Thus, effectively understanding correlation among different features Pearson's correlation coefficient is utilized to evaluate relationship between each feature in terms of target variable (Air Quality Index). This further illustrates correlation based on collinearity among target value and other features. The range of Pearson's correlation coefficient varies between -1 and $+1$. However, it is observed that if two variables don not share any correlation, then this coefficient is zero. On the other hand, if Pearson's correlation coefficient range is -1 to $+1$, it illustrates that association among two variables is stronger [23].

The heatmap in Fig. 4 represents correlation among various features, illustrated through specific colour patterns. The flamboyant colours represent direct association among every feature. It is observed that $PM_{2.5}$ and PM_{10} have the highest correlation with AQI, which involves a value of 0.98 and 0.97 although it depicts a minute correlation with relative humidity. Further, relative humidity also possesses an

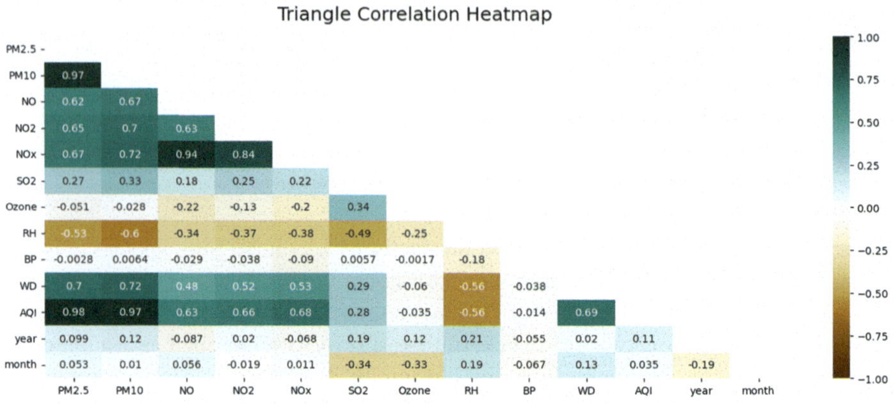

Fig. 4 Correlation among various features in the form of a heatmap

extremely minute correlation with wind direction and AQI. On the other hand, other features depict moderate association with each other as portrayed in the heatmap above.

3.2 *Variation in Pollutant Concentration*

For illustrating scenario of data utilized, the four principal air pollutant concentrations in Jadavpur area were plotted in the form of a line graph, and acquired results are illustrated in the table above. Moreover, these pollutants depicted apparent periodicity, i.e. $PM_{2.5}$, PM_{10}, NO_x and SO_x portrayed a seasonal trend of decrease in pollutant concentration levels during summer and monsoon months although a significant increase in pollutant concentration is observed during winters. $PM_{2.5}$, PM_{10}, NO_x and SO_x concentration attained their maximum values during December and minimum values were observed during September every year, illustrating a 'U' structure. Thus, during summer, due to powerful solar radiation, surface temperature increases distinctly and heats the surface air. This generates enhanced precipitation and transmission that hasten deposition and diffusion of atmospheric pollutants [24]. On the other hand, during winters, surface temperature is low, resulting in surface inversion, and meteorological conditions are unfavourable for vertical transmission. Thus, air pollution is significantly greater near the surface, as all four pollutants possess similar trends and influencing factors, while remaining pollutants such as ozone and benzoyl pyrene are required to be consolidated for predicting the overall Air Quality Index.

Figure 5 illustrates that the overall $PM_{2.5}$, PM_{10} and NO_x levels tend to decrease during April and further remain within standard limit till September. However, a significant increase in levels of $PM_{2.5}$, PM_{10} and NO_x is observed during September and it exceeds the standard concentration level during October, November,

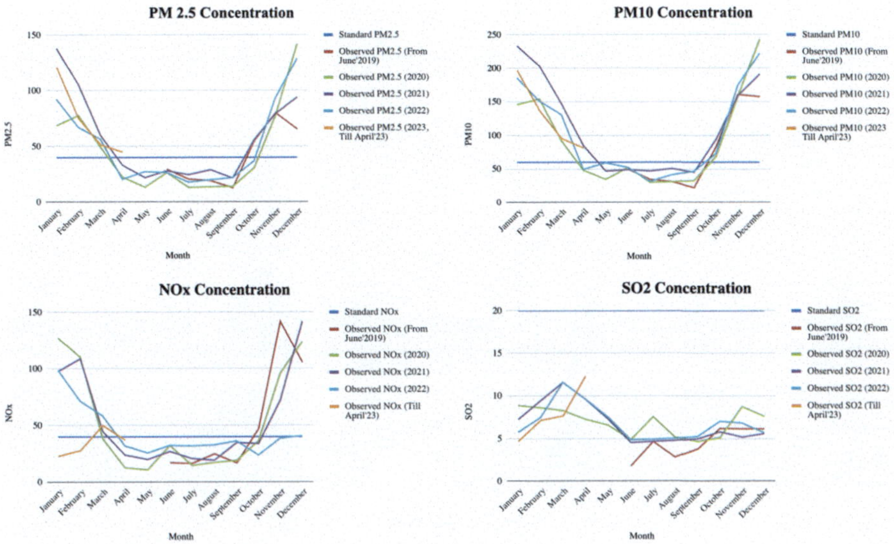

Fig. 5 Yearly (June 2019–April 2023) variation of various pollutant (PM2.5, PM10, NO_x, SO_2) concentration with respect to standard values recommended by the CPCB

December, January, February and March for all years taken into consideration. Moreover, SO_2 concentration in this region remains within limits of standard concentration value recommended by the CPCB, while a significant increase in SO_2 concentration was observed during March in the years 2021, 2022 and 2023.

3.3 Lichen Data Evaluation

Lichen species were predominantly collected from barks of tree species found in this region. The dominant tree specimens observed in this site were Royal Palm (*Roystonea regia*). The crustose and foliose variety of lichen were majorly found in this location, which involved Caperat lichen (*Parmelia caperata*), Wreath lichen (*Cryptothecia subnidulans*) and *Diploicia canescens* on tree barks as shown in Fig. 6.

Based on the growth aspects of lichen depicted in Fig. 7, it is observed that with respect to time, several varieties of lichen species found in Jadavpur area show a constant increase in their overall surface area. Growth rates in 20 lichen species were summarized for a span of 9 months. Direct lichenometry method has been considered as direct measurements of lichen growth were taken [25]. The *Parmelia caperata* variety of foliose lichen species develop consistently in this region at a growth rate of 0.30 mm^2 every 15 days. This crustose variety of lichen is observed to have a significant increase in growth rate during September, October and November.

Fig. 6 Images of Caperat lichen (*Parmelia caperata*) and Wreath lichen (*Cryptothecia subnidulans*)

Certain lichen species showed faster growth rate as compared to other lichen species found in the region.

From monitoring lichen species found in Jadavpur area, near major traffic intersections it was observed that initial growth was steady although it was accompanied by a period in which a linear growth of thallus area was observed for certain lichen species [26]. Thus, a direct association among size and age of lichen species has been depicted above through graphical representation (Fig. 7). This enabled reporting major change in growth pattern of certain lichen species based on seasonal aspects during monitoring tenure. Therefore, due to major change in surface area of specific lichen species during post-monsoon season, it could be stated that the overall pollution level prevalent in this area significantly decreased during these months, enabling sustenance of lichen species.

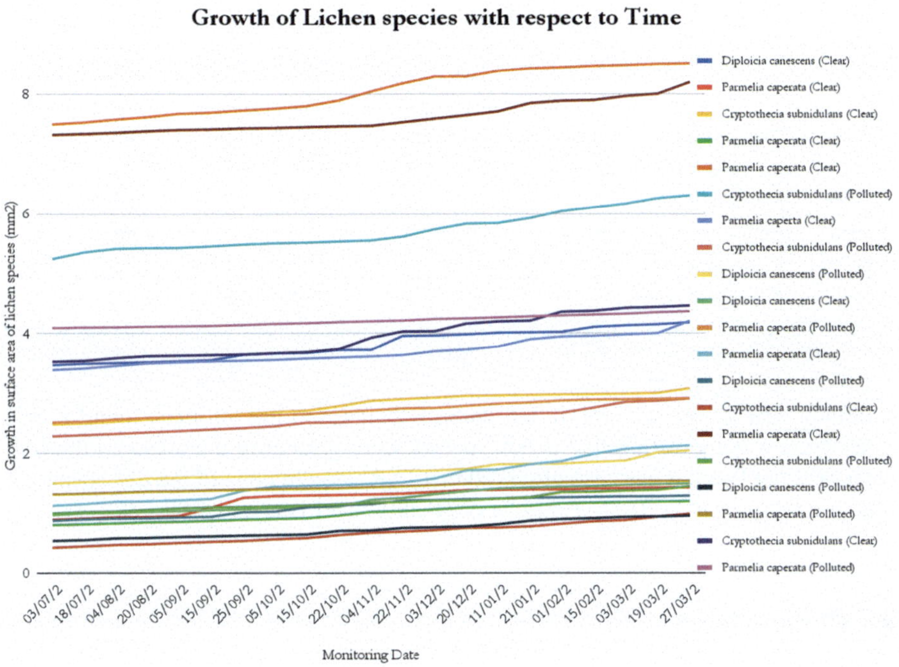

Fig. 7 Growth of lichen species surface area for monitoring time (July '22 to March '23)

3.4 *Evaluation of Index of Atmospheric Purity*

The compositional transformation in lichen communities were coordinated with transformation in atmospheric pollution levels. Moreover, the ecological index (Q) for specific species was evaluated through number of prevalent species (growing) at the study area and further dividing the sum through number of sampling sites in which the species have occurred. The frequency or cover of every species have been evaluated utilizing a transparent quadrant, which was also required for lichen sampling [18]. Therefore, in Jadavpur traffic intersection area, the value of number of species (n), factor of accompanying species (Q) and frequency of every species (f) was utilized for calculating IAP value and it is represented in Table 2.

From Table 2, the Index of Atmospheric Purity is calculated and the permissible range of IAP in terms of pollution levels has been analysed. Thus, based on the results it is observed that in study area, the overall pollution level in terms of tree specimens studied for IAP calculation is moderate. Mostly, the values obtained from *Roystonea regia* and *Wodyetia bifurcata* tree specimen belongs to quality level C (25<IAP<37.5) and quality level D (37.5<IAP<50) and predominant lichen species in these trees were observed to be Caperat lichen (*Parmelia caperata*) and Wreath lichen (*Cryptothecia subnidulans*). However, major lichen growth has been observed in tree J-1037 as its IAP range exceeded Level E (IAP>50), indicating

Table 2 Calculation for Index of Atmospheric Purity (IAP) lichen species in *Roystonea regia* and *Wodyetia bifurcata* trees

Tree no.	Tree name	Number of species (n)	Factor of accompanying species (Q)	Frequency of each species (f)	Index of Atmospheric Purity (IAP)
J-1037	*Roystonea regia*	2	12	61	73.2
J-1036	*Roystonea regia*	2	11	32	35.2
J-1035	*Roystonea regia*	3	9	52	46.8
J-1029	*Roystonea regia*	1	13	29	37.7
J-1016	*Roystonea regia*	2	17	15	25.5

extremely low pollution level. In the study, lichen desert has not been found in this area which was indicating that lichen species were present in the area and the IAP value was not 0 and it was also observed that IAP value in none of the tree specimen was in quality level A ($0<IAP<12.5$) or Quality level B ($12.5<IAP<25$).

4 Conclusion

This study takes into consideration lichen as a biomonitoring agent of air pollution and implements a systematic approach for monitoring lichen biomass and cover in Jadavpur area of Kolkata, which possess two significant traffic intersections nearby. It further depicts that there has been a significant increase in growth of lichen species in terms of surface area in roadside trees throughout a nine-month monitoring period. However, variety and quantity of lichen species observed in this region was less, indicating a moderate air quality. The lichen species observed in this region were Caperat lichen (*Parmelia caperata*), Wreath lichen (*Cryptothecia subnidulans*) and *Diploicia canescens*, and these species were predominantly observed more in barks of palm trees. Moreover, the coverage of lichen species was identified by employing a grid mapping method and further based on coverage and frequency of lichen species assisted in calculating Index of Atmospheric Purity. Therefore, IAP range calculated in significant palm trees has depicted a moderate pollution level for our study at the range 25–37.5. Furthermore, prevalent air quality in this region was also assessed through acquiring data from continuous monitoring stations of Central Pollution Control Board to determine concentration of particulates, SO_x and NO_x concentration and its impact on overall air quality. Based on this study, lichen diversity and growth in this region is hindered due to increasing levels of pollutant concentration as observed through the CPCB data. The roadside plants are constantly exposed to enhanced levels of vehicular traffic. The seasonal variation in pollutant concentration and lichen growth depicts that during winter months the growth observed in lichen is negligible due to degrading air quality, while lichen growth was observed to be at a faster pace during pre-monsoon and post-monsoon months. Prevalent impacts on lichen growth have also been closely observed due to changes in wind direction during various seasons. Therefore, this study illustrates the scope of AQI and its possible menaces on abundance and diversity of lichen species found near significant traffic intersections of Kolkata region.

References

1. I.R. Chowdhury, Scenario of vehicular emissions and its effect on human health in Kolkata city. Int. J. Hum. Soc. Sci. Invent. **4**(5), 1–9 (2015)
2. M. Das, S.K. Maiti, U. Mukhopadhyay, Distribution of PM 2.5 and PM 10-2.5 in PM 10 fraction in ambient air due to vehicular pollution in Kolkata megacity. Environ. Monit. Assess. **122**, 111–123 (2006)
3. National Ambient Air Quality Status and Trends, Central Pollution Control Board (2019)
4. D. Majumdar, P. Purohit, A.D. Bhanarkar, P.S. Rao, P. Rafaj, M. Amann, et al., Managing future air quality in megacities: Emission inventory and scenario analysis for the Kolkata Metropolitan City, India. Atmos. Environ. **222**, 117135 (2020)
5. CPCB, Parivesh Newsletter. Auto Emissions **6**(1) (1999)
6. S. Majumdar, S.S. Ram, N.K. Jana, S. Santra, A. Chakraborty, M. Sudarshan, Accumulation of minor and trace elements in lichens in and around Kolkata, India: An application of X-ray fluorescence technique to air pollution monitoring. X-Ray Spectrom. Int. J. **38**(6), 469–473 (2009)
7. S. Loppi, Lichens as sentinels for air pollution at remote alpine areas (Italy). Environ. Sci. Pollut. Res. **21**, 2563–2571 (2014)
8. S. Munzi, O. Correia, P. Silva, N. Lopes, C. Freitas, C. Branquinho, P. Pinho, Lichens as ecological indicators in urban areas: Beyond the effects of pollutants. J. Appl. Ecol. **51**(6), 1750–1757 (2014)
9. S. Majumder, D. Mishra, S.S. Ram, N.K. Jana, S. Santra, M. Sudarshan, A. Chakraborty, Physiological and chemical response of the lichen, Flavoparmelia caperata (L.) Hale, to the urban environment of Kolkata, India. Environ. Sci. Pollut. Res. **20**, 3077–3085 (2013)
10. S. Gupta, R. Khare, O. Bajpai, H. Rai, D.K. Upreti, R.K. Gupta, P.K. Sharma, Lichen as bioindicator for monitoring environmental status in Western Himalaya, India. Int. J. Environ. (2016) R. Honegger, 15 the symbiotic phenotype of lichen-forming Ascomycetes and their endo-and epibionts, Fungal Assoc., 287–339 (2012)
11. T.A. Green, L.G. Sancho, A. Pintado, Ecophysiology of desiccation/rehydration cycles in mosses and lichens. Plant Desiccation Tolerance, 89–120 (2011)
12. A. Aptroot, C.M. Van Herk, Further evidence of the effects of global warming on lichens, particularly those with Trentepohlia phycobionts. Environ. Pollut. **146**(2), 293–298 (2007)
13. A. Abas, A systematic review on biomonitoring using lichen as the biological indicator: A decade of practices, progress and challenges. Ecol. Indic. **121**, 107197 (2021). https://doi.org/10.1016/j.ecolind.2020.107197
14. L. Silberstein, B.Z. Siegel, S.M. Siegel, A. Mukhtar, M. Galun, Comparative studies on Xanthoria parietina, a pollution-resistant lichen, and Ramalina duriaei, a sensitive species. II. Evaluation of possible air pollution-protection mechanisms. Lichenologist **28**(4), 367–383 (1996)
15. A.Z.L.A.N. Abas, A.Z.A.H.A.N. Awang, Air pollution assessment using lichen biodiversity index (LBI) in Kuala Lumpur, Malaysia. Pollut. Res. **36**(2), 242–249 (2017)
16. S. Sharma, M. Zhang, J. Gao, H. Zhang, S.H. Kota, Effect of restricted emissions during COVID-19 on air quality in India. Sci. Total Environ. **728**, 138878 (2020)

17. P. Das, S. Joshi, J. Rout, D.K. Upreti, Lichen diversity for environmental stress study: Application of index of atmospheric purity (IAP) and mapping around a paper mill in Barak Valley, Assam, northeast India. Tropical ecology, 54(3) (2013)
18. F. LeBlanc, G. Robitaille, D.N. Rao, Biological response of lichens and bryophytes to environmental pollution in the Murdochville copper mine area, Quebec. J. Hattori Bot. Lab. (Japan) **38**, 71 (1974)
19. N. Muhammad, N.H. Hashim, N.A. Khairuddin, H. Yusof, S. Jusoh, A. Abas, et al., Identification of most tolerant lichen species to vehicular traffic's pollutants: A case study at Batu Pahat. J. Adv. Res. Fluid Mech. Therm. Sci. **42**(1), 57–64 (2018)
20. A. Mežaka, G. Brūmelis, A. Piterāns, The distribution of epiphytic bryophyte and lichen species in relation to phorophyte characters in Latvian natural old-growth broad leaved forests. Folia Cryptogamica Estonica **44**, 89–99 (2008)
21. K. Das, B.P. Nikita, A. Rani, P.L. Uniyal, Lichens as bioindicators and biomonitoring agents. Environ. Int. J. Sci. Technol. **15**, 18–25 (2021)
22. Github.io, Mutual information-based feature selection (2023). https://thuijskens.github.io/2017/10/07/feature-selection/
23. Cohen, I., Huang, Y., Chen, J., Benesty, J., Benesty, J., Chen, J., ... & Cohen, I. (2009). Pearson correlation coefficient. Noise reduction in speech processing, 1–4. M.E. Conti, G. Cecchetti, Biological monitoring: Lichens as bioindicators of air pollution assessment – A review, Environ. Pollut., 114(3), 471–492 (2001)
24. Z. Zhang, L. Ye, H. Qin, Y. Liu, C. Wang, X. Yu, J. Li, Wind speed prediction method using shared weight long short-term memory network and Gaussian process regression. Appl. Energy **247**, 270–284 (2019)
25. R. Armstrong, T. Bradwell, Growth of crustose lichens: A review. Geogr. Ann. Ser. A Phys. Geogr. **92**(1), 3–17 (2010)
26. R.A. Armstrong, T. Bradwell, Growth of foliose lichens: A review. Symbiosis **53**, 1–16 (2011)

Assessment of Pesticides Contamination in Surface Water and Development of Nature-Based Treatment Technology: A Case Study of Rural Area of West Bengal, India

Nilanjan Saha, Gourab Banerjee, Ronjon Heim, and Asis Mazumdar

Abbreviations

ACF	Activated carbon filter
BHC	Benzene hexachloride
CWT	Clear water tank
DIGF	Dynamic intake gravel filter
DO	Dissolved oxygen
EC	European Commission
GC-MS	Gas chromatography mass spectrometry
GHG	Greenhouse gas
GP	Gram Panchayat
GW	Groundwater
HRF	Horizontal roughing filter
IS	Indian Standards
IV	Iodine value
KLD	Kiloliters per day
LPH	Liters per hour
MPN	Most probable number
MSF	Multistage filtration
NTU	Nephelometric turbidity unit
OHT	Overhead tank

N. Saha (✉) · G. Banerjee · A. Mazumdar
School of Water Resources Engineering, Jadavpur University, Kolkata, India
e-mail: nilanjans.swre.rs@jadavpuruniversity.in

R. Heim
adelphi Research GmbH, Berlin, Germany

D. Mazumder (ed.), *Sustainable Advanced Technologies for Environmental Management*, Springer Proceedings in Earth and Environmental Sciences,
https://doi.org/10.1007/978-3-031-64006-3_17

SSF Slow sand filter
TDS Total dissolved solids

1 Introduction

West Bengal, India, is severely affected by increasing climate change impacts such as temperature rise and droughts [1, 2]. In addition, the growth of the urban and rural development and its economic development have led and will further lead to increased emission of greenhouse gas (GHG), making it also responsible for climate change. The rural population also suffers more and more from water scarcity due to declination of water table for extensive groundwater (GW) withdrawal and its concluding aggravated quality issues during the dry period in spring and summer. Former uncontaminated surface water sources have been contaminated in the last decades by wastewater discharges, effluents and seepage of pesticides from agricultural activities in the rural areas and may affect the human health [3–6]. Regional GW level declines due to overexploitation have affected surface water-based ecosystems, such as the declining or vanishing of wetlands and decreasing of vegetation coverage [7]. Around 2.5 million populations are directly affected by arsenic contamination in North 24 Parganas district of West Bengal, India [8].

Research conducted across Asia, Europe, and the United States of America revealed that pesticides and the breakdown products they produce are frequently found in surface water, sometimes at concentrations higher than the maximum individual limit (0.1 μg/L) set by the European Commission (EC) for drinking water [9–11]. In order to reduce the negative effects of these compounds on the environment and safeguard the quality of water resources, several nations have established monitoring programs to assess the degree of pesticide contamination in water bodies [12].

Water quality issues with the current conventional treatment procedure have led to the requirement of upgrading the treatment processes for which energy intensive processes such as advanced oxidation, adsorption processes and membrane treatment such as reverse osmosis are being discussed. While the current conventional drinking water treatment processes including abstraction are estimated to use around 0.35 kWh/m^3, advanced processes for the treatment of emerging pollutants such as pesticides and pharmaceuticals might range up to 3 kWh/m^3 [13, 14]. The other currently mostly practiced alternative of buying bottled water for drinking water purposes leads to 2–3.7 kg CO_2eq/L (1.55–2.83 kWh/L) [15], making a safe drinking water supply a relevant climate change mitigation measure. Considering all adverse situations in water sectors, multistage filtration (MSF) is adopted where geogenic contaminants' free surface water has been treated and distributed to the rural population. MSF technology has notable advantages as it does not use chemicals, mechanical equipment, or not locally available material [16–18]. It can be easily mangled, maintained, and operated by local people after basic training.

Different running projects in successfully operated for more than a decade have proven that the impact of implementing MSF plants in piped systems has a negligible effect on tariffs. The objective of the study is to analyze the water quality of the river named Padma situated in arsenic-affected village in Baduria Block, North 24 Pgs, West Bengal, India, while concurrently evaluating the performance of a pilot water treatment plant including pesticide parameters implemented in the region.

2 Materials and Methods

The present study was conducted for monitoring of water quality in three different sources of river water, pond water, and outlet of newly implemented nature-based surface water treatment plant in village Rasui, which is located at Chatra GP, Baduria Bock, North 24 Parganas District, West Bengal. In the village, 150 households of socially backward and economically poor were identified for implementing the pilot study. Based on the needs assessment survey, 10–12 kiloliters per day (KLD) water demand was considered for designing the capacity of treatment units.

The entire treatment process (modified multistage filtration) consists of dynamic intake gravel filter (DIGF), horizontal roughing filter (HRF), slow sand filter (SSF), and activated carbon filter (ACF). For this system, only photovoltaic solar pumps are used to withdraw pond water (raw) for surface water treatment process. The DIGF was fed from the raw water from pond and stored in raw water tank. After passing from HRF and cascade aeration process, pretreated water flowed into the SSF followed by ACF and collected in the clear water tank (CWT). In this tank the water is disinfected by chlorination and further pumped to the overhead tank (OHT) for supplying treated water through standposts (Fig. 1). For the filtration, different sizes of stone chips are used in DIGF and HRF. In SSF, effective sand size of 0.35 mm and uniformity coefficient of 3.5 were considered. In ACF, activated carbon of 800 iodine value (IV) was used for final treatment. The continuous flow rate of 500 liters per hour (LPH) is maintained and two parallel chambers of each treatment unit were constructed for incessant supply water 24 × 7 in cases of maintaining different treatment system. For 6 months, treated water was recirculated to the pond as the plant was in a biological ripening phase.

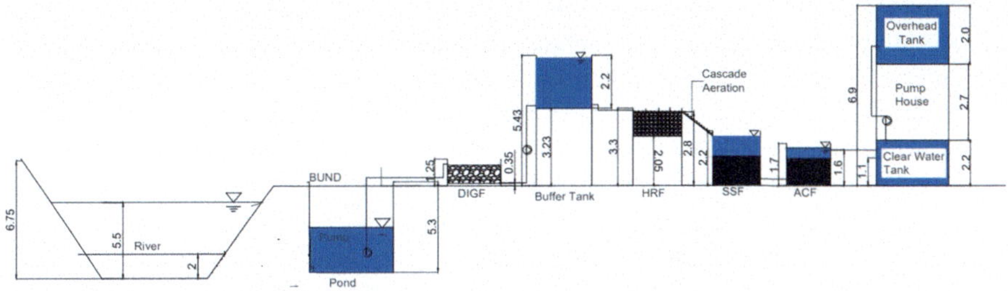

Fig. 1 Schematic flow and elevation diagram of multistage filtration system

A carefully designed sampling of the water is critical to ensure accuracy and reliability of the water quality results. For evaluation of treatment plant and river water quality, samples were taken over 1 year. For assessing the water quality parameters, water samples were collected directly from the OHT and fill the container to maximum level to minimize the impact with air. If oxygen is mixed into the water, it lowers the iron content of water samples as well as other organic parameters. Collected samples were always kept out of the reach of direct sunlight because the latter can accelerate oxidation processes. For the sampling of surface water, water was collected below the water level to avoid air contact. For bacteriological testing, water samples were strictly prohibited from human or animal contact so that external bacteria could not contaminate the water sample. For the pesticides test, the water is collected in 1000 mL amber-colored bottle to prevent the entry of direct sunlight and preserve it at 5 °C.

For the water quality analysis, pH, TDS, and dissolved oxygen (DO) were analyzed in HACH multimeter (HQ40d) with pH, TDS, and DO sensor in the field. For measuring turbidity, HACH Turbidity meter (2100Q) is used. For analysis of nitrate, phosphate, ammonia-N, iron, sulphate, Spectrophotometer (HACH DR2800) was used using HACH standard methods by powder pillow. The system was calibrated with standard solution for each method. For the determination of pesticide concentration, 500 mL of water sample, 50 mL of methylene chloride (Merck, Germany) as solvent, and 5 g sodium chloride were shaken into a 1000-mL separating funnel for about 10 min to thoroughly mix. Then mix solution was kept standing for 15 min to settle down and collect the lower aqueous solution filtered in conical flask passing through with anhydrous Na_2SO_4. The collected extract was then concentrated using a water bath with constant temperature of 60 °C. The concentrated extract was then transferred to a vial of 1 mL and tested in GC-MS (gas chromatography mass spectrometry) with standard solution of 17 organochlorine pesticides.

3 Results and Discussion

3.1 Analysis of Water Quality Parameters in River Water

As per water quality monitoring (Table 1), it is observed that during post-monsoon, the overall water quality is getting deteriorated with several pollutants such as pesticides, ammonia, and phosphate including turbidity. For the permissible limit, IS 10500:2012 is used for identifying the contamination level in the river water. Based on the field observations, during the summer, paddy (boro) are planted in the surrounding agricultural fields. The fields are irrigated with ground water/surface water by using submersible pump. At that time, the application of pesticides is higher and possibly will reach to surface water bodies (river, stream, pond) through surface runoff. As a result, higher concentration of β-BHC (0.72 μg/L), α-endosulfan

Table 1 Water quality characteristics of River Padma

Sl. no.		Desirable	Permissible					
Parameters				Min	Max	Average	SD	Median
No. of samples		***N* = 12**						
Organoleptic and physical parameters	pH value		6.5–8.5	7.25	8.20	7.49	0.26	7.405
	Turbidity, NTU	1	5	12.1	37.0	21.5	9.5	18.1
	TDS (mg/L)	500	2000	365.0	427.0	394.0	17.9	392
General parameters concerning substances undesirable in excessive amounts (mg/L)	Ammonia (as total ammonia-N)	0.5	0.5	0.22	3.20	1.85	1.19	2.055
	Dissolved oxygen			0.55	5.40	2.83	1.87	2.5
	Iron (as Fe)	1	1	0.53	1.46	0.75	0.27	0.655
	Nitrate (as NO_3)	45	45	1.00	5.00	2.89	1.37	2.75
	Phosphate (as PO_4)			0.37	2.34	1.22	0.54	1.175
	Sulphate (as SO_4)	200	400	25.0	34.0	29.5	3.3	29
Water quality parameters (μg/L) (pesticides)	α-BHC		0.01	0.012	0.089	0.040	0.029	0.028
	β-BHC		0.04	0.013	0.720	0.173	0.216	0.158
	γ-BHC		2	0.011	1.810	0.735	0.717	0.606
	Heptachlor		0.4	0.012	4.470	0.837	1.593	0.084
	Aldrin		0.03	0.049	0.430	0.135	0.139	0.08
	Heptachlor epoxide		0.2	0.043	1.000	0.246	0.306	0.119
	α-Endosulfan		0.4	0.110	2.237	0.933	0.944	0.272
	p,p′-DDE		1	0.020	0.590	0.248	0.236	0.26
	Endrin		0.6	0.374	0.407	0.391	0.023	0.3905
	p,p′-DDD		1	0.055	0.107	0.081	0.037	0.081
	Endrin aldehyde		0.6	0.066	0.192	0.145	0.060	0.161
	4,4′-DDT		1	0.106	0.312	0.209	0.146	0.209
	Methoxychlor		0.4	0.047	0.618	0.406	0.251	0.4785
	Endosulfan sulphate		0.4	0.059	1.110	0.370	0.497	0.1555
	Endosulfan-II		0.4	0.833	0.940	0.887	0.076	0.8865

(2.237 μg/L), endosulfan sulphate (1.11 μg/L), and endosulfan-II (0.940 μg/L) is observed at the mid-summer time.

During the monsoon, the water quality of the river is also deteriorated with jute rotting practices. Therefore, high concentration of phosphate (2.34 mg/L) and ammonia-N (3.2 mg/L) is observed, which causes massive algal bloom during and after the monsoon period. At the pre-monsoon, water quality gets slightly improved as the human activity has reduced on that time. It is also noted that dissolved oxygen level is reduced to less than 1 mg/L in certain events of the season, which may indicate the degradation and imbalance of aquatic life and ecosystems.

3.2 *Water Quality Analysis of Treatment Plant*

3.2.1 pH and TDS Removal

The pH and TDS varied at an average of 8.1 and 365.7 mg/L for both the raw and treated water of the water treatment plant. The usage of neutral media in the filtration plant ensures that the pH and TDS does not significantly vary. It is observed that during monsoon, the TDS value of raw water in the pond decreases up to 270 mg/L due to dilution with rain water.

3.2.2 Turbidity Removal

The turbidity values at the treated water were at an average of 0.6 NTU for the timeline, whereas for the raw water the average was around 14 NTU with a standard deviation of 3.16. It can be noted from the recorded data that during the monsoon and post-monsoon, i.e., from September to November, the turbidity values slightly increase. This could be associated with the runoff water from the bund area, which has increased the turbidity of the raw water (surface water pond). The highest turbidity was noted at 20.2 NTU in the month of October, immediately after the monsoon season. The treated water at no instance rose above the acceptable limit (1 NTU) suggested by IS 10500:2012, except for the initial month, which exceeded the accepted value. The filtration plant had an average removal efficiency within the range of 96–97% as depicted in Fig. 2. The removal efficiency of the plant can be higher if the turbidity of raw water is also higher. The overall system can ensure that with high turbidity, the plant is able to supply drinking water in which the turbidity should remain below the acceptable limit of IS 10500:2012.

3.2.3 Ammonia-N Removal

The average removal efficiency of the filtration plant was around 83.5%. The raw water had an average ammonia-N (as total ammonia-N) value of 0.75 mg/L, with a

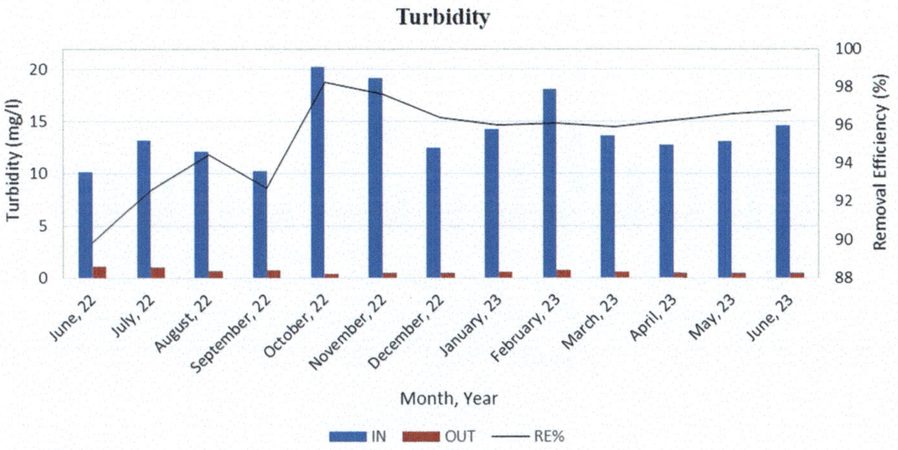

Fig. 2 Observed turbidity removal efficiency of treatment plant

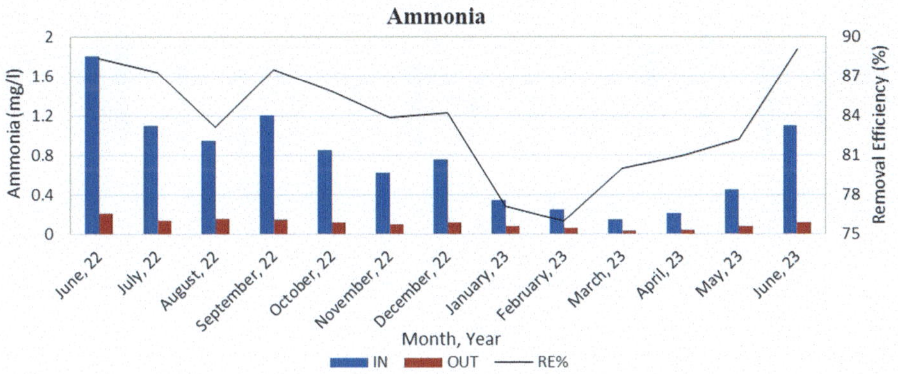

Fig. 3 Observed ammonia-N removal efficiency of treatment plant

standard deviation (SD) of 0.5 mg/L and a maximum of 1.8 mg/L. The treated value never surpassed the acceptable limit of 0.5 mg/L (IS 10500:2012), with an average of 0.11 mg/L (Fig. 3).

3.2.4 Iron Removal

The filtration plant achieved an average removal efficiency of approximately 68%. The raw water had an average iron content (as Fe) of 0.33 mg/L, with a standard deviation of 0.17 mg/L and a maximum value of 0.8 mg/L. The treated iron concentration consistently remained below the permissible limit of 0.3 mg/L, as specified in IS 10500:2012, with an average treated value of 0.09 mg/L (Fig. 4).

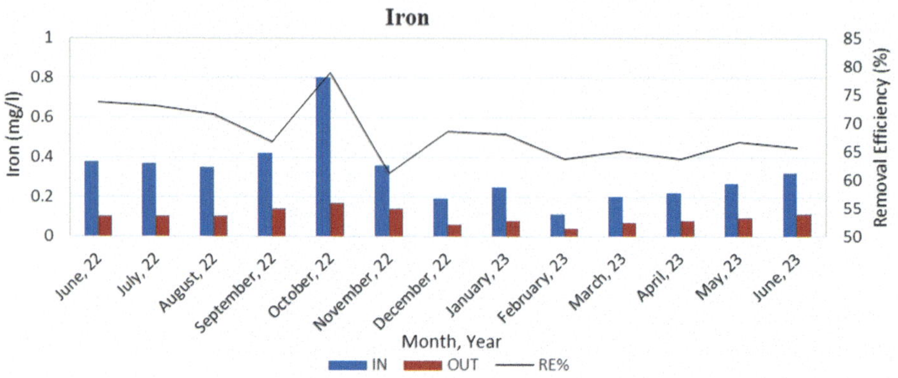

Fig. 4 Observed iron removal efficiency of treatment plant

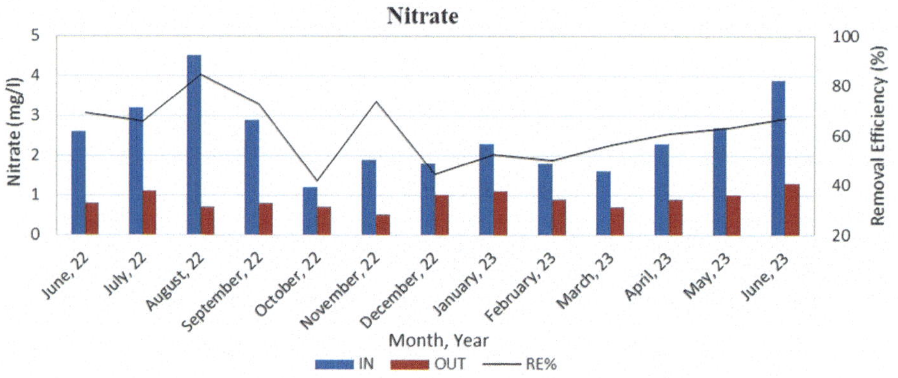

Fig. 5 Observed nitrate removal efficiency of treatment plant

3.2.5 Nitrate Removal

The filtration plant exhibited an average removal efficiency of roughly 62%. The incoming water had an average nitrate (as NO_3) concentration of 2.52 mg/L, accompanied by a standard deviation of 0.94 mg/L and a maximum level of 4.5 mg/L. It is noteworthy that the taw (raw water) and treated (treated water) nitrate concentration consistently remained below the acceptable threshold of 45 mg/L, as outlined in IS 10500:2012. With an average treated value of 0.88 mg/L, nitrate concentration has never surpassed the acceptable limit (Fig. 5).

3.2.6 Phosphate Removal

The filtration plant demonstrated an average removal efficiency of approximately 81%. In terms of the raw water, it featured an average phosphate level (as PO_4) of 1.59 mg/L, accompanied by a standard deviation of 0.84 mg/L and a maximum

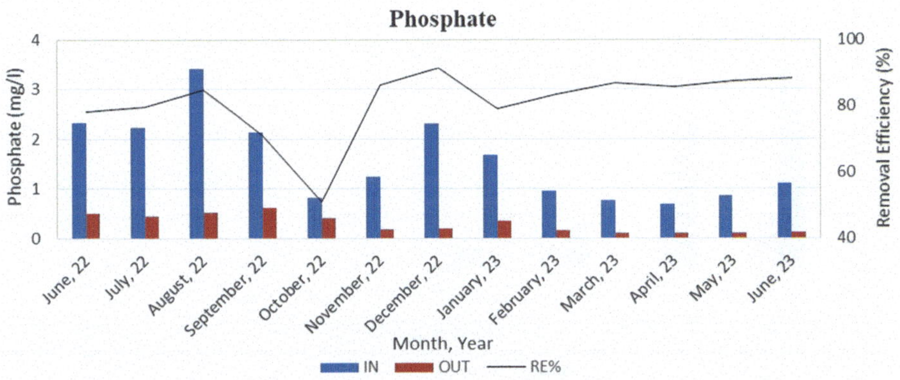

Fig. 6 Observed phosphate removal efficiency of treatment plant

reading of 3.42 mg/L. In contrast, the maximum phosphate concentration in the treated water was 0.62 mg/L, with an average concentration of 0.29 mg/L (Fig. 6).

3.2.7 Total Alkalinity and Total Hardness Removal

The filtration plant displayed an average removal efficiency of about 28%. The raw water featured an average total alkalinity (as $CaCO_3$) value of 121.84 mg/L, with a standard deviation of 3.02 mg/L and a maximum of 127 mg/L. Importantly, the treated water never exceeded the acceptable limit of 200 mg/L (according to IS 10500:2012), maintaining an average concentration of 87.46 mg/L. The same illustration occurred in case of hardness where removal efficiency varied 20–35%.

3.2.8 Removal of Pesticides Parameters

A total of 17 pesticides parameters are analyzed during the period of water quality monitoring. It is observed that four to five pesticides are relevant, which are shown below. In the initial period of treatment plant, the pesticides concentration is higher in the raw water. The circumstances can be possibly caused by mixing pollutants during excavation and refilling the pond with river water. During the time period of operation and maintenance, the observed initial concentrations of pesticides are higher and get significantly reduced due to recirculation and dilution with rain water.

3.2.8.1 β-BHC

The filtration plant achieved an average removal efficiency of approximately 33%. Raw water exhibited an average β-BHC concentration of 0.145 μg/L, with a standard deviation of 0.25 μg/L and a maximum reading of 0.72 μg/L. Notably, the treated

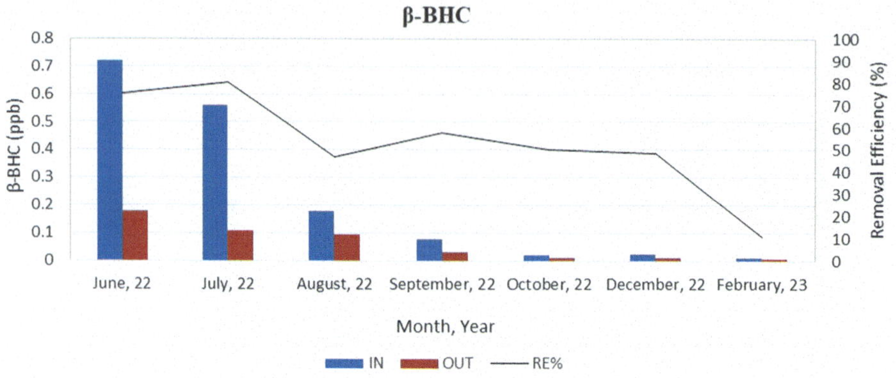

Fig. 7 Observed β-BHC removal efficiency of treatment plant

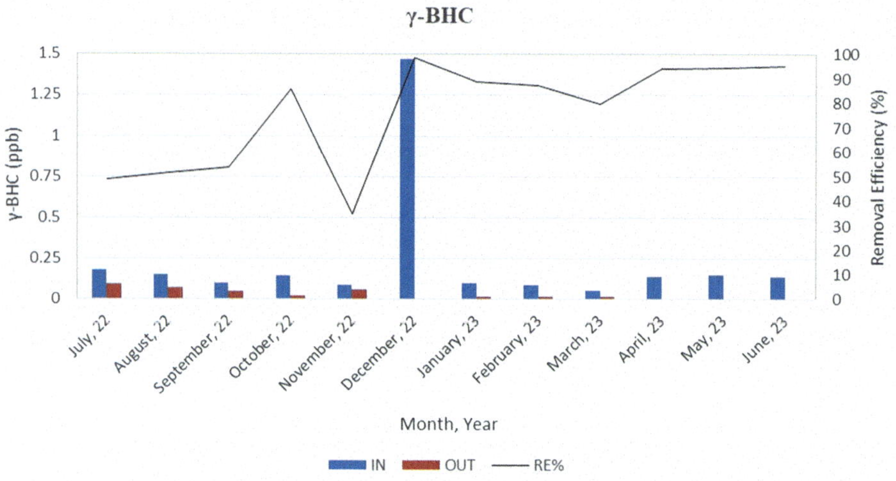

Fig. 8 Observed γ-BHC removal efficiency of treatment plant

water consistently remained below the acceptable limit of 0.04 μg/L, as stipulated by IS 10500:2012, maintaining an average concentration of 0.04 μg/L (Fig. 7). After February 2023, the concentration is not detected in the system.

3.2.8.2 γ-BHC

The filtration plant recorded an average removal efficiency of roughly 77.43%. The raw water featured an average γ-BHC concentration of 0.23 μg/L, with a standard deviation of 0.39 μg/L and a maximum of 1.47 μg/L. Significantly, the treated water consistently stayed within the acceptable limit of 2 μg/L, as defined by IS 10500: 2012, with an average concentration of 0.03 μg/L (Fig. 8).

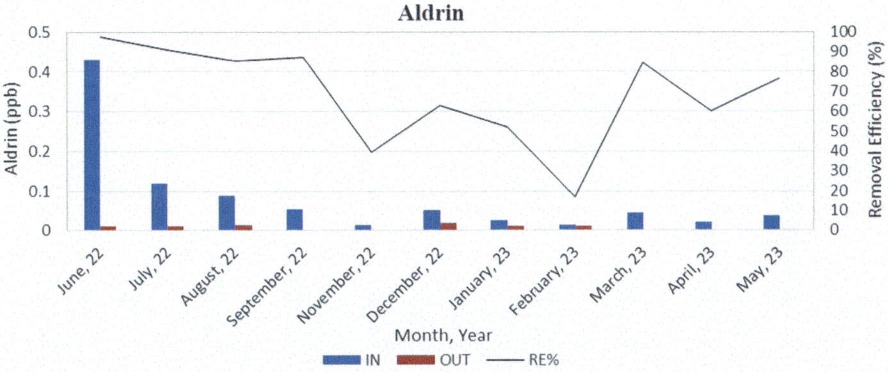

Fig. 9 Observed aldrin removal efficiency of treatment plant

3.2.8.3 Aldrin

The filtration plant, on average, achieved a removal efficiency of roughly 82%. In the water entering the system, the average aldrin concentration measured 0.08 mg/L, with a standard deviation of 0.12 μg/L and a maximum value of 0.43 μg/L (Fig. 9). It is noteworthy that aldrin concentration in the water leaving the plant consistently stayed below the acceptable limit of 0.03 μg/L, as outlined in IS 10500:2012, with an average reading of 0.01 μg/L.

3.2.8.4 Heptachlor Epoxide

The filtration plant achieved an average removal efficiency of about 83%. The raw water had an average concentration of heptachlor epoxide at 0.08 μg/L, with a standard deviation of 0.10 μg/L and a maximum reading of 0.34 μg/L. Notably, the treated water consistently remained below the maximum permissible limit of 0.098 μg/L, as defined by IS 10500:2012, with an average concentration of 0.02 μg/L (Fig. 10). After February 2023, the concentration is not detected in the system.

3.2.9 Bacteria (Total Coliform and Fecal Coliform) Removal

The filtration plant achieved an average removal efficiency of about 95% (Fig. 11). The raw water had an average concentration of Total Coliform at 754 MPN/100 mL, with a standard deviation of 461.8 and a maximum reading of 2000 MPN/100 mL at the starting period of treatment plant. It is noted, in raw water, that Total Coliform is gradually decreased due to recirculation and dilution with rainwater. A slight increase in Total Coliform is observed during summer time, which indicated the contamination from animal sources. It may be possible as the pond is not protected with the high bund and entry in the pond area is not prohibited during this time. After treatment, average Total Coliform is counted as 38 MPN/100 mL and maximum

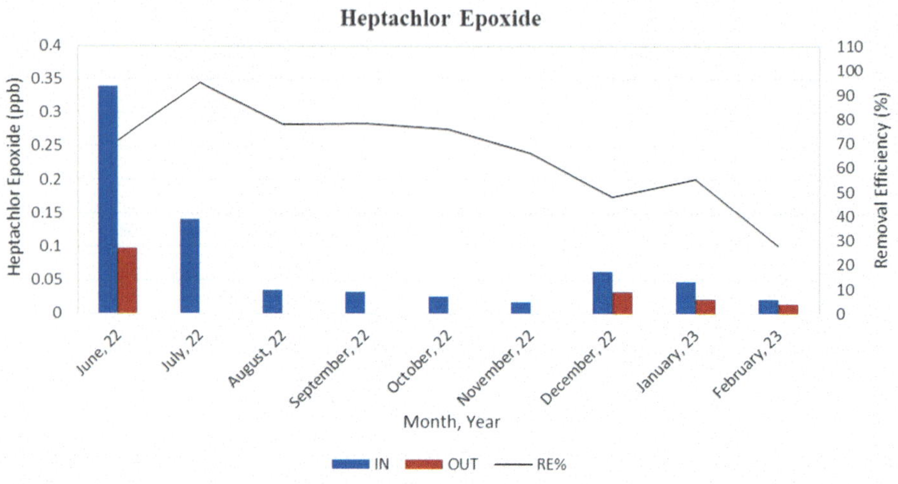

Fig. 10 Observed heptachlor epoxide removal efficiency of treatment plant

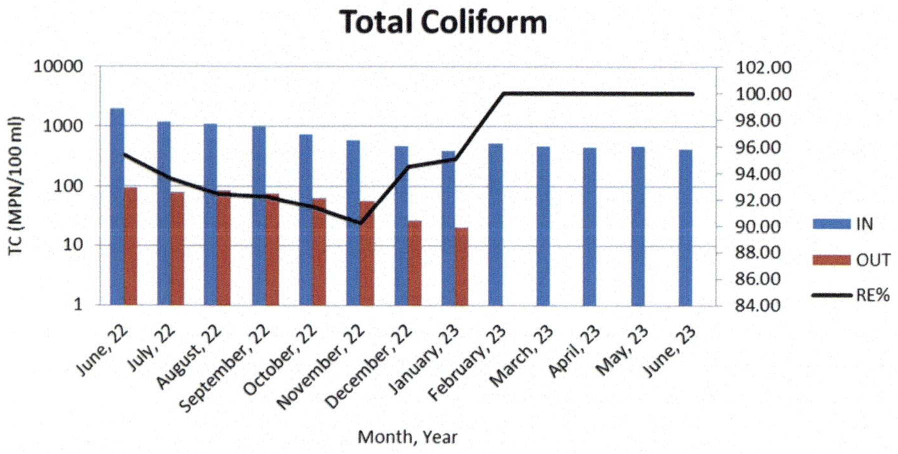

Fig. 11 Observed Total Coliform removal efficiency of treatment plant

reading of 94 MPN/100 mL at the starting period of treatment plant. The performance efficiency should be higher but due to less count of total coliform in raw water, efficiency is reduced. The disinfection process is conducted from February 2023 for which the Total Coliform is absent in the treated water (CWT). The same type of removal efficiency and the pattern is observed for Fecal Coliform.

The multistage filtration (MSF) technique, utilized in the surface water treatment plant described, distinguishes itself from conventional treatment methods through its innovative and comprehensive approach to addressing water quality challenges. This technique integrates multiple stages of filtration, each targeting specific contaminants, thus enhancing the overall effectiveness of water treatment [19].

A notable feature of this technique is the inclusion of a dynamic intake filter, which prescreens raw water to remove large particles before entering the treatment process, thereby improving system efficiency. Additionally, the horizontal roughing filter, designed to remove suspended solids and some chemical compounds, contributes to improved particle removal efficiency through its longer contact time between water and filter media [20].

The incorporation of slow sand filtration is crucial, as it allows for biological, physical, and chemical processes to effectively remove contaminants, including pathogens and turbidity [21]. Similarly, the activated carbon filter targets organic compounds, taste, odor-causing substances, and certain chemicals and pesticides, owing to its high surface area and adsorption capacity [22].

Moreover, the MSF technique demonstrates adaptability to handle varying water quality conditions, maintaining effectiveness despite challenges such as high pollutant loads and seasonal fluctuations [23]. Continuous monitoring and optimization of the treatment process further ensure maximum efficiency and effectiveness in water treatment.

4 Conclusion

The result of the treatment process of the surface water-based treatment plant comprising of a modified MSF technology such as dynamic intake filter, horizontal roughing filter, slow sand filtration, and activated carbon filter is fully successful. The filtration rate is considered 0.5 m^3/h for the pretreatment and treatment units and thus produces around 10–12 KLD. At the initial period of running of plant, pollutants' load was higher in the pond (raw water), which was due to refilling the pond with river water and having conducted excavation work in pond that remobilized the pollutants from the alluvial soil layer. During the operation and maintenance period, the quality of raw water improved due to recirculation of filtered water. All observed water quality parameters (except Total Coliform and Fecal Coliform due to the absence of disinfection) were below the permissible limit as per Indian Drinking Water Standards (IS 10500:2012) after the initial 3 months of treatment. Over the course of a year, the treatment process consistently achieved higher removal efficiencies, ranging from 90 to 98% for turbidity, 80 to 90% for ammonia, 70 to 80% for iron, about 85% for nitrate, and 60 to 90% for phosphate. In case of pesticides contamination, average removal efficiencies vary from 70 to 85% for different types of observed pesticides. A significant amount of bacterial removal is observed. These percentages indicate the effectiveness of the treatment in improving water quality. Even with elevated loads of pollutants, the treatment plant can ensure that treated water will follow IS guidelines for drinking water.

The river water is also contaminated with a high load of ammonia and pesticides that exceeds the permissible limit of drinking water due to agricultural effluents and jute rotting practices in the catchment. The river water is covered with algal blooms, which indicates a high nutrient load in the water and, as a result, the dissolved oxygen level decreases. The mentioned impacts can cause detrimental effects on water quality and ecosystem health if not properly managed. During May and June, the water quality improves, and river water can be used to refill the pond after pretreatment during that time. The treatment performance showed that the plant can handle even the most difficult season of the year; nevertheless, catchment area management for the river is recommended to reduce pollution loads. For long-term operation of the plant, a proper river intake system has to be setup to ensure pretreatment of the river water prior to entering the pond.

Acknowledgments This work was funded by the Science and Engineering Research Board, Department of Science and Technology, Government of India under Start-Up research Grant (SRG/2021/002021). The authors would like to thank School of Water Resources Engineering, Jadavpur University, India, Adelphi Research GmbH, Gemeinde Herrsching, and Indienhilfe e.V. Herrsching, Germany, for their support in laboratory and field data collection.

References

1. P.H. Gleick, H.S. Cooley, Energy implications of bottled water. Environ. Res. Lett. **4**(1) (2009). https://doi.org/10.1088/1748-9326/4/1/014009
2. JU, *Vulnerability and Risk Analysis 40 Revenue Villages Bankura and Purulia District, WB* (School of Oceanographic Studies, JU, DRCSC, 2019)
3. M.P. Ormad, N. Miguel, A. Claver, J.M. Matesanz, J.L. Ovelleiro, Pesticides removal in the process of drinking water production. Chemosphere **71**(1), 97–106 (2008) ISSN 0045-6535
4. M.A. Matin, M.A. Malek, M.R. Amin, S. Rahman, J. Khatoon, M. Rahaman, M. Aminuddin, A.J. Mian, Organochlorine insecticide residues in surface and underground water from different regions of Bangladesh. Agric. Ecosyst. Environ. **69**, 11–15 (1998)
5. M. Hasanuzzaman, M.A. Rahman, M.A. Salam, Identification and quantification of pesticide residues in water samples of Dhamrai Upazila, Bangladesh. Appl Water Sci **7**, 2681–2688 (2017). https://doi.org/10.1007/s13201-016-0485-1
6. R. Loos, B.M. Gawlik, G. Locoro, E. Rimaviciute, S. Contini, G. Bidoglio, EU-wide survey of polar organic persistent pollutants in European river waters. Environ. Pollut. **157**(2), 561–568., ISSN 0269-7491 (2009). https://doi.org/10.1016/j.envpol.2008.09.020
7. J. Liu, C. Zheng, Towards integrated groundwater Management in China. Integr. Groundwater Manag., 455–475., ISBN 978-3-319-23576-9 (2016). https://doi.org/10.1007/978-3-31923576-9_18
8. ADB, West Bengal Drinking Water Sector Improvement Project: Report and Recommendation of the President, Project Number (2018). https://www.adb.org/sites/default/files/project-documents/49107/49107-006-rrp-en.pdf
9. Y. Qian, H. Matsumoto, X.Y. Liu, S.Y. Li, X. Liang, Y.N. Liu, G.N. Zhu, M.C. Wang, Dissipation, occurrence and risk assessment of a phenyturea herbicide tebuthiuron in sugarcane and aquatic ecosystems in South China. Environ. Pollut. **227**, 389–396 (2017)
10. N.I. Rousis, R. Bade, L. Bijlsma, E. Zuccato, J.V. Sancho, F. Hernandez, S. Castiglioni, Monitoring a large number of pesticides and transformation products in water samples from Spain and Italy. Environ. Res. **156**, 31–38 (2017)
11. E. Herrero-Hernández, M.S. Rodríguez-Cruz, E. Pose-Juan, S. Sánchez-González, M.S. Andrades, M.J. Sánchez-Martín, Seasonal distribution of herbicide and insecticide residues in the water resources of the vineyard region of La Rioja (Spain). Sci. Total Environ. **609**, 161–171 (2017)

12. M. Bach, M. Diesner, D. Grossmann, D. Guerniche, U. Hommen, M. Klein, R. Kubiak, A. Muller, T.G. Preuss, J. Priegnitz, S. Reichenberger, K. Thomas, M. Trapp, Pesticide exposure assessment for surface waters in the EU. Part 2: Determination of statistically based run-off and drainage scenarios for Germany. Pest Manag. Sci. **73**(5), 852–861 (2017)
13. K. Smith, S. Liu, Energy for conventional water supply and wastewater treatment in urban China: A review. Glob. Chall. **1**(5) Special Issue: Water-Energy Nexus (2017). https://doi.org/10.1002/gch2.201600016
14. J.L. Vanderzalm, P.J. Dillon, S. Tapsuwan, P. Pickering, N. Arold, E.B. Bekele, K.E. Barry, M.J. Donn, D. McFarlane, *Economics, and Experiences of Managed Aquifer Recharge (MAR) with Recycled Water in Australia* (Australian Water Recycling Centre of Excellence, 2015)
15. International Energy Agency (IEA), *Water-Energy Nexus* (IEA, Paris, 2017) https://www.iea.org/reports/water-energy-nexus
16. G. Ochieng, F. Otieno, T. Ogada, S.M. Shitote, D. Menzwa, Performance of multistage filtration using different filter media against conventional water treatment systems. Water SA **30** (2004). https://doi.org/10.4314/wsa.v30i3.5085
17. P.K. Roy, S. Pal, G. Banerjee, A. Majumder, R. Chakraborti, M. Banik, A. Mazumder, Integrated, sustainable and eco-friendly surface and groundwater management in an arsenic affected rural area of West Bengal, India. Int. J. Environ. Eng. IJEE **3**(1) (2016) ISSN: 2374-1724
18. P.K. Roy, S. Pal, A. Majumder, G. Banerjee, A. Mazumdar, Decentralized integrated approach of water and wastewater management in rural West Bengal, in *Handbook of Environmental Materials Management*, ed. by C. Hussain, (Springer, Cham, 2019). https://doi.org/10.1007/9783-319-73645-7_2
19. D. de Jong, J. Smet, Multistage filtration: An innovative water treatment technology, in *Water encyclopedia*, ed. by J.H. Lehr, J. Keeley, (2005). https://doi.org/10.1002/047147844X.mw303
20. C. Mushila, G. Ochieng, F. Otieno, S.M. Shitote, C. Sitters, Performance of Horizontal-Flow Roughing Filter Units (HRF) under varied raw water conditions in Kenya. **2**, 136–144 (2016) ISSN: 2395-3160
21. K. Abdiyev, S. Azat, E. Kuldeyev, D. Ybyraiymkul, S. Kabdrakhmanova, R. Berndtsson, B. Khalkhabai, A. Kabdrakhmanova, S. Sultakhan, Review of slow sand filtration for raw water treatment with potential application in less-developed countries. Water **15**(11), 2007 (2023). https://doi.org/10.3390/w15112007
22. I.A. Saleh, N. Zouari, M.A. Al-Ghouti, Removal of pesticides from water and wastewater: Chemical, physical and biological treatment approaches. Environ. Technol. Innov. **19**, 101026 (2020). https://doi.org/10.1016/j.eti.2020.101026
23. S. Anbalagan, P. Kumar, S. Jeevanantham, K. Suresh, B. Tajsabreen, P.R. Yaashikaa, B. Reshma, Effective water/wastewater treatment methodologies for toxic pollutants removal: Processes and applications towards sustainable development. Chemosphere **280**, 130595 (2021). https://doi.org/10.1016/j.chemosphere.2021.130595

Methodology for Development of Cost Response Maps for Membrane Bioreactor

Bhaskar Sengupta, Abhisek Roy, and Somnath Mukherjee

Abbreviations

BOD	Five-day biochemical oxygen demand
BOD_u	Ultimate biochemical oxygen demand
CAPEX	Total bare construction cost
COD	Chemical oxygen demand
MAPE	Mean absolute percentage error
MBR	Membrane bioreactor
mld	Million litres per day
MLSS	Mixed liquor suspended solids
MLVSS	Mixed liquor volatile suspended solids
OPEX	Levelized cost based on energy requirement, operation and maintenance
TSS	Total suspended solids
VSS	Volatile suspended solids
WWTP	Wastewater treatment plant
3-D	Three dimensional

1 Introduction

Membrane bioreactor (MBR) is a recent but proven option for biological treatment of wastewater. This technology is an integration of conventional activated sludge process and the process for separation of biological sludge by the use of polymeric

B. Sengupta (✉) · A. Roy · S. Mukherjee
Environmental Engineering Division, Department of Civil Engineering, Jadavpur University, Kolkata, India
e-mail: aroy.civil@jadavpuruniversity.in; somnath.mukherjee@jadavpuruniversity.in

D. Mazumder (ed.), *Sustainable Advanced Technologies for Environmental Management*, Springer Proceedings in Earth and Environmental Sciences,
https://doi.org/10.1007/978-3-031-64006-3_18

membranes. There is no requirement of settling tanks either primary or secondary in case of use of membrane bioreactor. The use of membrane bioreactor appears as a preferred option where there is a requirement for construction or augmentation of wastewater treatment plant (WWTP) within a limited space. However, economic issues should also be taken into consideration for comparison with other technologies in respect of long-term sustainability with reference to costs for construction, operation and maintenance. In this context, stakeholders will require rapid cost data for MBR-based WWTP for comparison with other alternatives to arrive at suitable selection.

Several research studies have been made in last five decades in respect of development of cost functions to establish relationship between capacity and cost for biological treatment systems with several technologies [1–6]. Such studies were carried out for rapid cost estimation and commercial comparisons among several alternative technologies. However, these studies were made with collected cost history for different plants with conventional biological treatment technologies such as continuously mixed activated sludge, oxidation ditches, aerated lagoons and others. In some cases, the use of software has been exercised for comparison of costs of WWTPs with specific conventional technologies [7, 8]. Y. Tian adopted improved analytic hierarchy process for economic evaluation of wastewater treatment technologies such as MBR and others [9]. M. Zaerpour made a study to compare capital cost as well as cost for operation of MBR and conventional activated sludge technology based on similar inlet conditions and treated water quality for a particular WWTP [10]. For few case studies, the US National Research Council found that a logarithmic cost function is applicable for activated sludge, MBR and few other processes [11]. The Ministry of Urban Development–Government of India highlighted that small scale plants with MBR technology have been set up at several places in India. Thus, the techno-economic viability of MBR may be considered as proven one [12]. C. Visvanathan et al. addressed that due to broad domain of application and the functional features of membrane systems, MBR has gained significant importance in wastewater treatment [13]. Cost function for membrane bioreactor (MBR) based on engineering design as well as estimation is available in published literature [14]. An inlet BOD load of 250 mg/L has been considered to develop these cost functions. No other publication has come under notice related to cost functions for biological treatment with MBR. Cost implications for a specific design capacity of wastewater treatment plant (WWTP) with variation in inlet BOD level are an important issue. No such response study has been reported in published literatures. To bridge such a gap, an initiative has been adopted to develop response maps for rapid estimation of WWTP with MBR technology.

This chapter brings out the methodology followed to map in 3-D environment the cost response of biological treatment with membrane bioreactor for variation in inlet flow rate and inlet BOD. The use of such maps will assist stakeholders for rapid estimation of cost of MBR for any capacity and BOD as pertinent to a specific project.

Survey of literatures published in last five decades does not bring out any publication on 3-D response maps developed for rapid estimation of costs of WWTPs. The novelty of this research to explore 3-D cost response maps for WWTP with MBR prevails upon the approach, which is based on engineering rationale and delineated in detail hereinafter. The approach does not utilize any cost data library for determination of cost to ensure high degree of accuracy in estimated cost. Maps will project instant overall cost figure (inclusive of capital costs and operating costs over 25 years) for any capacity in between 5 million litres per day (mld) to 150 mld and BOD in between 100 milligrams per litre (mg/L) to 250 mg/L. The objective of this chapter is to present 3-D cost response surfaces as well as contour maps and illustrate the methodologies followed for development of the same. These maps may be utilized to predict the overall cost for WWTPs with MBR technology—a space economic technology for wastewater treatment. It is believed that these maps would be useful to stakeholders for evaluation purpose prior to construction of new project for WWTP.

Costs of primary treatment system and sludge collection as well as treatment system are not subject to significant variation with change in type of technology for biological treatment in WWTPs. In practice, the kind of technology selected for biological treatment in WWTPs in conjunction with required equipment and accessories will drive the height of capital expenses and operating expenses. For this reason, the costs of primary treatment system and sludge collection as well as treatment system are not necessary to be considered for comparison of alternative technologies for new construction of WWTPs. In this chapter the development of cost functions and thereafter 3-D maps are based upon the detailed engineering and cost estimations for biological treatment of wastewater and is focused upon MBR technology. Novel approach for construction of 3-D cost response map is delineated below.

2 Materials and Methods

Capacity

The capacities for WWTPs that were taken into consideration for the study are as follows:

1. A capacity in low range as 5 mld.
2. A capacity in medium range as 50 mld.
3. A capacity in large range as 150 mld.

Characteristics of Raw Wastewater

With a water delivery level of 135 litres per person in a day, the quality of wastewater as anticipated is presented in Table 1 [15].

Table 1 Characteristics of raw wastewater

Impurities	Unit	Content
TSS (total suspended solids)	mg/L	375.00
BOD (biochemical oxygen demand)	mg/L	250.00
COD (chemical oxygen demand)	mg/L	425.00

Table 2 Characteristics of treated wastewater

Impurities	Unit	Content
TSS	mg/L	10.00
BOD	mg/L	10.00

However, in actual situations, the BOD in wastewater will not remain constant. Six different BOD contents (100, 130, 160, 190, 220 and 250 mg/L) have been selected for this study. However, the ratio of BOD to COD has been kept same in all cases.

Characteristics of Treated Wastewater

The quality of treated wastewater as considered is presented in Table 2.

The characteristics provided in Table 2 are in compliance with the requirements specified by the Ministry of Environment & Forests of the Government of India.

Process Design Criteria

The process design rationale of biological treatment with MBR technology is presented in Table 3 [16].

Equipment and Accessories for WWTPs with MBR Technology

The equipment and accessories required for secondary treatment may be enlisted as follows:

1. Aerobic reactor basins and accessories.
2. Membrane chambers and accessories.
3. Reactor basin waste transfer pumps and pump-house.
4. Internal recirculation pumps and pump-house.
5. Mixed liquor recirculation pumps and pump-house.
6. Blowers and blower building.

MBR technology does not call for primary as well as secondary clarifiers.

Development of Model for Design and Estimation of Cost

A model as illustrated hereinafter has been developed by the use of Microsoft Excel for comprehensive design and estimation of bill of quantities as well as cost:

1. Process design to determine number and sizes of equipment and accessories [16]. Numbers of treatment chain and different equipment has been adopted as per the CAPDET—USEPA [17].
2. Quantity estimate for construction of different equipment has been worked out by the use of pertinent algorithms as illustrated in CAPDET—USEPA [17].

Table 3 Process design rationale of biological treatment with MBR technology

Parameters	Unit	Value
Design factor for lean condition		0.45
Design factor for peak condition		2.25
Content of BOD in thickener overflow return	mg/L	500.00
Fraction of plant flow as thickener overflow return		0.15
Content of BOD in centrate from sludge dewatering return	mg/L	380.00
Fraction of plant flow as sludge dewatering centrate		0.0060
Ratio of BOD to BOD_u		0.67
Ratio of BOD_u to VSS		1.42
Ratio of VSS to TSS		0.70
Temperature within reactor basin	°C	12.00
MLSS	mg/L	8000.00
MLVSS	mg/L	5600.00
Elevation at site	m	9.00
Atmospheric pressure at elevation of site	kPa	95.60
Design effective liquid height in reactor basin	m	4.07
Percentage clean water oxygen transfer efficiency for fine bubble ceramic diffusers	%	35.00
Point of air release for ceramic diffusers from bottom of reactor basin	m	0.50
Fouling factor for diffuser		0.90
Aeration α factor for BOD removal		0.50
Standard temperature	°C	20.00
Concentration of dissolved oxygen at standard temperature and pressure of 101,325 N/m^2	mg/L	9.08
Salinity and surface tension correction factor for BOD removal		0.95
Density of air	kg/m^3	1.20
Percentage of oxygen in air (by weight)	%	23.20
Oxygen transfer efficiency	%	8.00
Oxygen consumption	mg/mg of cell	1.42
Factor of safety		2.00
Kinetic data for removal of BOD		
Temperature	°C	20.00
Maximum specific bacterial growth rate	(g VSS/g VSS)/d	6.00
Half velocity constant	g bs COD/m^3	20.00
True yield coefficient	g VSS/g b COD	0.3125
	g VSS/g BOD	0.5000
Endogenous decay coefficient	(g VSS/g VSS)/d	0.06
Fraction of biomass as cell debris		0.15
Temperature activity coefficient		
Coefficient for μ_m		1.07
Coefficient for K_s		1.00
Coefficient for k_d		1.04

3. Cost estimation for different units complete with accessories.
4. Estimation of overall construction cost.
5. Estimation of operation and maintenance costs for 25 years of operation.
6. Estimation of area required for installation of the WWTP with MBR.

The above model will bring out costs for land, construction, operation and maintenance based on inputs for capacity, inlet BOD and allowable outlet BOD.

Quotes from vendors have been utilized for non-scheduled mechanical as well as electrical items. Schedule of rates published in 2021 has been taken into consideration for civil items [18].

The above approach avoids the use of cost data library to arrive at real-time cost with reasonable accuracy.

3 Results and Discussions

Construction cost, operation as well as maintenance cost for 25 years and cost of land as estimated by use of the developed model for different capacities and inlet BODs have been analysed by trend-line technique. Selection of most suitable cost response curves among several available alternatives has been made based on maximum value of coefficient of determination (R^2) to explore the nature of relationship between overall cost for the biological treatment with MBR and inlet BOD. A summary of results obtained for 150 mld capacity of WWTP is presented in the Appendix for information.

The cost response curves (which corresponds to maximum value for coefficient of determination) with BOD [as mg/L] along abscissa and total costs of WWTPs [as ₹ (crore)] along ordinate for each of 5, 50 and 150 mld capacities are presented in Figs. 1, 2 and 3.

Cost response function has been validated by estimation of mean absolute percentage error (MAPE).

Table 4 presents the identified most suitable cost response functions.

For 5 mld WWTP with MBR, cost vs. BOD plot is apparently linear. However, the R^2 value as obtained for polynomial trend is 1.000 and the same for linear trend is 0.9999. Therefore, the second-order polynomial function has been selected as the respective cost response function. Negative coefficient of x^2 in cost response function indicates that the rate of change of cost data is reduced by a factor as the BOD value in raw wastewater (shown along X-axis) increases. However, the overall cost remains always positive for any inlet BOD value.

Cost response function is represented by linear form for 50 mld WWTP with MBR. The respective R^2 value is 0.9999.

For capacity of 150 mld, cost response function is represented by polynomial function. The R^2 value in this case is 0.9997.

Mean absolute percentage errors for the cost response functions are 0.02%, 0.02% and 0.43%, respectively, for 5, 50 and 150 mld. In each case, mean absolute percentage error is within the accurate range (<10%).

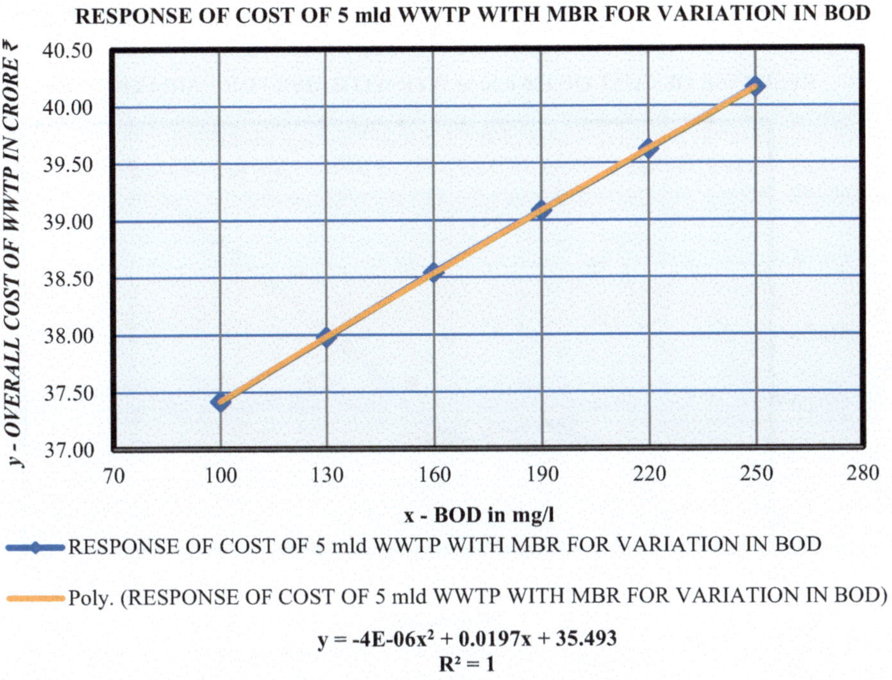

Fig. 1 Cost response curve—5 MLD WWTP with MBR technology

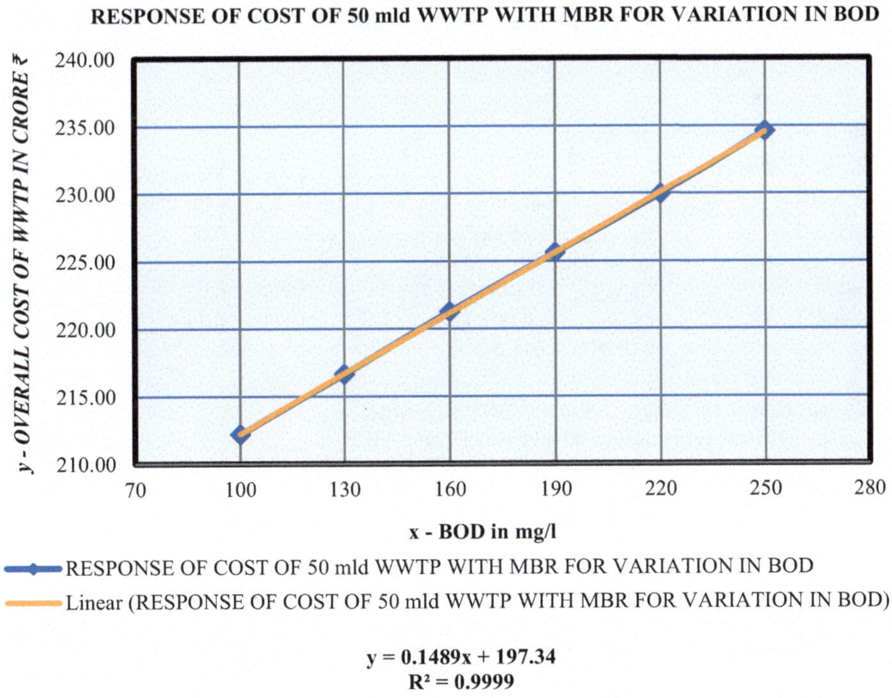

Fig. 2 Cost response curve—50 MLD WWTP with MBR technology

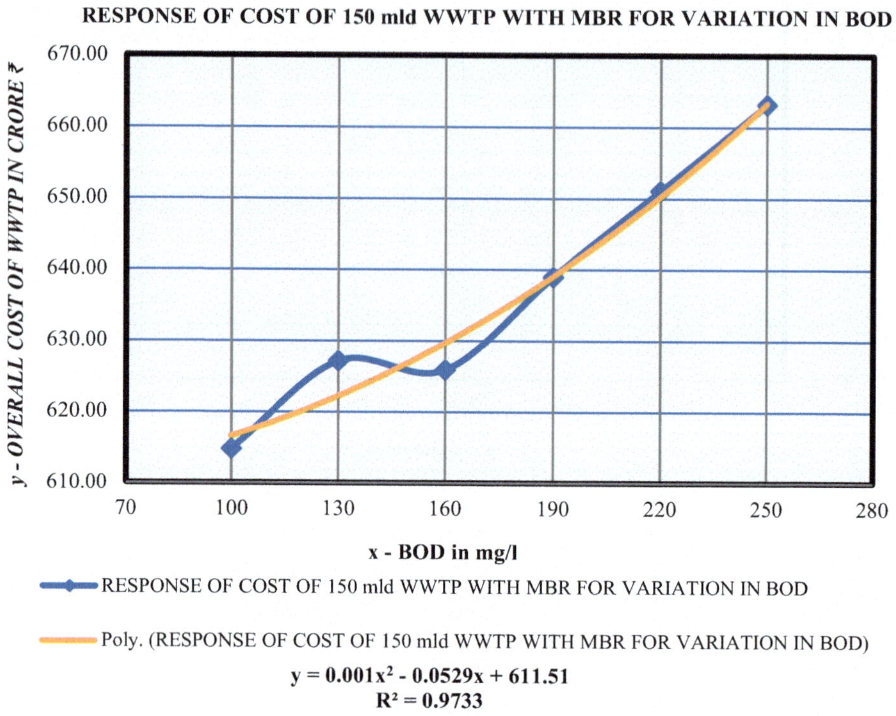

Fig. 3 Cost response curve—150 MLD WWTP with MBR technology

Table 4 Response functions of cost

Capacity of WWTP with MBR	Response function	R^2	Mean absolute percentage error
5 mld	$C_{R\text{-}5\ mld} = -4E\text{-}06*(x^2) + 0.0197*x + 35.493$	1.0000	0.02
50 mld	$C_{R\text{-}50\ mld} = 0.1489*x + 197.34$	0.9999	0.02
150 mld	$C_{R\text{-}150\ mld} = 0.001*(x^2) - 0.0529*x + 611.51$	0.9733	0.43

where,

$C_{R\text{-}5\ MLD}$ Cost [₹ (in crore)] of 5 mld WWTP with MBR
$C_{R\text{-}50\ MLD}$ Cost [₹ (in crore)] of 50 mld WWTP with MBR
$C_{R\text{-}150\ MLD}$ Cost [₹ (in crore)] of 150 mld WWTP with MBR
R^2 Coefficient of determination
x BOD [in mg/L] in raw wastewater

The datasets for variation of costs with change in inlet BOD for 5, 50 and 150 mld WWTPs with MBR are presented in Table 5.

The above datasets are used in 3-D environment to develop cost response surface map and contour map as shown in Figs. 4 and 5. Any of these maps may be used to predict a rapid cost for WWTP with MBR for a capacity range (5–150 mld) and inlet BOD range (100–250 mg/L).

Table 5 Datasets for variation of costs with change in BOD in raw wastewater

Predicted costs for WWTPs with MBR							
BOD removal							
Capacity (mld)	Description	BOD (mg/L)	BOD (mg/L)	BOD (mg/L)	BOD (mg/L)	BOD (mg/L)	BOD (mg/L)
		100.00	130.00	160.00	190.00	220.00	250.00
5	Predicted cost value in crore ₹ (as per polynomial cost function)	37.42	37.99	38.54	39.09	39.63	40.17
50	Predicted cost value in crore ₹ (as per linear cost function)	212.23	216.70	221.16	225.63	230.10	234.57
150	Predicted cost value in crore ₹ (as per polynomial cost function)	616.22	621.53	628.65	637.56	648.27	660.79

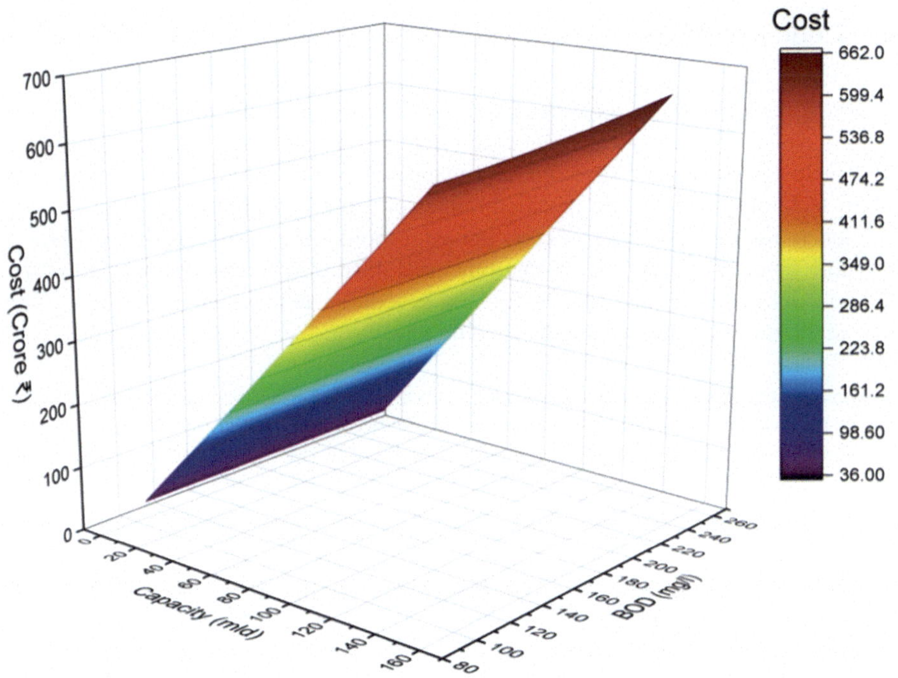

Fig. 4 3-D cost response surface map for WWTP with MBR technology

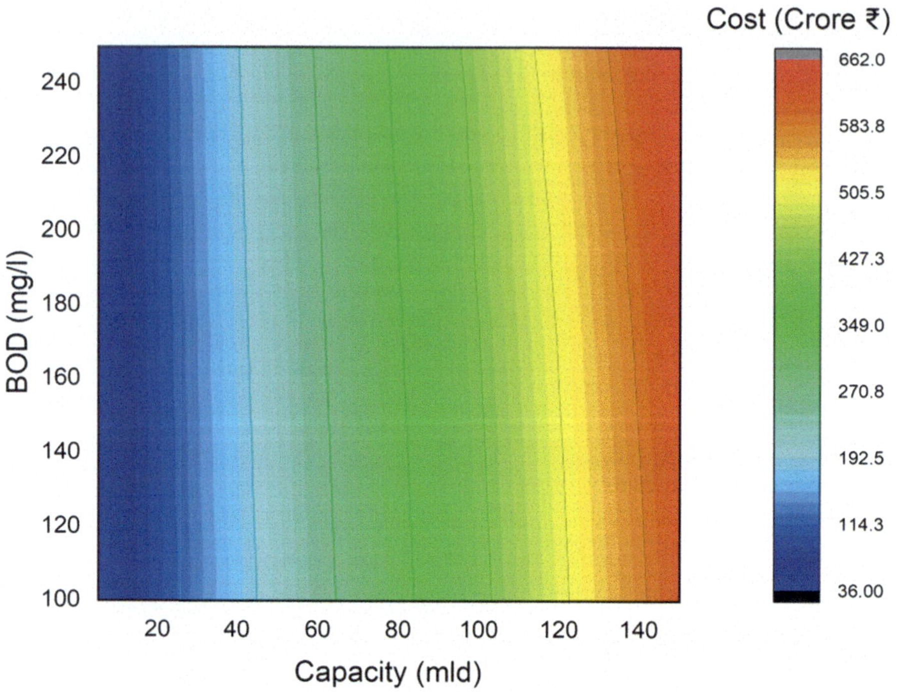

Fig. 5 3-D contour map for WWTP with MBR technology

4 Conclusions

3-D cost response maps have been developed with applicable engineering design criteria and do not include any historic reference. These maps will predict overall cost inclusive of capital, operation and maintenance expenditure for WWTPs with MBR technology within the specified ranges of capacity and inlet BOD.

Generation of 3-D surface or contour cost response map and its use at any place is possible by development of the cost functions illustrated in this chapter. However, the schedule of rates and quotes for non-scheduled items need to be taken into consideration as applicable for the specific place.

Acknowledgements The authors are thankful to several vendors for their support in respect of compilation of cost data for MBR media, pumps, blowers and motors that were used in the preparation of this chapter. The authors are also thankful to Environmental Engineering Division, Jadavpur University, Kolkata, India, for giving the opportunity for this specific research work.

Appendix: Summary of Cost Response Study for 150 mld WWTP with MBR

Design summary for BOD removal							
Inlet conditions							
		BOD (mg/L)	**BOD (mg/L)**	**BOD (mg/L)**	**BOD (mg/L)**	**BOD (mg/L)**	**BOD (mg/L)**
		100.00	**130.00**	**160.00**	**190.00**	**220.00**	**250.00**
Design parameter	**Unit**	**Data**	**Data**	**Data**	**Data**	**Data**	**Data**
Average flow rate for wastewater	m^3/d	150,000.00	150,000.00	150,000.00	150,000.00	150,000.00	150,000.00
Flow rate to MBRs	m^3/d	173,400.00	173,400.00	173,400.00	173,400.00	173,400.00	173,400.00
Average BOD load	kg/d	26,592.00	31,092.00	35,592.00	40,092.00	44,592.00	49,092.00
Average TKN load	kg/d	4930.37	5932.16	6933.95	7935.97	8937.76	9939.55
Removal of BOD							
Number of MBRs	Number	12	12	12	12	12	12
Aerobic solids residence time—Design value	d	5.00	5.00	5.00	5.00	5.00	5.00
Total volume of each MBR	m^3	1356.11	1592.85	1829.60	2066.34	2303.08	2539.83
Hydraulic detention time of each MBR	h	2.25	2.65	3.04	3.43	3.83	4.22
MLVSS (X_{MLVSS})	g VSS/m^3	3439.71	3427.09	3417.74	3410.53	3404.80	3400.14
Food to mass ratio	(g BOD/d)/g b VSS	0.48	0.47	0.47	0.47	0.47	0.47
Volumetric BOD loading	(kg $BOD/d)/m^3$	1.63	1.63	1.62	1.62	1.61	1.61
TSS purged per day	kg TSS/d	26,037.34	30,582.80	35,128.27	39,673.73	44,219.19	48,764.65
Observed yield (based on VSS)	g b VSS/g b COD	0.29	0.29	0.29	0.29	0.29	0.29
	g b VSS/g BOD	0.46	0.46	0.47	0.47	0.47	0.47
Observed yield (based on TSS)	g TSS/g b COD	0.62	0.62	0.62	0.62	0.62	0.62
	g TSS/g BOD	0.99	0.99	0.99	0.99	1.00	1.00
Overall oxygen demand	kg oxygen/h	1099.58	1286.80	1474.02	1661.23	1848.45	2035.67
Flow rate of air	m^3/min	937.27	1096.85	1256.44	1416.02	1575.61	1735.19
BOD	g/m^3	9.66	9.66	9.66	9.66	9.66	9.66
TSS	g/m^3	10.00	10.00	10.00	10.00	10.00	10.00
NH_4-N	g/m^3	19.19	23.54	27.89	32.25	36.60	40.95
NO_3-N	g/m^3	≤2	≤2.6	≤3.2	≤3.8	≤4.4	≤5

(continued)

Equipment summary for BOD removal

		BOD (mg/L)	BOD (mg/L)	BOD (mg/L)	BOD (mg/L)	BOD (mg/L)	BOD (mg/L)
		100.00	**130.00**	**160.00**	**190.00**	**220.00**	**250.00**
Description	**Unit**	**Data**	**Data**	**Data**	**Data**	**Data**	**Data**
MBRs and accessories							
Number		12	12	12	12	12	12
Length (each)	m	36.42	42.78	49.14	55.49	61.85	68.21
Width (each)	m	9.14	9.14	9.14	9.14	9.14	9.14
Depth (each)	m	4.57	4.57	4.57	4.57	4.57	4.57
Number of swing arm headers (each MBR)		12	14	16	18	20	22
Membrane chambers & accessories							
Number		12	12	12	12	12	12
Length (each)	m	25.65	25.65	25.03	25.03	25.03	25.03
Width (each)	m	3.26	3.26	3.26	3.26	3.26	3.26
Depth (each)	m	4.57	4.57	4.57	4.57	4.57	4.57
Number of membrane modules (each MBR)		10,080	10,080	9792	9792	9792	9792
Mixed liquor recirculation pumps and pump-house							
Number of pumps required		2	2	2	2	2	2
Capacity (for each pump)	m^3/h	14,270.00	14,240.00	14,210.00	14,180.00	14,150.00	14,120.00
Area required for pump house	m^2	226.00	225.00	225.00	225.00	224.00	224.00
Blowers and blower building							
Number of blowers required		4	4	4	5	5	5
Capacity (for each pump)	scfm	12,913.00	14,842.00	16,726.00	13,992.00	15,439.00	16,886.00
Area required for blower building	m^2	178.00	185.00	190.00	196.00	201.00	205.00

Estimated cost summary for BOD removal							
		BOD (mg/L)	**BOD (mg/L)**	**BOD (mg/L)**	**BOD (mg/L)**	**BOD (mg/L)**	**BOD (mg/L)**
		100.00	**130.00**	**160.00**	**190.00**	**220.00**	**250.00**
Description	**Unit**	**Data**	**Data**	**Data**	**Data**	**Data**	**Data**
MBRs and accessories							
Capital expenditure (CAPEX)	₹	2,110,174,723.08	2,123,523,038.33	2,079,189,133.19	2,092,301,110.56	2,105,319,012.44	2,118,271,723.00
Operation expenditure for 25 years (OPEX)	₹	778,144,593.05	864,119,909.41	942,500,245.17	1,028,231,329.85	1,113,869,503.35	1,199,430,190.80
Overall cost inclusive of CAPEX and OPEX	₹	2,888,319,316.13	2,987,642,947.74	3,021,689,378.36	3,120,532,440.41	3,219,188,515.79	3,317,701,913.80
Membrane chambers & accessories							
Capital expenditure (CAPEX)	₹	2,043,857,037.83	2,043,857,037.83	1,985,687,435.91	1,985,687,435.91	1,985,687,435.91	1,985,687,435.91
Operation expenditure for 25 years (OPEX)	₹	394,609,805.70	394,609,805.70	383,590,358.73	383,590,358.73	383,590,358.73	383,590,358.73
Overall cost inclusive of CAPEX and OPEX	₹	2,438,466,843.53	2,438,466,843.53	2,369,277,794.64	2,369,277,794.64	2,369,277,794.64	2,369,277,794.64
Mixed liquor recirculation pumps and pump-house							
Capital expenditure (CAPEX)	₹	165,029,871.27	164,644,548.18	164,259,342.92	163,874,255.71	163,489,286.75	163,104,436.26
Operation expenditure for 25 years (OPEX)	₹	373,701,111.07	372,875,821.63	372,050,534.39	371,225,249.35	370,399,966.51	369,574,685.88
Overall cost inclusive of CAPEX and OPEX	₹	538,730,982.34	537,520,369.81	536,309,877.31	535,099,505.06	533,889,253.27	532,679,122.14
Blowers and blower building							
Capital expenditure (CAPEX)	₹	152,611,901.39	164,944,606.07	176,356,933.80	196,675,539.38	207,879,371.40	218,654,512.47
Operation expenditure for 25 years (OPEX)	₹	0.00	0.00	0.00	0.00	0.00	0.00
Overall cost inclusive of CAPEX and OPEX	₹	152,611,901.39	164,944,606.07	176,356,933.80	196,675,539.38	207,879,371.40	218,654,512.47

(continued)

Estimated cost summary for BOD removal

		BOD (mg/L)	BOD (mg/L)	BOD (mg/L)	BOD (mg/L)	BOD (mg/L)	BOD (mg/L)
		100.00	**130.00**	**160.00**	**190.00**	**220.00**	**250.00**
Description	**Unit**	**Data**	**Data**	**Data**	**Data**	**Data**	**Data**
Estimated consolidated costs							
Capital expenditure (CAPEX)	₹	4,471,673,533.58	4,496,969,230.41	4,405,492,845.82	4,438,538,341.56	4,462,375,106.50	4,485,718,107.64
Operation expenditure for 25 years (OPEX)	₹	1,546,455,509.82	1,631,605,536.74	1,698,141,138.28	1,783,046,937.93	1,867,859,828.59	1,952,595,235.41
Overall cost inclusive of CAPEX and OPEX	₹	6,018,129,043.40	6,128,574,767.15	6,103,633,984.10	6,221,585,279.48	6,330,234,935.09	6,438,313,343.05
Cost of land							
Cost of land	₹	130,114,303.55	142,978,106.94	155,292,867.67	168,140,171.07	180,979,974.47	193,815,277.87
Overall cost							
Overall cost	₹	6,148,243,346.95	6,271,552,874.09	6,258,926,851.77	6,389,725,450.55	6,511,214,909.56	6,632,128,620.92
Overall cost	Crore ₹	614.82	627.16	625.89	638.97	651.12	663.21
Overall cost	Million USD	76.85	78.39	78.24	79.87	81.39	82.90

Estimation of MAPE (for BOD removal)

		BOD (mg/L)	BOD (mg/L)	BOD (mg/L)	BOD (mg/L)	BOD (mg/L)	BOD (mg/L)
		100.00	**130.00**	**160.00**	**190.00**	**220.00**	**250.00**
Description	**Unit**	**Data**	**Data**	**Data**	**Data**	**Data**	**Data**
Overall cost (as per exponential cost function)	Crore ₹	614.55	623.84	633.27	642.84	652.55	662.42
R^2		0.9547					
Absolute percentage error (APE)		0.04	0.53	1.18	0.61	0.22	0.12
Mean absolute percentage error (MAPE)		0.45					
Overall cost (as per linear cost function)	Crore ₹	613.52	622.86	632.20	641.55	650.89	660.23
R^2		0.9529					
APE		0.21	0.68	1.01	0.40	0.04	0.45
MAPE		0.47					
Overall cost (as per logarithmic cost function)	Crore ₹	611.30	624.37	634.71	643.27	650.57	656.94
R^2		0.9026					
APE		0.57	0.44	1.41	0.67	0.08	0.95
MAPE		0.69					
Overall cost (as per polynomial cost function)	Crore ₹	616.22	621.53	628.65	637.56	648.27	660.79
R^2		0.9733					
APE		0.23	0.90	0.44	0.22	0.44	0.37
MAPE		0.43					
Overall cost (as per power cost function)	Crore ₹	611.58	624.24	634.45	643.02	650.43	656.95
R^2		0.9070					
APE		0.53	0.46	1.37	0.63	0.11	0.94
MAPE		0.67					

References

1. A. Koul, S. John, A life cycle cost approach for evaluation of sewage treatment plants. Int. J. Innov. Res. Adv. Eng. (IJIRAE) **2**(7) (2015) ISSN:2349-2163
2. A.W. Sekandari, Cost comparison analysis of wastewater treatment plants. Int. J. Sci. Technol. Eng. **6**(1) (2019)
3. A. Giovanna, et al., Water Treatment Emergency: Cost Evaluation Tools, MDPI—07 (2019)
4. E. Friedler, E. Pisanty, Effects of design flow and treatment level on construction and operation costs of municipal wastewater treatment plants and their implications on policy making. Water Res. **40**(2), 3751–3758 (2006)
5. Gautam et al., Cost-effective treatment technology for small size sewage treatment plants in India. J. Sci. Ind. Res. **76**, 249–254 (2017)
6. M.K. Gratziou, M. Tsalkatidou, N.E. Kotsovinos, Economic evaluation of small capacity sewage processing units. Global Nest J. **8**(1), 52–60 (2006)
7. A.U.A. Arif et al., Cost analysis of activated sludge and membrane bioreactor WWTPs using CapdetWorks simulation program, Case study of Tikrit WWTP (middle Iraq). Alexandria Eng. J., Hosted by ELSEVIER (2020)
8. S. Jafarinejad, Cost estimation and economical evaluation of three configurations of activated sludge process for a wastewater treatment plant (WWTP) using simulation. Appl. Water Sci. Springer (2016)
9. Y. Tian, Selection of municipal wastewater treatment process based on improved analytic hierarchy process. IOP Conf. Ser. Earth Environ. Sci. (2019)
10. M. Zaerpour, Design, Cost & Benefit Analysis of a Membrane Bioreactor, Thesis Submitted in Partial Fulfilment of the Requirements for the Master's Degree – Politecnico di Milano – Department of Environmental and Geomatic Engineering (2014)
11. T. Guo et al., *Review of Cost Versus Scale: Water and Wastewater Treatment and Reuse Processes* (IWA Publishing – Water Science & Technology, 2014), p. 69.2
12. Ministry of Urban Development – Government of India, Recent Trends in Technologies in Sewerage System (2012)
13. C. Visvanathan et al., Membrane separation bioreactors for wastewater treatment. Crit. Rev. Environ. Sci. Technol. **30**(1), 1–48 (2010)
14. B. Sengupta, S. Mukherjee, Development of cost functions for biological treatment by membrane bioreactor. Global NEST J. **25**(8), 146–155 (2023)
15. Central Public Health and Environmental Engineering Organization in collaboration with JICA, Manual on Sewerage and Sewage Treatment Systems (2013)
16. Metcalf & Eddy, Inc., *Wastewater Engineering – Treatment and Reuse*, 4th edn. (McGraw Hill)
17. USEPA, Process Design and Cost Estimating Algorithms for the Computer Assisted Procedure for Design and Evaluation of Wastewater Treatment Systems (CAPDET) (1982)
18. Public Works Department (PWD), Government of India, Schedule of Rates (latest publication) of 2021

A Simplistic Method for Measurement of Bioaerosol in Indoor Air

Abhra Saha and Debabrata Mazumder

Abbreviations

CFU	Colony-forming units
LPM	Liter per minute
STP	Standard temperature and pressure
TSA	Tryptic soya agar

1 Introduction

Indoor air quality is a pivotal determinant of human health and comfort, especially considering substantial amount of time individuals spend within enclosed spaces. Amidst various factors influencing indoor air quality, the presence of bioaerosols emerges as a critical yet often overlooked aspect. Bioaerosols encompass minuscule airborne particles loaded with living microorganisms, encompassing a spectrum of entities such as bacteria, viruses, fungi, pollen, and allergens [1]. The significance of monitoring and managing bioaerosols in indoor environments cannot be overstated, particularly in settings where the well-being of occupants is paramount, including homes, offices, schools, and healthcare facilities [2]. Microorganisms suspended in the air contribute significantly to airborne pollutants, accounting for 7–35% of the total pollutants present [3]. In-depth understanding of the composition and concentration of bioaerosols within indoor air is crucial for multiple reasons, ranging from controlling the spread of diseases to reducing allergic reactions and ensuring an

A. Saha (✉) · D. Mazumder
Department of Civil Engineering, Indian Institute of Engineering Science and Technology, Shibpur, Howrah, West Bengal, India
e-mail: debabrata@civil.iiests.ac.in

D. Mazumder (ed.), *Sustainable Advanced Technologies for Environmental Management*, Springer Proceedings in Earth and Environmental Sciences,
https://doi.org/10.1007/978-3-031-64006-3_19

overall comfortable indoor environment [4]. Consequently, a diverse array of sophisticated bioaerosol-monitoring techniques has been developed, aiming to detect, quantify, and analyze these airborne particles effectively. Airborne indoor bacteria are mostly collected by using impinger and impactor for culture-based examination [5].

The microbial parameters in indoor environments are influenced by factors such as the number of individuals present, their activities, and the movement of dust, contributing to elevated microbial levels [6]. Individuals residing in densely populated areas face an increased likelihood of exposure to potentially harmful biological pollutants [7]. The transmission of microbes in communal settings such as schools, colleges, and hospitals has been associated with health concerns. The microbiological quality of air and ventilation within educational institutions has been shown to impact the health and academic performance of children. Elevated airborne fungus levels have also been correlated with persistent coughing and wheezing [8]. Moreover, there is a heightened risk of lower respiratory diseases in children compared to adults, with increased susceptibility observed in those attending daycare centers and primary schools. The culturable bioaerosol sample may be increased by 43% for *Staphylococcus epidermidis*, 27% for *Pseudomonus fluoresecens*, and 12% for *Escherichia coli* condensational growth compared to non-condensational growth [9].

Notably, this study aims to present a novel approach to measuring bioaerosols using a low-cost experimental setup. By doing so, we can gain valuable insights into the presence and concentration of bioaerosols in indoor environments, ultimately contributing to the improvement of air quality standards and public health. This novel method offers a practical and cost-effective way to gather important data on bioaerosols, which are often overlooked in indoor air quality assessments. By implementing this approach, we can enhance our understanding of the potential health risks associated with indoor bioaerosols and develop targeted strategies to mitigate their impact. The study also contributes to this field by introducing a novel and simplistic mechanism based on membrane filtration technique. This innovative approach enables the separation of bioaerosols from indoor air by means of membrane filtration. Subsequently, the bacterial species present in the bioaerosols are cultured on Tryptic Soy Agar, a widely used growth medium. The filter paper is strategically placed over the Tryptic Soy Agar within a Petri dish, and incubated at a controlled temperature of 35 ± 2 °C. The resulting number of colony units formed, in conjunction with the total volume of airflow, serves as a quantitative measure to express the density of bioaerosols. This study also aims to contribute valuable insights into the prevalence and characteristics of bioaerosols within indoor environment, with potential implications for selecting strategies related to ventilation, air purification, and infection control measures. Previous literature has highlighted various challenges in accurately measuring bioaerosols in indoor air, including the complexity and high cost of traditional sampling equipment. Additionally, the need for specialized training and technical expertise has been a barrier to widespread adoption of bioaerosol measurement techniques. In contrast, the simplistic method introduced in this study aims to address these challenges by providing a cost-effective and user-friendly alternative.

2 Materials and Methods

2.1 Chemicals

Tryptic Soy Agar (TSA, HIMEDIA, M1968) was used as a growing medium for bioaerosols during the production of bioaerosol food [10]. In order to prevent bacterial development in fungal culture medium, a 20 mL stock solution of cycloheximide (C81040) was made in this procedure [11]. An amount of 10 mg of pure cycloheximide was dissolved in 96% pure ethanol to create the cycloheximide stock solution. Cycloheximide, a crucial ingredient in fungal culture media, is a fungicide that specifically prevents the development of fungus. This stock solution is useful for establishing a fungus-free environment that is favorable for facilitating bacterial development.

2.2 Instruments

Throughout the course of the study, diverse tools were utilized to fulfill the particular demands of the study. The air movement was measured and managed using an airflow meter (JAPSIN), which has the capacity to handle up to 100 liters per minute (LPM). A Shimadzu weighing equipment with a maximum capacity of 220 g was used to quantify compounds precisely. A filter paper with pore sizes of 0.45 micrometers (μ) was used as the particle collection medium in order to efficiently catch bioaerosols. The membrane filter configuration, which included an air suction pump (RESUN ACO-003) was employed to enable the gathering of bioaerosols on the filter paper's surface. This setup was crucial for capturing and analyzing airborne particles in the study, ensuring accurate and controlled conditions for the investigation of bioaerosols.

2.3 Bioaerosol-Measuring Kit

A rectangular-shaped kit was used for measuring bioaerosols. The dimensions of the kit are kept as 40 cm × 25 cm × 15 cm (length × breadth × height). The monitoring unit contained one air suction pump, a membrane filter, and a rotameter, which are used to measure the airflow rate. The maximum capacity of the rotameter was 100 LPM. A membrane filter unit has two sections: the upper section sucks the air passing through the lower section by means of air suction pump. The maximum capacity of the air suction pump was 65 LPM. It also includes ethanol as sterilizer, tissue paper, and glass tube in addition to all of the devices.

2.4 Preparation of Agar

First, a requisite amount of agar was dissolved in boiled water completely. Thereafter, the agar solution was sterilized by autoclaving at 15 lbs (121 °C) for 15 min. After thorough mixing and cooling down to 45–50 °C, the agar solution was poured into sterile Petri dishes.

2.5 Sampling Procedure

A bioaerosol-measuring kit has been developed to detect the presence of microbial contamination in indoor air. Bioaerosol from different parts of the institute (Indian Institute of Engineering Science and Technology (IIEST), Shibpur) building was collected periodically. Tryptic soya agar was prepared by suspending 0.45 g in 10 mL of distilled water and then heated it to boiling for dissolving the medium completely. The media was subsequently sterilized by autoclaving at 15 pounds of pressure at 121 °C for 15 min. Thereafter, it was cooled to 45–50 °C before mixing well and pouring into a sterile Petri dish. After partial settlement, an air suction pump (ACO 003) was used with a suction rate of 30 LPM. A membrane filter of pore size 0.45 μ facilitates the suction air to pass through it and thereby to capture the bioaerosol on the surface of the filter paper. A flow meter (i.e., max capacity 100 LPM) was used to detect the flow of the pump at the time of suction. Sampling was continued for 5 min after placing the kit on a stool at a height of 1 m above the ground while remaining at least 1 m away from any physical barriers such as walls, windows, and entrance doors. The upper surface of the filter paper was observed to identify the presence of bioaerosol when it was kept in the Petri dish for 3 days at a temperature of 37 °C in a bacteriological incubator. Schematic diagram and photo image of the bioaerosol-measuring unit are shown in Fig. 1a, b, respectively.

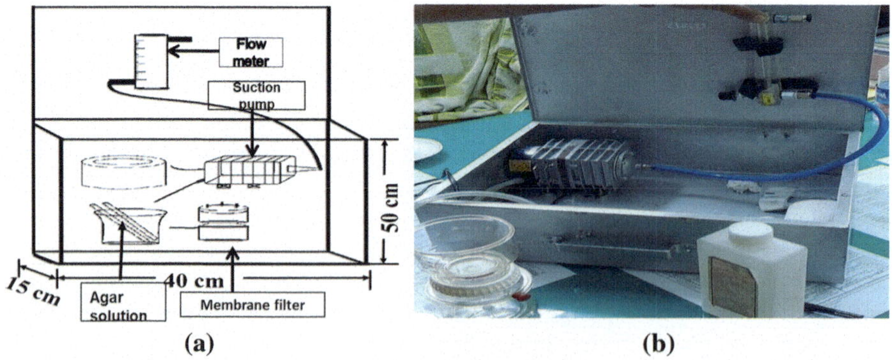

Fig. 1 (**a**) Schematic diagram of bioaerosol-measuring kits, (**b**) image of bioaerosol-monitoring unit

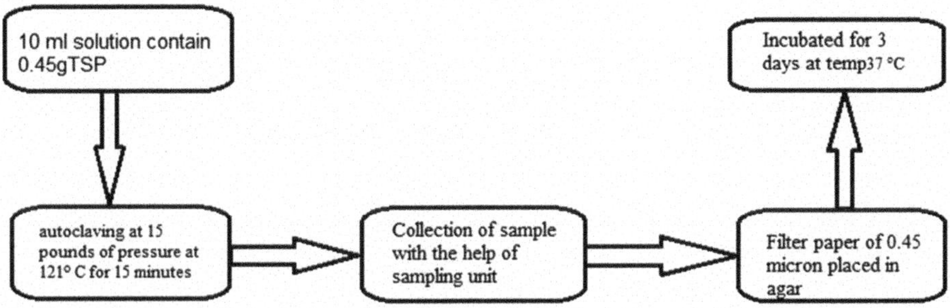

Fig. 2 Flow chart of preparation of agar and operation of bioaerosol-measuring unit

The step-by-step procedure for measuring bioaerosol from indoor air is depicted through a flow chart as shown in Fig. 2.

3 Results and Discussion

Samples were systematically collected from various classrooms within IIEST, Shibpur to conduct a thorough characterization of bioaerosol colonies. The outcomes of the bioaerosol-monitoring process are comprehensively presented in Table 1. Upon visual examination of the Petri dishes containing the sampled bioaerosols, it became evident that the colonies displayed distinct features, including varying colors, shapes, and orientations. This visual observation facilitated the identification of specific bacterial species, namely *Staphylococcus aureus*, *Escherichia coli*, and *Pseudomonas aeruginosa* as illustrated in Fig. 3.

The findings highlighted that the predominant bioaerosol colonies in the sampled indoor air were composed primarily of bacterial species as mentioned above. Importantly, the quantity of these identified bacteria exhibited variations across various locations within the building. These variations can be attributed to several environmental factors such as ventilation efficiency, airflow rates, and temperature disparities. The detailed data in the table further elucidate the differences in colony-forming units per cubic meter at standard temperature and pressure (CFU/m^3 at STP) between various locations. This study not only emphasizes the diversity and prevalence of bioaerosol colonies but also underscores the influence of environmental conditions on microbial concentrations within different areas of a typical technical institute building.

Figure 3 clearly exhibits that there is formation of different colors as well as shapes depending on the characteristics of bacterial species. On the basis of different colors, shapes, and orientations, the presence of *Staphylococcus aureus*, *Escherichia coli*, and *Pseudomonas aeruginosa* has been identified. The numbers of colony-forming units (CFU) are expressed per m^3 volume of sampled air at the standard temperature of 25 °C and 1 atm. pressure (STP) by applying necessary correction in volume. The total colony-forming units per m^3 at STP estimated at various locations

Table 1 Results of bioaerosol monitoring in various class rooms of IIEST, Shibpur

Sampling location	Ambient temperature (° C)	Duration of sample (min)	No. of colonies	CFU/ m^3	Identification of colonies
Block 93s (F)	32	5	4	27	*S. aureus* = 2 *E. coli* = 2 *P. aeruginosa* = 0
Block 93s (R)	32	5	10	68	*S. aureus* = 6 *E. coli* = 2 *P. aeruginosa* = 1
Environmental Lab Instrument Room	37	5	8	56	*S. aureus* = 4 *E. coli* = 2 *P. aeruginos* = 1
Air Quality Lab	32	5	5	34	*S. aureus* = 2 *E. coli* = 3 *P. aeruginos* = 0
Geotechnical Lab	34	5	23	158	*S. aureus* = 7 *E. coli* = 10 *P. aeruginos* = 6
Concrete Lab Class Room	35	5	15	103	*S. aureus* = 7 *E. coli* = 3 *P. aeruginos* = 5
Survey Lab	30	5	9	61	*S. aureus* = 3 *E. coli* = 1 *P. aeruginos* = 3
Carpentry Shop	30	5	12	81	*S. aureus* = 4 *E. coli* = 3 *P. aeruginos* = 5

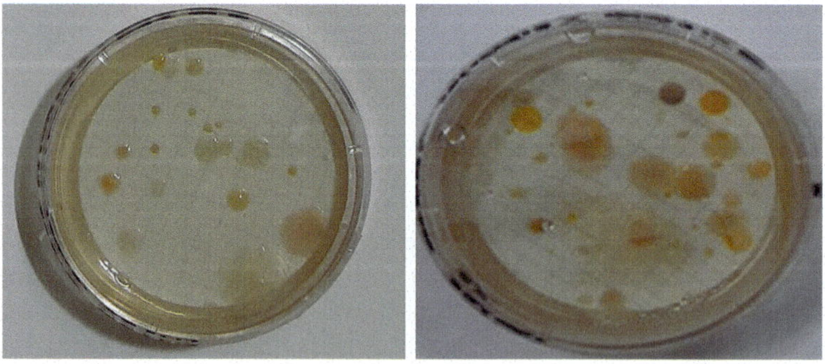

Fig. 3 Typical bioaerosol colonies in the Petri dish

are shown with a scatter diagram as shown in Fig. 4. On the other hand, colony-forming units of *Escherichia coli*, *Staphylococcus aureus*, and *Pseudomonas aeruginosa* per m^3 estimated at various places are shown with scatter diagram as shown in Fig. 5a–c, respectively.

Fig. 4 Total CFU/m^3 at STP at various places

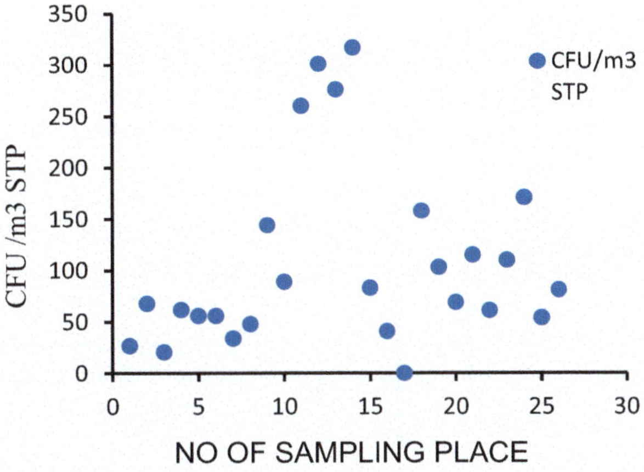

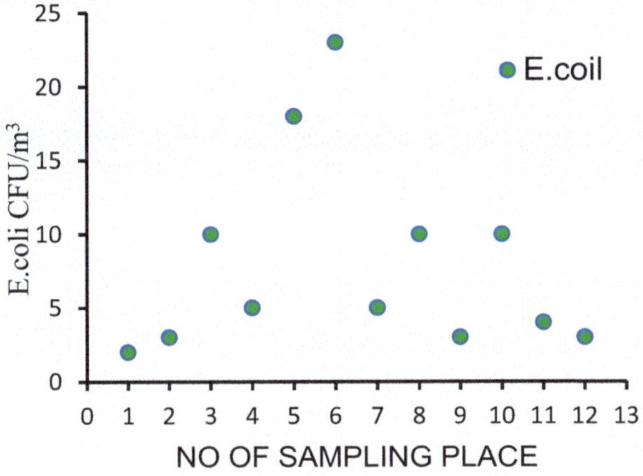

Fig. 5 (**a**) *E. coli* CFU/m^3 at various places, (**b**) *S. aureus* CFU/m^3 at various places, (**c**) *P. aeruginosa* CFU/m^3 at various places

The results indicate a lack of strong correlation between the sampling location and colony-forming units (CFU), underscoring the influence of diverse factors such as room size, percentage of openings, and ventilation rates on bioaerosol concentrations. As a whole, the presence of bioaerosols is more or less observed in each sampled area. The absence of a strong correlation suggests that the microbial concentrations are not solely determined by the geographical location within the institute building. Instead, variables such as room size and ventilation rates contribute significantly to the observed variations in CFU. Larger rooms or areas with higher ventilation rates may exhibit different bioaerosol concentrations compared to smaller spaces with limited airflow.

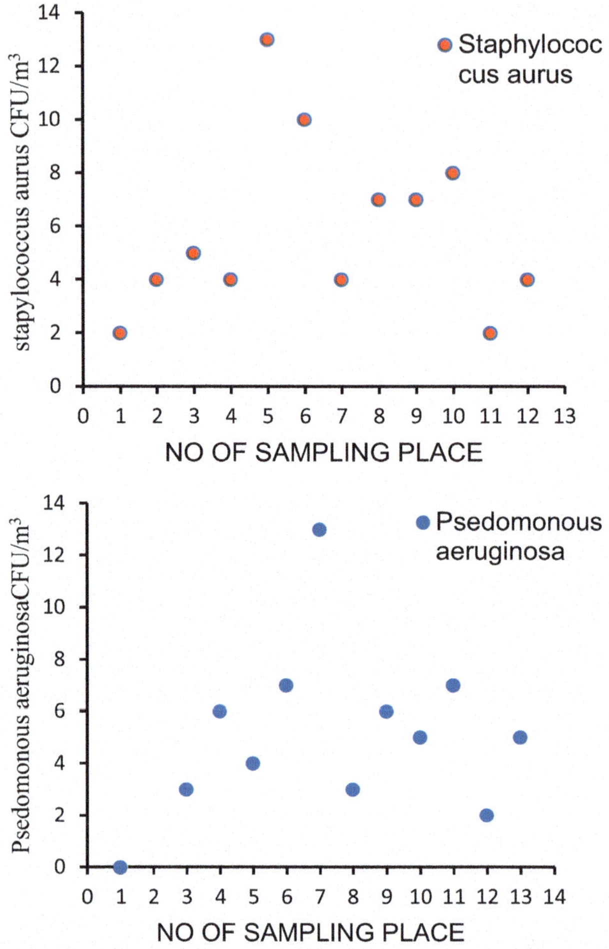

Fig. 5 (continued)

Despite the lack of a strong correlation, the consistent observation of bioaerosols in each location emphasizes the ubiquitous presence of these airborne particles. *Staphylococcus aureus*, *Escherichia coli*, and *Pseudomonas aeruginosa* are evidently present, showcasing the diversity of microbial colonies within the indoor environment. The results also highlighted the complexity of indoor air quality dynamics, where multiple factors interact to influence the concentration and distribution of bioaerosols. Understanding these intricacies is crucial for implementing effective strategies to manage and improve indoor air quality in various settings.

4 Conclusion

In conclusion, the novel method employed in this study demonstrated effectiveness in estimating bioaerosols in indoor air, shedding light on the prevalent microbial species, including *Staphylococcus aureus*, *Escherichia coli*, and *Pseudomonas aeruginosa*. Despite conducting the sampling under controlled conditions with a consistent airflow rate and an average room temperature of 35 °C, variations in bacterial concentration were evident. This suggests that other influencing factors, such as ventilation, airflow rate, and temperature played a substantial role in shaping the final bioaerosol composition. The visual identification of bacterial colonies through Petri dish examination, considering unique characteristics such as color, shape, and orientation, added depth to the understanding of microbial presence in indoor air. Future research endeavors should delve deeper into the intricacies of indoor air quality and its potential implications for occupant health and well-being. By gaining a more rational understanding of all associated factors, effective strategies can be developed to monitor bioaerosol in indoor air.

Acknowledgments This research work is supported by the research fellowship program of the Ministry of Human Resource Development, Government of India.

References

1. A.R.K. Gollakota, S. Gautam, M. Santosh, H.A. Sudan, R. Gandhi, V. Sam Jebadurai, C.M. Shu, Bioaerosols: Characterization, pathways, sampling strategies, and challenges to geo-environment and health. Gondwana Res. **99**, 178–203 (2021). https://doi.org/10.1016/j.gr.2021.07.003
2. T. Salthammer, E. Uhde, T. Schripp, A. Schieweck, L. Morawska, M. Mazaheri, S. Clifford, C. He, G. Buonanno, X. Querol, M. Viana, P. Kumar, Children's well-being at schools: Impact of climatic conditions and air pollution. Environ. Int. **94**, 196–210 (2016). https://doi.org/10.1016/j.envint.2016.05.009
3. E. Bragoszewska, A. Mainka, J.S. Pastuszka, Concentration and size distribution of culturable bacteria in ambient air during spring and winter in Gliwice: A typical urban area. Atmosphere (Basel) **8** (2017). https://doi.org/10.3390/atmos8120239
4. Y. Wen, J. Leng, X. Shen, G. Han, L. Sun, F. Yu, Environmental and health effects of ventilation in subway stations: A literature review. Int. J. Environ. Res. Public Health **17** (2020). https://doi.org/10.3390/ijerph17031084
5. J. Cox, H. Mbareche, W.G. Lindsley, C. Duchaine, Field sampling of indoor bioaerosols. Aerosol Sci. Technol. **54**, 572–584 (2020). https://doi.org/10.1080/02786826.2019.1688759
6. S.W. Kembel, J.F. Meadow, T.K. O'Connor, G. Mhuireach, D. Northcutt, J. Kline, M. Moriyama, G.Z. Brown, B.J.M. Bohannan, J.L. Green, Architectural design drives the biogeography of indoor bacterial communities. PLoS One **9**, 1–10 (2014). https://doi.org/10.1371/journal.pone.0087093
7. S.G. Al-Kindi, R.D. Brook, S. Biswal, S. Rajagopalan, Environmental determinants of cardiovascular disease: Lessons learned from air pollution. Nat. Rev. Cardiol. **17**, 656–672 (2020). https://doi.org/10.1038/s41569-020-0371-2

8. J. Hu, N. Li, Y. Lv, J. Liu, J. Xie, H. Zhang, Investigation on indoor air pollution and childhood allergies in households in six Chinese cities by subjective survey and field measurements. Int. J. Environ. Res. Public Health **14**, 1–21 (2017). https://doi.org/10.3390/ijerph14090979
9. J.Y. Park, D.S. Yoon, S. Li, J. Park, J. Il Bang, M. Sung, J.W. Jeong, Empirical analysis of indoor air quality enhancement potential in a liquid-desiccant assisted air conditioning system. Build. Environ. **121**, 11–25 (2017). https://doi.org/10.1016/j.buildenv.2017.05.011
10. A.U. Uzoechi, T.N.N. Obi, P.C. Nnagbo, C.N. Ohalete, V.I. Anyiam, Microbiological evaluation of indoor air quality of state university library. Asian J. Appl. Sci. **05**, 2321–0893 (2017). www.ajouronline.com
11. N. Rouhizohrab, F. Mohammadipanah, Suppression of predominant interfering bacteria in the purification process of myxobacteria, Iran. J. Microbiol. **14**, 721–729 (2022). https://doi.org/10.18502/ijm.v14i5.10968

Adsorptive Capacity of Surface-Modified Red Soil-Based AAB for Removal of Anionic Dye

S. Das, B. Mondal, D. Adak, S. P. Swain, and S. Shome

Abbreviations

AAB	Alkali-activated binder
CTAB	Cetyltrimethylammonium bromide
MO	Methylene orange
rpm	Revolutions per minute

1 Introduction

The inorganic polymer alkali-activated binder (AAB) is formed by aluminate and tetrahedral silicate units joined by common oxygen atoms. Alumina–silicate materials such as slag, fly ash, biomass ash, and kaolin can be used to make alkali-activated concrete [1]. Alkali-activated concrete is also known as geopolymer concrete. Geopolymer is a common adsorbent with three-dimensional structures ranging from amorphous to semicrystalline. Numerous studies have demonstrated the ability of geopolymers to absorb dyes and heavy metals. Tetrahedral aluminates in geopolymers have a negative charge, but substitutable cations can balance this charge. The ability of the geopolymer to exchange anion can be enhanced by surface modification using a cationic surfactant [2, 3].

S. Das (✉) · S. Shome
Haldia Institute of Technology, Haldia, India

B. Mondal
Guru Ghasidas Vishwavidyalaya, Bilaspur, India

D. Adak · S. P. Swain
National Institute of Technology, Meghalaya, Shillong, India

D. Mazumder (ed.), *Sustainable Advanced Technologies for Environmental Management*, Springer Proceedings in Earth and Environmental Sciences,
https://doi.org/10.1007/978-3-031-64006-3_20

The printing and dyeing industry has undergone substantial expansion, which has resulted in the rise of dye wastewater as a significant contributor to the pollution of water sources. Methyl orange (MO) is one of the chemicals that can be found in the wastewater produced during the dyeing process. This wastewater contains a substantial quantity of chemicals that are hazardous to both human health and the environment. The MO dye is an anionic dye that is not only toxic but also carcinogenic due to its toxicity. Consequently, the elimination of dye from wastewater in a timely and efficient manner has grown to become an urgent environmental concern. The elimination of dye from wastewater has been the subject of extensive research that has been conducted utilizing a variety of methods, such as photocatalytic degradation, biological degradation, electrochemical oxidation, and adsorption. When it comes to the removal of MO dye, adsorption is a method that is recommended owing to the fact that it is simple to apply, performs exceptionally well, and is economical. Many different types of adsorbents are used to remove MO dyes. These include activated carbon, biosorbent, charcoal, clays, minerals, polymers, nanoparticles, resins, and composites [4, 5].

The current study aims to remove MO dye by altering the surface of red soil-based AAB using the cationic surfactant cetyltrimethylammonium bromide (CTAB). We are interested in to conduct trials with varying contact times and dosages in order to determine the effectiveness of the surface-modified red soil-based AAB as an adsorbent for MO dye. The study will focus on adsorption kinetics and isotherm models.

2 Materials and Methods

2.1 Chemical Reagents

Chemicals used for the study are as follows:

1. Methyl orange (MO) dye was bought from Sigma Aldrich, with a molecular weight of 327.33 g/mol and a formula of $C_{14}H_{14}N_3NaO_3S$.
2. Laboratory grade liquid sodium silicate (Na_2Si0_3) was bought from LobaChemie Pvt. Ltd.
3. Laboratory grade sodium hydroxide (NaOH) pellets was bought from Merck Pvt. Ltd.
4. Merck Pvt. Ltd. provided laboratory grade cetyltrimethylammonium bromide (CTAB) with a molecular weight of 364.45 g/mol and a formula of $C_{19}H_{42}BrN$.

2.2 Preparation of Adsorbent

Samples of red soil were collected from the Shillong region in the state of Meghalaya, India. The components, which included sand, red soil, GGBS, and fly ash, were mixed together in a dry manner for a period of 2 min. This was followed by the addition of silicate solutions (Na_2SiO_3/K_2SiO_3) and alkali sources (NaOH/KOH), which were thoroughly mixed for an additional 5 min to ensure that the mixture was uniform. Finally, it was placed into a cube mold measuring 70.6 mm in size. Following a period of 24 h of casting, the samples were extracted from their respective molds and left to cure in the laboratory at room temperature for a further 7 days [6, 7]. Then, the samples were ground into powder and sieved into three distinct size ranges: 150, 300, and 600 μm [6]. In order to produce the modified AAB, a 0.01 M CTAB surfactant solution was combined with the crushed binder and processed for a period of 2 h. Then, the modified binder was washed with distilled water and then dried in an oven at a temperature of 100 °C for 1 h [8].

2.3 Instruments Used

A Shimadzu AUX220 high precision electronic balance was used for the task of weighing. A spectrophotometer was used to measure absorbance (Thermo-Scientific-Visible_SPECTRONICTM 200).

2.4 The Analytical Approach

A spectrophotometric method at a wavelength of 465 nm was used to determine the concentrations of MO within 0–12 mg/L. To determine the unknown MO concentration, a calibration curve was created in between absorbance and dye amount (Absorbance = 0.0709 Concentration (mg/L) − 0.0114, with the value of $r^2 = 0.9989$).

2.5 Experimentation

In batch investigations, an MO sample spiked with distilled water was utilized at 27 ± 2 °C. 150 rotations per minute were applied to the samples using a mechanical shaker. The adsorbent dosage and beginning MO concentration were 6 and 50 mg/L, respectively, for each experiment. A 50 mL MO sample was utilized to examine the effects of different parameters on MO removal. The effects of contact duration (0–60 min), adsorbent amount (1–10 g/L), and adsorbent size (150–600 μm) were investigated. The following equation [9] represents the reduction of MO (%):

$$R = \frac{C_0 - C_t}{C_0} \times 100 \quad (1)$$

where C_0 is the opening concentration of methyl orange and C_t is the MO percentage at any given time t.

2.6 Adsorption Kinetics

Four alternative kinetic models were examined in an effort to develop a greater understanding of the solvent adsorption process by the adsorbent. Among the models that were investigated were the first-order model [10], which is dependent on the concentration of the solution; the pseudo-first-order model [11], which is dependent on the capacity of the solid; the second-order model [10], which is dependent on the concentration of the solution; and the pseudo-second-order model [12], which is dependent on the sorption of solid phase. A comparison and analysis of these four models was carried out in order to determine which sorption method was the most viable option.

Here are the mathematical equations for four different models:

$$\text{First} - \text{order kinetic model}: \ \ln C_t = \ln C_0 - k_1 t \quad (2)$$

$$\text{Pseudo} - \text{first} - \text{order kinetic model}: \frac{dq_t}{dt} = k_{S1}(q_e - q_t) \quad (3)$$

$$\text{Second} - \text{order kinetic model}: \frac{1}{C_t} - \frac{1}{C_0} = k_2 t \quad (4)$$

$$\text{Kinetic model of pseudo} - \text{second} - \text{order}: \frac{1}{q_t} = \frac{1}{k_{S2} q_e^2} + \frac{1}{q_e} t \quad (5)$$

where

C_t = concentration of MO at any given time t
C_o = MO initial concentration
q_t = quantity of MO adsorbed at time t/the adsorbent's weight at time t
q_e = MO adsorbed per unit weight of adsorbent at equilibrium
k_1 = first-order reaction constant
k_{S1} = pseudo-first-order reaction model rate constant
k_2 = second-order reaction model constant
k_{S2} = pseudo-second-order reaction model rate constant

3 Results and Discussion

Performance of Red Soil-Based AAB and Surface-Modified Red Soil-Based AAB as MO Dye Adsorbent

Performance of red soil-based AAB and surface-modified red soil-based AAB as an adsorbent for MO dye was studied by varying particle size.

3.1 The Significance of Size of the Particles

The removal procedure in a batch study may be influenced by particle size. The removal efficiency will rise when the particle size is reduced [13, 14]. Three distinct ranges were used to study the impact of size: 150, 300, and 600 μm. The initial MO concentration was 50 mg/L, and 6 g/L of each adsorbent was used. The result is depicted in Fig. 1.

The elimination of MO is seen to vary from 32% to 84% for surface-modified red soil-based AAB and 6–18% for normal red soil-based AAB. The particle with the smallest size produced the maximum removal. This is because smaller particles have more surface area and, as a result, more adsorption sites. It can be seen that surface-modified AAB has more removal efficiency than normal AAB. Therefore, the surface-modified AAB with the smallest particle size (150 μm) was selected as the study's sole adsorbent.

Figure 2 illustrates the CTAB cationic surfactant's presence on the negatively charged red soil-based AAB surface. This modification would improve the adsorption capability of red soil-based AAB on MO [15]. The negatively charged geopolymer surface can be altered to a positive charge by using a cationic surfactant called CTAB. Methyl orange (MO) is an anionic dye that is absorbed through

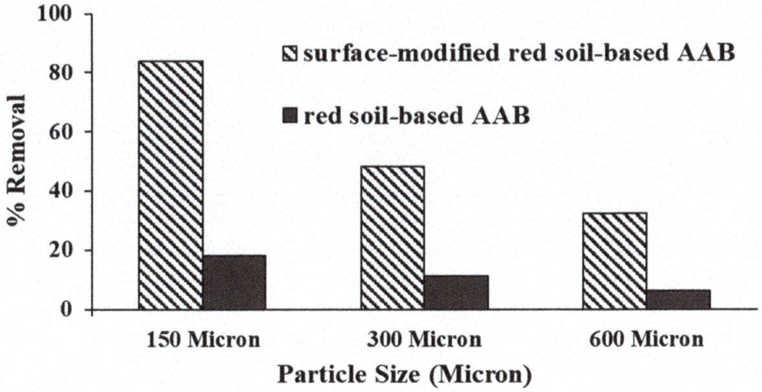

Fig. 1 Effect of particle size on MO adsorption onto surface-modified red soil-based AAB and red soil-based AAB

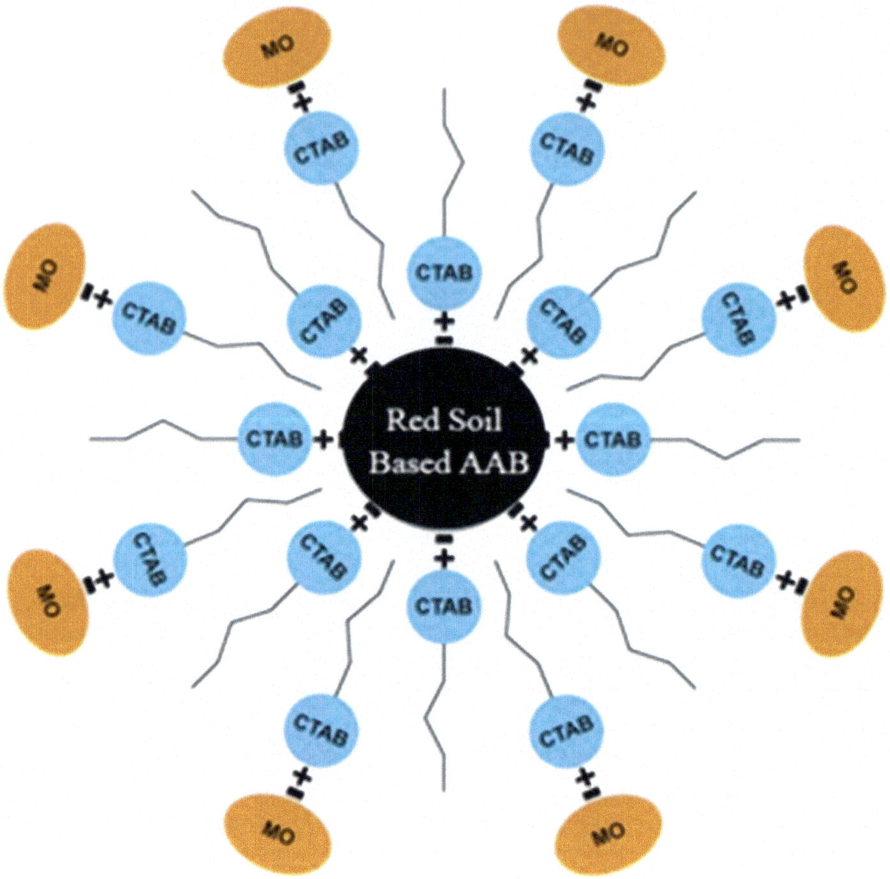

Fig. 2 A typical instance of how surface-modified red soil-based AAB adsorbs MO

electrostatic interaction with the positively charged head groups of cationic surfactants (CTAB).

3.2 The Effect of Time

The effect of time on the efficiency of MO dye removal by surface-modified red soil-based AAB is depicted in Fig. 3, which shows the relationship between time and adsorption. A solution of MO dye with a concentration of 50 mg/L was used for the adsorption technique, which was carried out using a volume of 50 mL. Adsorbent was applied at a dosage of 6 g/L. When using surface-modified red soil-based AAB as an adsorbent, the effectiveness of the removal of MO dye increased over time and reached a level that was typically consistent after 45 min. The presence of a large

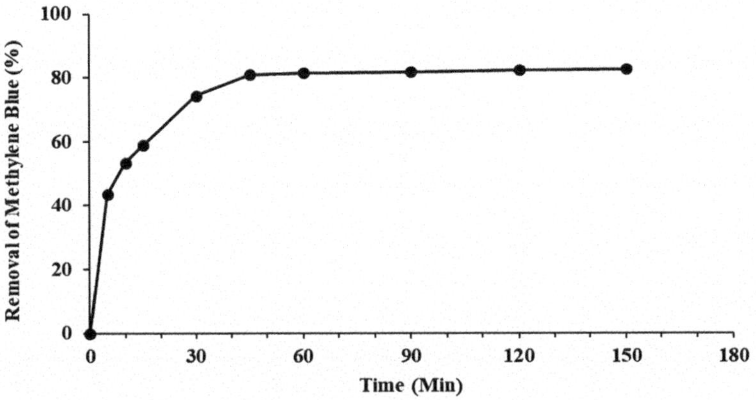

Fig. 3 Influence of time on the efficacy of MO removal

number of active sites on the surface of the adsorbent contributed to the rapid efficiency of the removal of MO dye during the early phase of the adsorption process. As soon as the system reached the equilibrium stage, the number of active sites on the surface of the adsorbent decreased. This led to a reduction in the rate at which the MO dye was removed from the system.

3.3 Adsorbent (Surface-Modified Red Soil-Based AAB) Dosage Effect

During the course of the adsorption process, the quantity of adsorbent that is utilized (surface-modified red soil-based AAB) has an effect on the number of sites that are available for the purpose of absorbing pollutants deposited on the surface. For the aim of this experiment, the dose of the adsorbent, which was comprised of surface-modified red soil-based AAB, was changed between 1 and 9 g/L. This is illustrated in Fig. 4.

Furthermore, during the whole of the procedure, the concentration of the adsorbate was maintained at a level of 50 mg/L throughout the entire operation. During the course of our investigation, we discovered that the most effective quantity of adsorbent to use was 6 g/L.

3.4 MO Dye Adsorption Kinetics Studies Using Surface-Modified Red Soil-Based AAB

The key components of kinetic investigations in batch experiments are the calculation of reaction coefficients and the contact duration between the solid and solution.

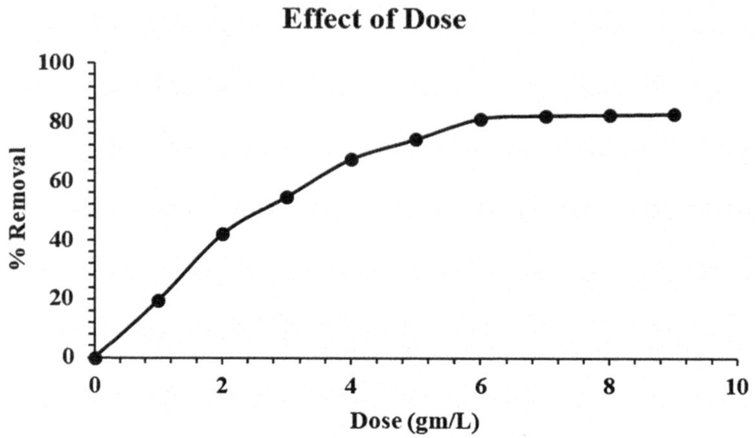

Fig. 4 Adsorbent (surface-modified red soil-based AAB) dosage effect on MO removal efficiency

Table 1 r^2 data and kinetics model equations for MO removal by surface-modified red soil-based AAB

Models	Linear-fit line equations	r^2
Kinetic model of first order	$\ln C_t = 3.1952 - 0.0093t$	0.6224
Model of pseudo-first-order kinetics	$\ln(q_1 - q_t) = 2.056 - 0.0584t$	0.9697
Pseudo-first-order kinetic model	$\frac{1}{C_t} - 0.0403 = 0.0008t$	0.8038
Pseudo-second-order kinetic model	$\frac{t}{q_t} = 0.2706 + 0.07075t$	0.9987

To perform kinetic tests on MO-spiked distilled water samples, a maximum adsorbent dosage of 6 g/L was used. The agitation time was set between 0 and 60 min, and the amount of MO removed by surface-modified red soil-based AAB over time was recorded and shown in Fig. 3. The outcomes showed a 45-min equilibrium time and a high initial rate of adsorption.

Section 2.6 applies four distinct reaction models to analyze the MO adsorption reaction kinetics data via surface-modified red soil-based AAB. Table 1 presents the summaries of the four kinetic models' formulations as well as the stated values of the regression model fitting lines' determining coefficients (r^2). The best-fitting model was the pseudo-second-order model, as shown by a comparison of the appropriate graphs and their r^2 values. The process is the outcome of chemical adsorption or chemisorption, according to this theory.

Table 2 Surface-modified red soil-based AAB for MO removal using Langmuir and Freundlich isotherm model

Isotherm	Parameters	Values
Langmuir isotherm model	Equation	$\frac{1}{q_e} = 0.02161\left(\frac{1}{C_e}\right) + 0.0315$
	The maximum capacity for adsorption, $q_{\max}$ (mg/g)	31.75
	Sorption system energy-related constant, b	1.46
	r^2	0.997
Freundlich isotherm mode	Equation	$\ln q_e = 1.573 \ln C_e + 0.831$
	Freundlich constant's capacity of adsorption, k_f [(mg g^{-1}) (L mg^{-1})$^{1/n}$]	13.14
	Efficacy of adsorption, $1/n$	1.031
	r^2	0.957

3.5 *MO Dye Adsorption Isotherm Studies Using Surface-Modified Red Soil-Based AAB*

The isotherm represents the adsorption process's equilibrium state. The MO dye adsorption by surface-modified red soil-based AAB was investigated using isotherms using Langmuir and Freundlich models. The Freundlich isotherm model takes into consideration multilayer adsorption since there are heterogeneous adsorbent surfaces with different adsorption capabilities, while the Langmuir isotherm model takes into consideration monolayer adsorption as of homogeneous adsorbent surfaces. Comparisons were made between the experimental results and Freundlich and Langmuir isotherm models, which both showed linear trends. However, based on their relative r^2 values, Langmuir isotherm appeared to be much more precise when compared with Freundlich isotherm. The constants were calculated and are shown in Table 2. The surface-modified red soil-based AAB's efficacy for MO absorption was determined to be 31.75 mg/g.

4 Conclusion

It was found that surface-modified red soil-based AAB was an excellent adsorbent for MO removal, with a removal rate of more than 80%. This was in comparison to the natural red soil-based AAB. A particle size of 150 μm was shown to be responsible for the largest quantity of removal that was achieved, according to the findings of the study. According to the findings of the current investigation, the surface-modified AAB adsorbent performed noticeably better than the unmodified samples in terms of the removal efficiency of the samples. As demonstrated by the r^2 value of 0.9987, the pseudo-second-order kinetic model was able to provide a precise illustration of the adsorption process during its development. Based on the findings, it was revealed that the Langmuir isotherm model is followed by the adsorption process. Furthermore, the surface-modified red soil-based AAB adsorbent has a capacity for adsorption, i.e., 31.75 mg/g. After studying the results, it was found that using surface-modified red soil-based AAB as an absorbent is a cheap and effective way to get rid of methyl orange (MO).

References

1. J. Davidovits, Geopolymers: Ceramic-like inorganic polymers. J. Ceram. Sci. Technol. **8**(3), 335–350 (2017)
2. J. Xu, M. Li, D. Zhao, G. Zhong, Y. Sun, X. Hu, J. Sun, X. Li, W. Zhu, M. Li, Z. Zhang, Y. Zhang, L. Zhao, C. Zheng, X. Sun, Research and application progress of geopolymers in adsorption: A review. Nanomaterials **12**, 3002 (2022)
3. T. Selkala, T. Suopajarvi, J.A. Sirvio, T. Luukkonen, P. Kinnunen, A.L.C.B. de Carvalho, H. Liimatainen, Surface modification of cured inorganic foams with cationic cellulose nanocrystals and their use as reactive filter media for anionic dye removal. ACS Appl. Mater. Interfaces **12**, 27745–27757 (2020)
4. A.A. Siyal, M.R. Shamsuddin, M.I. Khan, N.E. Rabat, M. Zulfiqar, Z. Man, J. Siame, K.A. Azizli, A review on geopolymers as emerging materials for the adsorption of heavy metals and dyes. J. Environ. Manag. **224**, 327–339 (2018)
5. K.O. Iwuozor, J.O. Ighalo, E.C. Emenike, L.A. Ogunfowora, C.A. Igwegbe, Adsorption of methyl orange: A review on adsorbent performance. Curr. Res. Green Sustain. Chem. **4**, 10017 (2021)
6. S. Das, B. Mondal, S.P. Swain, D. Adak, S. Shome, Removal of methylene blue from wastewater by red soil-based alkali activated binder, in *International Conference on Multidimensional Sustainability: Advanced Technologies for Industrial Pollution Control*, (Springer Nature Switzerland, Cham, 2022), pp. 105–115
7. D. Adak, S. Mandal, Study on the modified process for the development of fly ash based geopolymer mortar cured at ambient temperature. Indian Concr. J. **89**(1), 31–40 (2015)
8. B. Abbou, I. Lebkiri, H. Ouaddari, A. El Amri, F.E. Achibat, L. Kadiri, A. Ouass, A. Lebkiri, E.H. Rifi, Improved removal of methyl orange dye by adsorption using modified clay: Combined experimental study using surface response methodology. Inorg. Chem. Commun. **155**, 111127 (2023). https://doi.org/10.1016/j.inoche.2023.111127. ISSN 1387-7003
9. L. Wu, X. Liu, G. Lv, R. Zhu, L. Tian, M. Liu, Y. Li, W. Rao, T. Liu, L. Liao, Study on the adsorption properties of methyl orange by natural one dimensional nano-mineral materials with different structures. Sci. Rep. **11**, 10640 (2021)
10. Y.S. Ho, G. Mckay, Pseudo second order model for sorption process. Process Biochem. **34**, 451 (1999)
11. L.D. Benefield, C.W. Randall, *Biological Process Designs for Wastewater Treatment* (Prentice-Hall Inc., 1980), p. 1
12. S. Lagergren, B.K.S. Ventenskapsakad, Handl. 24 (1898), as cited by Wasay et al. Water Res. **30**, 1143 (1996)
13. Y.-S. Shim, Y.-K. Kim, S.-H. Kong, S.-W. Rhee, W.-K. Lee, The adsorption characteristics of heavy metals by various particle sizes of MSWI bottom ash. Waste Manag. **23**(9), 851–857 (2003). https://doi.org/10.1016/s0956-053x(02)00163-0
14. B. Huang, Z. Yuan, D. Li, M. Zheng, X. Nie, Y. Liao, Effects of soil particle size on the adsorption, distribution and migration behaviors of heavy metal(loid)s in soil: A review. Environ. Sci. Process Impacts **22**, 1596 (2020)
15. A. Purbasari, D. Ariyanti, E. Fitriani, J. Ecol. Eng. **24**(3), 90–98 (2023). https://doi.org/10.12911/22998993/157541. ISSN 2299–8993

Part IV
Advanced Wastewater Treatment Technologies

Treatability Study of Fish Market Wastewater in a Batch Reactor Using Suspended Growth Microbial Culture

Mainak Maiti, Abhisek Roy, and Somnath Mukherjee

Abbreviations

ASP	Activated sludge process
BOD	Biochemical oxygen demand
COD	Chemical oxygen demand
Fig	Figure
FMWW	Fish market wastewater
MBBR	Moving bed biofilm reactor
MLSS	Mixed liquor suspended solids
MLVSS	Mixed liquor volatile suspended solids
SBR	Sequential batch reactor
Tab	Table
TDS	Total dissolved solids
TSS	Total suspended solids

1 Introduction

Wastewater released from retail fish and wet markets in municipal and commercial shopping areas contributes a significant pollution load in the urban wastewater collection system. Due to high organic load (BOD/COD) and other nutritional elements such as N, P, etc., the refuse liquid waste turns into a septic condition resulting in a malodorous and obnoxious water environment. Very few literatures are

M. Maiti (✉) · A. Roy · S. Mukherjee
Civil Engineering Department, Jadavpur University, Kolkata, India
e-mail: aroy.civil@jadavpuruniversity.in

D. Mazumder (ed.), *Sustainable Advanced Technologies for Environmental Management*, Springer Proceedings in Earth and Environmental Sciences,
https://doi.org/10.1007/978-3-031-64006-3_21

available in the specific context for appropriate treatment of such wastewater. It is felt that there is an urgent need globally regarding the reduction of such wastewater at its source, especially on-site treatment to make a sustainable solution to such problem. Though the amount of wastewater generated from an organized formal municipal fish market may not be huge, considering the area of a town or city, the total generation of FMWW may be a great concern from the pollution point of view. Also, in the context of the Indian subcontinent, especially, many fish markets comprise a few chicken and meat shops. Hence, significantly high organic waste generated from those shops is also to be considered. FMWW consists of blood, flesh particles, scale, and other different organic components which means it imparts high BOD or COD loading [1]. It also contains high NH_4^+-N, PO_4^{3-}-P, TSS, TDS, etc. like fish processing wastewater [2, 3]. High nutrients generated from fish-related wastewater sources pose threats to aquatic environments [4]. It has been observed that bioreactors are very efficient in removing organic loads from fish processing wastewater [5, 6]. The present study aims to search for a feasible solution for treating FMWW. Moreover, the sustainability issue along with the "Reuse, Recycle, and Recovery" concept is becoming popular due to the global climate and environmental crisis. With this view, a research investigation has been undertaken to examine the possible biodegradability of organic loads containing FMWW in aerobic conditions. A necessary treatability study of FMWW has been done at the Environmental Engineering Laboratory, Civil Engineering Department, Jadavpur University to evaluate the performance and develop a suspended growth bioreactor system.

2 Materials and Methods

2.1 Collection of Real-Life FMWW

FMWW was collected at regular intervals on different days of the month including holidays from the local fish market.

The market generates approximately 1.2–1.5 KLD of wastewater which includes fish washing wastewater, fish cutting wastewater, and chicken cutting wastewater generated from a few chicken shops within the market. The market consists of 50 fish shops and 15 chicken shops approximately. At the time of each collection, 10 random fish shops and 3 different chicken shops were selected. Among the fish shops, different types of fish shops, i.e., shrimp/prawn shops, rohu/katla shops, sea fish shops, dedicated fish cutting shops, etc., were selected. Weekend holidays (Sunday and Saturday), working days, and even the holidays at the time of festival were chosen as collection dates so that the variation would be taken into consideration. Also, random collections from different shops were selected to take care of the probable high variability and obtain parameter values as realistic as possible. From every collection point, 1 L of sample wastewater was collected. After each collection, these samples were brought into the laboratory. Thereafter, a composite was prepared from the wastewater samples, and necessary filtration was carried out as

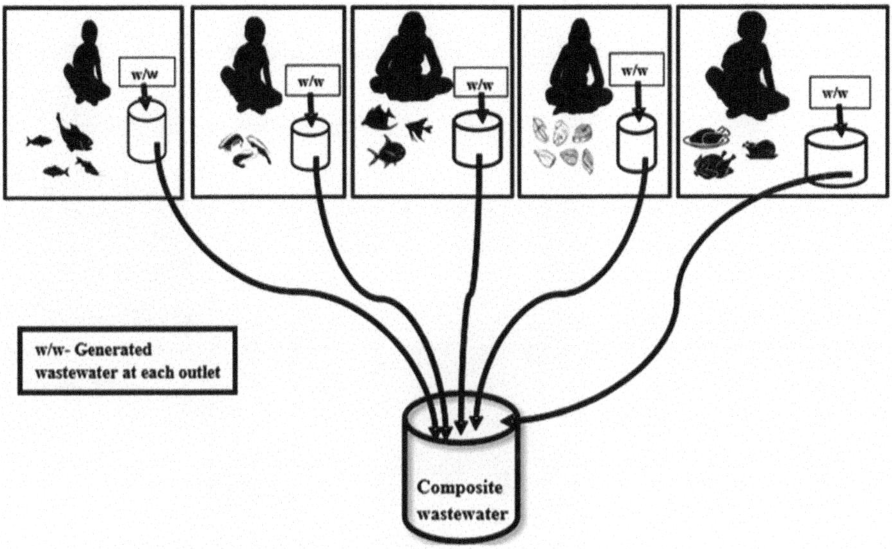

Fig. 1 Schematic sketch of wastewater collection and composite preparation after each collection from all collection points

pretreatment before doing the necessary characterization. Altogether, 20 no. of samples were collected for characterization study. A schematic sketch is presented below in Fig. 1 to give an idea about the collection and composite preparation procedure. Samples were preserved in a deep freezer below −4 °C. pH, TSS, TDS, COD, BOD, NH_4^+-N, and PO_4^{3-}-P were estimated in accordance with Standard Methods (23rd Edition) [7].

2.2 *Seed Acclimatization for Organic Carbon and Ammonia Oxidation Study*

Necessary microbial seeding and seed acclimatization was carried out with stock microbial seed (developed earlier) acclimated with synthetic sample simulated to FMWW. The microbial culture that is needed for the kinetic study was developed separately as shown in Fig. 2. The source of the earlier developed microbial stock seed was derived from the extraction of the sediment collected from the drainage system of fish market. The reactor was filled up with 1800 ml of substrate sample and 200 ml of previously acclimated seeds. A trace amount of conventional synthetic micronutrient solution was being added weekly. When a steady performance is noticed by an increment of mixed liquor volatile suspended solids (MLVSS), the synthetic sample was replaced with real-life wastewater, and corresponding to that, an acclimatization study with real-life FMWW is performed further.

Fig. 2 Seed acclimatization study

In the acclimatization period, the following parameters were maintained as follows: COD as 3000 ± 200 mg/ L, NH_4^+-N as 100 ± 20 mg/ L, and pH as 7.5–8. Air was supplied continuously through a system of aquarium compressors attached with ball-shaped diffusers. Five such air inlet lines were activated through compressors in which the maximum air flow was rated as 4 L/min for each line. It was purposefully done so that the reactor would remain in a completely mixed condition, and the dissolved oxygen level would maintain a level of 7–8 mg/L approximately. The air pumps were being rested intermittently so that their efficiency would not be affected too much. The seed acclimatization took 4 months approximately.

2.3 Reactor Setup and Experimental Method

Experiments have been done through batch mode with real-life FMWW collected from a local fish market situated near the university. The experimental setup used for the study is shown in Fig. 3. Subsequently, the kinetic study was performed in a reactor as shown in Fig. 3. The reactor is 45 cm in height and 14 cm in diameter. Along the different heights of reactors, sampling ports are provided (Fig. 3). Each test run was done for a total volume of 5 L. Initial pH was maintained at 6.8–7.4. Experiments were run at ambient temperatures (30–35 °C). At different sets, diluted wastewater samples were poured with different MLVSS.

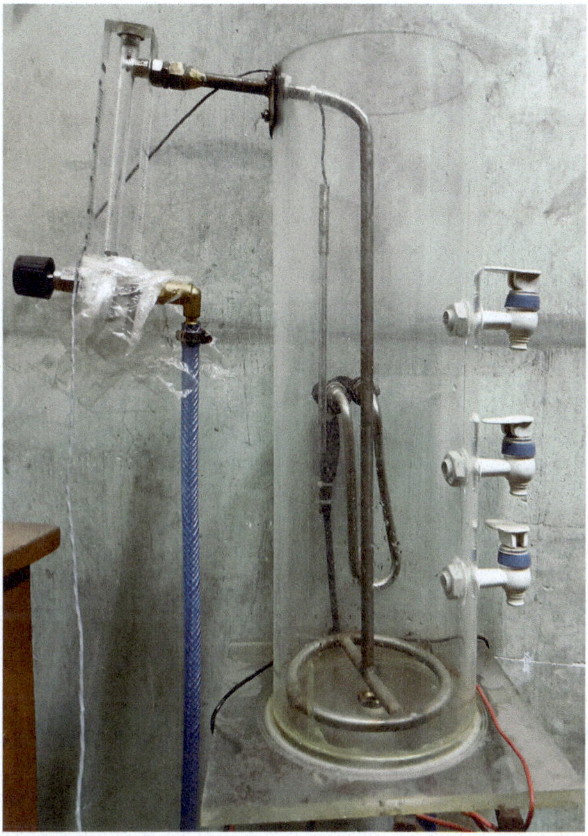

Fig. 3 Experimental setup

Batch studies were conducted for different contact times, and samples were withdrawn accordingly and tested for different parameters. Kinetic studies were conducted by varying COD concentration, contact time, MLVSS, and varying NH_4^+-N concentration. The experimental results were used for determining kinetic constants. All the chemicals used for analytical methods were of AR-grade quality. Kinetic constants, i.e., Y, k_d, k, and K_s, were estimated with the help of Lineweaver-Burk plots. All chemical analyses were done thrice and mean values of them have been taken. A schematic diagram of the reactor setup is given in Fig. 4.

3 Results and Discussion

3.1 Characterization of Real-Life FMWW

The wastewater collected from fish market was characterized analytically both before and after pretreatment. Pretreatment was done by vacuum filtration majorly

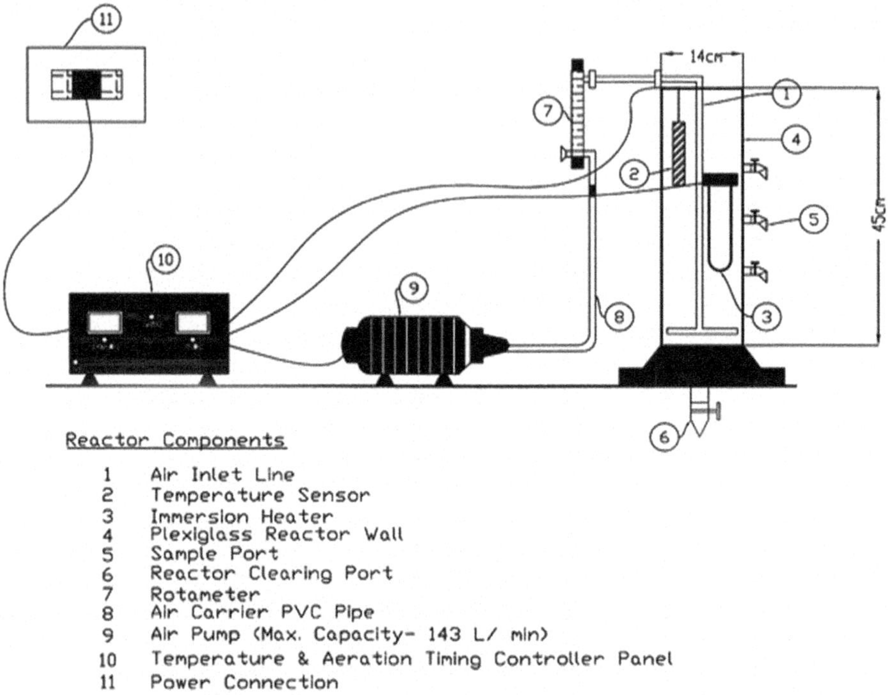

Fig. 4 Schematic diagram of reactor setup for kinetic study

to get rid of the suspended solids. Further, this pretreated wastewater was used for kinetic study. The parameters which were analyzed along with their values before and after the pretreatment are tabulated in Table 1. Higher values of COD, NH_4^+-N, and PO_4^{3-}-P indicate the presence of biodegradable organic and nutrients in larger amount in the collected real-life FMWW.

3.2 Biodegradation Study of Real-Life FMWW in Batch Reactor

Two sets of experiments were run to conduct time-concentration study by investigating organic carbon degradation. In both sets, COD decreased to a certain extent rapidly with respect to time. After that, an almost steady state was observed. 24 hours of contact time was selected. In both sets, the initial pH was set up at 7.5 approximately. In this period, the ambient temperature fluctuated between 30 and 35 °C. The rate of aeration was fixed at 25 L/min.

Table 1 Characteristics of real-life FMWW

Parameters	Levels	Before pretreatment	After pretreatment
COD (mg/L)	Maximum	4550	4280
	Minimum	4380	4090
	Average	4460 ± 52.8	4202.5 ± 60.62
BOD_5 (mg/L)	Maximum	3210	3100
	Minimum	3130	2945
	Average	3165 ± 48.98	3005 ± 50.43
NH_4^+-N(mg/L)	Maximum	510	497
	Minimum	439	411
	Average	472.5 ± 30.52	450.25 ± 29.59
TSS (mg/L)	Maximum	5263	12
	Minimum	5135	6
	Average	5201.75 ± 50.77	8 ± 0.58
TDS (mg/L)	Maximum	4280	4165
	Minimum	4192	4047
	Average	4230.42 ± 35.8	4105.62 ± 40.61
pH	Maximum	6.8	6.78
	Minimum	6.28	6.21
	Average	6.47 ± 0.18	6.44 ± 0.2
PO_4^{3-}-P (mg/L)	Maximum	185	150
	Minimum	102	91
	Average	150.8 ± 28.56	120.75 ± 29.31

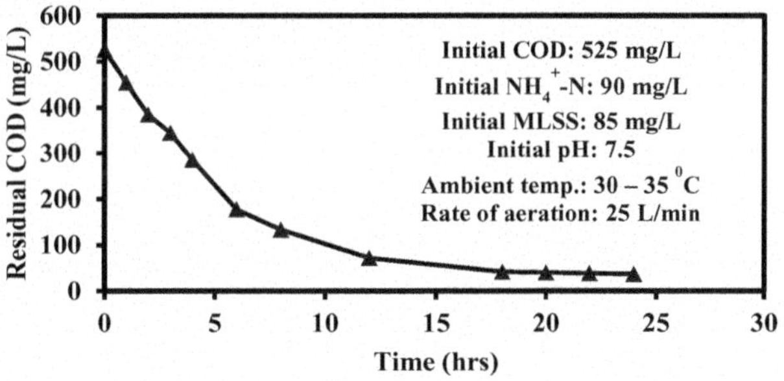

Fig. 5 Residual COD vs time (1st set)

3.2.1 COD Removal Study

In the first set, the initial COD concentration 525 mg/L was reduced almost to 37 mg/L (Fig. 5), i.e., approximately 93% COD was removed after a period of 24 hours.

The percentage removal of COD vs time has been plotted and shown in Fig. 6.

In the second set, 1050 mg/L of initially taken COD was reduced almost to 43.5 mg/L after 24 hours retention time (Fig. 7). Approximately 96% COD was removed in this period (Fig. 8).

Fig. 6 COD removal percentage vs time (1st set)

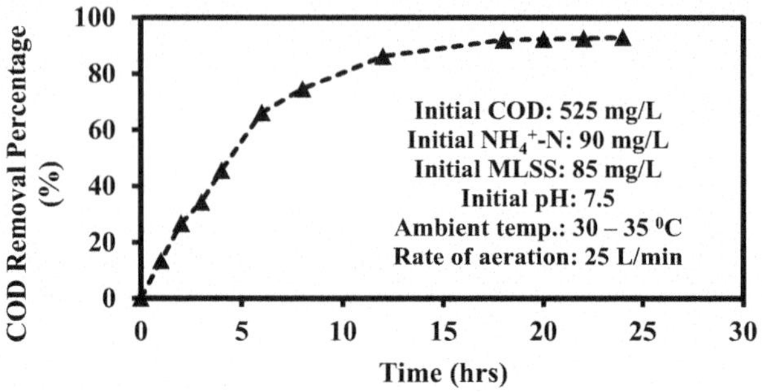

Fig. 7 Residual COD vs time (2nd set)

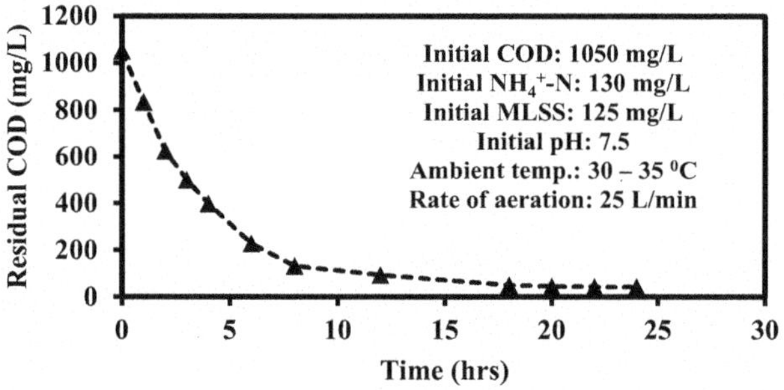

Fig. 8 COD removal percentage vs time (2nd set)

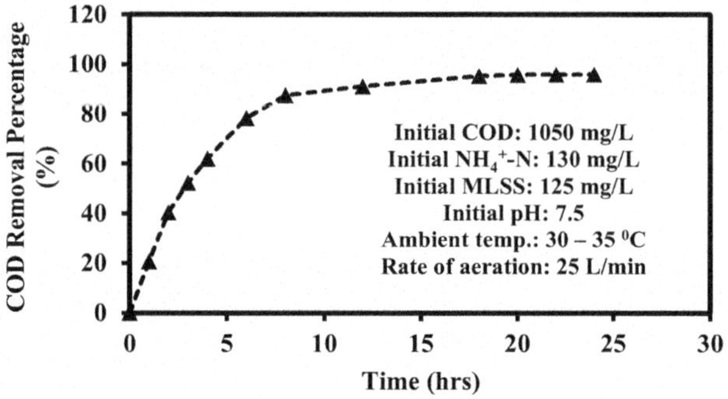

3.2.2 NH_4^+-N Removal Study

Combined COD and NH_4^+-N degradation were studied experimentally in the batch reactor. In the first set, initial NH_4^+-N concentration was taken as 90 mg/L along with the initial COD of 525 mg/L. Generally, if high COD is available in the wastewater, it creates hindrance to biological nitrification as the potential nitrifiers remain passive as the active carbonaceous microorganisms eat up the COD. After a significant decrease in the COD level, biological nitrification may gear up as the carbonaceous bacteria almost cease to work. Before that, the ammonia removal may happen due to the necessary nutrient uptake by the acting carbonaceous bacteria. Perhaps due to this reason, the rate of ammoniacal nitrogen removal may be found to

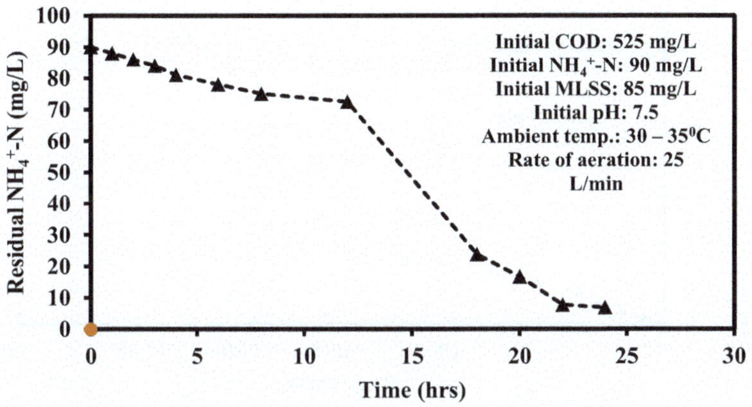

Fig. 9 Residual NH_4^+-N vs time (1st set)

be much less before the increasing removal rate under nitrification. In our study, perhaps this was the reason behind the observation that ammoniacal nitrogen removal steepened up after 12 hours, i.e., biological nitrification started to take place significantly. Also, the retention time was enhanced because the initial COD value was more and probably the predominant nitrifier was comparatively less. Also, substantial presence of reaction rate impeders like chloride, blood pigment, plasma, preservatives, etc. may not be overruled during the treatment of ammoniacal nitrogen.

Up to 12 hours, no significant degradation of NH_4^+-N happened. After this contact time, NH_4^+-N degraded to approximately 72.5 mg/L, i.e., almost 19.5% removal was achieved. This trend perhaps indicates that when there is enough biodegradable organic carbon present in the system, the potential carbonaceous bacterial community is very much active, and consequently, the activity of nitrifiers is largely suppressed. After 12 hours, a steep descendance in NH_4^+-N concentration was observed almost up to 22 hours. It indicates that the potential nitrifying microorganism colony got activated in this phase after most of the removal of organic carbon had happened. After 24 hours of contact time, NH_4^+-N concentration reduced to almost 7 mg/L (Fig. 9). The percentage removal of NH_4^+-N with respect to time has been plotted in Fig. 10. It was observed that approximately 92.5% removal of NH_4^+-N was achieved after 24 hours contact period. In the second set, the trend remained the same as described in the case of the first set. After 12 hours contact time, NH_4^+-N degraded to approximately 112.5 mg/L from the initial concentration value of 130 mg/L, i.e., almost 13.5% removal was achieved. After 12 hours, a similar type of steep declination in NH_4^+-N concentration was observed almost up to 20 hours. After 24 hours, NH_4^+-N concentration reduced to almost 7.7 mg/L (Fig. 11). The percentage removal of NH_4^+-N with respect to time has been plotted in Fig. 12. It was observed that approximately 94% NH_4^+-N was removed after 24 hours.

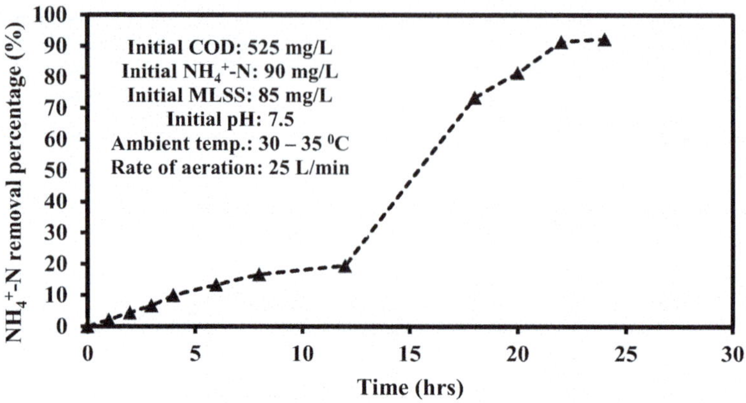

Fig. 10 NH_4^+-N removal percentage vs time (1st set)

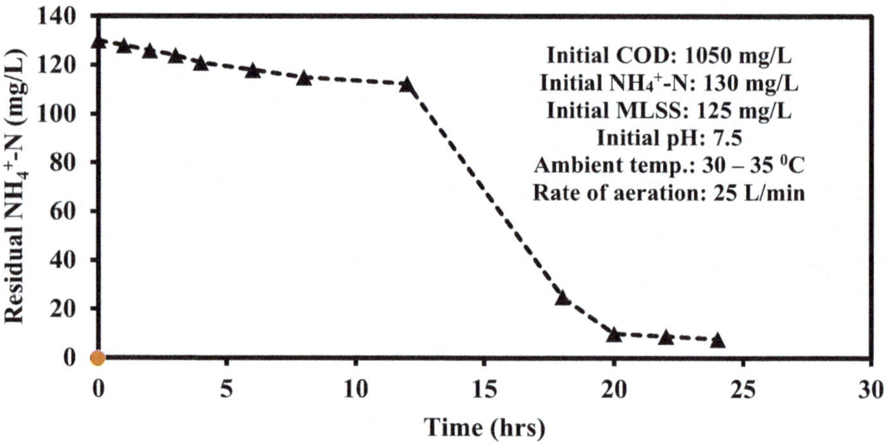

Fig. 11 Residual NH_4^+-N vs time (2nd set)

3.2.3 Study of Incremental Microbial Concentration

An initial microbial concentration, analyzed as mixed liquor suspended solids (MLSS), of 85 mg/L increased to 350 mg/L rapidly during 12 hours time period, but thereafter, it almost attained a steady-state condition. After 24 hours contact time in this conventional suspended culture system, MLSS increased almost up to 364 mg/L (Fig. 13). In the second set, MLSS concentration increased rapidly to 690 mg/L from an initial value of 125 mg/L after 12 hours. Afterwards, a steady-state condition was observed. After 24 hours contact time, MLSS increment was observed almost up to 723 mg/L (Fig. 14).

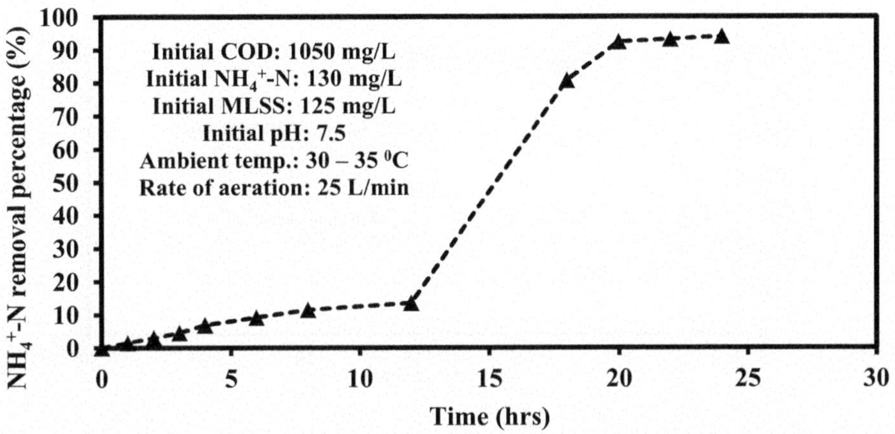

Fig. 12 NH_4^+-N removal percentage vs time (2nd set)

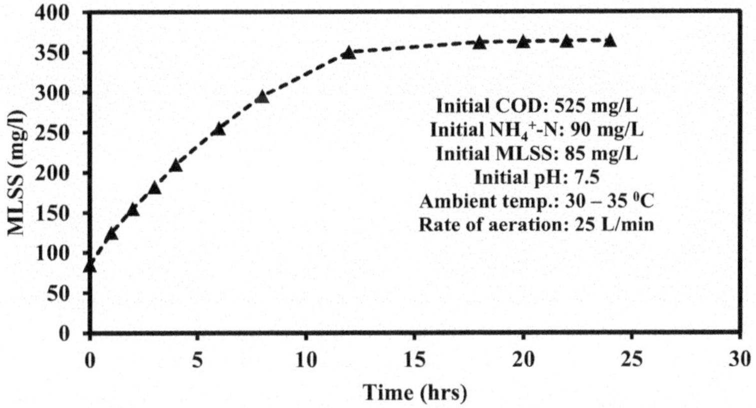

Fig. 13 MLSS vs time (1st set)

3.3 Determination of Kinetic Constants for COD Removal

A line of best fit has been drawn keeping $1/S$, i.e., reciprocal of residual COD concentration in X axis and $1/U_C$ as reciprocal of specific substrate (COD) utilization rate in Y axis with the experimental data. The corresponding graph is plotted in Fig. 15. The slope and intercept of best fit line have been used in the equation $1/U_C = [(K_s/k)\ (1/S)] + 1/k$ corresponding to Lineweaver-Burk plot [8]. The "k" (maximum substrate utilization rate) value has been evaluated as 23.47/day, and the value of "K_s: (half velocity constant) has been evaluated as 545.96 mg/L. The graph of $1/\theta_c$ has been plotted against U_C (Fig. 16). The "Y" (synthesis yield coefficient) value and "k_d" (endogenous decay constant) value have been determined from Fig. 16. Y and k_d have been evaluated from the conventional equation $1/\theta_c = YU_C$

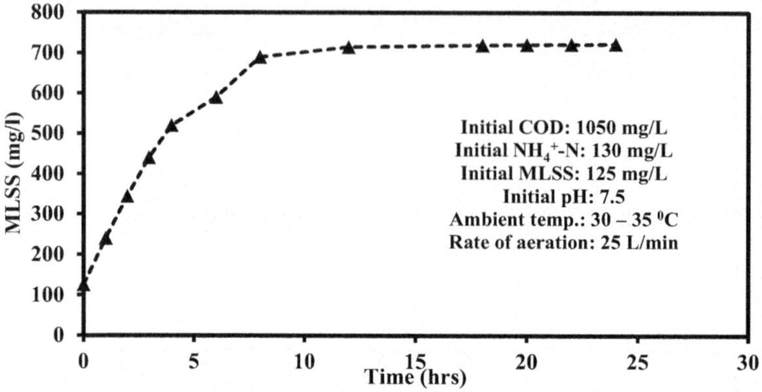

Fig. 14 MLSS vs time (2nd set)

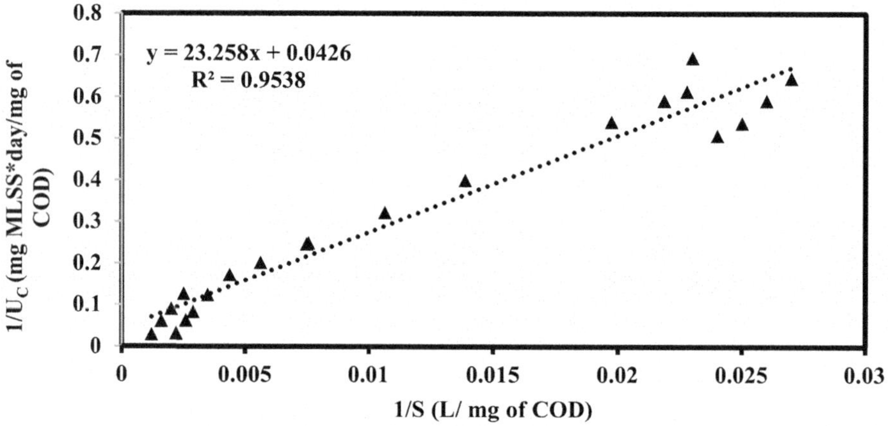

Fig. 15 Substrate utilization kinetic for COD removal study

$- k_d$. Y value has been found to be 0.737 mg of MLSS/mg of COD and k_d has been evaluated as 0.063 day.

3.4 Determination of Kinetic Constants for NH_4^+-N Removal

Another line of best fit has been drawn keeping $1/N$, i.e., reciprocal of residual NH_4^+-N concentration in X axis and $1/U_N$ as reciprocal of specific substrate (NH_4^+-N) utilization rate in Y axis with the experimental data. The corresponding graph is plotted in Fig. 17. The slope and intercept of best fit line have been used in the equation $1/U_N = [(K_s/k)\ (1/N)] + 1/k$ corresponding to Lineweaver-Burk plot in the similar manner as earlier. The "k" (maximum substrate utilization rate) value has

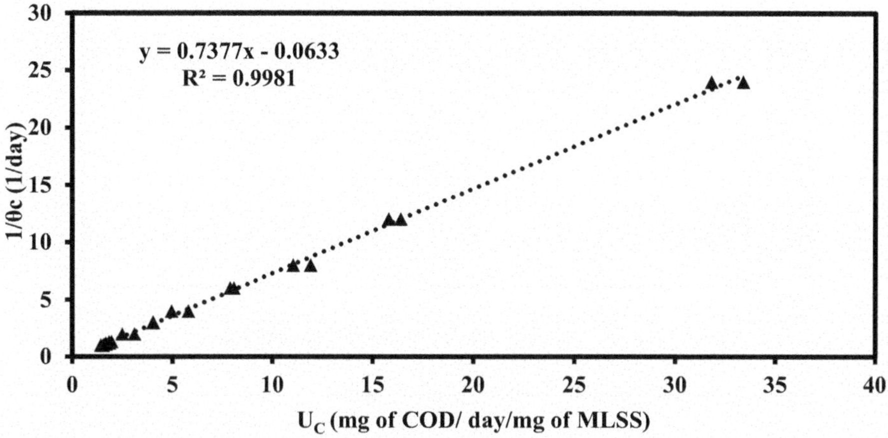

Fig. 16 Microbial growth kinetic for COD removal study

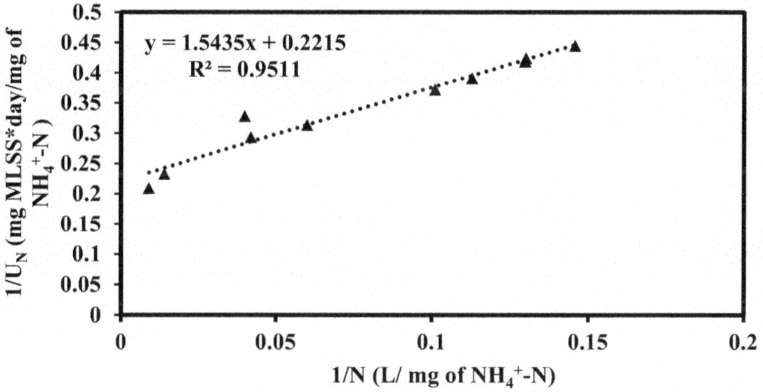

Fig. 17 Substrate utilization kinetic for NH_4^+-N removal study

been evaluated as 4.51/day, and the value of "K_s" (half velocity constant) has been evaluated as 6.97 mg/L. The graph of $1/\theta_c$ has been plotted against U_N (Fig. 18). The "Y" (synthesis yield coefficient) value and "k_d" (endogenous decay constant) value have been determined from Fig. 18. Y and k_d have been evaluated from the conventional equation $1/\theta_c = YU_N - k_d$. Y value has been found to be 0.448 mg of MLSS/mg of NH_4^+-N and k_d has been evaluated as 0.033/day.

All the evaluated kinetic constants for organic carbon and ammonia oxidation are given in Table 2.

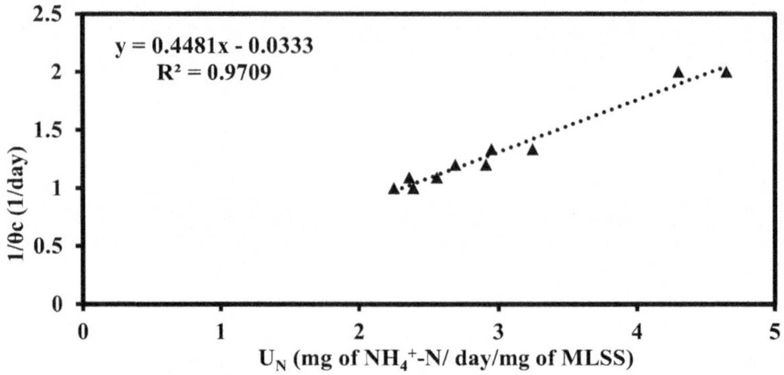

Fig. 18 Microbial growth kinetic for NH_4^+-N removal study

Table 2 Kinetic constants

Kinetic constants	Organic carbon oxidation	Ammonia oxidation	Unit
k	23.47	4.51	day^{-1}
K_S	545.96	6.97	mg/L
Y	0.737	0.448	mg MLSS/ mg COD (or NH_4^+-N)
k_d	0.063	0.033	day^{-1}

4 Conclusion

Fish and other animal meat selling establishments in the marketplace release wastewater containing significant amounts of organic and nutrient loads. However, appropriate treatment that is techno-economically feasible is seldom available in the existing literature. Specially in the Indian subcontinent or southeast Asia, there are innumerous retail fish markets discharging unattended and untreated high strength wastewater. It needs due attention to be paid to reduce the pollution load at source in line with the global convergence of source reduction of pollutants. It may be concluded from the present investigation that biological treatment in aerobic conditions is feasible in the case of fish market wastewater for removing organic carbon and ammoniacal nitrogen simultaneously. Also, the kinetic constants as evaluated for both the organic carbon and ammonia oxidation are favorable and corroborating the early research [9, 10].

However, aerobic treatment is commonly found where initial COD or BOD value is not too high. However, in the case of FMWW, initial COD was found to have values over 4000 mg/L which was significantly higher. Hence, the holistic target is to treat the wastewater, firstly, by anaerobic means and then by the aerobic way. Likewise, the kinetic study was carried out for relatively lower values of initial COD. Also, the feasibility of anaerobic treatment is underway in the laboratory. It is expected that anaerobic treatment will reduce the organic load to some extent which may enable the subsequent aerobic treatment to take place effectively.

As the real-life FMWW was diluted to achieve the target initial COD values, the consequent descent in initial NH_4^+ -N concentrations also took place. In this study, our primary objective was to study the feasibility of real-life FMWW treatment by suspended growth aerobic method and find the kinetic constants by the laboratory-grown mixed microbial culture derived from fish market drainage sediment. The primary feasibility was observed under experimental conditions. Further the performance study with undiluted real-life FMWW will be conducted in the coming days by different aerobic methods such as ASP, MBBR, SBR, etc. to observe the viability of aerobic treatments for real-life FMWW containing high organic and nutrient load.

References

1. M.T. Jamal, A. Pugazhendi, Treatment of fish market wastewater and energy production using halophiles in air cathode microbial fuel cell. J. Environ. Manag. **292**, 112752 (2021)
2. G.D. Bhowmick, B. Neethu, M.M. Ghangrekar, R. Banerjee, Improved performance of microbial fuel cell by in situ methanogenesis suppression while treating fish market wastewater. Appl. Biochem. Biotechnol. **192**, 1060–1075 (2020)
3. A. Palenzuela-Rollon, G. Zeeman, H.J. Lubberding, G. Lettinga, G.J. Alaert, Treatment of fish processing wastewater in a one- or two-step upflow anaerobic sludge blanket (UASB) reactor. Water Sci. Technol. **45**(10), 207–212 (2002)
4. M.S. Islam, S. Khan, M. Tanaka, Waste loading in shrimp and fish processing effluents: Potential source of hazards to the coastal and nearshore environments. Mar. Pollut. Bull. **49**(1), 103–110 (2004)
5. P. Chowdhury, T. Viraraghavan, A. Srinivasan, Biological treatment processes for fish processing wastewater–A review. Bioresour. Technol. **101**(2), 439–449 (2010)
6. A. Nowaka, R. Mazurb, E. Paneka, E. Dacewiczc, K. Chmielowskic, Treatment efficiency of fish processing wastewater in different types of biological reactors. Phys. Chem. Earth **109**, 40–48 (2019)
7. American Public Health Association, APHA, *Standard Methods for the Examination of Water and Wastewater*, 23rd edn. (American Public Health Association/American Water Works Association/Water Environment Federation, Washington, D.C., 2017)
8. H. Lineweaver, D. Burk, The determination of enzyme dissociation constants. J. Am. Chem. Soc. **56**(3), 658–666 (1934)
9. J.-S. Huang, C.-G. Jih, T.-J. Sung, Performance enhancement of suspended-growth reactors with phototrophs. J. Environ. Eng. **125**(6), 501–507 (1999)
10. Y.C. Ching, G. Redzwan, Biological treatment of fish processing saline wastewater for reuse as liquid fertilizer. Sustainability **9**(7), 1062 (2017)

Comparative Assessment and Phytoremediation Potentials of *Ipomoea carnea* and *Ricinus castor* Species for Wastewater Treatment

Sagar M. Gawande and Dilip D. Sarode

Abbreviations

BOD	Biochemical Oxygen Demand
COD	Chemical Oxygen Demand
GOI	Government of India
IMD	Indian Meteorological Department
NSSO	National Sample Survey Office
PVC	Polyvinyl Chloride
SDG	Sustainable Development Goals

1 Introduction

As per the 76th National Sample Survey Office (NSSO), around 42% of rural communities travel to fetch drinking water. The traveling ranges from 0.2 to 5 km, with an average time of 15 hours a month to get sufficient drinking water [1]. The depletion of natural resources, environmental pollution, and climate change will pose possible health problems and lack of food and potable water [2]. It is now widely acknowledged that the twenty-first century marks the era of environmental challenges worldwide. All nations are now confronted with the growing dangers of climate change, pollution, deforestation, disasters, land use change, biodiversity and ecosystem service losses, global hunger, loss of agriculture, environmental health diseases and epidemics, and encroaching issues of inequality in socioeconomic

S. M. Gawande (✉) · D. D. Sarode
Department of General Engineering, Institute of Chemical Technology, Mumbai, India
e-mail: civ16sm.gawande@pg.ictmumbai.edu.in

D. Mazumder (ed.), *Sustainable Advanced Technologies for Environmental Management*, Springer Proceedings in Earth and Environmental Sciences,
https://doi.org/10.1007/978-3-031-64006-3_22

status and development [3]. The scale of these ecological losses has increased by human intervention. We are living through the most catastrophic environmental crisis in human history. We must solve these growing ecological concerns; otherwise, we may face an existential life crisis on our planet [4, 5].

The United Nations recently approved the 2030 Agenda for Sustainable Development, which had 169 interconnected targets and 17 Sustainable Development Goals (SDGs). The 2030 Agenda's aims and targets include eradicating poverty; reducing pollution; boosting dependability and sustainability; assuring access to clean air, water, and energy; sanitation and hygiene; sustainable cities and urbanization; and biodiversity protection [6].

By eradicating gender inequality, conflicts, poverty, and environmental dangers, a sharp focus should be placed on achieving economic, social, cultural, civic, political, and human rights under the SDGs. Additionally, 197 nations ratified the Paris Agreement 2015 in 2015, a worldwide framework for tackling and mitigating climate change [7].

According to the Paris Agreement, wealthy and impoverished nations agreed to take steps to slow global warming while also achieving the SDGs in a broader sense. However, more progress has yet to be made toward reaching the Sustainable Development Goals' objectives or the Paris Agreement targets [8]. Every country today faces a massive challenge in meeting the energy and employment needs of an ever-increasing population while also combating climate change by lowering its emissions of greenhouse gases. To achieve sustainable development, we must make educated and scientific decisions [9].

There is an urgent need for collaboration among multiple stakeholders in developing new and improved infrastructures, innovation, and research, creating environmentally friendly technologies, and adopting a shared minimum sustainable development strategy. Phytoremediation is a solution and an expanding field. In this process of phytoremediation, inorganic and organic contaminants are absorbed by different parts of the plants. The primary sources of organic and inorganic pollutants are chemicals, pesticides, herbicides, spills, and other human-made environmental releases. Excessive emission of these toxins negatively impacts people's and the environment's health. These pollutants must be removed from the environment at great financial and operational expense by employing phytoremediation. The adopted plants can absorb and break down these contaminants at low operation and maintenance costs with environmental benefits.

2 Material and Methods

2.1 Selection and Cultivation

Ipomoea carnea stems are collected from dry land, semiarid, and waterlogged areas. Collected stems are wrapped in damp jute bags to preserve and prevent the loss of natural cell and plant qualities throughout the transfer from site to laboratory. Before

Map of Maharashtra state and Location of Pune city	Location map of Study area (Gogalwadi Village) near Pune City

Fig. 1 Showing the geographical location of the implementation of the treatment system [11]

being moved to the planting pot or experimental unit, each collected sample species is submerged for 24 hours in water with a pH of 6–7.5 at a depth of about 12 inches. The stem's vertical immersion helps to prevent the shock phenomena when the species is supposed to transfer to soil media by allowing it to become habituated to the laboratory and site environment. It is possible to prepare the *Ipomoea carnea* stems by using the soil medium produced with a moisture level of about 67%. The stems are prepared such that each 6- to 8-inch high stem has a cut angle at the bottom of around 45°. In the rectangular container, roughly 4–6 stems of sufficient height are planted in prepared media, spaced about 3–4 inches apart. The experimental setup is installed and maintained at a house situated at Gogalwadi Village near Pune City of Maharashtra (Fig. 1) [10].

The research experiment setup to treat the wastewater by phytoremediation process and phyto units are shown in Fig. 2. The prepared media that provide plant support (phyto unit) used locally available materials from the setup area, such as sand, brickbats, coarse- and medium-sized aggregate media, and soil. This supporting media is poured into a PVC rectangular container about 560 × 470 × 470 mm. The bottom 125 mm thick layer has brickbats of sizes ranging from 80 mm to 120 mm and 60 mm to 85 mm wide. A 105 mm thick blend layer of 20 mm and 5 mm aggregate was placed on the previous brickbat layer.

The sandy soil is spread out and coated with 165 mm thick. For free circulation and the accumulation of volumetric changes in the media placed, a freeboard of 75 mm is provided. The circular openings of 55 mm diameter are positioned 100–200 mm from the edges and 85–90 mm from the bottom of the 7 mm thick PVC plates as a baffle wall. The pilot plant setup and matured plant species are shown in Fig. 3.

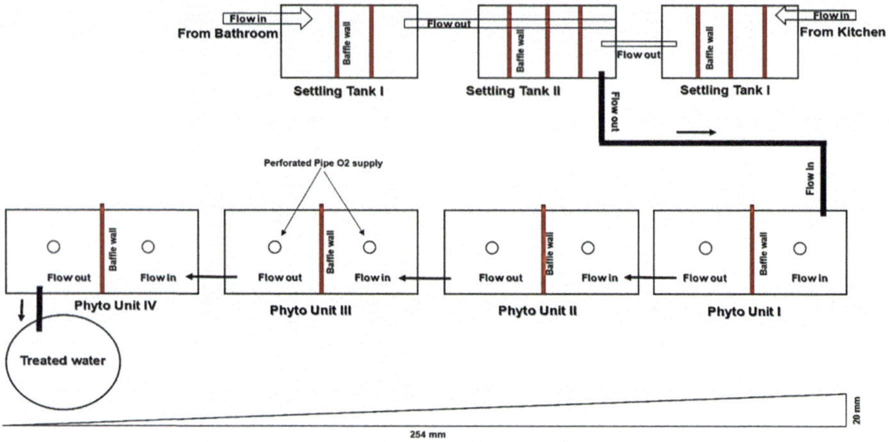

Fig. 2 Showing the different units and purpose of designed treatment system [11, 12]

Fig. 3 Showing the physical status and purpose of the on-site treatment system [12]

3 Results and Discussion

The *Ipomoea carnea* and *Ricinus communis* plant species are considered variables. Each plant contributed to obtaining dependable results. The plants were fully grown in the PVC rectangular tanks on-site conditions and fed with wastewater and nutrients. Table 1 shows the performance data for BOD, COD, ammonium-nitrogen ($NH_3^{-}N$), and phosphate $[PO_4]^{3-}$.

The adopted plant species *Ipomoea carnea* and *Ricinus communis* successfully treated the supplied wastewater during the treatment run. The system successfully lowered the BOD by about 80.14% with an average inlet concentration of 310.73 mg/l reduced to 61.73 mg/l at the outlet. The inlet COD was found to be 394.07 mg/l which reduced towards an outlet of 93.00 mg/l with the removal efficiency of 76.37%, as shown in Fig. 4.

Table 1 Performance and removal efficiency observed during the experimentation run

Treatment run	BOD			COD			NH_3^-N			$[PO_4]^{3-}$		
	In	Out	Removal %	In	Out	Removal %	In	Out	Removal %	In	Out	Removal %
1	320	71	77.81	393	92	76.59	31.1	7.1	77.17	5.7	0.92	83.86
2	310	62	80.00	389	87	77.63	30	7.5	75.00	5.5	0.9	83.64
3	298	59	80.20	428	98	77.10	30.3	7.6	74.92	5.2	1	80.77
4	300	54	82.00	385	93	75.84	31.6	7.5	76.27	4.9	0.93	81.02
5	310	57	81.61	405	93	77.04	31.2	7.5	75.96	5.8	0.9	84.48
6	310	64	79.35	389	90	76.86	30.2	7.2	76.16	5.5	1	81.82
7	308	64	79.22	428	92	78.50	30	7	76.67	5.2	1	80.77
8	300	56	81.33	388	94	75.77	30.8	7.2	76.62	4.9	0.92	81.22
9	310	60	80.65	380	94	75.26	34.1	7.3	78.59	5.8	0.9	84.48
10	322	62	80.75	410	96	76.59	30.8	7.2	76.62	5.2	1	80.77
11	332	66	80.12	373	92	75.34	31.3	7.5	76.04	5	0.9	82.00
12	318	68	78.62	385	92	76.10	31	7.1	77.10	5.1	0.9	82.35
13	297	59	80.13	377	93	75.33	30.2	7	76.82	5	0.9	82.00
14	316	62	80.38	389	93	76.09	33.2	7.1	78.61	5.3	1	81.13
15	310	62	80.00	392	96	75.51	31.2	7.1	77.24	5.5	0.92	83.27
Average	**310.73**	**61.73**	**80.14**	**394.07**	**93.00**	**76.37**	**31.13**	**7.26**	**76.65**	**5.31**	**0.94**	**82.24**

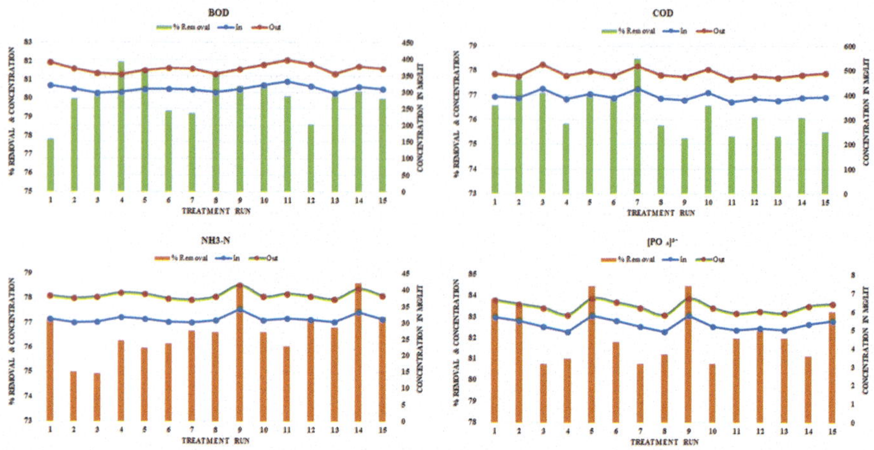

Fig. 4 Showing removal efficiency and concentration for BOD, COD, NH_3^-N, and $[PO_4]^{3-}$

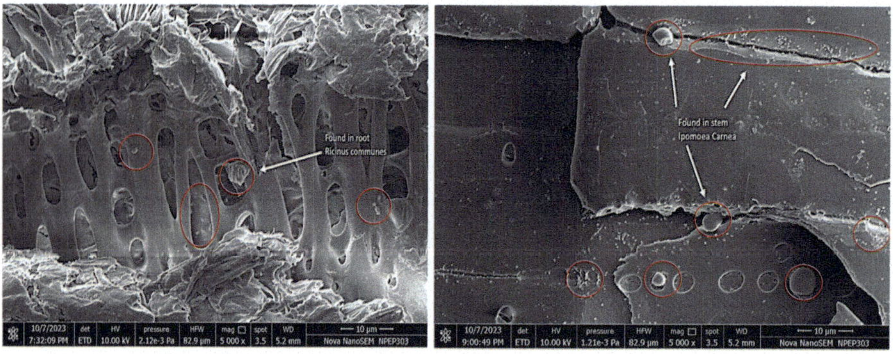

Fig. 5 FESEM Analysis of *Ipomoea carnea and Ricinus communis*

The average concentration of ammonium-nitrogen (NH_3^-N) at inlet was found to be 31.13 mg/l and that of outlet was 7.26 mg/l with around 76.65% of removal during the experimentation run. The average phosphate $[PO_4]^{3-}$ concentration at inlet was around 5.31 mg/l with an outlet of 0.94 mg/l showing the removal efficiency of 82.24% for the time run of experimentation.

3.1 Field Emission Scanning Electron Microscopy (FESEM)

Additionally, Fig. 5 shows that the biological investigation and dimension features of the accumulated matter present in the cells of plants were analyzed by Field Emission Scanning Electron Microscopy (FESEM) examination. The internal details of *Ipomoea carnea's* stem and *Ricinus communis'* roots for their internal structural arrangement are needed and could help better understand the selected species at magnification level.

4 Conclusion

This developed system is independent of electricity, maintenance, chemicals, skilled supervision, and workmanship for effective running and treating of the wastewater. Each phyto unit has a capacity of 80.64 l, and the total available capacity is 322.56 l. The developed system's total one-time investment and construction cost was around INR 28000=00, including piping, PVC containers, fixtures, etc. This developed system produces around 250–300 l of treated wastewater daily. During the experimental run, the adopted plant species and developed phyto media showed a notable performance in reduction of targeted parameters. The average inlet concentration of BOD was 310.73 mg/l reduced to 61.73 mg/l at the outlet, and the removal efficiency for BOD was 80.14%. A 76.37% removal efficiency of COD was observed with an inlet concentration of 394.07 mg/l and outlet of 93.00 mg/l. With 76.65% of removal efficiency, ammonium-nitrogen (NH_3^-N) was found to have an inlet concentration of 31.13 mg/l and outlet concentration of 7.26 mg/l, whereas for phosphate $[PO_4]^{3-}$, the average inlet concentration was around 5.31 mg/l with an outlet of 0.94 mg/l showing the removal efficiency of 82.24%.

The internal features of the particle present in these plants were further illuminated by Field Emission Scanning Electron Microscopy (FESEM) examination. For the internal voids of dried stem of *Ipomoea carnea*, internal structural arrangement of stem of *Ipomoea carnea*, and layer interlocking of cut section of stem, the opening size ranges from 24.23 to 41.87 μm and solid particles size ranges from 48.53 to 194.6 μm. The structural details shows the stem of *Ricinus castor*, the structural details of stem internal openings of *Ricinus castor*, structural details of stem shapes of internal openings of *Ricinus castor*, the details of roots of *Ipomoea carnea* @ 3 μm, 5 μm and 20 μm.

This research study will help rural and semi-urban communities and local governing bodies reuse the generated wastewater for non-drinking purposes such as gardening, constructing water-bound macadam (WBM) roads, flushing toilets and bathrooms, etc., without much environmental, economic, and financial burden. Compared to other plant species used and adopted by other researchers in the phytoremediation process, very few provide the opportunity to utilize the biomass produced during or after treatment. In this system, the stems attain a height of around 180–210 cm and a root length of about 45–75 cm. The biomass produced will provide an alternative income source to the system adopter as the *Ipomoea carnea* not only has medicinal properties but also increases the fertility of the farmland, and its extensive root system can help protect against soil erosion. The oil from the seeds of *Ricinus communis* can be produced and sold to the biodiesel, soaps, and lubrication industries. The oil cake produced as a byproduct of oil extraction makes a fantastic organic fertilizer and enhances soil fertility.

References

1. Unit Level Data & Report on NSS 76th Round for Schedule 1.2, July-December, 2018 (Drinking Water, Sanitation, Hygiene and Housing Condition), https://mospi.gov.in/unit-level-data-report-nss-76th-round-schedule-12-july-december-2018-drinking-water-sanitation
2. *Executive Summary, Water Pollution from Agriculture: A Global Review* (Food and Agriculture Organization of the United Nations Rome, 2017)
3. E.W. Chu, J.R. Karr, Environmental impact: Concept, consequences, and measurement, in *Reference Module in Life Sciences*, (2017). https://doi.org/10.1016/B978-0-12-809633-8.02380-3
4. L. Laybourn-Langton et al., *Initial Report, This Is a Crisis-Facing up to the Age of Environmental Breakdown* (Institute for Public Policy Research, 2017)
5. P. Horton, B.P. Horton, R.-d. Sustainability, Living in harmony with life on earth. One Earth **1**(1), 86–94 (2019). https://doi.org/10.1016/j.oneear.2019.08.019
6. A. Breuer, H. Janetschek, D. Malerba, Translating sustainable development goal (SDG) interdependencies into policy advice. Sustainability **11**(7), 2092 (2019). https://doi.org/10.3390/su11072092
7. F. Shayan, M.A. Zahed, et al., Sustainable development goals (SDGs) as a framework for corporate social responsibility (CSR). Sustainability **14**(3), 1222 (2022). https://doi.org/10.3390/su14031222
8. J. Mensah, Sustainable development: Meaning, history, principles, pillars, and implications for human action: Literature review. Cogent Soc. Sci. **5**, 1 (2019). https://doi.org/10.1080/23311886.2019.1653531
9. J. Charles Rajesh Kumar, M.A. Majid, Renewable energy for sustainable development in India: Current status, future prospects, challenges, employment, and investment opportunities. Energy Sustain. Soc. **10**, 2 (2020). https://doi.org/10.1186/s13705-019-0232-1
10. S. Gawande, D.D. Sarode, Insight of *Ipomoea carnea* and its use in wastewater treatment as ecofriendly, sustainable solution. World J. Adv. Res. Rev. **13**(02), 483–491 (2022). https://doi.org/10.30574/wjarr.2022.13.2.0185
11. S. Gawande, D.D. Sarode, A case study on comprehensive investigation of a sustainable wastewater treatment system: Promoting environmental sustainability. Water Energy Int. **66r**(5), 32–36 (2023)
12. S.M. Gawande, D.D. Sarode, Treatment of rural domestic wastewater with artificial constructed wetlands using Ipomoea carnea and Ricinus castor plants. Environ. Eng. Manag. J. **22**(9), 1523–1534 (2023). https://doi.org/10.30638/eemj.2023.128

Adsorption of Fluoride from Wastewater with the Activated Carbon Derived from Coconut Shell

Firdous Ahmad Dar, Akshit Kumar, Muzammil Hussain, Jatin Sotra, and Swamy Kurella

Abbreviations

ASTM	American Society for Testing and Materials
BET	Brunauer-Emmett-Teller
CS	Coconut shell
CSAC	Coconut shell activated carbon
FTIR	Fourier transform infrared
IR	Infrared
NaF	Sodium fluoride
NEERI	National Environmental Engineering Research Institute
SEM	Scanning electron microscope
WHO	World Health Organization

1 Introduction

The current state of contamination of water throughout India is precarious. These days, pollution has reached dangerous levels in majority of river systems throughout India. Experts from India's National Environmental Engineering Research Institute (NEERI) in Nagpur have determined that around 70% of the country's water supply is contaminated [1]. Studies indicate that every nation is grappling with the issue of water contamination, indicating that India does not stand alone in this regard [2]. According to most studies, the main causes of pollution in rivers are sewage

F. A. Dar · A. Kumar · M. Hussain · J. Sotra · S. Kurella (✉)
Department of Chemical Engineering, National Institute of Technology Srinagar, Srinagar, Jammu and Kashmir, India
e-mail: kurellaswamy@nitsri.ac.in

D. Mazumder (ed.), *Sustainable Advanced Technologies for Environmental Management*, Springer Proceedings in Earth and Environmental Sciences,
https://doi.org/10.1007/978-3-031-64006-3_23

that is dumped into them, untreated industrialized wastewater that is dumped into rivers, as well as runoff that comes through farms which utilize synthetic fertilizers, pesticides, insecticides, as well as animal waste. Freshwater constitutes 3% of the overall water content across the planet, owing to its critical nature. A negligible fraction, precisely 0.01%, of that potable water is accessible to humanity to meet all needs [3]. Fluoride is identified as a prevalent pollutant in water as well as soil sources. When it comes to anionic species that inhabit the outermost layer of the Earth, fluoride is 13th among them. Its concentration and quantity in the world's water supply are in varying concentrations and proportions. The topography of the areas where these water sources flow influences the fluoride concentration that might be present in lakes, rivers, and canals [4]. The concentrations of fluoride within underground waters and reservoirs were due to the mineral constitution of the bedrock through which water penetrates. If we consider the instance of ocean waters, fluoride levels typically range from 1.2 to 1.4 mg/L [5–7]. The elementary state of fluoride does not occur naturally due to its severe reactive nature. The main diverse sources of fluoride include soil, water, and volcanic activity. Concentration of fluoride in groundwater, surface water, and wastewater can diverse considerably. The fluctuation in fluoride concentrations depends on various factors as mentioned above. Below is a comparative analysis of fluoride level in such water sources:

Wastewater: The fluoride levels in wastewater can exhibit considerable variation based on the origins of contamination. Industrial sector is the primary source that contaminates the water with fluoride and generates fluoridated wastewater. The usual fluoride level in industrial wastewater may span from 1 to 100 ppm or even exceed this range in certain instances.

Groundwater: Fluoride levels in groundwater are predominantly shaped by the underlying geological and hydrogeological processes. In areas characterized by fluoride-rich geological formations, the fluoride level in groundwaters may surpass the recommended limit of WHO. Conversely, in areas with low natural fluoride levels, groundwater fluoride level may be below 0.1 mg/L. Human activities such as industrial outlets, agricultural outlets, and inadequate waste disposal practices can also raise fluoride levels in groundwater.

Surface Water: The levels of fluoride in surface water can fluctuate based on natural sources, human contributions, and the closeness to industrial operations. In regions characterized by substantial industrial operations or geological formations rich in natural fluoride, the fluoride concentrations in surface water in such areas may vary from 0.1 to 10 mg/L or even exceed this range. In regions with low industrial sector and less natural fluoride sources, surface aquifers may exhibit fluoride content below 0.1 mg/L.

Fluoride content in soil can arise from various factors, encompassing both natural phenomena and human events. The presence could be ascribed to external elements, like the flow from industrial facilities or runoff from residential and household areas. Soil's complete fluoride levels typically fall between 150 and 400 mg/kg; yet in the instance of clay, these measurements experience notable variation. Fluoride content in clay is quite high—1000 mg/kg. Fluoride contamination in soil is predominantly attributed to phosphorus fertilizers, which typically ranges from 1% and 1.5% [6, 8,

9]. In the realm of water, human-made activities such as industrial discharges from coal-burning industries and extensive fluoridation of communal water are accountable for the fluoride occurrence in aquatic environment. They constitute among the most important contributions. The contamination of water systems by fluoride may be attributed to several factors such as the formulation of rocks, the characteristics of water, the soil profile, the groundwater infiltration, and the chelation of elements, along with the depth of wells [10]. A volcanic explosion not only throws enormous amounts of ash into the air; however, it generates a great deal of dirt including dust. Due to its extremely significant fluoride levels, ash from volcanoes represents a prominent cause for fluoride contamination. Heterogeneous constituents, which include fluoride, are released during such explosions. Whenever they reach the grounds surface (the lithosphere), they can persist there over quite a while. The degradation process causes fluoride to seep into underground watercourses [11, 12].

India is home with approximately 12 million tons fluoride stored underground, compared to a total of 85 million tons presumed to be prevalent worldwide. According to the recommendations established by the World Health Organization (WHO), the highest permitted amount for fluoride that may be found in water for consumption are limited to 1.5 mg/L. Fluoride within specific quantity has become an essential feature for optimal bone development as well as tooth enamel growth. If it exceeds the permissible limit, then it causes fluorosis [3]. Fluorosis has afflicted approximately 25 nations worldwide, and the disease continues spreading at alarming pace [13]. Within India, 19 states have significant fluoride concentrations in their subsurface water [14–18]. India ranks first in the world with regard to the number of people possibly impacted by fluorosis because the levels of fluoride within water from ground sources are in excess of 1.5 mg/L [19]. Harnessing the collective brilliance of countless experts and dedicated researchers, who relentlessly strive to conquer the formidable challenge of purifying water, numerous conventional and unconventional methods are presently employed to alleviate people's suffering. Currently various defluoridation approaches are employed for fluoride removal from water. However, there are certain drawbacks concerning technological feasibility, operational/maintenance ease, and socioeconomic factors [2, 20]. Different techniques for defluoridation are reverse osmosis [21], adsorption [22], ion-exchange [23], nanofiltration [24], and precipitation/coagulation [25]. The drawbacks associated with some of the techniques are the coagulations/precipitation has found to be expensive, pH-dependent efficacy, presence of interfering ions in the sample, necessitating pH adjustments, excessive residual aluminum concentration, sludge formation containing high levels of harmful aluminum fluoride complexes, and the requirement for extensive dehydration before disposal, while ion exchange is costly, prone to the impact of co-existing ions, necessitating frequent material regeneration, leading to the production of dangerous materials, harmful aqueous lavish produced during regenerating media. Moreover, the efficiency of the approach is significantly altered by pH; reverse osmosis is found to be having high initial investment costs, significant operational and maintenance cost, and hazardous leftover material. Among the defluoridation techniques, the famous Nalgonda technique is a revolutionary one which was developed by NEERI, Nagpur, India [20]. Under such approach, the chemical dose includes a makeup of alum as well as lime. Experience the remarkable superiority of coagulation over aluminum hydroxide flocs in the captivating pH of 6–9 for enhanced defluoridation efficiency [26]. Compared to the above technologies, adsorption possesses functional efficacy in the elimination of redundant fluoridation from water owing its economical nature, user-friendly functioning, and substantial capability during adsorption, along with reusability of material [13]. Coconut shell-based activated carbon exhibits superior removal efficiency when contrasted with both the Nalgonda method and activated alumina. This superiority can be attributed to its elevated adsorption capability, area, and economic nature. Furthermore, its reusability and environment friendly characteristics render it a preferred choice for water treatment purpose. Thus, an attempt is made to adsorb fluoride on activated carbon derived from coconut shell.

2 Materials and Methods

2.1 Materials

During the experimentation, excellent-quality analytical-grade chemicals were employed. The "fluoridated water mixture" was created using anhydrous sodium fluoride (NaF) sourced from Merck. To prepare a standard sample mixture with a level of 100 mg/L, 2.210 g of sodium fluoride were mixed with 1000 mL of distilled water. Subsequently, the initial solution was diluted to achieve the desired fluoride content in the solutions.

2.2 Synthesis of Adsorbent

The study discovered that the adsorbent is obtained by grinding shells of coconuts needed for impregnation to be effective, following a procedure outlined in published research [27]. Initially, the coconut shells were cleaned using purified water for elimination of adhering debris and other unwanted materials and then dried overnight at approximately 100 °C in order to lower moisture levels. The dried product was crushed and sieved until the particle diameter was uniformly about 0.45–0.15 mm, resulting in a material designated as CS (coconut shell). Subsequently, 20 g of dried chestnut shells were soaked in 250 mL of 2 M potassium hydroxide solution as the activating agent over a period of 24 h at ambient environment. After filtration, the material was placed into a furnace. The precursor was activated using a pyrolyzer; carbonization occurred at 750 °C and was maintained for 2 h in an inert atmosphere to get at least 20% yield, with a consistent flow of 99% pure nitrogen supply for the entire process at 5 °C per min with 100 cm^3/min rate. Once cooled, the carbonized leftover remnants were cleaned using Milli-Q distilled water until a neutral pH was achieved and maintained (monitored with an HP, pH measuring device, model CRISON 602). The corresponding material, which underwent carbonization, was then baked in an oven at 100 °C for 24 h to obtain the final activated carbon product (CSAC750).

2.3 Stock Solution Preparation

The fluoride residual content was determined using the SPADNS spectroscopic approach, which relies on the interaction between F^{-} ions and zirconium. This interaction separates a portion of F^{-} ions, forming a colorless ($ZrF62^{-}$) anionic compound as well as a colored agent. When compared to a conventional fluoridated sample, the color generated becomes progressively lighter with an increase in fluoride quantity. The zirconium-SPADNS complex combined compound was chosen due to its high stability in the dark and its ease of preparation by combining two distinct chemicals/reagents [28].

1. Incorporate 958 mg of SPADNS (2-(4-Sulfophenylazo)-1, 8-dihydroxy-3, 6-naphthalene disulfonic acid trisodium salt) from Sigma in 500 mL of pure water.
2. Mix 133 mg of zirconyl chloride hydrate from Sigma with approximately 25 mL of distilled water; add 350 mL of HCl and then dilute the mixture with distilled water to reach a total volume of 500 mL.
3. Prepare the Zr-SPADNS complex mixture by combining equal proportions of the above two reagents. This complex mixture can be stored for approximately 2 years when kept in the dark.

Following this, add 2 mL of the Zr-SPADNS solution to 10 mL of deionized water, serving as a blank, and then add it to 10 mL of specimens that have been diluted with water. Thoroughly mix the solution after each addition. The fluoride concentrations of the papered samples were analyzed using the spectrophotometer, and the calibration curve has been plotted for the corresponding values. Later following the adsorption treatment the samples were again checked so as to check the leftover concentration of fluoride in them by taking the reference of the already plotted calibration curve through the spectrometer. The posttreatment analysis for fluoride level has been performed for 2, 5, and 10 ppm using the spectrophotometer, and the corresponding remaining levels of fluoride were determined.

3 Results and Discussion

3.1 *Characterization of Activated Carbon*

The characterization of the obtained value-added products would be done by various techniques such as proximate analysis, CHNS, Fourier Transform Infrared (FTIR), SEM, Brunauer-Emmett-Teller (BET), and spectrophotometer.

3.1.1 Proximate Analysis

The outcomes of proximal evaluation are summarized in Table 1. Moisture content is the water content stored between the gaps surrounding dead cells and cellular membranes. The weight of the empty crucible without lid was 18.3149 g and the weight of the sample was 1 g. Therefore, the weight of crucible without lid but with the sample was 19.3149 g which was put into the oven at 105 °C for 1 h. After 1 h, the weight of crucible without lid with sample after operation was 19.2549 g and the weight of the sample after the operation was 0.94 g. Therefore, the moisture content was 0.06 g and the percentage of moisture content is 6%. The volatile matter corresponds to the fraction that has been pushed out as gas (which encompasses moisture) through bringing over 950 degrees Celsius for 7 min. Initially, the weight of empty crucible with lid was 26.2184 g and the weight of the sample was 1 g. Therefore, the weight of the crucible with lid and with sample was 27.2184 g, and it was operated in Muffle furnace at 950 °C for 7 min. At the end of the operation, the weight of crucible with lid and with sample was 26.394 g and the final sample weight was 0.824 g. Thus, the volatile matter was 0.176 g and the percentage of volatile matter was 17.6%. Ash content is the inorganic residue that remains after combustion in air at specific temperature. The weight of empty crucible without lid was 18.3149 g and the weight of the sample was 1 g; as a result, the weight of the crucible without lid and with sample was 19.3149 g, and it was operated in Muffle furnace at 700 °C for 1 h. Later, the weight of the crucible without lid and with sample became 18.4049 g and the weight of the final sample (ash) was found to be 0.09 g. Therefore, the percentage ash content was 9%. Fixed carbon is the residual weight that refers to the leftover weight following the expulsion of volatile substances, omitting ash and

Table 1 Proximate analysis of activated carbon derived from coconut shell

Proximate analysis	Temperature °C	Time (min)	Content %
Moisture content (ASTME871)	100	60	6.0
Volatile matter (ASTME872)	950	7	17.6
Ash content (ASTMD1102)	700	60	9.0
Fixed carbon	–	–	67.4

moisture, and it can be estimated by subtracting the sum of the percentages of volatile matter, moisture content, and ash content from 100 percent. Therefore, the fixed carbon percentage was in the sample 67.4%.

3.1.2 Scanning Electron Microscope (SEM)

SEM represents a kind of microscopy that creates images using electrons rather than luminescence. That has rendered it possible for analysts to look at a great deal with wider range of specimens. Compared to conventional microscopy instruments, the scanning electron microscope offers several benefits. As it can be seen in Fig. 1, the SEM's broad scope of vision enables much of a material to appear under frame at once. Additionally, the SEM offers significantly greater resolution, allowing for considerably greater magnification of specimens that are relatively close together. The investigator gets significantly greater control over amplification level of the SEM since it employs electromagnetic fields instead of lenses. The real remarkably detailed pictures along with those benefits make the scanning electron microscope among the greatest valuable tools in study nowadays.

3.1.3 Fourier Transform Infrared (FTIR)

This represents the predominant method about infrared spectrum. The fundamental principle behind infrared spectroscopy is that the material captures a portion of infrared (IR) light when it transmits within the material during the analysis. The energy which traverses through specimen is confirmed. FTIR is the most selected technique during infrared spectroscopy due to multiple advantages. First and foremost, it never damages the specimen in question . Secondly, it is far quicker than previous methods. Thirdly, it has far greater accuracy along with sensitivity. Waves

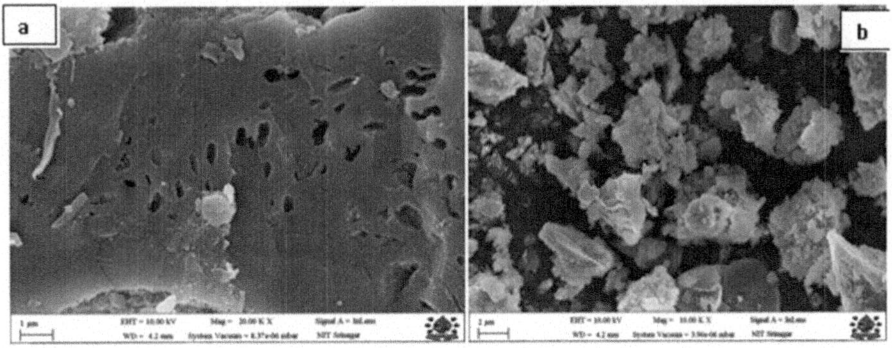

Fig. 1 SEM images of coconut shell activated carbon (**a**) before adsorption (**b**) after fluoride adsorption

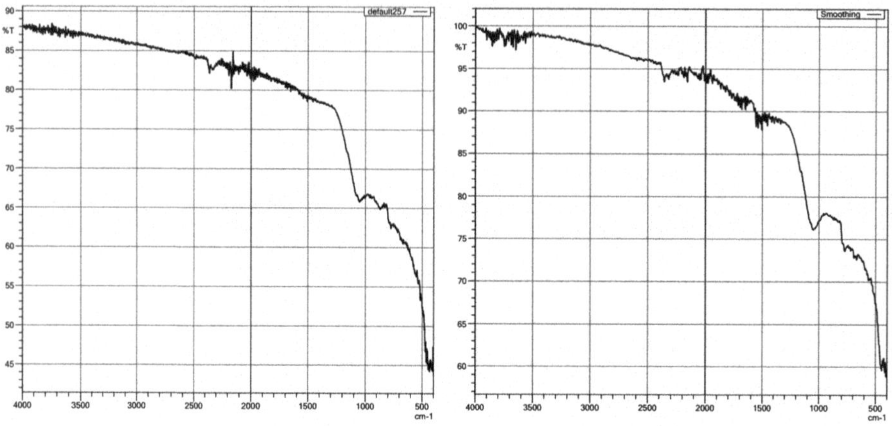

Fig. 2 FTIR spectra of activated carbons derived from coconut shell

may be broken down mathematically using the Fourier transform, which then yields the wavelength of the time-dependent wave.

The graphs referred to by the term "interferogram" represent precise interferometer's "output," nor the spectroscopic spectra which are actually used. By applying the Fourier transformation, the interferogram of light is transformed onto the familiar plot of infrared spectroscopic spectra. In Fig. 2, the distinct phases of the material appear prior to and following fluoride treatment. Some of the fields that make use of FTIR spectroscopy include the pharmaceuticals sector, food inspection, and synthetic organic polymeric research, along with petrochemicals engineering. In a nutshell, its versatility makes it useful for many things, including operational monitoring, chemical identification, and a mixture composition determination. This is employed for recognizing the given specimen's functional groupings.

3.2 *Batch Adsorption Experiment*

3.2.1 Effect of pH and Initial Fluoride Concentration

In this experimental work, the experiments were performed in a pH range of 5–8. Also, adsorbent dose was 1 g in 250 mL of fluoride solution. Initial fluoride concentration was taken as 2–10 ppm. We found that the pH and the commencing fluoride concentration play a major role in the defluoridation approach. The maximum fluoride elimination adsorption efficiency that was achieved after 24 h was 85% at pH 7 having the initial fluoride concentration of 10 ppm. The experimental results of different samples with different parameters are listed in the Table 2 and shown in Fig. 3.

Table 2 Fluoride adsorption efficiencies of activated carbon derived from coconut shell at adsorbent dose (1 g/ 250 mL)

pH	% Adsorption		
	At 2 ppm	At 5 ppm	At 10 ppm
5	25	24	25
6	54	52	60
7	68	74	85
8	65	72	78

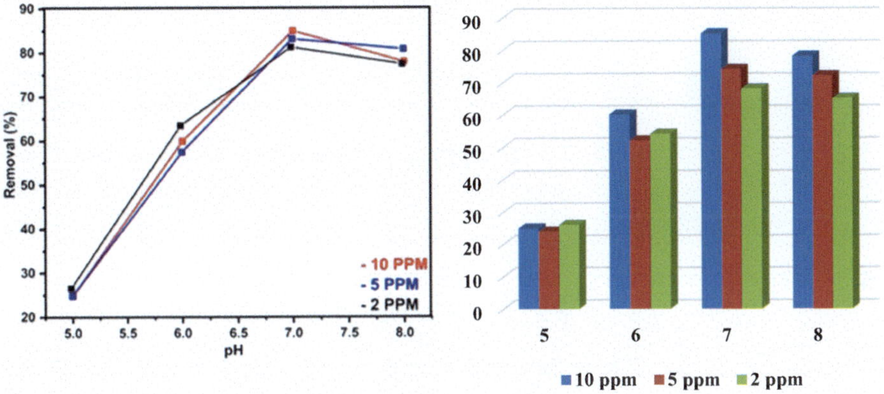

Fig. 3 Effect of pH and initial fluoride concentration on fluoride adsorption by coconut shell activated carbon

3.3 *Adsorption Comparison Among Different Adsorbents*

Different adsorbents used for the defluoridation purpose are compared and their corresponding results are summarized in Table 3.

4 Conclusion

Activated carbon from coconut shells was used for the removal of F^- from aqueous solutions. SEM, FTIR, and BET methods had been used to characterize those produced products. The BET surface area was measured to be 91 m^2/g. The material's homogeneous as well as porous morphology was shown through SEM data, with the activated carbon specimen displaying additional cavities. Based on the acquired data, the highest level of adsorbed fluoride was of 85% experience and recorded around pH 7 during a 24-h interaction with an adsorbent dosage of 1 g/ 250 mL. It has also been compared the maximum adsorption capacity observed in the present work with the other adsorbents capacities reported in the literature.

Table 3 Adsorption comparison among different adsorbents for defluoridation

Adsorbent	Initial concentration (mg/L)	Adsorbent dose	pH	Equilibrium time (min)	Surface area (m^2/g)	Adsorption efficiency	References
Magnesium amended activated carbons obtained using coconut shells	10	1 g/L	7	1440 min	42.5	92%	[28]
Alumina impregnated activated carbon	10	6 g/L	6.1	180 min	68	89%	[29]
Rice husk coated iron oxyhydroxide	10	0.8 g/L	4	45 min	53.7	83%	[30]
Sawdust impregnated ferric hydroxide and activated alumina	5–50	0.1 g/ 10 mL	6.5	240 min	22.5	74%	[31]
Coconut shells activated carbon	13.2	1 g/L	7	300 min	78.7	68.2%	[32]
Fluoride removal by adsorption using coconut shells	2–10	1 g/ 250 mL	7	1440 min	91	85%	Present work

Acknowledgment The authors thankfully acknowledge National Institute of Technology Srinagar, Jammu and Kasmir, India.

References

1. A.R. Chinchmalatpure, B. Gorain, S. Kumar, D.D. Camus, S.D. Vibhute, Groundwater pollution through different contaminants: Indian scenario, in *Research Developments in Saline Agriculture*, ed. by J.C. Dagar, R.K. Yadav, P.C. Sharma, (Springer Singapore, Singapore, 2019), pp. 423–459. https://doi.org/10.1007/978-981-13-5832-6_15
2. F.A. Dar, S. Kurella, Recent advances in adsorption techniques for fluoride removal – an overview. Groundw. Sustain. Dev. **23**, 101017 (2023). https://doi.org/10.1016/j.gsd.2023.101017
3. J. Singh, P. Yadav, A.K. Pal, V. Mishra, Water pollutants: origin and status, in *Sensors in Water Pollutants Monitoring: Role of Material*, (2020). https://doi.org/10.1007/978-981-15-0671-0_2
4. M. Yousefi, S. Ghalehaskar, F.B. Asghari, A. Ghaderpoury, M.H. Dehghani, M. Ghaderpoori, A.A. Mohammadi, Distribution of fluoride contamination in drinking water resources and health risk assessment using geographic information system, northwest Iran. Regul. Toxicol. Pharmacol. **107**, 10440 (2019). https://doi.org/10.1016/j.yrtph.2019.104408
5. S. Ali, S.K. Thakur, A. Sarkar, S. Shekhar, Worldwide contamination of water by fluoride. Environ. Chem. Lett. **14**, 291–315 (2016). https://doi.org/10.1007/s10311-016-0563-5
6. S. Ayoob, A.K. Gupta, Fluoride in drinking water: a review on the status and stress effects. Crit. Rev. Environ. Sci. Technol. (2006). https://doi.org/10.1080/10643380600678112
7. S. Peckham, N. Awofeso, Water fluoridation: a critical review of the physiological effects of ingested fluoride as a public health intervention. Sci. World J. **2014**, 1–10 (2014). https://doi.org/10.1155/2014/293019
8. F. Ahmad Dar, S. Kurella, Fluoride in drinking water: an in-depth analysis of its prevalence, health effects, advances in detection and treatment. Mater. Today Proc. (2023). https://doi.org/10.1016/j.matpr.2023.05.645
9. H.W.F. Bunce, Fluoride in air, grass, and cattle. J. Dairy Sci. **68**, 1706–1711 (1985). https://doi.org/10.3168/jds.S0022-0302(85)81016-X
10. G. Biswas, M. Kumari, K. Adhikari, S. Dutta, A critical review on occurrence of fluoride and its removal through adsorption with an emphasis on natural minerals. Curr. Pollut. Rep. **3**, 104–119 (2017). https://doi.org/10.1007/s40726-017-0054-8
11. N. Nabbou, M. Belhachemi, M. Boumelik, T. Merzougui, D. Lahcene, Y. Harek, A.A. Zorpas, M. Jeguirim, Removal of fluoride from groundwater using natural clay (kaolinite): optimization of adsorption conditions. C. R. Chim. **22**, 105–112 (2019). https://doi.org/10.1016/j.crci.2018.09.010
12. A. Rasool, A. Farooqi, T. Xiao, W. Ali, S. Noor, O. Abiola, S. Ali, W. Nasim, A review of global outlook on fluoride contamination in groundwater with prominence on the Pakistan current situation. Environ. Geochem. Health **40**, 1265–1281 (2017). https://doi.org/10.1007/s10653-017-0054-z
13. J. He, Y. Yang, Z. Wu, C. Xie, K. Zhang, L. Kong, J. Liu, Review of fluoride removal from water environment by adsorption. J. Environ. Chem. Eng. **8**, 104516 (2020). https://doi.org/10.1016/j.jece.2020.104516
14. S.K. Jha, A.K. Nayak, Y.K. Sharma, Response of spinach (Spinacea oleracea) to the added fluoride in an alkaline soil. Food Chem. Toxicol. **46**, 2968–2971 (2008). https://doi.org/10.1016/j.fct.2008.05.024
15. I. Hussain, M. Arif, J. Hussain, Fluoride contamination in drinking water in rural habitations of Central Rajasthan, India. Environ. Monit. Assess. **184**, 5151–5158 (2012). https://doi.org/10.1007/s10661-011-2329-7

16. S. Saxena, U. Saxena, Study of fluoride contamination status of ground water in Bassi Tehsil of district. Int. J. Environ. Sci. **3**, 2251–2260 (2013). https://doi.org/10.6088/ijes.2013030600040
17. J.M. Sim, K.M. Leong, Feasibility study on fluoride removal in drinking water in Mehsana, India. Int. NGO J. **6**, 224–228 (2011). https://doi.org/10.5897/NGOJ11.023
18. R.A. Duraiswami, U. Patankar, Occurrence of fluoride in the drinking water sources from Gad River Basin, Maharashtra. J. Geol. Soc. India **77**, 167–174 (2011). https://doi.org/10.1007/s12594-011-0020-9
19. J. Podgorski, M. Berg, Global analysis and prediction of fluoride in groundwater. Nat. Commun. **13**, 1–9 (2022). https://doi.org/10.1038/s41467-022-31940-x
20. N. Gangani, V.C. Joshi, S. Sharma, A. Bhattacharya, Fluoride contamination in water: remediation strategies through membranes. Groundw. Sustain. Dev. **17**, 100751 (2022). https://doi.org/10.1016/j.gsd.2022.100751
21. S. Dubey, M. Agarwal, A.B. Gupta, Recent developments in defluoridation of drinking water in India, in *Environmental Pollution*, ed. by V. Singh, S. Yadav, R. Yadava, (Springer, 2018), pp. 345–356. https://doi.org/10.1007/978-981-10-5792-2_28
22. M. Habuda-Stanić, M. Ravančić, A. Flanagan, A review on adsorption of fluoride from aqueous solution. Materials (Basel) **7**, 6317–6366 (2014). https://doi.org/10.3390/ma7096317
23. S.V. Jadhav, E. Bringas, G.D. Yadav, V.K. Rathod, I. Ortiz, K.V. Marathe, Arsenic and fluoride contaminated groundwaters: a review of current technologies for contaminants removal. J. Environ. Manag. **162**, 306–325 (2015). https://doi.org/10.1016/j.jenvman.2015.07.020
24. M. Tahaikt, R. El Habbani, A. Ait Haddou, I. Achary, Z. Amor, M. Taky, A. Alami, A. Boughriba, M. Hafsi, A. Elmidaoui, Fluoride removal from groundwater by nanofiltration. Desalination **212**, 46–53 (2007). https://doi.org/10.1016/j.desal.2006.10.003
25. M.M. Emamjomeh, M. Sivakumar, Fluoride removal by a continuous flow electrocoagulation reactor. J. Environ. Manag. **90**, 1204–1212 (2009). https://doi.org/10.1016/j.jenvman.2008.06.001
26. W.X. Gong, J.H. Qu, R.P. Liu, H.C. Lan, Effect of aluminum fluoride complexation on fluoride removal by coagulation. Colloids Surf. A Physicochem. Eng. Asp. **395**, 88–93 (2012). https://doi.org/10.1016/j.colsurfa.2011.12.010
27. A.K. Tolkou, M. Mitrakas, I.A. Katsoyiannis, M. Ernst, A.I. Zouboulis, Fluoride removal from water by composite Al/Fe/Si/Mg pre-polymerized coagulants: characterization and application. Chemosphere **231**, 528–537 (2019). https://doi.org/10.1016/j.chemosphere.2019.05.183
28. A.K. Tolkou, S. Trikalioti, O. Makrogianni, D.G. Trikkaliotis, E.A. Deliyanni, G.Z. Kyzas, I.A. Katsoyiannis, Magnesium modified activated carbons derived from coconut shells for the removal of fluoride from water. Sustain. Chem. Pharm. **31**, 100898 (2023). https://doi.org/10.1016/j.scp.2022.100898
29. A. Saini, P.H. Maheshwari, S.S. Tripathy, S. Waseem, S.R. Dhakate, Processing of rice straw to derive carbon with efficient de-fluoridation properties for drinking water treatment. J. Water Process Eng. **34**, 101136 (2020). https://doi.org/10.1016/j.jwpe.2020.101136
30. P. Pillai, Y. Lakhtaria, S. Dharaskar, M. Khalid, Synthesis, characterization, and application of iron oxyhydroxide coated with rice husk for fluoride removal from aqueous media. Environ. Sci. Pollut. Res. **27**, 20606–20620 (2020). https://doi.org/10.1007/s11356-019-05948-8
31. P. Dhanasekaran, O. Sahu, Arsenate and fluoride removal from groundwater by sawdust impregnated ferric hydroxide and activated alumina (SFAA). Groundw. Sustain. Dev. **12**, 100490 (2021). https://doi.org/10.1016/j.gsd.2020.100490
32. M. Said, R.L. Machunda, Defluoridation of water supplies using coconut shells activated carbon: batch studies. Int. J. Sci. Res. **3**, 2327–2331 (2014)

Removal of Suspended Solids and COD from a Textile Industry Wastewater Using Electrocoagulation and Advanced Treatment Process

Tamal Kanti Das, Ashim Kumar Bhattacharya, and Sailendra Nath Mandal

Abbreviations

APHA	American Public Health Association
AWWA	American Water Works Association
BIS	Bureau of Indian Standards
BOD	Bio-chemical Oxygen Demand
COD	Chemical Oxygen Demand
DC	Direct Current
EC	Electrocoagulation
IS	Indian Standards
TDS	Total Dissolved Solids
TSS	Total Suspended Solids
UV	Ultraviolet
UV VIS Spectrophotometer	Ultraviolet Visible Spectrophotometer
WPCF	Water Pollution Control Federation

T. K. Das (✉) · S. N. Mandal
Civil Engineering Department, N.I.T.T.T.R, Salt Lake City, Kolkata, India
e-mail: snmandal@nitttrkol.ac.in

A. K. Bhattacharya
N.I.T.T.T.R., Academy of Water Technology and Environ Management, Kolkata, India

D. Mazumder (ed.), *Sustainable Advanced Technologies for Environmental Management*, Springer Proceedings in Earth and Environmental Sciences,
https://doi.org/10.1007/978-3-031-64006-3_24

1 Introduction

Water is one of the lifelines for the survival of mankind and there is no alternative available so far. Though most of the earth surface is covered with water, but unfortunately, only 1% of it is for human consumption. In India, the necessity of water increased manyfold, as the population is increased about more than four times since 1947, the year of the independence. In present, the rate of production from different industries also increased; simultaneously, the wastewater generation also increases. It is compulsory to treat the wastewater before discharging it in nature [1]; otherwise it may contaminate the resources of water [2]. The World Water Assessment Program (WWAP), 2012, organized by the United Nations, published the World Water Development Report 4 where it is predicted that by 2025 most of Northern India may face economic water scarcity, whereas the Southern part may suffer from physical water scarcity. In last decades, coagulation came to be the most approachable and cost-effective physiochemical process, vastly adopted by the industries [3]. In coagulation, the application of different coagulants was involved to turn the impurities, especially the colloidal impurities, in the form of flocs and settle it down as sludge [4]. Now the large-scale treatment produces large amount of sludge, and its disposal becomes a big challenge for the industries.

As the solution [5], electrocoagulation is introduced to separate the suspended impurities by utilizing the concept of electrolysis. In this technique, the application of chemicals becomes almost nil; consequently, the production of sludge becomes very much less [5]. The performance of the technique was already reviewed for the treatment of effluent from various industries like paper mills, textile, pharmaceutical, etc. [6–8]. Here, the study has been carried out on textile wastewater which is one of the biggest threats to the environment [9]. In this research, the removal efficiency of two significantly selected parameters, total suspended solids (TSS) and chemical oxygen demand (COD) from textile industry wastewater by electrocoagulation technique, was investigated and reported [10]. The experimental trials have been executed with constant DC supply, and the removal of TSS and COD has been investigated with the time variation of electrolysis, until the removal reached up to the optimum range.

2 Materials and Methods

The sample of the effluent was freshly collected from one of the textile manufacturing factory of Dhulagarh, West Bengal, before each trial. The wastewater contains high percentage of unfixed dyes. After the collection of effluent, with TSS and COD, some other significant parameters of the effluent such as pH, BOD, and conductivity were immediately determined with sophisticated instruments according to standard methods prescribed by APHA-AWWA-WPCF [11] and BIS. The chemical parameters are tabulated in Table 1. The total solids (TS) of effluent were measured by the

Table 1 Chemical parameters of raw textile wastewater

Chemical parameters	Concentration range
pH	9.25–9.5
Total suspended solids (mg/L)	580–597
BOD (mg/L)	350–390
COD (mg/L)	858–963
Conductivity (μS/cm.)	7400–7500

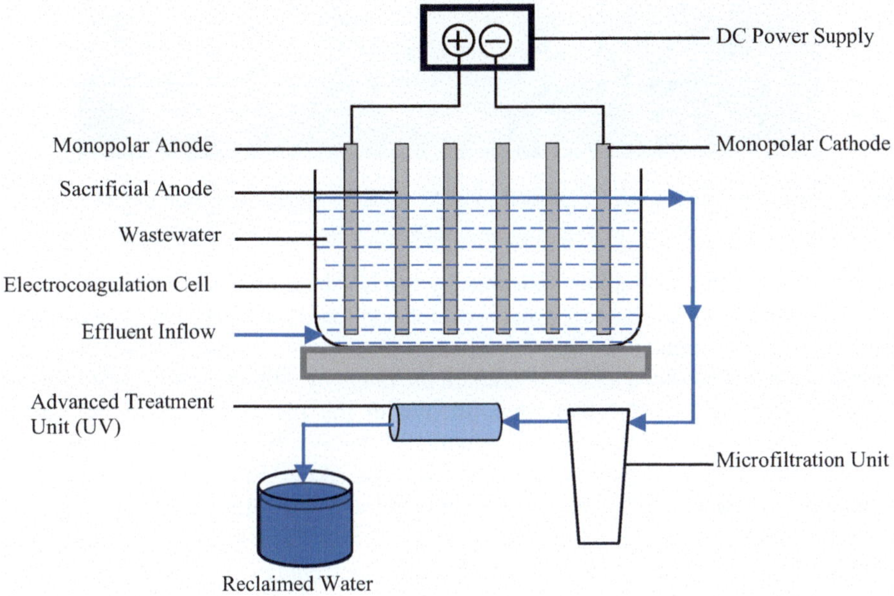

Fig. 1 Schematic flow diagram of treatment process

evaporating method as per section 2540 B (APHA 1998), and the total dissolved solids (TDS) were measured by the same evaporation process according to IS 3025 (Part 16): 1984. The samples were filtered through 42 Whatman filter papers. The evaporation was done on steam bath at 103–105 °C temperature. Finally, the TSS were determined by eliminating the TDS values from corresponding TS. The method was found comparatively suitable for this particular case and chosen on the basis of the results from a series of trials according to other standard methods. The COD values have been measured by DR6000 UV VIS spectrophotometer after the consumption of dichromate as per section 5220 A (APHA 1998).

The experimental setup of pilot reactor of electrocoagulation with advanced treatment was fabricated for the pilot study with a treatment capacity of 25 liters/hour (Fig. 1) [12]. A set of 40 numbers of iron (Fe) plates (125 mm × 200 mm × 3 mm), separated by inert material, were used as electrodes [7, 8] and installed in the reactor for electrolysis. The arrangement has been made for DC power supply of 220 volts, 20 amps to the monopolar electrodes. The EC-treated effluent was further

Fig. 2 Raw textile wastewater, EC-treated and UV-treated reclaimed water (left to right)

passed through microfiltration, followed by the UV-based advanced treatment process, and finally the reclaimed water was collected and analyzed. The working pH range for EC was found to be 5.5–7.5. The raw effluent was found alkaline (pH > 9); therefore the pH has been adjusted to attain the range by adding diluted H_2SO_4 in the effluent. The pilot reactor had been filled with the pH-adjusted effluent, and a constant DC power supply with the potential difference of 100 volts was supplied for each trials with varying time of electrolysis. After EC and especially after advanced treatment (UV), the color of the effluent was almost getting removed (Fig. 2). Finally, the same parameters of the reclaimed water after each trials were determined, and for each time of electrolysis, the percentage of removal of TSS and COD was observed for variation of the time of electrolysis, to determine optimum removal efficiency of the process.

3 Results and Discussion

After each trial, the determined TSS and COD values of the reclaimed water were compared with the TSS and COD values of the raw effluent, and the percentage of removal was calculated. The percentage of removal of TSS and COD, both with respect to the corresponding time duration of the electrolysis for the constant potential difference of 100 volts, is graphically represented in Figs. 3 and 4, respectively.

Initially, after just 25–30 minutes of electrolysis, the removal of TSS and COD was possible up to 75% and 60%, respectively. At the initial trials for 45 minutes to 1.5 hours of electrolysis, the outstanding improvement of the results has been observed. It was noticed that after 1.5 hour of electrolysis, the curves have become flatter and beyond 2 hours of electrolysis the removal has become saturated for both

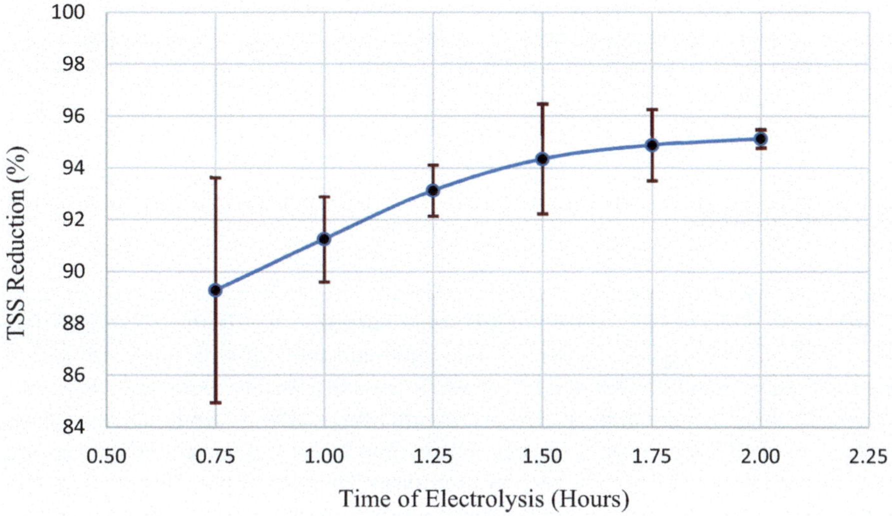

Fig. 3 Reduction of TSS (%) varies with the time of electrolysis (hrs)

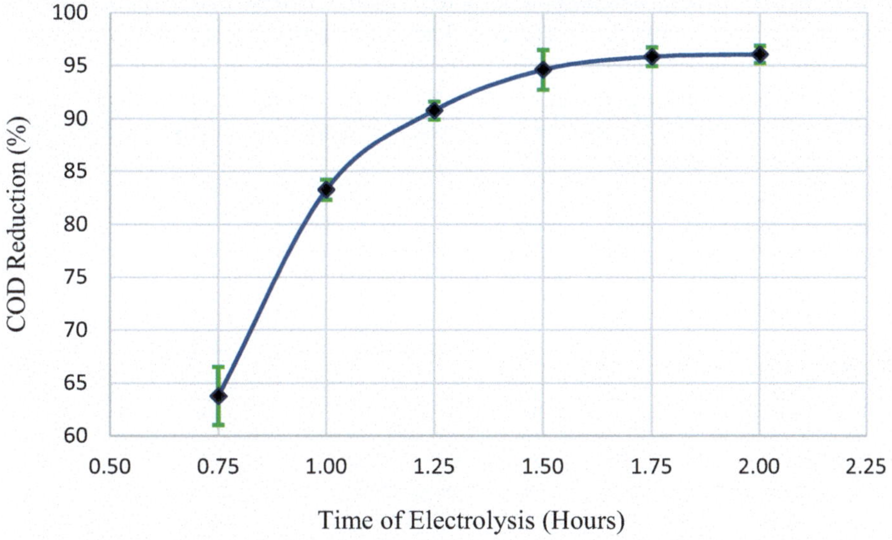

Fig. 4 Reduction of COD (%) varies with the time of electrolysis (hrs)

cases. Ultimately, the percentage of removal becomes constant, and particularly for this effluent, up to 95% of TSS and 96% of COD have been possible to remove.

The effluent from textile industry contains dyes of soluble and insoluble both. The color of the effluents get generated due the presence of the insoluble matters. One can realize by observing the EC-treated reclaimed water that certain amount of color was still left, but after passing through the UV treatment unit, the color almost got removed. The decoloring process has successfully happened only by UV treatment, even without any additional arrangement like ozonation or adding oxidants. Therefore, technically it may be not an oxidation process but purpose was served only with the UV treatment unit. Moreover, it also disinfects the wastewater.

4 Conclusion

The efficiency of the electrocoagulation basically depends upon the type of electrodes, time of electrolysis, potential difference (in volts), etc. In this paper, the time of electrolysis is analyzed. Generally, one may presume that with the increment of the time the removal of TSS and COD will be proportionately increased. However, practically such increment happened at the initial stage of 0.75–1.5 hours; especially in the case of COD removal, it improves from 63% to 94%. After a stipulated period of time, the removal efficiency becomes saturated. Hence, optimum removal of TSS and COD is achieved at the ultimate saturation stage of time. Beyond this percentage of removal, there will be no possibility of further improvement by increasing only the time of electrolysis with the constant DC power supply of 100 volts. There will be a possibility to enhance the percentage removal of the impurities by increasing the amount of power supply to the EC cell. The characteristics of the reclaimed water were observed and compared with the permissible values prescribed by Central Pollution Control Board. Based on the satisfactory results of the treated effluent parameters, it can be recommended to use for gardening, horticulture or other non-potable secondary uses.

Acknowledgments The present research work is supported by the PhD Program under Quality Improvement Program of Ministry of Education, Government of India.

References

1. D.A. Yaseen, M. Scholz, Textile dye wastewater characteristics and constituents of synthetic effluents: A critical review. Int. J. Environ. Sci. Technol. **16**, 1193–1226 (2019)
2. D. Rajkumar, K. Palanivelu, Electrochemical treatment of industrial wastewater. J. Hazard. Mater. **B113**, 123–129 (2004). https://doi.org/10.1016/j.jhazmat.2004.05.039
3. J.-Q. Jiang, The role of coagulation in water treatment. Curr. Opin. Chem. Eng. **8**, 36–44 (2015)
4. J. Gregory, V. Dupont, Properties of flocs produced by water treatment coagulants. Water Sci. Technol. **44**, 231–236 (2001)
5. Padmaja et al., A comparative study of the efficiency of chemical coagulation and electrocoagulation methods in the treatment of pharmaceutical effluent. J Water Process. Eng. **34**, 101153 (2020)
6. M.M. Emamjomeh, M. Sivakumar, Review of pollutants removed by electrocoagulation and electrocoagulation/flotation processes. J. Environ. Manag. **90**, 1663–1679 (2009)

7. R. Katal, H. Pahlavanzadeh, Influence of different combinations of aluminum and iron electrode on electrocoagulation efficiency: Application to the treatment of paper mill wastewater. Desalination **265**, 199–205 (2011)
8. M. Boroski et al., Combined electrocoagulation and TiO_2 photoassisted treatment applied to wastewater effluents from pharmaceutical and cosmetic industries. J. Hazard. Mater. **162**, 448–454 (2009)
9. N. Daneshvar, A. Oladegaragoze, N. Djafarzadeh, Decolorization of basic dye solutions by electrocoagulation: An investigation of the effect of operational parameters. J. Hazard. Mater. **B129**, 116–122 (2005)
10. B. Merzouk, K. Madani, A. Sekki, Using electrocoagulation-electroflotation technology to treat synthetic solution and textile wastewater, two case studies. Conference on membranes in drinking and industrial water production, 2010
11. APHA-AWWA-WPCF, *Standard Methods for the Examination of Water and Wastewater*, 20th edn. (Washington, DC, 1998)
12. P. Holt, G. Barton, C. Mitchell, Electrocoagulation as a wastewater treatment. The 3rd Annual Australian Environmental Engineering Research Event, 1999

Leachate Treatment Using Low-Cost Adsorbents: A Comparative Study

Potshangbam Janshi Devi, Akanksha Srivastava, and Potsangbam Albino Kumar

Abbreviations

AR	Analytical reagent
BOD	Biochemical oxygen demand
COD	Chemical oxygen demand
MSW	Municipal solid waste
NTU	Nephelometric turbidity unit
PAP	Pineapple peel
SSH	Sunflower seed husk
SWM	Solid waste management
TDS	Total dissolved solids
TS	Total solids
TSS	Total suspended solids
TW	Tea waste

1 Introduction

In recent times, due to rapid pace of industrialization and urban expansion, global concerns about environmental protection have escalated leading to environmental degradation. Over the past few years, the nature of domestic waste has undergone a dramatic shift, primarily influenced by the emergence of a more affluent society. Cities worldwide produce billions of tons of solid waste annually, and projections

P. J. Devi · A. Srivastava · P. A. Kumar (✉)
Civil Engineering Department, National Institute of Technology, Manipur, India
e-mail: albino@nitmanipur.ac.in

D. Mazumder (ed.), *Sustainable Advanced Technologies for Environmental Management*, Springer Proceedings in Earth and Environmental Sciences,
https://doi.org/10.1007/978-3-031-64006-3_25

indicate that this quantity is anticipated to rise in the coming years [1]. One of the by-products of the solid waste is leachate, formed from chemical and biological reactions occurring when rainwater permeates through the landfill. If not appropriately managed and treated, leachate can result in detrimental environmental effects, including unpleasant odors and groundwater contamination [2]. These leachates often contain elevated levels of soluble organic substances and inorganic ions. Solid waste and leachate management encounters substantial challenges globally. The World Bank estimates that municipal solid waste generation will surge from 2.01 billion tons in 2016 to a projected 3.40 billion tons by 2050, reflecting rapid urbanization and population growth [3]. Alarmingly, over 40% of waste worldwide is disposed of in open dumps, as highlighted in the Global Waste Management Outlook 2015 [4]. This underscores the urgent need for enhanced waste management infrastructure and practices. The World Health Organization reports that improper waste management contributes to 5–10% of all communicable diseases in developing countries, emphasizing the critical link between waste disposal and public health [4]. These statistics underscore the pressing need for comprehensive strategies integrating technology, policy reforms, and community engagement to establish sustainable waste management practices capable of addressing environmental, social, and health challenges effectively. Several methods and technologies have been explored, implemented, and suggested for treating landfill leachate. These approaches span from straightforward techniques like leachate recirculation through the landfill to more complex ones involving a combination of physical, chemical, and biological processes [4]. A challenge in treating leachate lies in the varying characteristics of the leachate over time and space, making the selection of suitable treatment methods quite challenging. The treatability of leachate is delineated into three primary categories: integration with domestic sewage for treatment in municipal plants, degradation through aerobic or anaerobic processes, and application of chemical or physical treatment techniques. Biological treatments work well for younger leachate with abundant volatile fatty acids but are less effective for older or stabilized leachates [4]. This study focuses on the most cost-effective method for treating landfill leachate.

In India, an annual generation of 30.3 million tons of municipal solid waste (MSW) equates to approximately 350 g of waste per person on average [4]. This volume is poised to escalate in the foreseeable future due to population expansion and industrial growth. According to estimates from the Central Pollution Control Board of India, the MSW generation rate is anticipated to soar to 299.3 million tons by 2047 [5], assuming a rise in India's urban population to 45% from the current 28%. This study assesses the existing solid waste management (SWM) infrastructure in Imphal, a town in Northeast India, and explores the viability of an SWM plan. Imphal presently generates about 120 tons/day of solid waste, with projections indicating an increase to approximately 170 tons/day by 2035 [6]. With this huge amount of waste being generated, there is a need for waste disposal. Among the various methods like recycling, recovery, incineration, and composting used for managing solid wastes, landfilling remains the primary approach [6]. However, landfilling faces significant challenges due to the sheer quantity of waste generated and its environmental impact. This necessitates exploring alternative approaches for handling solid waste. Adsorption present a promising avenue in the realm of waste management. Biomaterials like sunflower seed husk, tea waste, and pineapple peel exhibit remarkable adsorption properties that can aid in mitigating the challenges posed by leachate contaminants in landfills [7]. These adsorbents offer a sustainable and cost-effective solution for reducing the environmental impact of waste disposal. Their porous structures and extensive surface areas provide ample sites for trapping and adsorbing contaminants present in leachate, effectively reducing the risk of groundwater and soil contamination around landfill sites. Furthermore, the presence of functional groups in these adsorbents facilitates interactions with various pollutants, enhancing their adsorption efficiency.

2 Materials and Methods

2.1 Synthesis of Adsorbent

In this study, sunflower seed husk, pineapple peel, and tea waste, abundant in the Northeast region, were chosen as cost-effective adsorbents based on a comprehensive literature review, as their potential as adsorbents for leachate COD removal had not been explored. Availability of these materials in the region addresses the challenge of consistent raw material supply, promising solutions for organic contamination removal. The adsorbents—sunflower seed husk (SSH), tea waste (TW), and pineapple peel (PAP)—were procured locally near NIT Manipur Campus. After thorough cleaning with distilled water to remove dust, they underwent oven drying at 150 °C for 24 hours to eliminate moisture and volatile impurities. Chemical impregnation using H_3PO_4 followed, soaking the material for 24 hours. Subsequently, activation involved heating in a muffle furnace at 550 °C for 20 minutes. Washing with warm distilled water ensured removal of residues and impurities, followed by filtration and drying at 100 °C for 24 hours. The dried material was sieved for uniformity and stored airtight. Therefore, the obtained activated carbon is termed as sunflower seed husk activated carbon (SSH), tea waste activated carbon (TW), and pineapple peel activated carbon (PAP).

2.2 Leachate Sample Collection and Analysis

The study involved the collection of landfill leachate samples from the Lamdeng Khunou Solid Waste Management Plant, Manipur, India (24°84′24.7704″ N, 93°53′ 23.9244″ E). Two leachate ponds, the raw leachate pond and the sump pond for leachate evaporation and stabilization, are present at the treatment site. Stabilized leachate samples were initially collected solely from the sump pond every 2 weeks between July to September 2023.The samples were carefully gathered and placed in clean polypropylene containers, totaling 20 L. These containers were then transported to the laboratory for analysis and subsequent storage. Various parameters, including pH levels, chemical oxygen demand (COD), biochemical oxygen demand (BOD), total solids (TS), total dissolved solids (TDS), total suspended solids (TSS), as well as the presence of major ions and heavy metals like iron, were investigated. To ensure accuracy and reliability, all analyses and tests throughout this study adhered to standardized methods as per the guidelines [8].

2.3 Batch Experiment

The coagulation-flocculation experiments conducted at the laboratory bench scale were set up using a Phipps and Bird Jar-test apparatus [9]. This apparatus consisted of six jars, each with a capacity of 2 L, equipped with six spindle steel paddles for stirring. The experimental setup allowed for both rapid mixing at 200 rpm for 2 minutes to ensure sample homogeneity with the coagulant solution, followed by slow mixing at 50 rpm for 30 minutes to facilitate floc formation and settlement with the time and speed controlled automatically by the apparatus. Subsequently, sludge settling occurs for an additional 30 minutes. After settling, supernatant is withdrawn from beakers, taken 2 cm below the surface, using pipettes for chemical analysis. All experiments were conducted in triplicate for enhanced reliability. For each batch test, 1 L leachate samples were utilized in individual beakers to conduct the coagulation-flocculation experiments. Throughout these experiments, analytical reagent (AR) grade chemicals were employed to analyze and assess the various parameters under study [7, 9]. This ensured precision and accuracy in evaluating the effects of coagulation-flocculation on the leachate samples. The amount of adsorbate adsorbed was estimated using Eq. (1):

$$q_e = [(C_o - C_e)/m]\ V, \tag{1}$$

where C_o is the initial concentration and C_e is the concentration at equilibrium in mg/l, V is the volume of the sample, and m is the mass of adsorbent (g/L).

Different models have been used in kinetic studies—Lagergren's pseudo-first-order (Eqs. 2 and 3) and Lagergren's pseudo-second-order (Eqs. 4 and 5) [9].

Lagergren's pseudo-first-order:

$$\log(q_e - q_t) = \log\ q_e - (k_1/2.303)\ t \tag{2}$$

Further, equation can be expressed as a function of time as

$$q_t = q_e\left(1 - e^{-k_1 t}\right) \tag{3}$$

Lagergren's pseudo-second-order:

$$dq_t/dt = k_2(q_e - q_t)^2 \tag{4}$$

$$t/q_t = \left(1/k_2 q_e^2\right) + (t/q_e) \tag{5}$$

where q_e and q_t are the amounts of metal ions adsorbed on adsorbent (mg/g) at equilibrium and at time t, respectively, and k_1 and k_2 are the adsorption rate constant of pseudo-first-order equation (minutes^{-1}) and pseudo-second-order equation (g/mg. minutes), respectively. The adsorption isotherm elucidates the distribution of ions or molecules between the adsorbent and the bulk liquid phase once equilibrium is reached [9]. In this study, two commonly utilized models, namely, the Langmuir and Freundlich isotherms, were employed to investigate the relationship between adsorbed ions (q) and aqueous concentration (C). By utilizing the Langmuir isotherm (Eq. 6) and Freundlich isotherms (Eq. 7), this research aims to comprehend the equilibrium relationship between these variables:

$$q_e = (q_{max} b C_e)/(1 + b C_e), \tag{6}$$

$$q_e = k_f C_e^{1/n}, \tag{7}$$

where q_{max} is the maximum adsorption capacity or monolayer coverage capacity (mg/g), b is the Langmuir adsorption constant (L/mg), k is the indicator of adsorption capacity, and n represents the Freundlich model's affinity factor [9]. The acceptability of the isotherm equations was evaluated with respect to the correlation coefficients R, estimated from linear regression of the adsorption data.

2.4 Methodology

The analysis of the leachate samples encompassed a comprehensive study of several physicochemical parameters. These included pH levels, turbidity, alkalinity, total solids (TS), total dissolved solids (TDS), total suspended solids (TSS), as well as the assessment of major ions like calcium (Ca^{+2}), chloride (Cl^-), fluoride (F^-), nitrate (NO_3^-), and heavy metals such as iron. All characterization procedures were conducted at an ambient temperature of approximately 25 ± 2 °C to maintain consistency across the tests. The effect of pH was initially examined through experiments conducted at different pH levels (2, 4, 6, 8, 10, 12) employing an adsorbent dose of 4 g/L, in order to determine the optimal working pH. The pH of the sample was adjusted to the desired levels by the addition of appropriate amounts of lime ($Ca(OH)_2$) and H_2SO_4. Subsequently, the study of the influence of contact time on COD adsorption was carried out at an optimal pH of 4 (identified from pH effect study) by employing 4 g/L of sunflower seek husk, pineapple peel, and tea waste for 5–180 minutes. During the initial study, employing a dosage of 4 g/L adsorbent revealed notable variations in the removal percentage of the leachate across different contact times. Consequently, 4 g/L was selected for this study to highlight its higher efficiency across various contact times. The adsorption kinetics of COD onto powdered activated SSH, PAP, and TW was carried out by withdrawing and analyzing the sample at 5, 10, 15, 30, 60, 90, 120, and 180 minutes. Adsorption isotherm is carried out for different dose of adsorbent 2, 4, 6, 8, 10, and 12 g/L by using Langmuir isotherm and Freundlich model. The average values derived from these repeated analyses were then presented, providing a more reliable representation of the leachate sample characteristics.

3 Results and Discussion

3.1 Characterization of Landfill Leachate

The leachate collected at the Lamdeng landfill site in Imphal, Manipur, exhibited stabilized characteristics, with a pH of 8.01 ± 0.68. Notably, it displayed high TDS content of 4388 mg/L, COD of 3200 mg/L, and BOD_5 of 465 mg/L. The BOD_5/COD ratio of 0.15 indicates limited biodegradability, suggesting a significant conversion of organics to methane [9, 10]. With the low BOD_5/COD ratio, the leachate, identified as old or stabilized, is better suited for physicochemical treatment than biological methods. While the presence of heavy metal ions was found to be negligible, the COD concentration remains considerably high, and other parameters detected in the leachate are presented in Table 1.

3.2 Optimization of pH for Coagulation with Ferric Chloride

As per Fig. 1, it is found that coagulant ferric chloride exhibits optimal effectiveness at a pH of 4, utilizing a random dose of 4 g/L. At this pH level, the results indicate the

Table 1 Physicochemical characteristics of raw leachate

Sl. no.	Parameters	Concentration	Permissible limit (CPCB 2017)
1	pH	8.01	5.5–9
2	Conductivity, ms/cm	6.55	4.7
3	Turbidity, NTU	161	300
4	Total dissolved solids, mg/L	4388	1200
5	COD, mg/L	3200	100
6	BOD_5, mg/L	465	50
7	BOD_5/COD	0.15	0.5
8	Alkalinity, mg/L	2520	600
9	Total iron	25	3

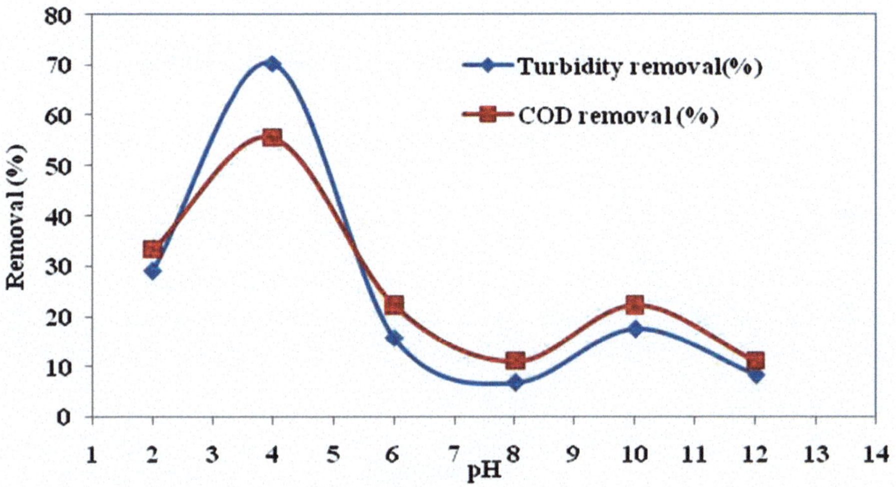

Fig. 1 pH optimization for coagulation with ferric chloride

highest removal efficiencies for turbidity (70.31%) and COD (55.55%). Beyond pH 4, both turbidity and COD removal efficiencies decline with increasing pH, which demonstrated higher COD removal by ferric chloride at pH 4 compared to pH 6 and 12. The optimum pH for ferric chloride coagulation is influenced by the water's organic matter concentration, with organic removal being more effective under slightly acidic conditions. In cases of higher organic content, the optimum pH tends to shift towards slightly more acidic values. Adequate alkalinity in leachate is essential during the flocculation process for two reasons: first, to maintain buffer capacity and prevent a drastic pH drop due to ferric chloride addition and, second, to ensure a sufficient supply of bicarbonate for the formation of metal hydroxide [9].

3.3 Optimization of Ferric Chloride Dose

Various coagulant dosages (1–12 mg/L) were employed at the optimum pH 4 as obtained from Fig. 1, and the result is shown in Fig. 2 with ferric chloride at an optimal dosage of 4 mg/L. The turbidity reduction increases from 15% to 70.93% at a dose of 4 mg/L and COD reduction from 15% to 55.55%. Further increases in ferric chloride dosage did not enhance removal efficiencies, aligning with the charge neutralization theory. Ferric chloride, through Fe^{3+} and its hydrolyzed products, destabilizes negative colloids by charge neutralization [9]. However, beyond the optimal dosage, colloids absorb cations, regaining stability through electrical repulsion. The parameters detected for leachate after coagulation are shown in Table 2.

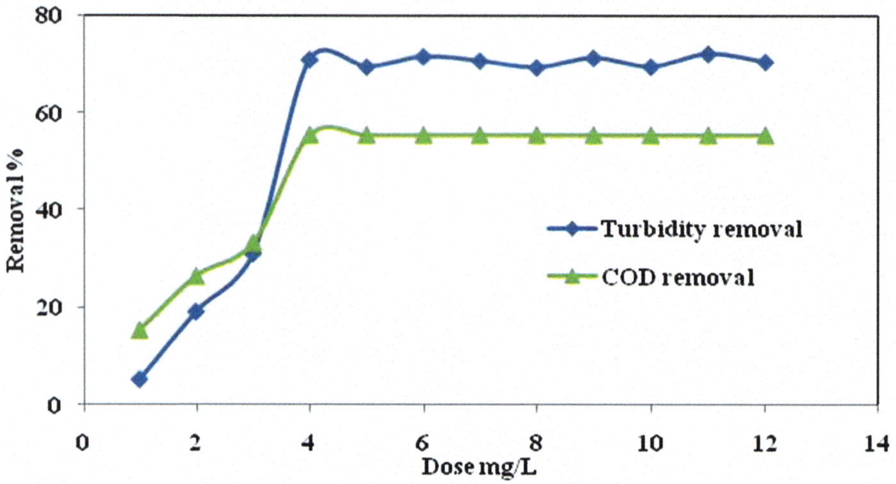

Fig. 2 Ferric chloride dose optimization

Table 2 Leachate parameter after coagulation

Sl. no.	Parameters	Concentration
1	pH	4
2	Conductivity (ms/cm)	6.5
3	COD (mg/L)	1600
4	BOD_5 (mg/L)	204
5	Turbidity (NTU)	161
6	TDS (mg/L)	4388
7	Alkalinity (mg/L)	2520
8	Iron (mg/L)	24.9

3.4 Leachate Parameters After Adsorption

In Table 3, the adsorbents SSH, PAP, and TW reduced COD further till 160, 290, and 320 mg/L, respectively, yielding efficiency of 95.4%, 91.7%, and 90.8%. On the other hand, BODs were reduced to 40.1, 34.9, and 55.6 mg/L by SSH, PAP, and TW, respectively, with removal efficiency 91.4%, 92.5%, and 88.04%. It is found that all three adsorbents can efficiently remove pollutants, not up to desired disposable standards of river but can be considered as sewage.

3.5 Effect of Contact Time

Effect on contact time on COD removal efficiency of leachate and adsorption capacity is shown in Fig. 3. Maximum COD removal efficiency for leachate using SSH was observed during the first 30 minutes and achieved equilibrium after

Table 3 Leachate parameters after adsorption

Sl. no.	Parameters	SSH	Removal %	PAP	Removal %	TW	Removal %
1	pH	4	–	4	–	2	–
2	Conductivity (ms/cm)	5.7	11.6	5.6	13.7	6.5	0.6
3	COD (mg/L)	160	90	290	81.8	320	80
4	BOD_5 (mg/L)	40.1	80.3	34.9	82.9	55.6	72.7
5	Turbidity (NTU)	6.7	95.8	4.3	97.3	2.3	98.5
6	TDS (mg/L)	1220	72.1	1800	58.9	1440	67.1
7	Alkalinity (mg/L) as $CaCO_3$	400	84.1	400	84.1	350	86.1
8	Iron (mg/L)	2.1	91.4	4.4	82.3	5.0	79.8

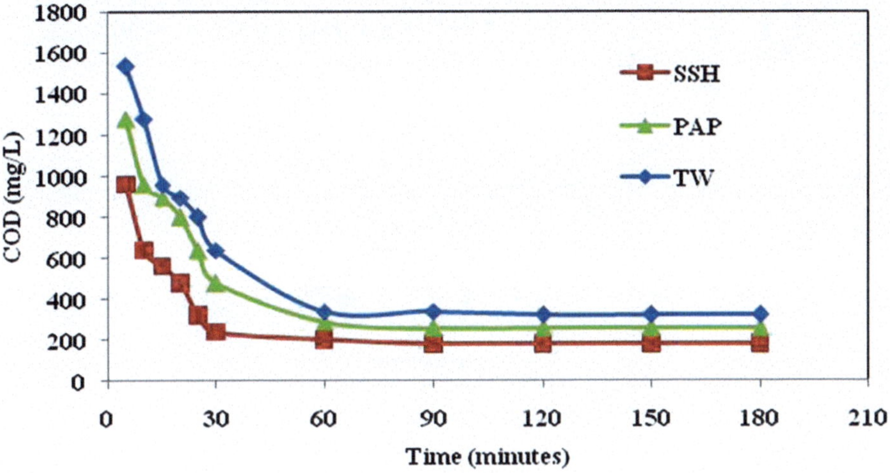

Fig. 3 Time chart for SSH, PAP, and TW

60 minutes. For PAP and TW, removal efficiency of COD increases rapidly during the first adsorption stage and then at lower rate, and equilibrium was attained at 90 minutes. At the end of 180 minutes observation period, COD effluent was 180, 320, and 256 mg/L for SSH, TW, and PAP, respectively. This slowdown stemmed from a rapid decline in immediate solute adsorption, indicating a reduction in available sites for COD uptake. Consequently, this suggests that the adsorption likely took place through film diffusion [10].

3.6 Determination of Adsorption Kinetics

The kinetics of adsorption of COD onto powdered activated SSH, PAP, and TW were carried out by withdrawing and analyzing the samples at the time 5, 10, 15, 30,

60, 90, 120, and 180 minutes. The adsorption kinetic experiments were carried out separately for three different adsorbents SSH, TW, and PAP. Figures 4 and 5 show the graph for Lagergren's pseudo-first-order adsorption kinetic with R^2 value of 0.724, 0.811, and 0.766 and Lagergren's pseudo-second-order adsorption kinetic with R^2 value of 0.999, 0.995, and 0.986 for SSH, PAP, and TW, respectively. The correlation coefficient (R^2) for the second-order equation significantly exceeds that of the pseudo-first-order counterpart, indicating a stronger fit. From Table 4, it is understandable that the predicted pseudo-second-order q_e is very close to the actual q_e values, in stark contrast to the pseudo-first-order equation, underscoring its

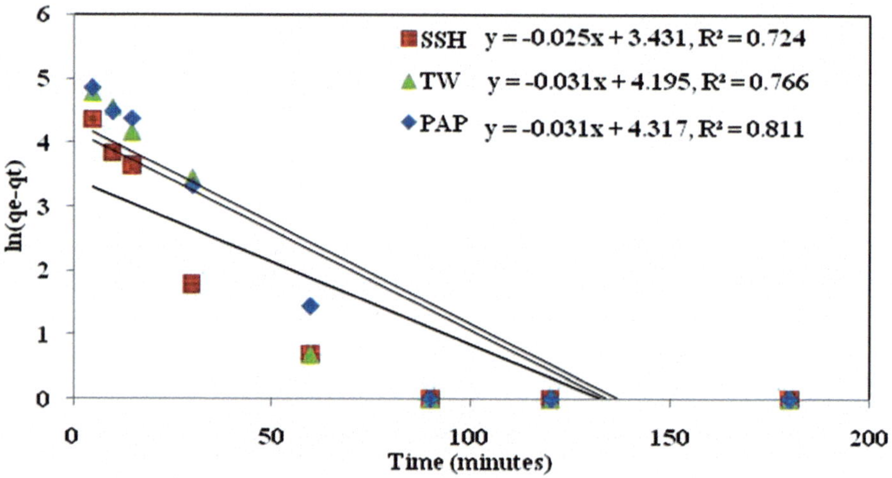

Fig. 4 Lagergren's pseudo-first-order adsorption kinetics

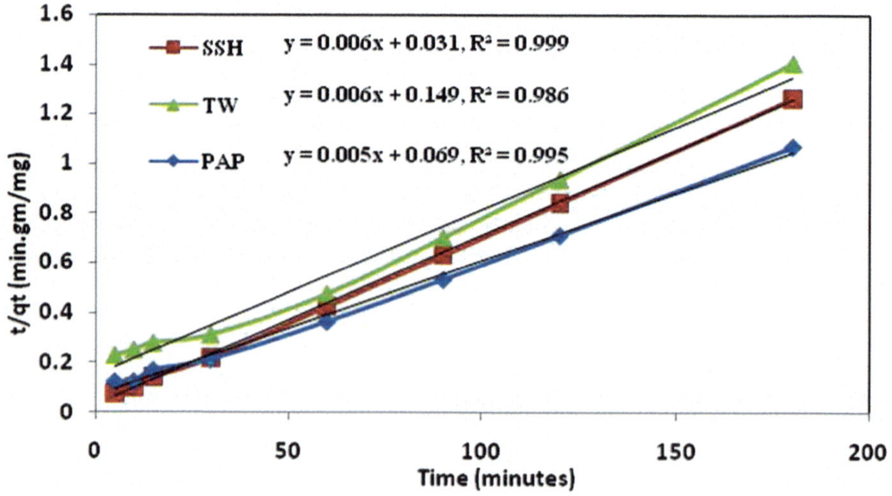

Fig. 5 Lagergren's pseudo-second-order adsorption kinetics

Table 4 Fitting parameters of pseudo-first-order and pseudo-second-order kinetic models

		Pseudo-first order			Pseudo-second order		
Adsorbent	$q_{e,exp}$ (mg/g)	K_1	R^2	$q_{e,cal}$ (mg/g)	K_2	R^2	$q_{e,cal}$ (mg/g)
SSH	158	−0.025	0.724	30.907	0.001161	0.999	166.666
TW	147	−0.031	0.766	74.963	0.000241	0.986	166.666
PAP	168	−0.031	0.811	66.353	0.000362	0.995	200

superior predictive accuracy. This data confirms the adherence of COD adsorption to a pseudo-second-order kinetic model, thereby affirming the precise chemical interactions between the adsorbents SSH, TW, and PAP and the constituents present in the leachate [11, 12].

3.7 pH Optimization for Adsorption

The optimum pH of COD removal percentage for all three adsorbents SSH, TW, and PAP are shown in Fig. 6. COD removal increases from pH 2 to pH 4 for SSH and PAP, respectively; however, for TW, removal efficiency of COD was found to the highest at pH 2. Thereafter, effectiveness of COD removal decreases with further increase in pH due to increase in the solubility of organic matter [12].

3.8 Dose Optimization for Adsorbent

The effect of dose was studied at a range of 2–12 g/L as shown in Fig. 7. The removal efficiency of COD increases with an increase in dosages of coagulant. The optimum ferric chloride dosages for the leachate sample were found to be 8 g/L for TW and PAP and 10 g/L for SSH. It can be observed that raising coagulant doses over the ideal levels did not considerably increase the leachate's turbidity removal efficiency, in accordance with the charge neutralization concept and the sweep flocculation phenomenon [11]. The combination of leachate and ferric chloride caused the negatively charged colloids to collapse because the hydrolyzed species of Al^{+3} reacted with them, neutralizing their charge. Electrical repulsion caused the colloids to acquire a positive charge once they had absorbed the cations over the ideal dosage, stabilizing them.

3.9 Adsorption Isotherms

The adsorbent dose study data were further treated using the most applied isotherm model Langmuir and Freundlich's isotherm (Eqs. 6 and 7, respectively). Figures 8 and 9 depict the graphical representations of the linear Langmuir (Eq. 6) and

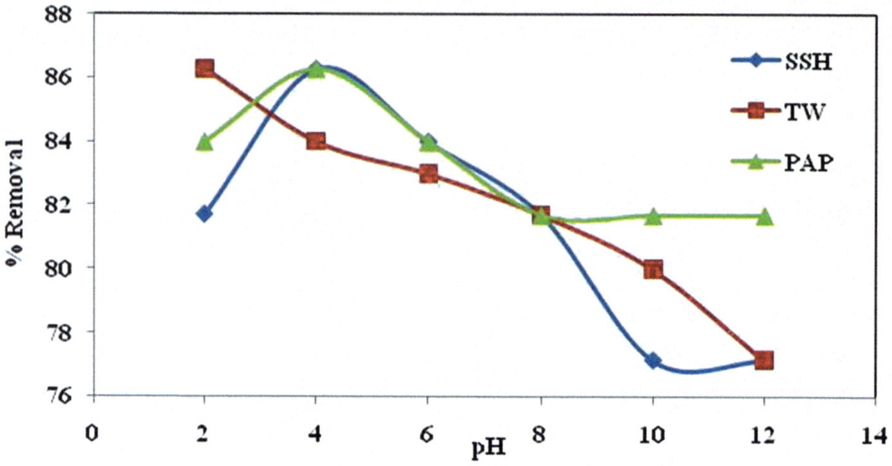

Fig. 6 pH optimization

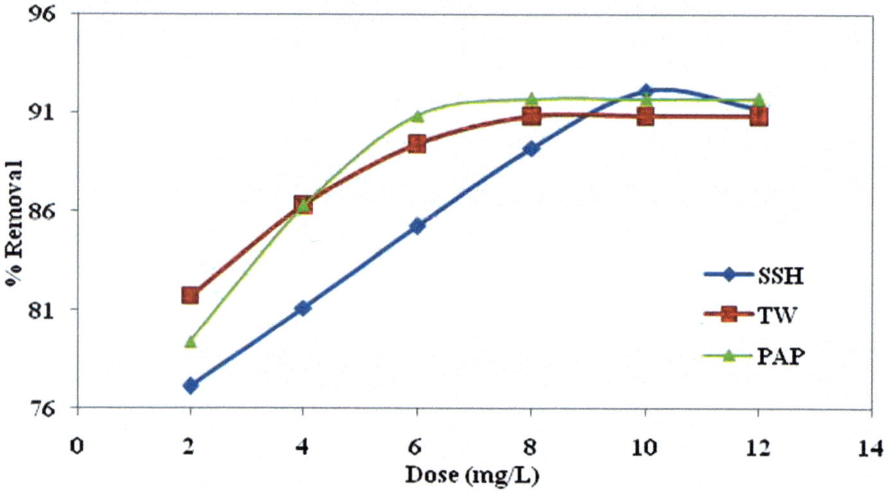

Fig. 7 Dose optimization

Freundlich (Eq. 7) equations, respectively, with $1/C_e$ vs $1/q_e$ and $\ln C_e$ vs $\ln q_e$ plots. Table 5 outlines the isotherm parameters and regression statistics for both models, facilitating the identification of the superior fit. Observations from the figures indicate that the adsorption of COD on the three adsorbents SSH, TW, and PAP aligns more suitably with the linear form of the Freundlich isotherm giving higher correlation coefficients (R^2 values of 0.955, 0.964, and 0.953, respectively) compared to the lower R^2 values associated with the Langmuir isotherm. Therefore, Freundlich model suggests a better fit model of COD adsorption as it accommodates

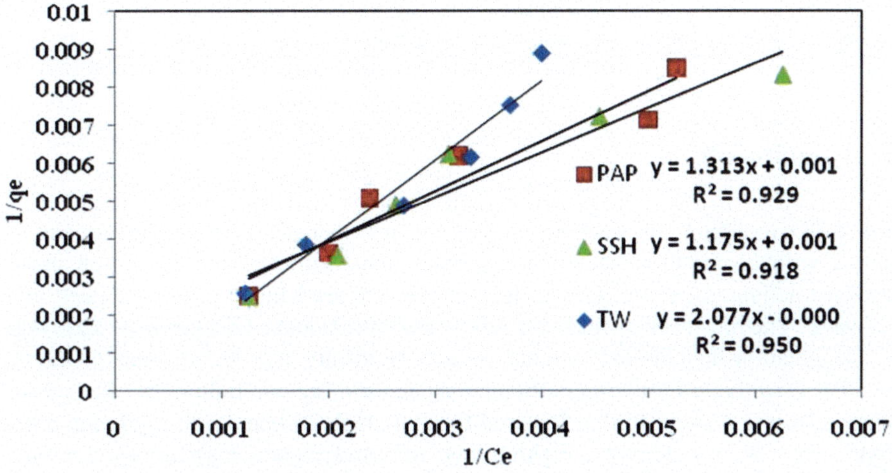

Fig. 8 Langmuir adsorption of SSH, PAP, and TW

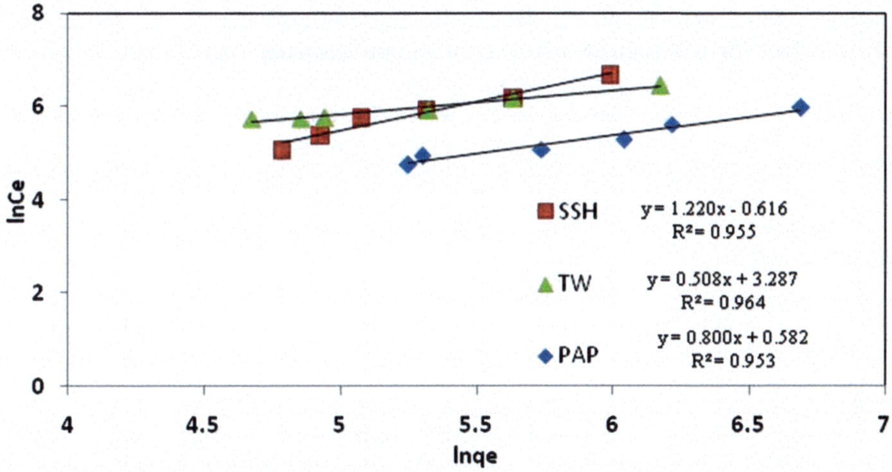

Fig. 9 Freundlich adsorption of SSH, PAP, and TW

Table 5 Langmuir and Freundlich isotherm parameters

	Langmuir isotherm parameters		Freundlich isotherm parameters	
Adsorbents	q_{max}	B	k_f	$1/n$
SSH	1000	1175	0.540101	1.22
TW	−250	−912.7	26.76246	0.508
PAP	1000	1313	1.789614	0.8

heterogeneity in the adsorption process making it more versatile for systems where adsorption may occur on sites with varying affinities [13].

4 Conclusion

The COD reduction in leachate underwent a systematic multistep treatment. It started with coagulation-flocculation pretreatment, followed by adsorption using low-cost activated carbon derived from sunflower seed husk, pineapple peel, and tea waste. The raw leachate had a turbidity of 543 NTU and a COD level of 3200 mg/L. Coagulation-flocculation notably reduced turbidity by 70.31% and COD by 55.55%, effectively removing particulate matter and organic pollutants. The subsequent adsorption stage further lowered COD to 160, 290, and 320 mg/L for SSH, PAP, and TW, respectively, highlighting activated carbon's ability to eliminate residual organic contaminants. Activated carbon from sunflower seed husk and pineapple peel showed superior adsorption properties, hinting at their potential as effective adsorbents for leachate treatment. This comprehensive treatment, blending coagulation-flocculation and activated carbon adsorption, offers a promising strategy to reduce the environmental impact of landfill leachate.

Acknowledgment The present research work is supported under the Research fellowship program of Ministry of Human Resource Development, Government of India.

References

1. S. Kaza, L. Yao, P. Bhada-Tata, F. Van Woerden, *What a Waste 2.0: A Global Snapshot of Solid Waste Management to 2050. Urban Development* (World Bank, Washington, DC, 2018), pp. 1–2
2. R.H. Hredoy, M.A.B. Siddique, M.M. Rahman, Impacts of landfill leachate on the surrounding environment: a case study on Amin Bazar Landfill, Dhaka (Bangladesh). Soil Syst., 1–2 (2022)
3. A. Valavanidis, Global municipal solid waste (MSW) in crisis. Refuse Disposal, 1–2 (2023)
4. D.C. Wilson, L. Rodic, P. Modak, *Global Waste Management Outlook* (UNEP, 2015), p. 4
5. G.R.Y. Perera, W.K.C. Neetha Dayanthi, Low-cost treatment method for organic matter and nutrients in landfill leachate. J. Water Health **21**, 1089–1101 (2023)
6. N.A. Rosli, M.A. Ahmad, T. Ubaidah, et al., Pineapple peel–derived carbon for adsorptive removal of dyes. Mater. Chem. Phys. **306**, 40–52 (2023)
7. X. Xin, X. Liu, O. Myounghak, J. Park, Y.F. Chen, Adsorption properties of heavy metal ions in landfill leachate by Na-bentonite. Mater. Test. **61**, 195–210 (2019)
8. APHA, *Standard Methods for the Examination of Water and Wastewater*, 23rd edn. (American Public Health Association, Washington, DC, 2017)
9. A. Gholizadeh, M. Kermani, M. Gholami, Kinetic and isotherm studies of adsorption and biosorption processes in the removal of phenolic compounds from aqueous solutions: comparative study. J. Environ. Health Sci. Eng. **11**, 07–15 (2013)

10. M. Verma, S. Chakraborty, S. Kumari, et al., Co-treatment of stabilized landfill leachate and municipal wastewater in a granular activated carbon-sequencing batch reactor (GAC-SBR). Process. Saf. Environ. Prot. **174**, 189–196 (2018)
11. E. Mohammad-pajooh, A.E. Turcios, G. Cuff, et al., Removal of inert COD and trace metals from stabilized landfill leachate by granular activated carbon (GAC) adsorption. J. Environ. Manag. **228**, 189–196 (2018)
12. V. Saxena, S.K. Padhi, P.K. Dikshit, L. Pattanaik, Recent developments in landfill leachate treatment: aerobic granular reactor and its prospects. Environ. Nanotechnol. Monit. Manage. **18**, 175–190 (2022)
13. M.A.M. Reshadi, A. Bazargan, G. McKay, A review of the application of adsorbents for landfill leachate treatment: focus on magnetic adsorption. Sci. Total Environ., 931–950 (2020)

Assessment of HRT's Impact on COD Reduction of Leachate Through Lab-Scale UASB Reactor

Malvika Thangjam, Potsangbam Albino Kumar, and Ankit Dubey Prakashvir

Abbreviations

BOD	Biological oxygen demand
COD	Chemical oxidation demand
GC	Gas chromatography
GLSS	Gas-liquid-solid separator
HRT	Hydraulic retention time
MSW	Municipal solid waste
OLR	Organic loading rate
UASB	Upflow anaerobic sludge blanket
VFAs	Volatile fatty acids

1 Introduction

The issue of municipal solid waste has become a significant concern worldwide with the amount of waste generated reaching staggering levels. Annually, the world produces 2.01 billion tons of municipal solid waste, and at least 33% of this waste is not managed in an environmentally safe manner [1]. The situation is projected to worsen in the coming years, as global waste generation is expected to reach a staggering 3.4 billion tons in 2050 [2]. The escalating waste production worldwide poses a dire risk to the environment and public health. India, a significant contributor to this crisis, ranks in the top 10 nations for municipal solid waste generation,

M. Thangjam · P. A. Kumar (✉) · A. D. Prakashvir
Civil Engineering Department, National Institute of Technology, Manipur, India
e-mail: albino@nitmanipur.ac.in

D. Mazumder (ed.), *Sustainable Advanced Technologies for Environmental Management*, Springer Proceedings in Earth and Environmental Sciences,
https://doi.org/10.1007/978-3-031-64006-3_26

annually producing more than 62 million tons of waste, exacerbating the global waste burden. Unfortunately, the management of municipal solid waste in India is far from ideal. Only 50% of the waste is treated, while 18.4% is simply landfilled, and a significant 31.7% remains unaccounted for [3]. This lack of proper waste management infrastructure exacerbates the environmental and health risks associated with uncontrolled waste disposal. Imphal City, Manipur, alone generates approximately 100–120 tons of municipal solid waste per day, highlighting the immense challenge faced by local authorities in managing and disposing of such a significant amount of waste typically disposed of in sanitary landfills [4]. The popularity of sanitary landfills as a municipal solid waste (MSW) disposal method stems from their straightforward disposal process and low cost. However, they also produce highly contaminated leachate, a significant drawback. Landfill leachate is rich in organic pollutants like chemical oxygen demand (COD), biochemical oxygen demand (BODs), ammonia, suspended solids, alkalinity, heavy metals, and inorganic salts [5]. It also carries elevated concentrations of phenol, nitrogen, and phosphorus. If inadequately managed, this hazardous leachate poses a significant risk of contaminating both surface and groundwater reservoirs, underscoring the critical importance of proper treatment and disposal protocols. The primary environmental issues associated with landfill sites include leachate infiltration, contamination of surrounding land and aquifers, and odor problems. Physicochemical treatment techniques such as ultrafiltration [6], nanofiltration [7], coagulation-flocculation [8], reverse osmosis [9], and chemical precipitation [10] are deployed to effectively eliminate refractory pollutants from leachate. Coagulation-flocculation followed by activated carbon adsorption is commonly employed in leachate pretreatment, often preceding physical-chemical or biological processes [11]. Given the stubborn nature of leachates, treatment typically necessitates a combination of techniques. Integrating biological and physicochemical processes can optimize treatment efficiency by capitalizing on each method's strengths, offering a potent solution for leachate remediation [12]. However, the two-step process is complicated, and biological process techniques are generally employed. Because of the elevated COD levels in leachates, high-rate anaerobic reactors like the Upflow Anaerobic Sludge Blanket are favored for treatment compared to aerobic methods [13]. UASB reactor is an effective method for treating leachate where these reactors utilize anaerobic bacteria to degrade organic pollutants present in the leachate. Furthermore, it has shown high efficacy and versatility, effectively treating diverse organic wastewater types with varying COD levels across a broad range of environmental circumstances. However, the performance of UASB reactors depends on various factors such as hydraulic retention time (HRT), temperature, organic loading rate (OLR), pH, alkalinity, granulation, wastewater composition, and mixing dynamics. This study focuses on the investigation of the effects of different HRTs for leachate treatment taken from Lamdeng Khunou, Imphal, Manipur, and also examines various parameters like pH, COD, alkalinity, VFAs, etc.

Investigating the effects of HRT on leachate treatment by UASB reactors is crucial in optimizing the system's performance. A longer HRT allows for better contact between the leachate and the anaerobic bacteria, promoting efficient degradation of organic pollutants. On the other hand, a shorter HRT may result in incomplete treatment and lower removal rates. Therefore, finding the optimal HRT for leachate treatment is essential for achieving maximum pollutant removal and maintaining a stable reactor performance.

2 Materials and Methods

2.1 Landfill Leachate

Sanitary waste landfill leachate was collected from the Lamdeng Khunou landfill site in Imphal [24.8170°N, 93.9368°E]. The site comprised two leachate ponds: one for raw leachate and another as a sump pond for evaporation and stabilization. Collection from the raw leachate pond was done once every 2 weeks from October 2022 to April 2023. Samples were preserved in 20 L clean propylene containers at 4 °C to prevent microbial activity that might alter leachate characteristics. The raw leachate exhibited a strong odor and a dark brownish hue, suggesting significant colloid and fine solid content. These samples served as feed for the reactor, initiating subsequent treatment processes effectively.

2.2 UASB Reactor Setup

The schematic diagram of UASB reactor is shown in Fig. 1. A laboratory-scale reactor was constructed from fiber plastic and had a total volume of 6 L, with an effective height of 160 cm and a diameter of 7 cm. It included features for feeding, gas collection, and residue draining and operated at ambient temperature. Gas production was tracked daily through the water displacement technique, ensuring accurate monitoring of the reactor's performance. Leachate, designated for treatment, entered the reactor from the bottom via a peristaltic pump (Cole-Parmer). Flow rates were adjusted accordingly: 6.25 mL/min for a 16-h hydraulic retention time (HRT), 9.375 mL/min for a 24-h HRT, and 18.75 mL/min for a 48-h HRT. The leachate ascended through a blanket of biologically activated sludge with an upflow velocity of 0.1 m/s, primarily composed of granular aggregates. These sludge aggregates, characterized by excellent stability, remained intact under practical conditions, ensuring efficient treatment as leachate interacted with the granules. Anaerobic conditions within the reactor facilitated the production of gases, predominantly methane and carbon dioxide, which promoted internal mixing, crucial for the development and maintenance of biological granules.

However, some gas remained trapped within the granules. To address this, a gas-liquid-solid separator (GLSS) was incorporated at the reactor's top to facilitate the effective separation of gas, liquid, and granules. Within the GLSS, gas-surrounded particles collided with the bottom of degassing baffles, causing them to settle back into the sludge blanket. Meanwhile, the treated water exited the reactor, ensuring the integrity of the treatment process. This meticulously designed system optimized gas-

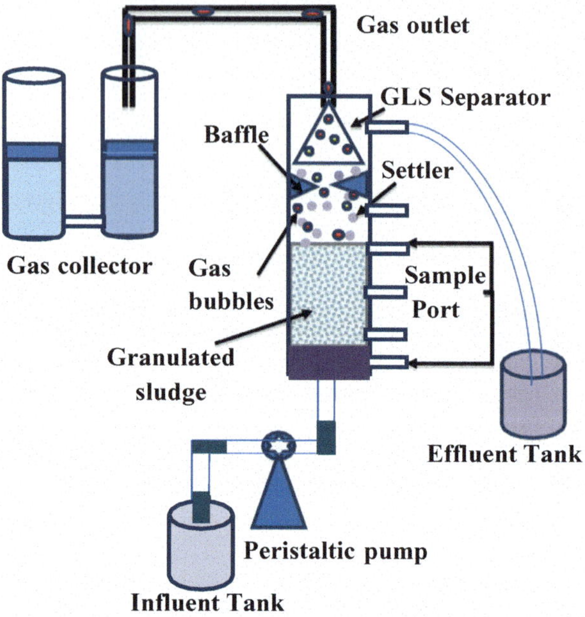

Fig. 1 Schematic diagram of the lab UASB reactor system

liquid-solid separation, enhancing the reactor's efficiency in treating leachate while minimizing potential losses of valuable granular aggregates [14].

2.3 *UASB Reactor Start-Up*

The anaerobic digestion process began with the dissolution of 300 g of cow dung in 3 L of water, maintaining uniformity throughout the setup. To scale for the reactor, cow dung was mixed with 20 L of water and sieved with a cotton cloth to ensure an optimal slurry consistency. Stored in capped bottles, this slurry contained methanogenic bacteria crucial for anaerobic digestion, abundant in cow rumen. Given the slow growth of methanogens, a 30-day period allowed for microbial development. The resulting slurry was combined with leachate in a 1:1 ratio and introduced into the 6-L reactor for stabilization through recirculation, promoting microbial growth [15]. Once stabilization occurred, marked by consistent COD removal and the formation of a sludge blanket, the reactor was deemed operational and ready for leachate input.

Operating at hydraulic retention times (HRTs) of 16, 24, and 48 h, leachate was introduced based on the reactor's volume and flow rate. HRT, crucial for organic matter degradation, optimizes anaerobic digestion efficiency. This systematic approach established an effective anaerobic digestion system, utilizing cow dung-derived methanogens for leachate treatment. Adhering to scientific principles and operational parameters showcased the reactor's potential for sustainable waste treatment and resource recovery.

2.4 Methodology

Various water quality parameters like pH, TSS, COD, BOD, etc. were analyzed as per the standard method [16]. Volatile fatty acids (VFAs) were analyzed by using the distillation method in which 500 mL water samples containing various concentrations of acetic acid were steam-distilled. The content of volatile acids in distillate was determined by titration with 0.1 N NaOH solution and expressed as acetic acid content [17]. In this procedure, the potentiometric titration system was used in place of the standard titration method, and also, the concentration of NaOH solution was matched with acetic acid content in samples every time. For biogas composition, gas chromatography (GC) analysis is used to calculate the content of a chemical product. In contrast to conventional chromatography techniques, gas chromatography uses a gas as the mobile phase and separates the constituents as vapors. It is therefore used to isolate and identify tiny molecular weight molecules in the gas phase. The sample is vaporized in the injection port and can be either a gas or a liquid. The mobile phase in gas chromatography is a carrier gas, commonly helium due to its low molecular weight and chemical inertness. The analyte is forced through the column by the mobile phase when pressure is applied. A stationary phase-coated column is used for separation.

3 Results and Discussion

3.1 Characteristics of Raw Leachate

The characteristics of raw leachates from the sanitary landfill site of Lamdeng Khunou Solid Waste Management Plant, Manipur, Imphal, are presented in Table 1.

3.2 Initial Phase of the Reactor: Effects of pH and Alkalinity

The culture/cells in the reactors were initially developed by feeding the reactor with glucose concentration having the same COD concentration as leachate, i.e., 2800 mg/L. During the period of acclimatization, the pH value of the reactor is maintained in the range of 6.8–7.2 using H_2SO_4 and NaOH (1N). With time, alkalinity was observed to increase from 2520 to 2900 mg/L largely due to the release of carbon dioxide (which is acidifying in nature) from the reactor, and so the pH of the system increased. Thereafter, leachate was injected gradually to replace the synthetic leachate of glucose. When leachate was injected, efficiency was in the range of 45–68.75% while starting, and after 7 days, whole synthetic leachate was replaced by leachate. The pH, COD, alkalinity, dissolved solids, and volatile fatty acids

Table 1 Initial physico-chemical characteristics of raw leachate

Sl. no.	Parameters	Raw leachate
1	pH	6.55–8.2
2	Temperature (°C)	25
3	Conductivity, ms/cm	6.55
4	Turbidity, NTU	161
5	Chloride, mg/L	1597.25
6	Total dissolved solids, mg/L	4388
7	BOD_5, mg/L	400
8	COD, mg/L	2800
9	BOD_5/COD, mg/L	0.14
10	Total alkalinity as $CaCO_3$, mg/L	2520

Table 2 pH values of influent and effluent throughout the process at 16-, 24-, and 48-h HRT

Weeks	16-h HRT		24-h HRT		48-h HRT	
	INF	EFF	INF	EFF	INF	EFF
1	6.75	7.6	6.89	8.11	7.03	7.62
2	6.55	7.68	6.96	7.7	6.89	7.7
3	6.88	7.8	7	7.4	6.95	8.2
4	6.85	8.1	6.8	8.75	6.95	8.05
5	6.75	7.83	6.96	7.75	6.91	7.89
6	6.75	7.8	7.01	8.1	7.1	8.85
7	6.8	8.1	6.98	7.5	6.8	7.9
8	6.85	7.8	6.95	7.55	7.11	7.99
9	6.88	7.75	6.95	7.88	7.03	8.2
10	6.9	8.08	7.01	8.55	6.7	7.99
11	6.74	8.18	6.95	8.08	7.02	7.85
12	6.83	8	6.98	8.26	6.81	8.03
13	6.81	7.92	6.8	8.4	6.85	8.05
14	6.8	7.65	6.87	7.75	6.91	8.1
15	6.88	7.9	6.95	8.1	7.03	7.85

(VFA) of both leachate feed and treated effluent from the reactors were analyzed daily as part of the assessment process.

The values of pH of different weeks of operation at different HRTs are summarized in Table 2. The pH variation in the influent ranging throughout the processes at different HRTs, i.e., 16-, 24-, and 48-h, respectively, are 6.55–6.9, 6.8–7.01, and 6.7–7.11. This concentration range is favorable since the pH for anaerobic digestion of most organisms is 6.0–9.0 [12]; beyond this range, digestion can proceed, but with less efficiency. Initially, the reactor's pH exhibited oscillations due to CO_2 stripping by air. The reduction in CO_2 resulted in the decrease of carbonic acid (H_2CO_3) and bicarbonate ion concentration (HCO^{3-}), thereby consuming H^+ ions. Alkalinity and pH are closely interlinked, so a drop in alkalinity typically corresponds to a decrease in pH. Table 3 illustrates the alkalinity values of influent and treated effluent at different hydraulic retention times (HRTs). Noticeably, influent alkalinity showed significant fluctuations during the initial stages, possibly due to

Table 3 Alkalinity values (mg/L of $CaCO_3$) of influent and effluent throughout the process at 16-, 24-, and 48-h HRT

Weeks	16-h HRT		24-h HRT		48-h HRT	
	INF	EFF	INF	EFF	INF	EFF
1	500	1000	1850	2900	750	1000
2	850	1000	1750	2250	900	1100
3	950	1350	2250	2450	700	800
4	675	1075	1600	2100	900	1000
5	800	1000	1500	1900	850	1000
6	1000	1175	1700	2300	900	1500
7	1100	1750	1300	1700	750	1250
8	1500	1750	1600	1900	1500	1700
9	800	1100	1800	2100	1250	1600
10	750	850	1000	1200	1200	1500
11	750	850	700	1000	1250	1600
12	700	800	1000	1400	1000	1350
13	650	800	1250	1750	1250	1550
14	600	750	1200	1650	900	1200
15	800	1000	900	1100	1300	1650

variations in leachate pH or composition. For instance, in the 16-h HRT operation, influent alkalinity ranged from 500 to 1500 mg/L, while effluent alkalinity ranged from 750 to 1750 mg/L. These fluctuations underscore the dynamic nature of leachate treatment processes. A minimum alkalinity of approximately 300 mg/L is essential to counteract the acidity of dissolved carbon dioxide gas and establish a buffer within the optimal pH range. Elevated alkalinity levels can be advantageous during startup phases to neutralize initial acidity, particularly when volatile fatty acid levels are higher than usual. Ammoniacal nitrogen, primarily released through cell lysis (the breakdown of cell protein), serves as a significant source of alkalinity in the system. During the 24-h hydraulic retention time (HRT) operation, influent alkalinity ranged from 700 to 2250 mg/L, while effluent alkalinity varied between 1000 and 2900 mg/L. For the 48-h HRT operation, influent alkalinity ranged from 700 to 1500 mg/L, with treated effluent ranging from 800 to 1700 mg/L. Notably, the 48-h HRT operation exhibited greater stability, with minimal fluctuation compared to the other HRTs. Moreover, its alkalinity values showed a continuous increase, indicating enhanced reactor efficiency.

3.3 *Effects of Ions*

The efficiency of dissolved solids removal varied significantly across different hydraulic retention times (HRTs). At a 16-h HRT, the reactor exhibited the highest removal percentage, reaching 81.82% in the third week, but decreased to a minimum of 3.66% by the ninth week. Conversely, for 24- and 48-h HRT reactors, maximum solid reduction percentages of 64.85% and 72.04% were achieved on the 14th and

15th weeks, respectively, as shown in Fig. 2. Hence, for dissolved solids removal, operation at a 16-h HRT is preferred.

Examining the influent chloride content removal efficiency across different HRTs revealed higher treatment efficacy at 48-h HRT, reaching 74.52%, compared to 67.19% at 24-h HRT and 63.00% at 16-h HRT as shown in Fig. 3. This underscores the superior performance of the 48-h HRT, followed by the 24-h HRT and then the 16-h HRT.

Moreover, analysis of chemical oxygen demand (COD) levels in influent and effluent at different hydraulic retention times (HRTs) revealed varying patterns, with

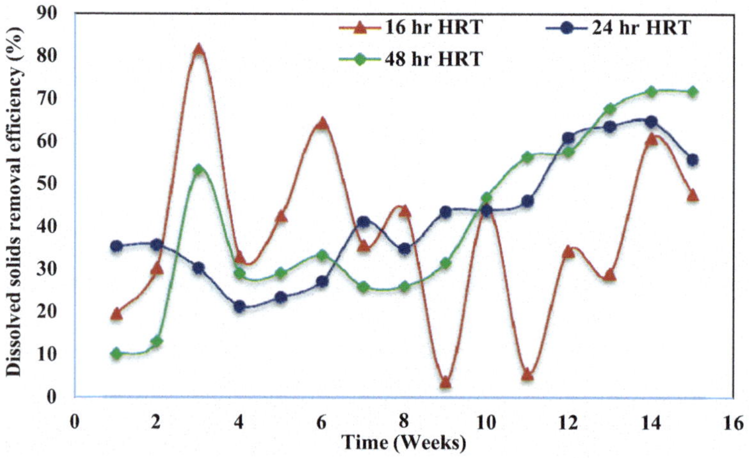

Fig. 2 Removal efficiency of dissolved solids at HRTs: 16-, 24-, and 48-h

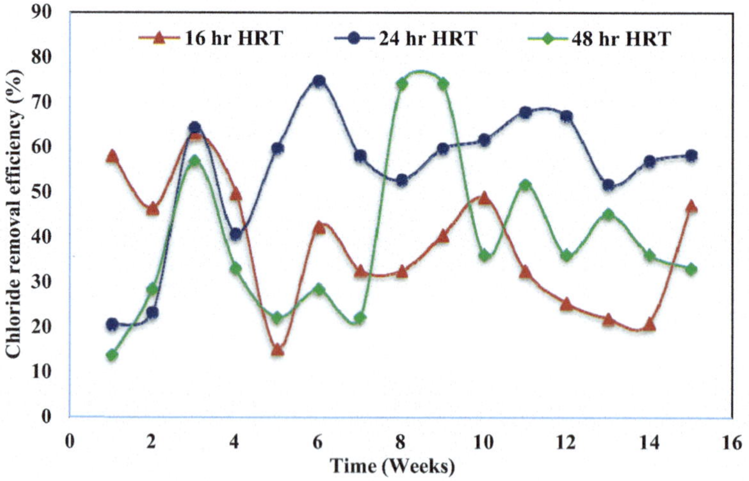

Fig. 3 Removal efficiency of chloride content at HRTs: 16-, 24-, and 48-h

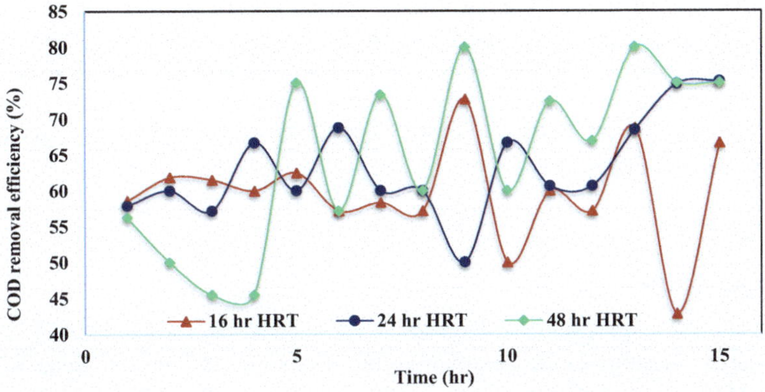

Fig. 4 COD removal efficiency at HRT: 16-, 24-, and 48-h Effects of VFAs and Biogas

notable fluctuations observed, especially noticeable at a 16-h HRT, as depicted in Fig. 4. Conversely, stability was observed at 24- and 48-h HRTs, with the reactor achieving a treatment efficiency of approximately 80% at 48-h HRT, compared to 75.19% at 24-h HRT and 68.75% at 16-h HRT. Thus, the reactor's efficiency was highest at a 48-h HRT, owing to its extended contact time facilitating effective organic matter degradation.

In summary, the analysis reveals the importance of HRT in influencing the treatment efficiency of the reactor. While shorter HRTs may yield higher dissolved solids removal initially, longer HRTs prove more effective in achieving consistent and superior treatment outcomes, particularly evident in chloride content removal and COD reduction. Therefore, the 48-h HRT emerges as the optimal operating condition, ensuring optimal treatment performance and maximizing organic matter degradation efficiency.

3.4 *Effects of VFAs and Biogas*

The analysis of volatile fatty acids (VFAs) serves as a crucial method for evaluating the efficacy of the digestion process within treatment facilities. In the study, influent leachate VFAs spanned from 201 to 384 mg/L across different HRTs. Notably, at a 16-h HRT, VFAs fluctuated within 202.4–303.29 mg/L due to varied leachate compositions, while effluent VFAs remained stable, ranging from 25 to 51 mg/L. Similarly, during the 24-h HRT, influent VFAs fluctuated between 189.4–267 mg/L due to composition variations, while effluent VFAs maintained stability at 28.6–35.2 mg/L. These findings highlight leachate composition's impact on VFA concentrations during treatment. Also, in the 48-h HRT, influent leachate VFAs concentration varies in the range of 176.8–325 mg/L throughout the processes, but the effluent conc. is much lower than influent, i.e., 29.54 mg/L. The difference between influent and effluent conc. is quite noticeable. Regarding the biogas, it was observed that maximum biogas production is at 48-h HRT, i.e. 840–925 mL, while at 16- and 24-h HRTs, biogas production is 360–450 and 600–650 mL, respectively. The biogas production consistently rises across various treatment processes. Additionally, the observed percentages of biogas constituents include nitrogen at 32%, methane at 26%, carbon dioxide at 21%, hydrogen at 19%, and carbon monoxide at 2%.

4 Conclusion

This research investigates the treatment of leachate sourced from Lamdeng Khunou, Imphal, Manipur, India, employing Upflow Anaerobic Sludge Blanket (UASB) technology across various hydraulic retention times (HRTs), with a particular emphasis on COD removal efficiency. A 6 L USAB reactor was set up, and an investigation was done on the treatment of leachate at varied HRTs, i.e., 16-, 24-, and 48-h in a continuous process where maximum reduction efficiency was obtained in the 48-h HRT. At the initial stage, the reactor started up for acclimatization and the glucose was replaced by leachate. Then, the pH starts fluctuating in the initial stages of the different HRTs taken, but in the case of 48-h HRT, the fluctuation decreases and then stays in the range of pH 6–9, which is favorable for anaerobic digestion. The pH of effluent variations throughout the processes at different HRTs, i.e., 16-, 24-, and 48-h, respectively, are 7.6–8.1, 7.4–8.75, and 7.62–8.85. Also, the alkalinity of 48-h HRT showed stability due to its low variation in both influent and effluent which was not the case in the 16- and 24-h HRTs. The value of alkalinity varies from 700-1500 mg/L at and 800–1700 mg/L at effluent for HRT-48 hours whereas that of HRT-24 hours varies from 700–2250 mg/L at influent and 1000–2900 mg/L at effluent. Higher alkalinity values imply a stronger resistance to pH fluctuations. The COD reduction was also maximum in 48-h HRT, i.e., 80%, whereas at 16- and 24-h HRTs, it was 68.75% and 75.19%, and biogas production at 16-, 24-, and 48-h were 450, 650, and 925 mL, respectively. Thus, it can be said that the best performance of the reactor in the treatment of landfill leachate was obtained at 48-h HRT with 80% COD reduction, VFAs reduced from 325 to 25 mg/L, alkalinity increased from 700 to 1700 mg/L, dissolved solids reduction and chloride content removal were found to be 72.04% and 74.52%, respectively, and biogas production was 925 mL. From this study, it can be also concluded that the UASB treatment method is effective for the leachate treatment of municipal solid waste landfills, in Imphal, Manipur.

Acknowledgment The present research work is supported under the Research fellowship program of the Ministry of Human Resource Development, Government of India.

References

1. S.E. Derdera, G.S. Ogato, Towards integrated, and sustainable municipal solid waste management system in Shashemane city administration, Ethiopia. Heliyon **9**, e21865 (2023)
2. S. Kaza, L.C. Yao, P. Bhada-Tata, F. Van Woerden, *What a Waste 2.0: A Global Snapshot of Solid Waste Management to 2050* (World Bank, Washington, DC, 2018)
3. Y. Pujara, J. Govani, K. Chabhadiya, H. Patel, K. Vaishnav, P. Pathak, Waste-to-energy: suitable approaches for developing countries, in *The Handbook of Environmental Chemistry (HEC)*, vol. 99, (Springer, 2020), pp. 173–191
4. I.C. Yadav, N.L. Devi, Municipal solid waste management in Imphal Town, Northeast India: a critical analysis of existing management practices and proposed action plans. International Journal of Waste Resources **6**, 2 (2016)
5. J. Zheng, X.-G. Wang, Y. Sun, Y.-X. Wang, H.-Q. Sha, X.-S. He, X.-J. Sun, Natural and anthropogenic dissolved organic matter in landfill leachate: composition, transformation, and their coexistence characteristics. J. Hazard. Mater. **465**, 133081 (2023)
6. S. Nazia, N. Sahu, V. Jegatheesan, S.K. Bhargava, S. Sridhar, Integration of ultrafiltration membrane process with chemical coagulation for proficient treatment of old industrial landfill leachate. Chem. Eng. J. **412**, 128598 (2021)
7. M. Thirumal, V. Srinivasan, R. Manibharathi, K. Vinoth, R. Thenmozhi, Treatment of landfill leachates by nanofiltration. International Research Journal of Engineering and Technology (IRJET) **91**, 1209–1217 (2019)
8. R. Bouyakhsass, S. Souabi, S.K. Rifi, A. Taleb, A. Pala, A. Madinzi, Optimization of coagulation-flocculation for landfill leachate treatment: an experimental design approach using response surface methodology. Environmental Nanotechnology, Monitoring & Management **20**, 100841 (2023)
9. K. Ushikoshi, T. Kobayashi, K. Uematsu, A. Toji, D. Kojima, K. Matsumoto, Leachate treatment by the reverse osmosis system. Desalination **150**, 121 (2003)
10. T. Zhang, L. Ding, H. Ren, Pretreatment of ammonium removal from landfill leachate by chemical precipitation. Journal of Hazardous Materials **166**, 911–915 (2008)
11. M.J. Puchana-Rosero, E.C. Lima, B. Mella, D. da Costa, E. Poll, M. Gutterres, A coagulation-flocculation process combined with adsorption using activated carbon obtained from sludge for dye removal from tannery wastewater. J. Chil. Chem. Soc. **63**, 3867 (2018)
12. T.H. Tsui, W. Hao, B. Song, S.S. Liu, A. Bhardwaj, J.W.C. Wong, Food waste leachate treatment using an Upflow Anaerobic Sludge Bed (UASB): effect of conductive material dosage under low and high organic loads. Bioresour. Technol. **304**, 122738 (2020)
13. M. Dhamsaniya, D. Sojitra, H. Modi, M.A. Shabiimam, A review of the techniques for treating the landfill leachate. Mater. Today Proc. **77**, 358–364 (2022)
14. A. Yasar, N. Ahmad, M.N. Chaudhry, A.A.A. Khan, Sludge granulation and efficiency of phase separator in UASB reactor treating combined industrial effluent. J. Environ. Sci. **19**, 553 (2007)
15. A. Patra, T. Park, M. Kim, Y. Zhongtang, Rumen methanogens and mitigation of methane emission by anti-methanogenic compounds and substances. J. Anim. Sci. Biotechnol. **8**, 13 (2017)
16. APHA, *Standard Methods for the Examination of Water and Wastewater*, 23rd edn. (American Public Health Association, Washington, DC, 2017)
17. M. Ghidotti, D. Fabbri, C. Torri, S. Piccinini, Determination of volatile fatty acids in digestate by solvent extraction with dimethyl carbonate and gas chromatography-mass spectrometry. Anal. Chim. Acta **1034**, 92 (2018)

Removal of COD in Leachates Through Coagulation and Adsorption Using Magnetic Iron Oxide-Activated Adsorbent

Lairenlakpam Helena, Suraj Chingkhamayum, and Potsangbam Albino Kumar

Abbreviations

AC	Activated carbon
COD	Chemical oxygen demand
IIAC	Magnetic iron oxide-activated adsorbent
MSW	Municipal solid waste

1 Introduction

The generation of large amounts of solid waste is one of the biggest environmental concerns as a result of rapid urbanization and industrialization. Municipal solid waste (MSW) landfilling results in leachates, which contain high quantities of heavy metals, inorganic salts, and toxic biodegradable organic debris. The typical stabilized landfill leachate has limited biodegradability, a high ammonium nitrogen content of 3000–5000 mg/L, and a moderate COD level between 5000 and 20,000 mg/L [1]. To eliminate refractory pollutants from leachate, several researchers have published studies on physicochemical treatment techniques, including reverse osmosis, coagulation-flocculation, adsorption, chemical precipitation, and nanofiltration [1, 2]. The most practical approach, given the inherent limits of each of these techniques, is the combination of coagulation-flocculation with a biological process [2]. However, a physicochemical process is a better suitable substitute for the biological process since it is less costly and requires a greater

L. Helena · S. Chingkhamayum · P. A. Kumar (✉)
Civil Engineering Department, National Institute of Technology, Manipur, India
e-mail: albino@nitmanipur.ac.in

D. Mazumder (ed.), *Sustainable Advanced Technologies for Environmental Management*, Springer Proceedings in Earth and Environmental Sciences,
https://doi.org/10.1007/978-3-031-64006-3_27

area [2]. In leachate pretreatment, coagulation-flocculation is usually employed either before biological or physical-chemical procedures to reduce the colloids thereby reducing COD and interfering with the next step treatment [3]. Among physicochemical treatment methods after coagulation-flocculation, adsorption is widely adopted as it is a low-cost and very feasible method [3]. The application of activated carbon-based adsorbent is generally replaced by nano-based adsorbent because of its larger surface area and higher effectivity [4]. Some of the nano-based adsorbents are iron oxide-based nanomaterials, chitosan-coated nanoparticles, inorganic-supported nanocomposites, organic polymer-supported nanocomposites, and magnetic nanocomposites. Nano-sorbents have several benefits, one of which is their high sorption capacity, which makes them more suitable and effective for purifying water. Because of their exceptional physiochemical qualities, nanomaterials as adsorbents have attracted a lot of attention in the explanation of ecological challenges like wastewater purification. These qualities set them apart from traditional adsorbents in several ways. Pollutants can be efficiently and rapidly treated with adsorbents with a high rate of adsorption having a large surface area [4]. Because of its numerous functional groups, vast surface areas, highly porous structure, and surface reactivity, the magnetic iron nanocomposite has been used [5]. Nano sorbents typically have a higher surface area and a well-defined pore structure due to their nanoscale dimensions. This can lead to enhanced adsorption capacity for COD molecules. Among these nano-adsorbents, biomaterial-based nanocomposite adsorbents are the most economical ones [4, 6]. The objective is to explore the potential of nano-activated carbon as a means to adsorb COD within leachates. The study involves a comprehensive assessment of its feasibility for adsorption.

2 Materials and Methods

2.1 Synthesis of Nano-adsorbent

For the synthesis of activated carbon (AC), the ash gourd was first peeled, then cleaned, and dried. Next, a miller was used to grind the dried ash gourd peels and sieved with a 300 μm sieve. It is subjected to 24-hour H_3PO_4 impregnation and a 30-minute carbonization process in a muffle furnace at 350 °C. After that, the ultimate form of activated carbon was sieved through a 125 μm screen [7, 8]. For the synthesis of nano-adsorbent, the required quantity of AC was mixed vigorously with a magnetic stirrer at 800–900 rpm in a 400 mL solution containing 9.8 g of $FeCl_3{\cdot}6H_2O$ (28 mM) and 3.8 g of $FeSO_4{\cdot}7H_2O$ (14 mM) [9, 10]. To precipitate the hydrated iron oxide, a 5.0 M NaOH solution was applied dropwise and left for 24 hours. The Fe_3O_4/AC nanocomposite was separated using a magnet, cleaned until the pH was neutral, and then allowed to dry in a hot air oven. Eventually, after drying and crushing in a mortar and pestle, magnetic iron oxide-activated adsorbent (IIAC)

Fig. 1 (**a**) Ash Gourd Peels (**b**) Activated carbon (**c**) Nanocomposite Synthesis (**d**) Final Magnetic iron composite

was produced [11, 15]. The peels from the ash gourd, the activated carbon, the synthesis of the nanocomposite, and the final IIAC are listed in order in Fig. 1a–d.

2.2 *Sample Collection*

Landfill leachate was sampled from the evaporation sump pond (stabilized leachate) on the Lamdeng Khunou Solid Waste Management Plant's sanitary landfill, Manipur, India (24°84′24.7704″ N, 93°53′ 23.9244″ E), during the month of February–March 2022. The samples were kept away from microbiological activity that could alter the leachate's properties in 20 L clean polypropylene containers at 4 °C

2.3 *Coagulation-Flocculation Experiment*

At room temperature, the coagulation-flocculation process was run in a Phipps and Bird Jar test device (PB-600). The mixture was subjected to quick blending for 2 minutes at 100 rpm to promote the interaction of coagulants with suspended particles, and to agglomerate the colloids and form flocs, slow stirring at 25 rpm for 30 minutes was conducted along with 1 hour of sedimentation. pH values, ranging from 2 to 10, were adjusted to the landfill leachate sample to identify the ideal pH at which the majority of the turbidity and COD are removed. The same was done in different alum dosages ranging from 2 to 10g/L.

2.4 Batch Adsorption Experiments

Adsorption studies were carried out with different parameters. The impact of pH on COD adsorption was initially studied by conducting experiments at different pH values (2–10) for 120 minutes at 300 rpm while using two IIAC dosages of 0.8 and 1 g/L. Small variations in pH were controlled with varying concentrations of H_2SO_4 and NaOH. Next, by contacting the leachate with IIAC 0.8 g/L for 120 minutes, the effects of contact time on COD adsorption by IIAC were investigated, and the adsorption kinetics were examined. The dose of 0.8 g/L adsorbent during the initial investigation showed a substantial variation in the leachate removal percentage at different contact times

Every adsorption study was carried out using a jar-test apparatus (Phipps and Bird PB-600 model). For every adsorption experiment, 1 L of pretreatment leachates was utilized in a beaker, and IIAC treated the leachates for 120 minutes at 300 rpm. Following a 30-minute settling period, the adsorbed supernatant was taken out of the beakers, filtered using Whatman No. 47 filter paper, and its contents were examined.

3 Results and Discussion

3.1 First-Stage Treatment by Coagulation-Flocculation

The characteristics of the leachate collected during the study period are pH 8–8.9, turbidity 78–80 NTU, and COD concentration of 3000–3100 mg/L. A coagulation-flocculation procedure was employed to lower the high concentration of colloids that cause turbidity in the leachate in the first stage of treatment . In addition to removing turbidity, coagulation-flocculation pretreatment resulted in partial removal of COD also [12, 13]. Coagulation-flocculation tests were conducted for the stabilized leachate sample by varying the pH of 1 L leachate samples with the same dose of alum to investigate the impact of pH on turbidity. This involved altering the pH of the leachate samples of 1 L volume while utilizing the same dose of aluminum sulfate. Leachate samples with the optimal pH were treated to varying coagulant dosages after pH optimization to investigate the effect of coagulant dosage on the elimination of turbidity from the leachate sample [14].

3.1.1 Effect of pH on Turbidity Removal by Coagulation-Flocculation Using Alum

Using alum, the influence of pH on turbidity reduction was investigated after leachate pH modifications ranging from 2 to 10. It is shown in Fig. 2 how pH affects a decrease in turbidity. Turbidity reduction increased from pH 2 to pH 4 with 70.31% removal at pH 4. Thereafter, the effectiveness of coagulation-flocculation decreases with a further increase in pH due to a decrease in the charge density of the flocculant concerning pH [13]. Correspondingly, the COD concentration of leachate decreases from 3084 mg/L to 1920 mg/L suggesting 37.7% of the colloids imparting COD.

3.1.2 Effect of Coagulant Dose on Turbidity Removal by Coagulation-Flocculation Using Alum

Further, the dose effect was studied at the ideal pH 4 at a range of 2–6 g/L, and the result is demonstrated in Fig. 3. As the dose of alum increases from 2 to 10 g/L, turbidity reduction increases from 18.9% to 70.93%. Thereafter, a further increase in

Fig. 2 The effect of pH on the removal of turbidity

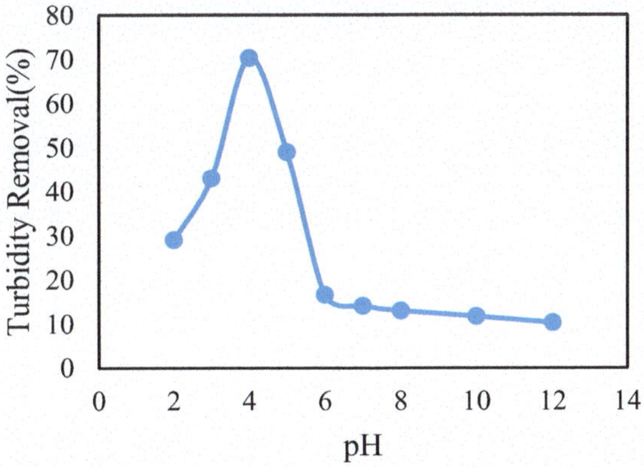

Fig. 3 The effect of coagulant dose on turbidity removal

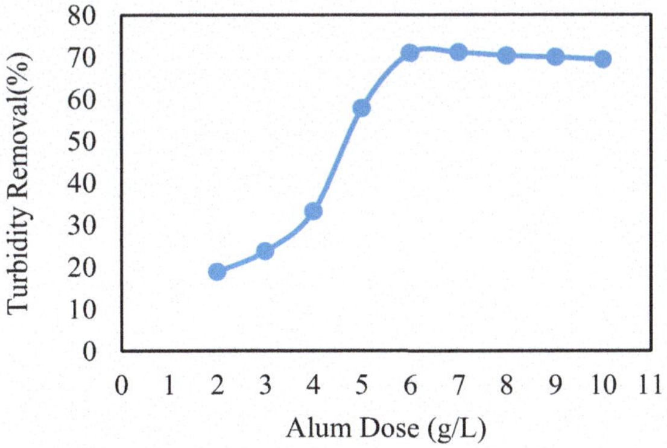

alum does not yield any significant removal of turbidity. Therefore, the optimum alum dose is obtained as 6 g/L. According to the principle of charge neutralization and the occurrence of sweep flocculation, it can be seen that increasing coagulant dosages above the optimal levels did not significantly boost the leachate's turbidity removal efficiencies. The hydrolyzed Al^{3+} species reacted with the negatively charged colloids in the leachate and alum mixture, neutralizing their charge and causing the colloids to collapse. Due to electrical repulsion, the colloids acquired a positive charge once they absorbed the cations over the ideal dosage, thereby rendering them stable [13].

3.2 Second-Stage Treatment by Adsorption

3.2.1 Effect of pH

The effect of initial pH on the COD adsorption capacities has been studied at various pH ranging from 2 to 8 and contacting with two IIAC doses of 0.8 and 1 g/L. Figure 4 shows the graph between COD removal versus the pH of the solution. The trend for the pH effect is the formation of the peak at pH 4 and a decrease in COD adsorption on both sides of pH. The highest possible adsorption percentage is 83.22% and 92.55% at IIAC doses of 0.8 and 1 g/L, respectively. IIAC also showed a positive surface charge at acidic pH, favoring anionic organic molecules in the leachate, such as aromatic carboxylic acids, hydrocarbons, phenolic compounds, etc. The IIAC surface developed a negative charge at basic pH, which made the organic molecules in the leachate repellent and resulted in less COD removal.

Fig. 4 Optimization of pH for removal of COD

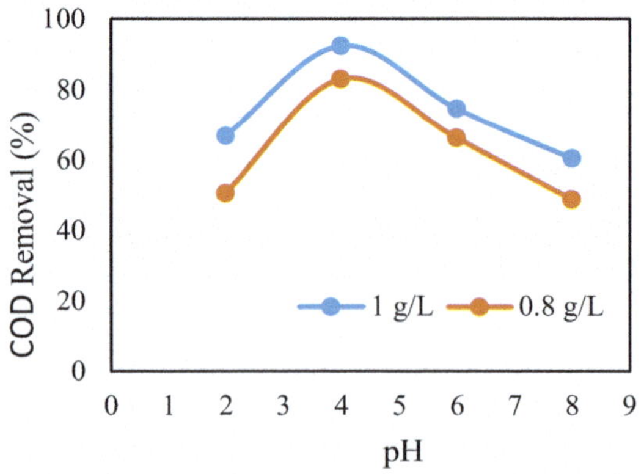

3.2.2 Effect of Time

The effectiveness of COD removal from the adsorbent at various time intervals is displayed in Fig. 5. The percentage removal of COD increases rapidly during the first adsorption stage up to 30 minutes and then at a lower rate up to 60 minutes and achieves equilibrium after 60 minutes. The rapid COD adsorption can probably be explained by the abundance of active sites on the adsorbent surface during the first 30 minutes, where the removal percentage is approximately 70.83% and within 60 minutes 88.5% removal. This corresponds to a q_t value increase from 580 mg/g at 5 minutes to 2160 mg/g at 120 minutes.

Lagergren's pseudo-first-order, pseudo-second-order, and intraparticle diffusion linearized kinetics equations (Eqs. 1–3, respectively) were applied once more to the equilibrium time experiment's adsorption data. The results are shown in Figs. 6, 7, and 8, respectively [16, 17].

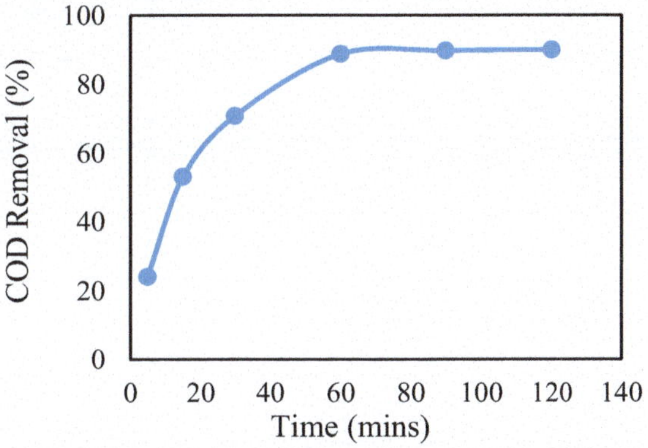

Fig. 5 Optimization of time for removal of COD

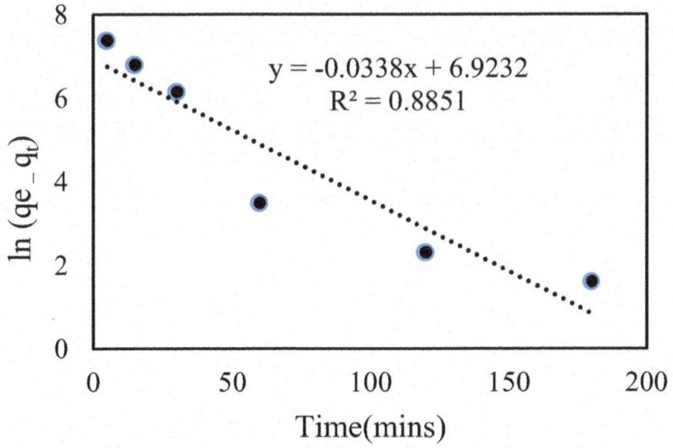

Fig. 6 Lagergren pseudo-first-order model

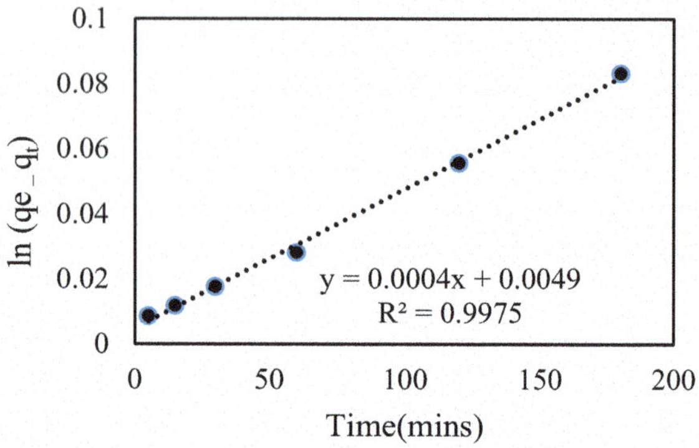

Fig. 7 Lagergren pseudo-second-order model

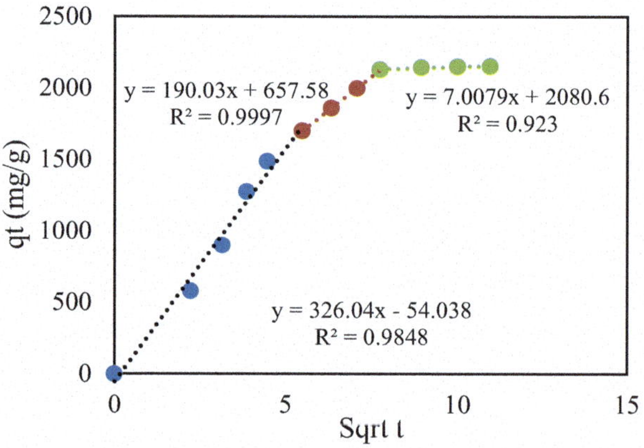

Fig. 8 Intraparticle diffusion model

Pseudo-First-Order Kinetics:

$$\ln(q_e - q_t) = \ln q_e - K_1 t \tag{1}$$

where q_e and q_t are the amounts of COD adsorbed per g of IIAC at equilibrium and at time t, respectively. K_1 is the first-order rate constant (min^{-1}).

Pseudo-Second-Order Kinetics:

$$\frac{t}{q_t} = \frac{1}{K_1} \cdot q_e^2 + \frac{1}{q_e} \cdot t \tag{2}$$

where K_2 ($\text{g mg}^{-1}\,\text{min}^{-1}$) is the adsorption rate constant for the pseudo-second-order reaction.

Intraparticle Diffusion Model:

$$q_e = K_d \cdot T_{1/2} + C \tag{3}$$

where K_d is the rate constant of the intraparticle transport ($\text{g mg}^{-1}\,\text{min}^{-1}$) and C is the constant that indicates the thickness of the margin layer, i.e., the larger the intercept, the greater the boundary layer effect.

A chi-square test was also done to compare all isotherms in the same axial setting. The chi-square value (χ^2) is calculated as shown in Eq. (1). χ^2 indicates the closeness between the experimental data with the various models [13].

$$\chi^2 = \sum \frac{(q_e - q_{em})^2}{q_{em}} \tag{4}$$

where q_{em} is the equilibrium capacity (mg/g) calculated from the model and q_e is the equilibrium capacity (mg/g) from the experimental data.

The second-order equation's correlation coefficient (R^2) is 0.9975, which is significantly greater than the 0.8851 of the pseudo-first order which can be seen in Table 1. The predicted pseudo-second-order q_e of 2500 mg/g is near to the experimental q_e of 2155 mg/g as compared to 1832 mg/g of pseudo-first order suggesting fixing the COD adsorption onto IIAC following pseudo-second order. This indicates that one of the mechanisms for COD adsorption on IIAC is the formation of covalent bonds through chemical interactions with the leachate chemicals [4]. The kinetic model parameters given in Table 1 substantiate this assessment with chi-square error for pseudo-second order. This means that the chemical adsorption process which is similarly comparable to the earlier study on biomaterial-activated carbons could be verified as one of the processes responsible for the majority of the organic matter removal from leachate [5, 7].

3.2.3 Optimization of Dose

Starting with a COD concentration of 1920 mg/L at pH 4, the impact of IIAC dose on COD was investigated by varying the number of dosages ranging from 0.2 to 1.2 g/L. As the IIAC dose increases, the amount of unadsorbed COD residual decreases, as seen in Fig. 9. The elimination of COD increased from 32.3% to 88.9% with a IIAC rise from 0.2 to 1.2 g/L. It is clear from this result that an increase in adsorbent dose causes a rise in IIAC's functional groups and total surface area, which increases the number of active sites available for COD adsorption from aqueous solution [16]. This could be explained by the fact that when IIAC dosage is increased, more unoccupied active sites for COD adsorption remain available and vacant. The IIAC dose of 1.2 g/L is regarded as the ideal IIAC dose in this investigation because negligible COD elimination was seen following it.

The experimental data were assessed using the Langmuir, Freundlich, and Redlich Peterson isotherms [16]. Equations (6)–(8), respectively, contain references to the linearized adsorption isotherm equations.

Table 1 Coefficients of adsorption kinetic models

S. N	Kinetic models	Parameters	R^2	χ^2	q_e, cal
1	Pseudo-first order	K_1(min^{-1}): −0.0528	0.8851	3.2956	1832
2	Pseudo-second order	K_2(g/mg.min): 0.00003265	0.9975	0.006432	2155

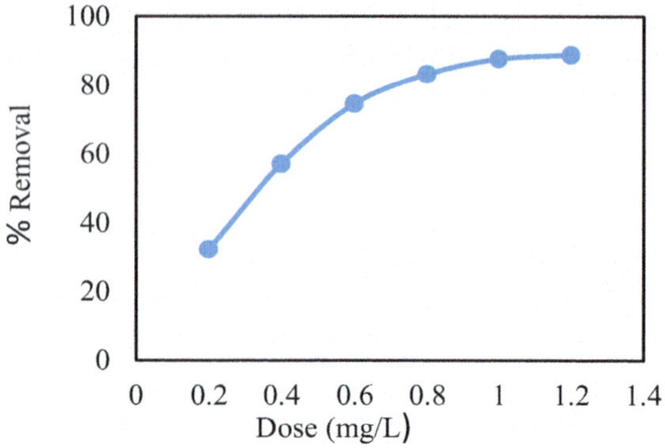

Fig. 9 Optimization of dose for removal of COD

Langmuir Model:

$$1/q_e = 1/K_L \cdot Q_m \cdot C_e + 1/Q_m \tag{6}$$

where K_L is the Langmuir constant, which denotes the energy of adsorption or the equilibrium constant of the adsorbate-absorbent equilibrium (L/g or L/mol depending on the unit of Q_m and C_e), and Q_m (mg/g) and C_e (mg/L) are the maximum adsorption capacity and the concentration at equilibrium, respectively.

Freundlich Model:

$$\log(q_e) = \log K_f + (1/n)\log C_e \tag{7}$$

where C_e is the equilibrium concentration of the adsorbate (mg/L). K_F and $1/n$ are the Freundlich constants that represent the adsorption capacity and the intensity of adsorption, respectively.

Redlich-Peterson Model:

$$\ln(C_e/q_e) = \beta \ln C_e - \ln A \tag{8}$$

where β is an exponent that lies between 0 and 1 and A is the Redlich-Peterson isotherm constant (Lg^{-1}).

As indicated in Figs. 10, 11, and 12, the correlation coefficient (R^2) of the Langmuir, Freundlich, and Redlich Peterson models are 0.9989, 0.9418, and 0.9770, respectively. With an R^2 of 0.9989, the Langmuir model provided the best fit.

The determination of coefficients and isotherm parameters for each model are listed in Table 2. With higher R^2 for Langmuir, the adsorption data obeyed Langmuir isotherm the best. However, Redlich-Peterson R^2 is also close to unity. The chi-square error was also evaluated and is incorporated in Table 2. Also, the experimental COD adsorption capacity of 3110 mg/g is close to the maximum adsorption capacity of 3333.3 mg/g predicted by the Langmuir model.

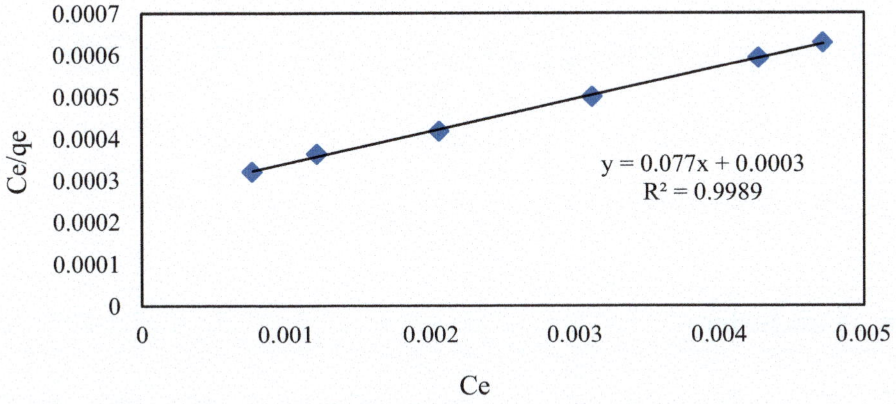

Fig. 10 Langmuir model

Fig. 11 Freundlich model

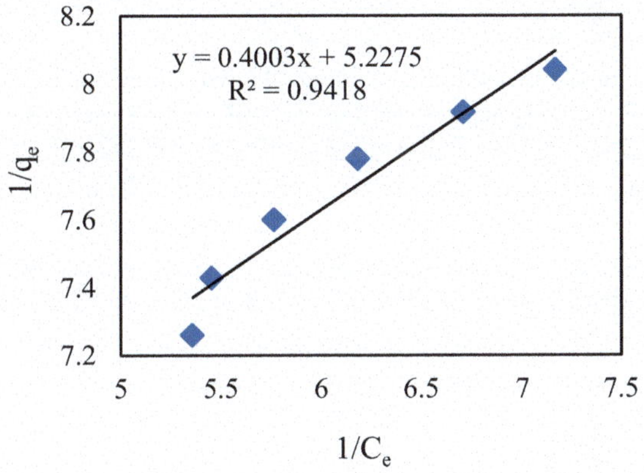

Fig. 12 Redlich-Peterson model

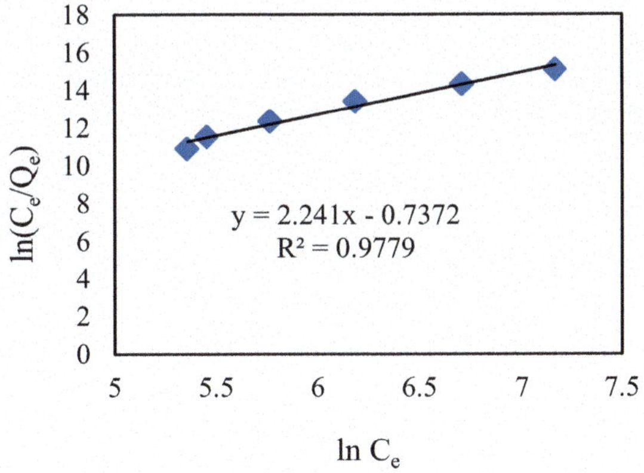

Table 2 Coefficients of adsorption isotherm models

S. N	Isotherm models	Parameters	R^2	χ^2
1	Langmuir	Q_m (mg/g): 3333.33 K_L: 0.00226	0.9989	0.001
2	Freundlich	K_f: 546.53 $1/n$: 0.2042	0.9418	40.940
3	Redlich-Peterson	β: 2.241 A (L/g): 2.09	0.9779	0.015

4 Conclusion

According to the present study, the coagulation-flocculation pretreatment method used on the leachate eliminated 70.93% of the turbidity and 37.4% of COD at pH 4 and an alum dose of 6 g/L. The efficacy of a novel landfill leachate treatment that coupled coagulation with leachate adsorption onto composite nanoparticle adsorbent was assessed in this study using the percentage of COD removal. The nanocomposite's base material was activated carbon derived from ash gourd peels. A nanocomposite of Fe_3O_4 and AC showed the highest COD removal efficiency when pH = 4 and a dose of 1.2 g/L was applied. In this case, equilibrium was attained after adsorption for 60 minutes. The elimination of COD proceeds quickly at first but eventually slows down reaching equilibrium. The explanation for this phenomenon is that there are a lot of vacant surface areas that are accessible for adsorption in the first phase. However, as time goes on, molecules of the solute on the bulk and solid phases exert repulsive forces, making it more difficult to occupy the remaining empty surface sites. The COD adsorption onto IIAC can be explained by the pseudo-second-order kinetics and Langmuir isotherm. Langmuir isotherm fits better showing that its suitability is prominent in idealized conditions with low concentrations and homogeneous surfaces. The increased characteristics of nano-activated carbon suggest that it is a promising material for leachate adsorption. In any case, before it is widely used as a leachate treatment method, more investigation and development are required to deal with issues of cost-effectiveness, scalability, environmental impact, safety concerns, and regulatory permissions.

References

1. S. De, T. Hazra, A. Dutta, Application of integrated sequence of air stripping, coagulation-flocculation, electrocoagulation and adsorption for sustainable treatment of municipal landfill leachate. Cleaner Waste Syst. **3**, 1–12 (2022)
2. X. Wang, S. Chen, X. Gu, K. Wang, Pilot study on the advanced treatment of landfill leachate using combined coagulation, fenton oxidation and biological aerated filter process. Waste Manage. **29**(4), 1354–1358 (2009)
3. V. Laishram, P.A. Kumar, Reduction of COD and turbidity in leachates through coagulation-adsorption. J. Indian Chem. Soc., 1–10 (2020)

4. M. Jain, M. Yadav, T. Kohoutc, M. Lahtinend, V.K. Garge, M. Sillanpääa, Development of iron oxide/activated carbon nanoparticle composite for the removal of Cr(VI), Cu(II) and Cd(II) ions from aqueous solution. Water Resour. Ind. **20**, 54–74 (2018)
5. P.N. Dave, L.V. Chopda, Application of iron oxide nanomaterials for the removal of heavy metals. J. Nanotechnol. **398569**, 1–4 (2014)
6. L.P. Lingamdinne, J.R. Koduru, R.R. Karri, Green synthesis of iron oxide nanoparticles for lead removal from aqueous solutions. Key Eng. Mater. **805**, 122–127 (2019)
7. Q.J.M. Slaiman, C.K. Haweel, Y.R. Abdulmajeed, Removal of heavy metals ions from aqueous solutions using biosorption onto bamboo. Iraqi J. Chem. Pet. Eng. **11**(3), 23–32 (2010)
8. S.M. Kakom, N.M. Abdelmonem, I.M. Ismail, A.A. Refaat, Activated carbon from sugarcane bagasse pyrolysis for heavy metals adsorption. Sugar Tech. **25**(3), 619–629 (2023)
9. J. Boguniewicz-Zablocka, I. Klosok-Bazan, V. Naddeo, C.A. Mozejko, Cost-effective removal of COD in the pre-treatment of wastewater from the paper industry. Water Sci. Technol. **81**(7), 1345 (2020)
10. K. Kaur, M. Kaur, M.K. Ubhi, N. Kaur, J.-M. Grenache, Composition optimization of activated carbon-iron oxide nanocomposite for effective removal of Cr (VI) ions. Mater. Chem. Phys. **258**, 1–13 (2021)
11. M. Kaura, S. Kumari, P. Sharma, Removal of Pb (II) from aqueous solution using nanoadsorbent of Oryza sativa husk: isotherm, kinetic and thermodynamic studies. Biotechnol. Rep. **25**, 1–8 (2020)
12. R.P. Chávez, E.C.C. Pizarro, Y.L. Galiano, Landfill leachate treatment using activated carbon obtained from coffee waste Technical Article. Eng. Sanit. Ambient. **24**(04), 833 (2019)
13. H.I. Abdel-Shafy, A.M. Ibrahim, A.M. Al-Sulaiman, R.A. Okasha, Landfill leachate: sources, nature, organic composition, and treatment: an environmental overview. Ain Shams Eng. J. **89**, 1–11 (2023)
14. C. Ayşenur Turan, M. Kobya, C. Iskurt, E. Gengec, A. Khatae, A techno-economical assessment of treatment by coagulation-flocculation with aluminum and iron-bases coagulants of landfill leachate membrane concentrates. Chemosphere **314**, 153–160 (2023)
15. N.A. Yazid, Y.C. Joon, Co-precipitation synthesis of magnetic nanoparticles for efficient removal of heavy metal from synthetic wastewater. AIP Conf. Proc. **2124** (2019)
16. Z. Chaouki, M. Hadri, M. Nawdali, M. Benzina, H. Zaitana, Treatment of a landfill leachate from Casablanca city by a coagulation-flocculation and adsorption process using a palm bark powder (PBP). Sci. Afr. **12**, 1–16 (2021)
17. N. Ayawei, A.N. Ebelegi, D.W. Hindawi, Modelling and interpretation of adsorption isotherms. J. Chem. **2017**, 3039817, 1–11 (2017)

Effect of Anammox Bacteria for Synthetic Wastewater Treatment Containing Ammonia and Nitrite

Indrani Chakraborty and Apurba Dey

Abbreviations

BR	Batch reactor
COD	Chemical oxygen demand
FBR	Fluidized-bed reactor
N-DN	Nitrification and denitrification
SBR	Sequencing batch reactor
WWTP	Waste water treatment plant

1 Introduction

Ammonia and nitrite are the harmful components in wastewater that get accumulated after the discharge from waste generation sites into the water bodies which affects aquatic as well as human life. These components need to be treated before being dispersed into water bodies. The relatively recent phenomenon known as anaerobic ammonium oxidation (anammox) was first proposed two decades ago [1]. The breakdown of ammonia and nitrite to nitrogen gas creates a nitrogen-anoxic shortcut in the presence of a recently discovered bacterial phylum belonging to the group Planctomycetes [2]. Although several species of the anammox group of bacteria were identified, none of them have been successfully grown in pure culture because of unknown cellular characteristics and growth requirements [3]. Traditional removal of ammonia and nitrite is being done by nitrifying and denitrifying bacteria,

I. Chakraborty (✉) · A. Dey
Department of Biotechnology, National Institute of Technology, Durgapur, Paschim Bardhaman, West Bengal, India
e-mail: ic.22bt1112@phd.nitdgp.ac.in; adey.bt@nitdgp.ac.in

D. Mazumder (ed.), *Sustainable Advanced Technologies for Environmental Management*, Springer Proceedings in Earth and Environmental Sciences,
https://doi.org/10.1007/978-3-031-64006-3_28

but after the discovery of the anammox group of bacteria in WWTP, it has found immense importance because of the property to oxidize ammonia with nitrite to form nitrogen gas. Despite being more cost-effective than the N-DN process, very little data is available to support the on-site development of anammox in real-time which involves the cultivation and optimization of conditions to support their growth in specific environments [4]. The "anammox" process, which oxidizes ammonia, occurs inside the anammoxosome where nitrite is present as an electron acceptor. This syntrophic metabolic activity is carried out inside the Planctomycetes by chemolithoautotrophic bacteria, which assemble into a monophyletic cluster. Anammox bacteria can only metabolize ammonium and nitrite using CO_2 or bicarbonate as their sole carbon source. The substances in lab-scale bioreactors are digested based on this general equation when they are operating in a steady state [5, 6].

$$NH_4 + 1.32NO^- + 0.066HCO^- + 0.13H^+ \rightarrow 1.02N_2 + 0.26NO_3^- \\ + 0.066CH_2O_{0.5}N_{0.15} + 2.03H_2O$$

Developing a strong strategy is crucial to opening up the possibility of an anammox start-up in India as the field is relatively new and more time-consuming. This will not only close a significant research gap but also contribute to the global achievement of sustainable nitrogen removal.

As the anammox group of bacteria grows slowly, thus researchers have looked into several reactor topologies, including fluidized bed reactors (FBR) [7], sequencing batch reactors [6], and membrane bioreactors [8]. Because of its ability to retain all of the biomass, the sequencing batch reactor (SBR) is thought to be the most ideal reactor for the growth of anammox bacteria among these [6, 9].

Herein, we started up a batch reactor (BR) using kitchen waste samples as seed by applying an unconventional start-up strategy. The term unconventional strategy means that the ratio of NH_4^+-N to NO_2^- -N in the influent was not predetermined but rather varied depending on the sample characteristics. Thus, the primary goal of this work was to maintain the NO_2^- -N levels below the inhibitory levels to initiate the anammox process using the available seed in a batch reactor [10].

2 Materials and Methods

2.1 *Materials*

A Scott bottle of 1 L volume was selected for the enrichment purpose. pH meter was used for measuring the changes in pH of the sample. Conical flask for reagent preparation, test tube, glass beaker, funnel, and measuring cylinder of Borosil make were taken for experimental analysis.

2.2 Sample Collection and Enrichment

Anammox bacteria thrive quickly in easily accessible sources due to their rapid enrichment. A sample was taken from the CSIR-CMERI Durgapur kitchen waste and placed in a 3 L PET jar before being brought into the laboratory. The sample consisted of water and a thick slurry mixture. It was first kept undisturbed for 6 hours to sediment; after that the supernatant water was discarded, and thick slurry samples were pre-enriched in synthetic autoclaved media in a sterile Scott bottle, and it was agitated under a magnetic field in a magnetic stirrer [11]; then 20% (v/v) of pre-enriched sample was transferred to a new synthetic medium where ammonium and nitrite source is provided with limited carbon source. The batch culture for a 1-L volume was operated at an optimized condition of 37 °C at 120 rpm with pH 7.5 for 45 consecutive days. Estimation of ammonia and nitrite was done using a multiparameter photometer equipped with a COD digester. Pre-enriched sample in the presence of ammonia and nitrite gets converted to nitrogen gas. For the enrichment, the experiment was carried out using a synthetic medium containing NH_4^+-N and NO_2^- -N. Synthetic wastewater was prepared following literature reports, and Table 1 lists its composition [12]. Ammonium sulphate ($(NH_4)_2SO_4$) and sodium nitrite ($NaNO_2$) were used as a source of ammonia and nitrite, respectively, in this study [12].

Table 1 Composition of synthetic medium

Component	Quantity (g/L)
$KHCO_3$	1.0
$(NH_4)_2SO_4$	0.942
$NaNO_2$	0.673
KH_2PO_4	0.140
$CaCl_2$	0.050
$MgCl_2$	0.050
Na_2S	0.250
***Trace element solution 1**	**1 mL**
$FeSO_4$	5
EDTA powder	5
****Trace element solution 2**	**1 mL**
EDTA	15.0
$ZnSO_4$	0.43
$CoCl_2$	0.24
$MnCl_2$	0.63
$CuSO_4$	0.25
Na_2MoO_4, $2H_2O$	0.22
$NiCl_2$	0.19
Na_2SeO_3	0.21
H_3BO_3	0.01
Na_2WO_4	0.05

2.3 Effect of Anammox Activity

To confirm the anaerobic ammonium, disappearance was studied using 400 mL pre-enriched sludge which was added to a 1-L Scott bottle that was fed with autoclaved synthetic medium (the composition of which is shown in Table 1). The entire setup was kept in a shaker bacteriological incubator at 37 °C, pH of 7.5, and 120 rpm. Literature study suggests it is the optimized parameter for the growth of anammox group of bacteria [13, 14]. 20 mL of sample was taken out aseptically after every 72 hours, and the concentrations of pH, NH_4, and NO_2 were recorded [4]. Every batch was run continuously for 60 days, and the experiment was repeated three times.

2.4 Chemical Assessment

A multiparameter photometer with a COD digester was used to assess the COD of the sample (HI 839800 2008 series), based on the dichromate method [15]. 0.2 mL sample was added to a COD kit (HI 93754A-0) COD low range EPA Reagent vial, a vial of 2 mL deionized water (DEIONIZED 120) taken as blank and then processed in a Hannah COD digester for 2 hours at 150 °C. By using the COD HR technique, the samples were monitored at 420 nm against digested distilled water as blank.

Ammonia nitrogen (NH_3-N) was measured through Nessler's reagent in the same multiparameter photometer [16, 17].

After diluting 1 mL of sample to a final amount of 10 mL, 4 drops (150 μL) of Nessler's reagents, one and two, were added, then thoroughly mixed, and allowed to stand for 3 minutes. Next, using the ammonia HR technique with a 1 mg/mL ammonium chloride standard solution, the absorbance at 420 nm was calculated [18].

Using the ferrous sulfate method, 10 mL of filtered material was added along with one packet of nitrite HR (H193708-0) reagent and allowed to react for 10 minutes, and nitrite was detected at 575 nm.

A digital pH meter was used to monitor each sample's pH regularly. A two-point calibration was performed using standard buffers with pH values of 4.0 and 7.0 [19].

2.5 Biological Assessment

Periodic evaluations and microscopic analysis were carried out to monitor the presence and growth of the anammox group of bacteria. Since anammox bacteria are autotrophic, they do not rely on organic carbon for energy; instead, they use ammonium as their energy source to oxidize nitrite to nitrogen gas. Ammonia and nitrite concentration is checked regularly within 3-day intervals for the growth and activity of anammox bacteria [4].

3 Results and Discussion

In the initial phase of pre-enrichment, there is low utilization of ammonia and nitrite around 20–30%; after 18–20 days of pre-enrichment, the utilization of ammonia with nitrite was found to be 50% [3]. The pre-enriched sample was fed with a low concentration of nitrite, and it has been found that there is complete utilization of ammonia along with nitrite within the reactor. Microscopic analysis showed the presence of red color beads in the enriched sample of a 60-day culture [20]. From the study, it was found that there is complete utilization of ammonia with nitrite by the enriched anammox culture within 60 days of batch culture from the kitchen waste sample. Through a physical and microbiological observation, the sludge was discovered to be light brown, primarily enriched with Gram-positive bacillus and Gram-negative coccus in the initial phase of the pre-enrichment. Following 1 month of enrichment, the majority of the bacteria in the sludge were Gram-negative coccus, and a small number of Gram-positive bacillus and actinomycetes were found [4]. After 60 days of operation, a significant change in the quality of the sludge was detected. A red cluster with a noticeable particle size of 4–6 mm was apparent in the microscopic view. The utilization of ammonia and nitrite inside the reactor was analyzed every 3-day interval (Table 2) and the data was plotted (Fig. 1). Figure 2 depicts the microscopic view of anammox bacteria in the enriched sample.

4 Conclusion

In this study, an attempt has been made to start up an anammox process in the best possible way to address the problem of ammonia and nitrite in the wastewater. A batch reactor was set up for the enrichment of the experiment from a kitchen waste sample. Anammox group of bacteria is slow growing and needs continuous

Table 2 Ammonia and nitrite utilization by anammox bacteria

No. of days	NH_4^+ (mg/L)	NO_2^- (mg/L)
1	107	160
4	99	149
7	87	131
10	8	11
13	3	4
16	0	0

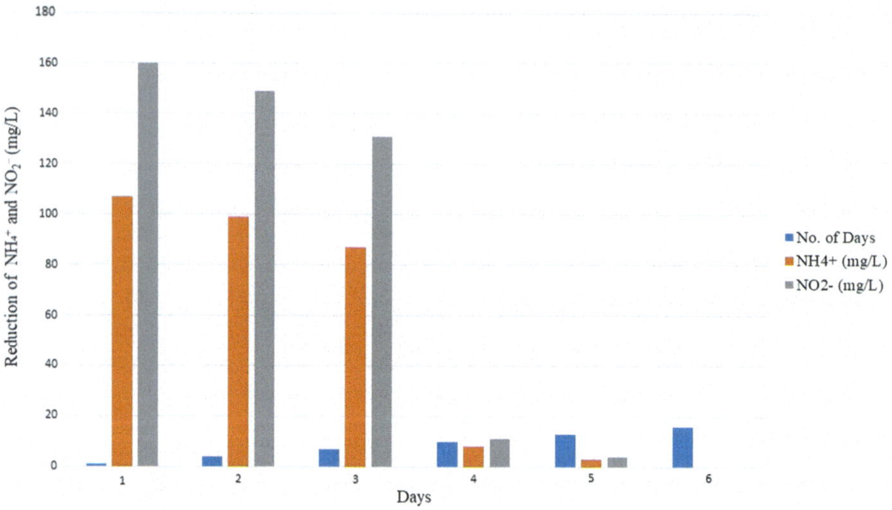

Fig. 1 Reduction of ammonia and nitrite in batch culture

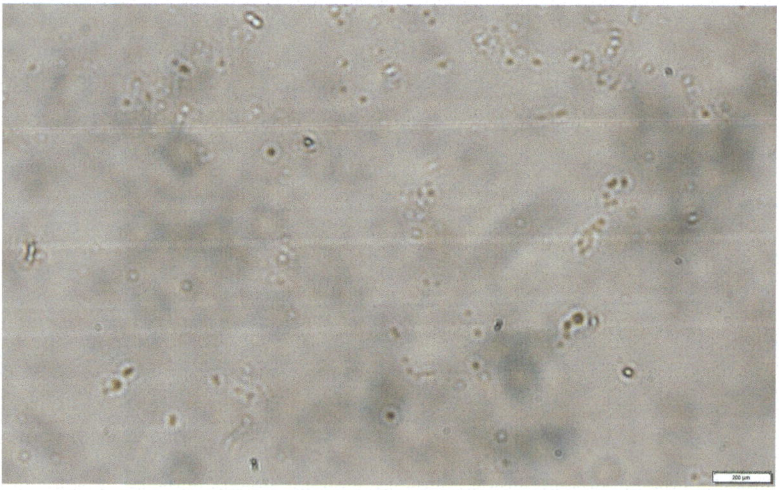

Fig. 2 Anammox bacteria under inverted microscope OLYMPUS IX73 at 100× resolution

monitoring as excess nitrite in the medium causes toxicity to the cells. Continuous monitoring of pH, ammonia, and nitrite in the system was done along with a regular supply of synthetic medium. During the initial days (first month) of enrichment, there was no change in sludge quality, and the color of the sludge was dark brown. After the enrichment of 60 days, the sludge becomes light brown. From the result, it can be concluded that ammonia and nitrite get utilized from the system within 16 days of supply. The microscopic visualization confirms the presence of an anammox group of bacteria by the appearance of the red cluster. Further study is required in the reactor to check the amount of nitrogen gas produced by the reactor during the initial and final period of enrichment. Additionally, for a large-scale operation to start up successfully, it is necessary to extensively explore the feasibility of carbon load tolerance, viable food-to-microbe ratio, and the reduction of volatile suspended solids generation inside the reactor with an inorganic carbon source.

Acknowledgment The current research work was done under the supervision of Dr. Apurba Dey, Professor, Department of Biotechnology, NIT Durgapur. Acknowledging Dr. Amit Ganguly, Chief Scientist, Director's Secretariat cum Technical Cell, CMERI Durgapur for providing the sample. The Research work is funded by the DST-SERB Government of India's fellowship program.

References

1. A.A. Van De Graaf, M. Arnold, P. De Bruijn, M.S.M. Jetten, L.A. Robertson, A.J. Gijs Kuenen, Anaerobic oxidation of ammonium is a biologically mediated process. Appl. Environ. Microbiol. **61**, 1246–1251 (1995)
2. S. Verma, A. Daverey, J.G. Lin, Successful start-up of anammox process from activated sludge and anaerobic sludge in a sequencing batch reactor using an unconventional strategy. Int. Biodeterior. Biodegradation **156** (2021). https://doi.org/10.1016/j.ibiod.2020.105132
3. M.C.M. Van Loosdrecht, X. Hao, M. Jetten, W. Abma, Use of Anammox in urban wastewater treatment. Water Sci. Technol. Water Supply **4**, 87–94 (2004)
4. S. Neogi, A. Dey, P.K. Chatterjee, Studies on rapid initiation of Anammox process for starch industry effluent treatment, in *Waste Water Recycling and Management: 7th IconSWM—ISWMAW 2017*, vol. 3, (Springer Singapore, Singapore, 2019), pp. 97–110
5. B. Kartal, L. van Niftrik, J.T. Keltjens, H.J.M. Op den Camp, M.S.M. Jetten, Anammox-growth physiology, cell biology, and metabolism, in *Advances in Microbial Physiology*, (Academic Press, 2012), pp. 211–262
6. M. Strous, J.J. Heijnen, J.G. Kuenen, M.S.M. Jetten, The sequencing batch reactor as a powerful tool for the study of slowly growing anaerobic ammonium-oxidizing microorganisms. Appl. Microbiol. Biotechnol. **50**, 589–596 (1998)
7. W.R.L. van der Star, A.I. Miclea, U.G.J.M. van Dongen, G. Muyzer, C. Picioreanu, M.C.M. van Loosdrecht, The membrane bioreactor: a novel tool to grow anammox bacteria as free cells. Biotechnol. Bioeng. **101**, 286–294 (2008)
8. C. Trigo et al., Start-up of the Anammox process in a membrane bioreactor. J. Biotechnol. **126**(4), 475–487 (2006)
9. A.A. van de Graaf, P. de Bruijn, L.A. Robertson, M.S.M. Jetten, J.G. Kuenen, Autotrophic growth of anaerobic ammonium-oxidizing micro-organisms in a fluidized bed reactor. Microbiology **142**, 2187–2196 (1996)
10. A. Daverey, S.H. Su, Y.T. Huang, S.S. Chen, S. Sung, J.G. Lin, Partial nitrification and anammox process: a method for high strength optoelectronic industrial wastewater treatment. Water Res. **47**, 2929–2937 (2013)
11. J. Wang, C. Yu, S. Zhang, T. Li, D. Zhang, X. Wang, Study on rapid start-up of anammox process under the influence of magnetic field, in *E3S Web of Conferences*, (EDP Sciences, 2018)
12. K.A. Third et al., Enrichment of anammox from activated sludge and its application in the CANON process. Microb. Ecol. **49**, 236–244 (2005)

13. A. Dey, S. Neogi, P.K. Chaterjee, Studies on optimization of physical growth factors for rapid initiation of anammox. Carbon **125**, 272 (2019)
14. L. Zhang, Y. Narita, L. Gao, M. Ali, M. Oshiki, S. Okabe, Maximum specific growth rate of anammox bacteria revisited. Water Res. **116**, 296–303 (2017). https://doi.org/10.1016/j.watres.2017.03.027
15. D. Fatta, A. Achilleos, A. Nikolaou, S. Meriç, Analytical methods for tracing pharmaceutical residues in water and wastewater. TrAC – Trends Anal. Chem. **26**, 515–533 (2007). https://doi.org/10.1016/j.trac.2007.02.001
16. Standard Test Methods for Ammonia Nitrogen in Water 1. https://doi.org/10.1520/D1426-08
17. Standard Methods for the Examination of Water and Wastewater (1999)
18. S. Neogi, A. Dey, P.K. Chatterjee, Microflora from leaf debris is suitable for treatment of starch industry wastewater. Eng. Life Sci. **16**, 683–689 (2016). https://doi.org/10.1002/elsc.201500086
19. Designation: D 698-07 Standard Test Methods for Laboratory Compaction Characteristics of Soil Using Standard Effort (12,400 ft-lbs/ft3 (600 kN-m/m3)) 1 (2012)
20. S. Uyanik, O.K. Bekmezci, A. Yurtsever, Strategies for successful ANAMMOX enrichment at laboratory scale. Clean (Weinh) **39**, 653–657 (2011)

Removal Capacity of *Lemna minor* as Biosorbent for Removal of Cadmium from Simulated Wastewater in Combination with Ceramic Membrane

Priyankari Bhattacharya, Sachin Mandal, Ashis Sadhukhan, Sk Jinnur Rahaman, and Sathi Banerjee

Abbreviations

AAS	Atomic absorption spectrophotometer
BET	Brunauer, Emmett, and Teller
FESEM	Field emission scanning electron microscopy
FTIR	Fourier transform infrared spectroscopy
ID	Inner diameter
LMH	Liter per meter square hour
LPM	Liter per minute
OD	Outer diameter
XRD	X-ray diffraction

1 Introduction

Heavy metal pollution and water contamination is one of the environmental concerns that still create concern for mankind. Cadmium among many toxic metals poses serious toxic effect as human body cannot excrete out completely this metal and it also has synergistic effect with other metals [1, 2]. Moreover, this heavy metal is not needed by either plants or animals or beneficial to soil and is toxic in nature thereby causing pollution [3]. Cadmium gets discharged in the environment from different activities like electroplating, cement, batteries, mining, etc. [4]. Different treatment technologies like chemical precipitation, electrochemical methods, etc. are widely

P. Bhattacharya · S. Mandal · A. Sadhukhan · S. J. Rahaman · S. Banerjee (✉)
Metallurgical and Materials Engineering Department, Jadavpur University, Kolkata, India

D. Mazumder (ed.), *Sustainable Advanced Technologies for Environmental Management*, Springer Proceedings in Earth and Environmental Sciences,
https://doi.org/10.1007/978-3-031-64006-3_29

used methods for cadmium removal [5, 6]. Adsorption-based treatment technologies are used for metal removal because of its effectiveness and regeneration [7]. Most commonly used adsorbent is activated carbon that is widely used for removal of various pollutants like metals, dyes, etc. However, due to cost, regeneration problem, and disposal issue, use of activated carbon is restricted [8, 9]. Low-cost biosorbents are prepared from fruit peels, leaves, agricultural wastes, etc. Biosorption is based on efficiency of any biosorbent to bind to metal ions either by chelation or surface adherence or chemisorption. Moreover adsorbents derived from waste materials bear the capacity of pollutant remediation. The presence of such materials like macrophytes, algae, agricultural wastes, etc. imposes load on the environment. Therefore, utilizing such wastes as biosorbent served dual role of pollution reduction. Duckweed has the capacity to double its biomass volume in 2 days causing eutrophication of water bodies. Live duckweed is capable of bioaccumulating toxic metals causing threat to aquatic life. Thick mats covering water bodies disrupt navigation. These factors contributed towards utilizing duckweed as low-cost biosorbents for removal of heavy metals [10–13].

Biosorbent prepared from duckweed showed promising results in removal of cadmium from simulated wastewater. Duckweed has high growth rate and is easy to culture, and they exhibit metal accumulating property.

In the present study, cadmium removal was observed using biosorbent prepared from duckweed followed by membrane study that served dual role of biosorbent separation and metal removal in single step. Adsorption isotherm data was fitted with Langmuir and Freundlich isotherm models. Membrane study was then carried out for separation of biosorbent and other contaminants present in wastewater.

2 Materials and Methods

2.1 Preparation of Biosorbent

Duckweed was collected from nearby pond at Jadavpur University campus. It was cleaned thoroughly and dried in hot air oven and then subjected to thermal activation in furnace at 550 °C.

2.2 Preparation of Cadmium Solution

50 mg/L stock solution was prepared from cadmium standard solution by Sigma-Aldrich (1000 mg/L). Working solutions of 10–40 mg/L were then prepared by dilution.

2.3 Removal Study

pH, temperature, initial concentration of Cadmium ions, biosorbent dosage, and contact time determine the biosorption process. To find the optimum concentration of above factors, batch scale biosorption study was carried out. The experiments were carried out in three replicated=s and standard deviation was applied to get the average data. The effect of pH (2–8), biosorbent dose (0.25–2 g/L), and metal concentration (10–50 mg/L) was studied. Batch biosorption study was carried out using 100 mL solution and biosorbent under stirring condition. Samples were drawn at regular intervals and subjected to AAS analysis. Kinetic study was carried out at pH 4, 1 g/L biosorbent dose, and 10 ppm cadmium solution.

2.4 Biosorbent Characterization

Biosorbent prepared from duckweed was characterized in terms of x-ray diffraction (XRD, SmartLabSE, Rigaku Corporation, Japan) at 2θ range: 20–80° using Cu anode material (λ = 1.541 A°), Fourier Transform Infrared Spectroscopy (FTIR, IRPrestige-21, Shimadzu, Japan), and Field Emission Scanning Electron Microscopy (FESEM, Inspect F50, FEI, USA).

2.5 Membrane Study

Membrane study was conducted in pilot scale prototype consisting of 20 L capacity feed tank with stirring arrangement. Ceramic membrane filtration module was made of stainless steel with rotameter (Range 0–7 LPM) and differential pressure showing at glycerine pressure gauge (upstream and downstream both). Single channel ceramic membrane of 300 mm L and 10/7 mm (OD/ID) was used. Simulated wastewater was taken in feed tank and recirculated along with biosorbent. The effect of time, transmembrane pressure, and cross-flow velocity were studied.

3 Results and Discussion

3.1 Characterization of Nanoparticles

Biosorbent was characterized by the following methods. BET surface area of biosorbent was determined to be 125 m^2/g.

3.2 XRD

Phase identification and cell dimensions in terms of crystal orientation, crystal defects, lattice parameters, crystallite size, etc. of biosorbent were analyzed by XRD. Figure 1 displays XRD analysis of duckweed biosorbent from which crystalline nature, lattice parameters, orientation, and phase is determined. Biosorbent was crystalline in nature with broad peak at $2\theta = 30$ that corresponded to planes (111), (100), (211), (110), (133), (011), (101), and (132). The planes corresponded to cellulose and lignin that were present in duckweed. Bragg angles 2θ corresponding to 30 represents face-centered cubic crystals [14]. Mean crystallite size was obtained from Scherrer Eq. (1) where D is the mean diameter of crystallite, 0.89 was taken as Scherrer Constant, wavelength $\lambda = 1.54056$ Å, and $\beta 2 = B2 - b2$ (B is the full width at half maxima and b instrumental broadening). Crystallite size thus obtained from Scherer equation was 22.08 nm.

$$D = \frac{K\lambda}{\beta \cos \theta} \tag{1}$$

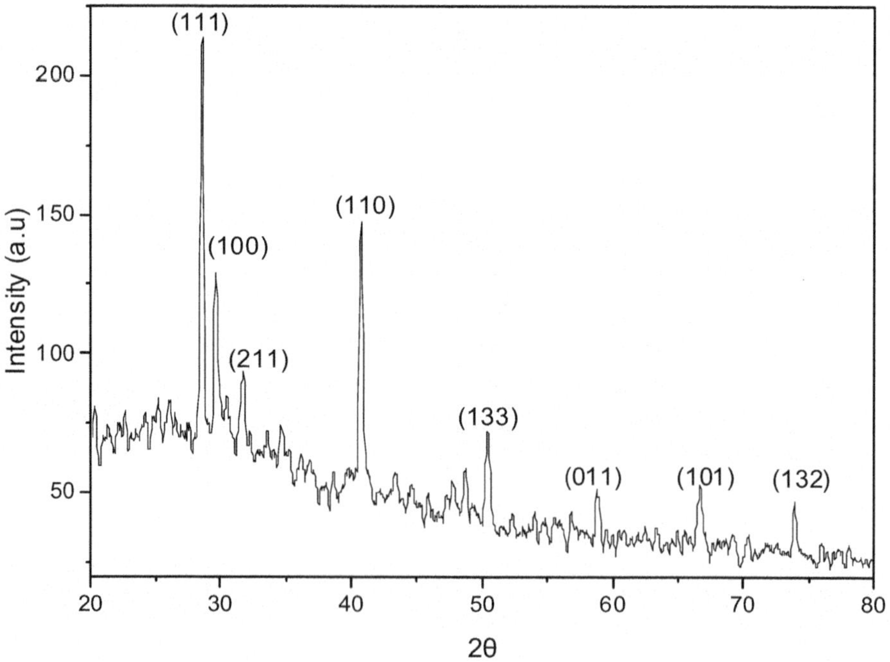

Fig. 1 XRD analysis of duckweed biosorbent

3.3 FTIR

Functional groups that are present and involved in adsorption were determined from FTIR analysis (Fig. 2). Peaks present at around 3284 cm^{-1} was mainly due to O-H stretch which after biosorption slightly shifted to 3260 cm^{-1}. Peaks at 2276 and 1601 cm^{-1} were due to presence of alkanes having 3284 cm^{-1} C-H stretch and -C=C- stretch of alkenes. A major shift in peak from 1601 to 1575 cm^{-1} was observed suggesting complexation of carboxylic groups with cadmium. Stretching vibrations at 1436 cm^{-1} (before) and 1438 cm^{-1} (after) might be due to N-H bond. A new peak was observed at 1074 cm^{-1} in FTIR of duckweed after biosorption which might be due to cadmium. Shift in peak from 680 cm^{-1} (before) to 445 and 581 cm^{-1} (after) suggests presence of amines and aromatics in biosorption process. Peaks at 880 and 807 cm^{-1} can be due to presence of C-O group [15].

3.4 FESEM

To observe the microstructure of the biosorbent before and after cadmium removal, FESEM was performed (Fig. 3a). Surface of unexposed biosorbent appeared smoother and porous with spherical structure whereas after exposure to cadmium solution surface appeared dense and irregular with frequent cavities and significant reduced pores. This behavior was observed in other studies also. The reason might be due to deposition of Cd^{2+} which were freely available to open pores of the biosorbent [16].

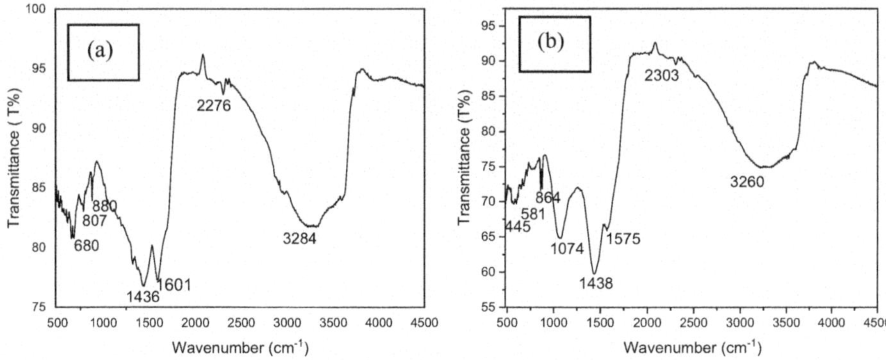

Fig. 2 FTIR analysis of biosorbent (**a**) before cadmium adsorption and (**b**) after cadmium adsorption

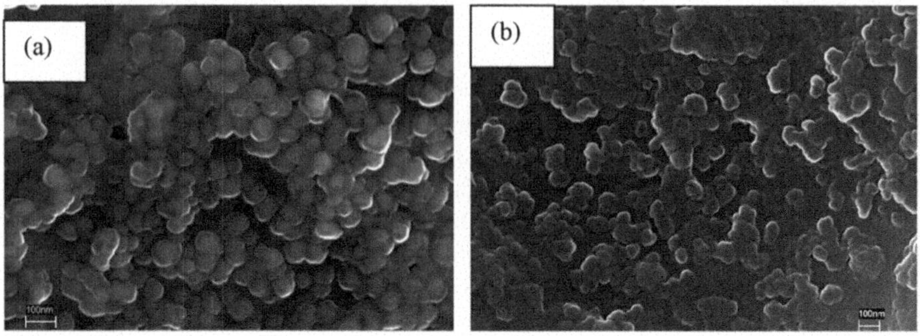

Fig. 3 FESEM analysis of biosorbent (**a**) before adsorption of cadmium (**b**) after adsorption of cadmium

Fig. 4 Effect of biosorbent dose on cadmium removal (metal concentration: 10 ppm; pH: 4)

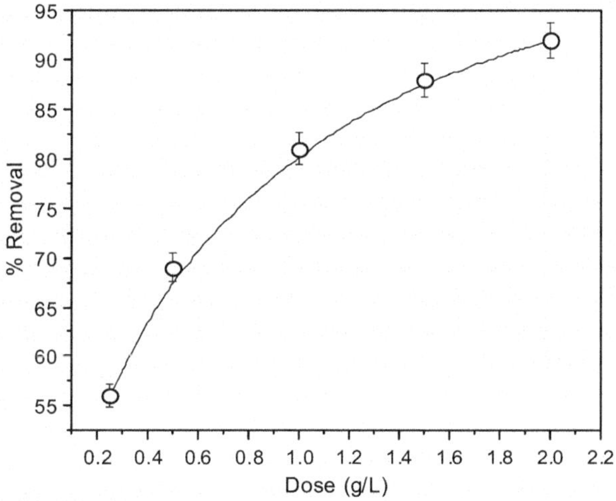

3.5 *Effect of Dose on Adsorption*

Effect of biosorbent dose on removal of cadmium is shown in Fig. 4 at an initial metal concentration of 10 ppm. Dose was varied from 0.25 to 2 g/L. Initially, with increasing dose, adsorption increased, i.e., from 0.25 to 1 g/L; an increase of 69–82% cadmium removal was observed. However, further increasing the dose showed no further significant metal removal. Therefore, for further studies 1.0 g/L biosorbent dose was selected. At high dose of biomass, electrostatic interaction causes decreased metal biosorption. Under low biomass concentration, biosorption takes place on surface as well as intracellular level causing greater biosorption.

3.6 Effect of Concentration on Biosorption

From Fig. 5, it was observed that increasing metal concentration from 10 to 50 ppm causes decrease in biosorption efficiency of duckweed. About 83% cadmium removal was observed for 10 ppm cadmium solution at pH 4 and 1 g/L biosorbent dose. At 50 ppm metal concentration, only 40% removal of cadmium was observed. At particular biomass dose, available sites get exhausted when metal concentration is increased thus causing reduced uptake by biosorbent. For reproducibility, the study was conducted in three replicates and data represented average of the experimental data. Standard deviation was shown in form of error bars.

3.7 Effect of pH on Adsorption

To observe the effect of pH on cadmium adsorption, pH of the solution was varied from 2 to 8 (Fig. 6). Initial cadmium solution is highly acidic (~2) which when diluted with water increases slightly. pH of the solution was increased with NaOH, and further increasing the pH resulted in precipitation of Cd $(OH)_2$. Maximum removal was obtained at pH 4 with about 86% which further decreased with increasing pH.

pH of solution is very important in determining removal capacity of adsorbent. In the present study, pH of the solution was varied from 2 to 8. Optimum pH obtained was pH 4 which was further used for study. Increasing the pH beyond 4 resulted in removal of cadmium but at decreased rate. This might be due to repulsion between H^+ and Cd^{2+} ions (pH 2–3). However, with increasing pH, precipitation of Cd^{2+} as cadmium hydroxide takes place (pH > 5). Therefore, pH 4 was optimum for this

Fig. 5 Effect of initial cadmium concentration on its removal (biosorbent dose 1 g/L; pH 4)

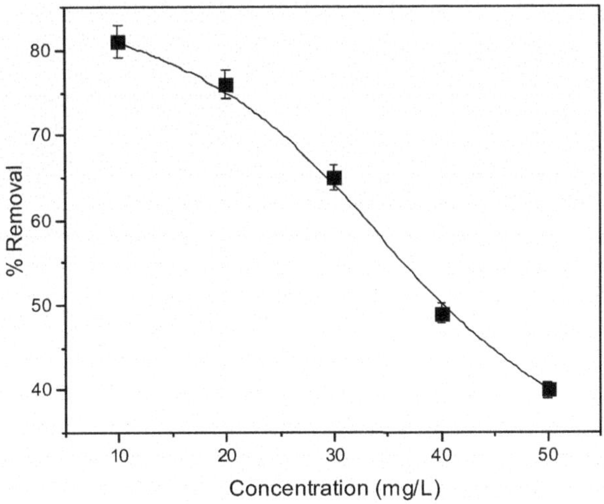

Fig. 6 Effect of pH on removal of cadmium (biosorbent dose 1 g/L; metal concentration: 10 ppm)

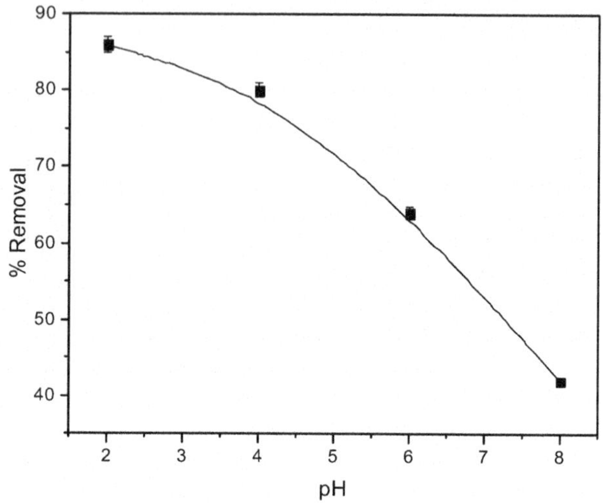

Fig. 7 Effect of contact time on removal (biosorbent dose 1 g/L; pH 4; 10 ppm cadmium concentration)

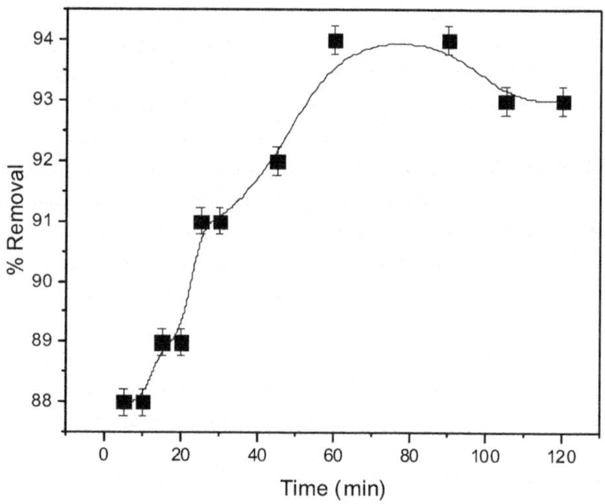

study. Similar phenomenon was also observed in some other studies conducted by other authors [17–19].

3.8 *Effect of Contact Time on Biosorption*

At constant biosorbent dose of 1 g/L, pH 4, and 10 ppm cadmium solution, maximum removal of 84% was obtained at 60 min of contact time (Fig. 7) which later became constant or further 120 min (84–87%). The resultant equilibrium time was found to be 90 mins. This might be attributed to the fact that attachment of more dye particles became available for binding on nanoparticles surface. However, after a certain time, equilibrium was reached as with increase in time all available adsorption sites were occupied by the dye particles.

3.9 Adsorption Modeling

Cadmium removal data by duckweed was computed by mathematical isotherm models of Langmuir and Freundlich and depicted in Fig. 8a, b. The Langmuir isotherm favors monolayer adsorption and is based on assumption that adsorption sites are limited and uniform, and after saturation sets in, no further adsorption occurs. Freundlich isotherm on the other hand is based on assumption that it is multilayer and adsorbent surface is heterogeneous [14]. The correlation parameters are tabulated in Table 1. The following equation explains Langmuir isotherm (Eq. 2):

$$\frac{1}{Q_e} = \frac{1}{K_L \cdot Q_{max}} \cdot \frac{1}{C_e} + \frac{1}{Q_{max}} \tag{2}$$

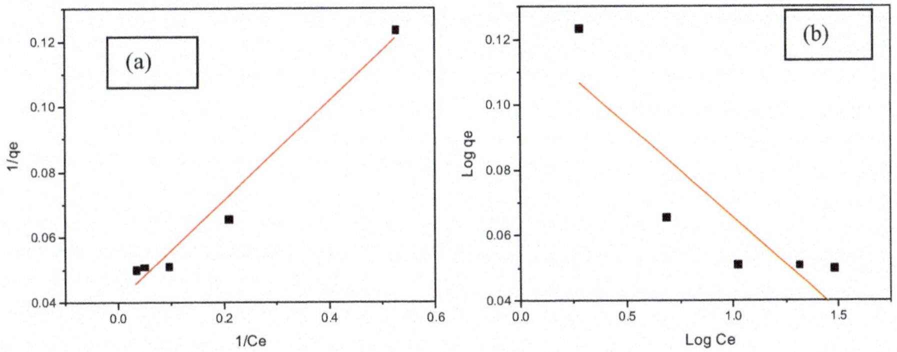

Fig. 8 Adsorption isotherm fitting for (**a**) Langmuir (**b**) Freundlich

Table 1 Langmuir and Freundlich isotherm parameter

Isotherm	Parameter	Value
Langmuir	Q_m (mg/g)	24.7
	K_L (L/mg)	0.26
	R^2	0.977
Freundlich	K_F (L/g)	1.327
	$1/n$	0.057
	R^2	0.768

where Q_{max} (mg/g) stands for maximum adsorption capacity of adsorbent, K_L (L/mg) is Langmuir constant, and C_e (mg/L) is maximum adsorption capacity at equilibrium.

Freundlich isotherm is represented by the following Eq. (3):

$$\text{Ln}Q_e = \frac{1}{N} \cdot \text{Ln}C_e + \text{Ln}K_F \quad (3)$$

where K_F (L/g) is the Freundlich constant and $1/n$ is adsorption intensity that determines surface heterogeneity.

Adsorption of cadmium by duckweed is well explained by Langmuir isotherm model ($R^2 = 0.97683$) than Freundlich model ($R^2 = 0.76851$). Also from Table 1, Q_{max} is 24.715 for Langmuir isotherm suggesting uniform adsorption sites and monolayer adsorption [14].

3.10 Membrane Study

Pilot scale membrane study was carried out using 10 ppm synthetic cadmium solution as feed (15 L) along with 1 g/L biosorbent dose and pH 4. The feed was recirculated for 15 min before onset of experiment. Membrane study was conducted at 2 bar transmembrane pressure for 120 min. Initial flux of 78 LMH (L^{-1} m^{-2} h^{-1}) was obtained which became 69 LMH within 60 min of operation. Further decrease in flux was not obtained, and it thus reached a steady state within 120 min (Fig. 9a). Removal of cadmium was measured by drawing permeate samples at regular intervals and subjected to analysis. Maximum removal of 94% was obtained within 60 min, and further increasing the time did not affect removal much (Fig. 9b). As biosorbent tends to form cake over membrane surface, so desorption might take place thereby reducing removal efficiency [20].

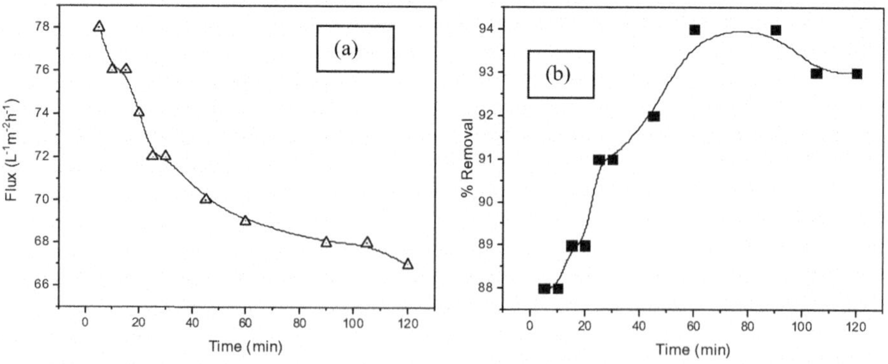

Fig. 9 Effect of time on (**a**) flux profile (**b**) cadmium removal at 2 bar transmembrane pressure using 15 L of 10 ppm cadmium solution as feed, pH 4, and 1 g/L biosorbent dose

4 Conclusion

Cadmium removal was achieved using biosorbent-based treatment where duckweed was used as biosorbent. About 80% removal was obtained which further enhanced to 91% after membrane study. Adsorption obeyed Langmuir isotherm suggesting biosorbent to be monolayer. Maximum adsorption occurred at pH 4 using 1 g/L dose and 10 ppm cadmium solution. FTIR analysis suggested involvement of amines, aromatics functional groups to be involved in biosorption process. The biosorbent had high surface area of 125 m^2/g determined from BET analysis. The combined process also removed other pollutants and separated biosorbent. The biomass can be regenerated for further use.

Acknowledgment The present research work is supported under the Department of Science and Technology, Government of India, vide Grant no. DST/WOS-B/WWM-2/2021 (G) dated 17.09.2021.

References

1. H. Yaacoubi, O. Zidania, M. Mouflih, M. Gourai, S. Sebti, Removal of cadmium from water using natural phosphate as adsorbent. Procedia Eng. **83**, 386–393 (2014)
2. R. Kumar, J. Chawla, I. Kaur, Removal of cadmium ion from wastewater by carbon-based nanosorbents: a review. J. Water Health **13**, 18–33 (2015)
3. R.S. Kislioglu, Removal of Ag, Au, and As from acid mine water using Lemna gibba and Lemna minor—a performance analysis. Water **15**, 1–10 (2023)
4. M.S. Patterer, I. Bavasso, J.E. Sambeth, F. Medici, Cadmium removal from aqueous solution by adsorption on spent coffee grounds. Chem. Eng. Trans. **60**, 157–162 (2017)
5. Z. Khan, A. Elahi, D.A. Bukhari, A. Rehman, Cadmium sources, toxicity, resistance and removal by microorganisms-a potential strategy for cadmium eradication. J. Saudi Chem. Soc. **26**, 1–18 (2022)
6. C. Chen, J.L. Wang, Removal of Pb2+, Ag+ , Cs+ and Sr2+ from aqueous solution by brewery's waste biomass. J. Hazard. Mater. **151**, 65–70 (2007)
7. S.S. Alquzweeni, R.S. Alkizwini, Removal of cadmium from contaminated water using coated chicken bones with double-layer hydroxide (Mg/Fe-LDH). Water **12**, 1–13 (2020)
8. X. Yang, Y. Wan, Y. Zheng, F. He, Z. Yu, J. Huang, H. Wang, Y.S. Ok, Y. Jiang, B. Gao, Surface functional groups of carbon-based adsorbents and their roles in the removal of heavy metals from aqueous solutions: a critical review. Chem. Eng. J. **366**, 608–621 (2019)
9. S. De Gisi, G. Lofrano, M. Grassi, M. Notarnicola, Characteristics and adsorption capacities of low-cost sorbents for wastewater treatment: a review. Sustain. Mater. Technol.., vol. **9**, 10–40 (2016)
10. L. Chen, Y. Fang, Y. Jin, Q. Chen, Y. Zhao, Y. Xiao, H. Zhao, Biosorption of Cd2+ by untreated dried powder of duckweed Lemna aequinoctialis. Desalin. Water Treat. **53**, 1–12 (2013)
11. J.L. Reyes-Ledezma, E. Cristiani-Urbina, L. Morales-Barrera, Biosorption of Co2+ ions from aqueous solution by K2HPO4-pretreated duckweed Lemna gibba. Processes **8**, 1–21 (2020)
12. B.M. Greenberg, X.-D. Huang, D.G. Dixon, Applications of the aquatic higher plant Lemna gibba for ecotoxicological assessment. J. Aquat. Ecosyst. Health **1**, 147–155 (1992)

13. P. Prakash, J.A. Kumar, B. Dhandapani, D. Vishnu, S.H. Sree, S. Madhumeena, Y. Lavanya, L. Inbathamizh, Utilization of eutrophicated Lemna minor for biosorption of acid blue dye. Biomass Convers. Biorefinery, 1–13 (2021)
14. S. Dey, S.R. Basha, G.V. Babu, T. Nagendra, Characteristic and biosorption capacities of orange peels biosorbents for removal of ammonia and nitrate from contaminated water. Cleaner Mater. **1**, 1–14 (2021)
15. F. Gourchane, F. Zaaboul, A. Bouyahya, M. Laaouan, A. El Hourch, Bioadsorption of CI Reactive Blue 203 dye by duckweed Lemna gibba's powder. Morrocan J. Chem. **11**, 674–688 (2023)
16. S.C.M. Rivas, R.I.A. Corral, M.C.F. Félix, A.R.I. Rubio, L.V. Moreno, G.R.-C. Montfort, Removal of cadmium from aqueous solutions by Saccharomyces cerevisiae–alginate system. Materials **12**, 1–14 (2019)
17. M. Dehvaria, B. Jamshidia, S. Jorfic, S. Pourfadakarid, Z. Skandarie, Cadmium removal from aqueous solution using cellulose nanofibers obtained from waste sugarcane bagasse (SCB): isotherm, kinetic, and thermodynamic studies. Desalin. Water Treat., 218–228 (2021)
18. M. Ugrina, N. Vukojević Medvidović, I. Penava, Influence of initial pH of solution on cadmium removal by natural and iron-modified zeolite, in *Proceedings of the 5th Serbian-Croatian-Slovenian Symposium on Zeolites*, (2013), pp. 76–79
19. M.H. Ehrampoush, M. Miria, M.H. Salmani, A.H. Mahv, Cadmium removal from aqueous solution by green synthesis iron oxide nanoparticles with tangerine peel extract. J. Environ. Health Sci. Eng. **13**, 1–7 (2015)
20. P. Bhattacharya, D. Mukherjee, S. Dey, S. Ghosh, S. Banerjee, Development and performance evaluation of a novel CuO/TiO2 ceramic ultrafiltration membrane for ciprofloxacin removal. Mater. Chem. Phys. **229**, 106–116 (2019)

Earthen Pot Membrane Electrodialysis (EPME) for the Removal of Reactive Red 141 (RR141) Dye from Aqueous Solution

Budhodeb Biswas and Chanchal Majumder

Abbreviations

ATR-IR	Attenuated total reflection infrared
COD	Chemical oxygen demand
ED	Electrodialysis
EPME	Earthen pot membrane electrodialysis
RR 141	Reactive red 141
UV-Vis	Ultraviolet visible

1 Introduction

In recent times, the rapid growth of industrialization has brought about unprecedented advancements across various sectors [1]; however, it has also led to the generation of vast volume of industrial effluents, laden with diverse pollutants that pose significant environmental and health challenges [2]. Among these pollutants, textile dyes have emerged as noteworthy contributors to environmental pollution due to their persistent nature, vibrant colors, and potential toxicity [3, 4]. Reactive dyes, in particular, represent a prominent category of textile dyes extensively utilized for their ability to form covalent bonds with textile fibers, resulting in vibrant and durable coloration [5]. However, the effluents discharged from textile industries, containing residues of these reactive dyes, often exhibit high color intensity and resistance to conventional treatment methods [6, 7]. Reactive red 141 (RR141), a

B. Biswas (✉) · C. Majumder
Department of Civil Engineering, Indian Institute of Engineering Science and Technology (IIEST), Shibpur, Howrah, West Bengal, India
e-mail: chanchal@civil.iiests.ac.in

D. Mazumder (ed.), *Sustainable Advanced Technologies for Environmental Management*, Springer Proceedings in Earth and Environmental Sciences,
https://doi.org/10.1007/978-3-031-64006-3_30

representative diazo reactive dye, exemplifies the challenges associated with dye-containing effluents [8]. This dye is extensively employed due to its vivid red hue and favorable colorfastness properties. Nevertheless, its presence in wastewater streams introduces substantial aesthetic, environmental, and health concerns. RR141 consists of complex molecular structures, often accompanied by reactive functional groups such as sulfonic acids, which hinder its easy degradation and removal from aqueous solutions [9]. The mitigation of dye pollution in industrial effluents has spurred extensive research efforts towards the development of efficient and sustainable treatment techniques [10–13]. Electrodialysis (ED) is an electrochemical separation process that has shown potential in addressing dye pollution by utilizing selective ion exchange membranes [14]. This technique has garnered attention due to its ability to selectively transport ions and polar compounds, thereby facilitating the separation and removal of contaminants from aqueous solutions [15–17]. Nevertheless, the implementation of conventional electrodialysis encounters challenges related to membrane fouling, limited transport rates, and high energy consumption [18]. In the pursuit of enhancing electrodialysis for dye removal, innovative approaches are being explored. One such approach involves the integration of an earthen pot membrane system into the electrodialysis process. The earthen pot, a low-cost and locally available material, has been used traditionally for water purification due to its unique porosity and ion-exchange properties [19, 20]. By incorporating the earthen pot as a membrane separator in electrodialysis, the limitations of conventional ion-exchange membranes could potentially be mitigated, leading to improved efficiency and reduced energy consumption.

This paper presents a comprehensive study on the application of earthen pot membrane electrodialysis (EPME) for the removal of reactive red 141 dye from aqueous solution. The unique advantages of the earthen pot as a membrane material, coupled with the electrochemical driving force of electrodialysis, offer a novel perspective on addressing the challenges posed by dye-contaminated wastewater. Through a systematic investigation of the operational parameters and performance indicators, this study aims to provide insights into the feasibility and efficacy of EPME as a sustainable and efficient approach for dye removal, contributing to the advancement of wastewater treatment technologies.

2 Materials and Methods

2.1 *Experimental Setup*

The batch experiments were conducted using a 2000 mL borosilicate beaker. Within the beaker, an earthen pot was placed to separate the cathode and anode chambers. This earthen pot had an inner diameter of 80 mm and a depth of 115 mm. The graphite electrodes employed in both the cathode and anode were 300 mm in length and 10 mm in diameter. The space between these electrodes was 30 mm, resulting in an effective electrode area of 17.28 square centimeters (cm^2). To apply the desired current for the treatment process, the electrodes were connected to a digital DC power supply known as the scientific 30 V, 5 A power supply PSD 3005. The Reactive red 141 dye used in the experiments was procured from Modi Dye Chem in Kolkata, West Bengal-700001. Each experimental run utilized a total sample volume of 1400 mL, distributed as 600 mL in the cathodic cell and 800 mL in the anodic cell. For a visual representation of the experimental setup, please refer to Fig. 1.

2.2 Analytical Procedure

The RR 141 dye was analyzed using a Smart UV-Vis Double Beam Spectrophotometer (Systronics, model 2203, India). A scanning process was performed on RR 141, covering concentrations spanning from 10 to 80 mg/L, to identify the wavelength at which it exhibited maximum absorption as shown in Fig. 2a. This

Fig. 1 Original experimental setup

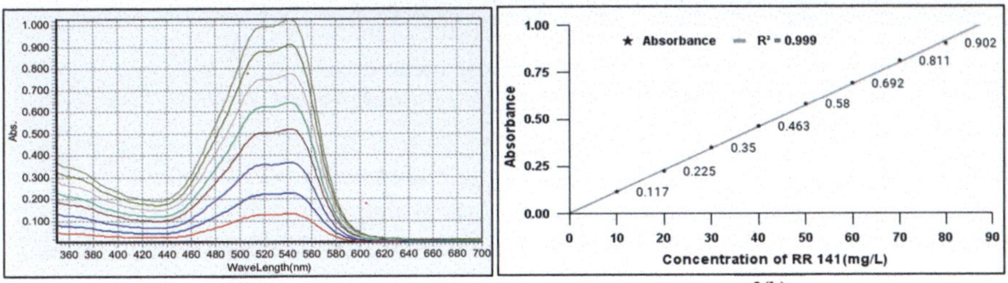

2(a) 2(b)

Fig. 2 (**a**) Spectrum scan of RR 141 for concentration ranging from 10 to 80 mg/L and (**b**) calibration curve for RR 141

maximum absorption peak was found at a wavelength of $\lambda = 543$ nm [21]. Subsequently, eight samples were prepared, each having different concentrations ranging from 10 to 80 mg/L, and a calibration curve was generated as shown in Fig. 2b. The relationship between concentration and absorbance was found to be linear up to a concentration of 80 mg/L, adhering to Beer Lambert's Law.

2.3 *Influence of pH on Degradation of RR 141*

The investigation was conducted across various pH levels (2.78, 4.83, and 8.92). Several parameters, including equilibrium pH, the percentage of color removal, and the percentage of COD removal, were examined and subsequently reported. Sampling was carried out at 10-minute intervals over a 60-minute ED operation, collecting 20–30 mL samples from both the cathode and anode chambers and placing them in glass vials. For each of these samples, measurements were taken for pH, absorbance, and COD values. Each run encompassed a total of 13 samples, comprising six from the cathode, six from the anode, and one initial raw sample. The applied current remained constant at 0.3 A throughout all experimental runs, and the distance between the cathode and anode was consistently maintained at 30 mm. The change in appearance for anode and cathode samples is shown in Fig. 3.

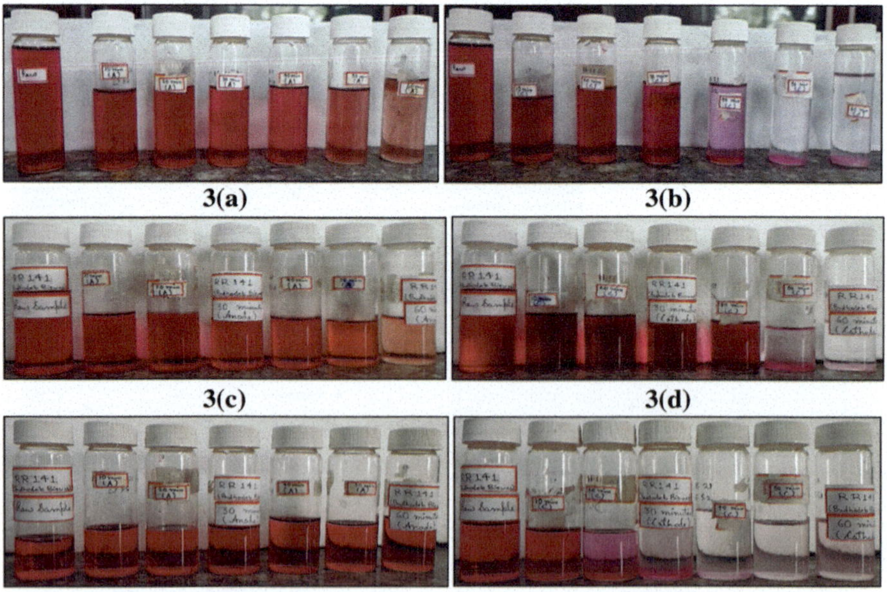

3(e) 3(f)

Fig. 3 The change in appearance of color for anolyte (left column) and catholyte (right column) for various initial pH. (**a, b**) Had initial pH 2.78. (**c, d**) Had initial pH 4.83. (**e, f**) Had initial pH 8.92

Figure 3a, b represents the anolyte and catholyte, respectively, for experimental run at pH 2.78. Similarly, Fig. 3c, d represents the anolyte and catholyte, respectively, for experimental run at pH 4.83. Again, Fig. 3e, f represents the anolyte and catholyte, respectively, for experimental run at pH 7.92. In all the figures, the raw sample is kept at extreme left for comparison.

2.4 *EPME on Real Textile Wastewater*

Wastewater originating from Kothari Processors Pvt. Ltd. in Dhulagori, Howrah, West Bengal-711313, was gathered for analysis. The initial characteristics of the untreated water included a pH of 9.8 $\pm$ 0.3, an absorbance of 0.451 $\pm$ 0.082 at 543 nm, and a concentration of RR 141 at 44 $\pm$ 8 mg/L. Electrodialysis was initially performed at a pH of 9.02, with an electrode distance of 30 mm and a current of 0.3 A, as depicted in Fig. 4a. Periodic samples were taken at 15-minute intervals to observe color changes, with noteworthy removal occurring exclusively in the anolyte, not the catholyte, as depicted in Fig. 4b. Following this, a comparable sample volume underwent electrodialysis at an initial current of 0.5 A (as illustrated in Fig. 4c), while keeping other conditions constant. Samples were collected every 15 minutes over a total duration of 90 minutes. The visible alteration in color is apparent in Fig. 4e. Subsequently, the equilibrium pH and absorbance at 543 nm were determined for the collected samples, as reported in Table 1.

3 Results and Discussion

3.1 *Influence of pH on Degradation of RR 141*

During electrodialysis carried out at pH 2.78, the data revealed a gradual decline in both absorbance and the concentration of the colored compound RR 141 over time. As the ED operation progressed, there was an observed increase in the equilibrium pH for cathode samples, while for anode samples, it decreased (as depicted in Fig. 5a). At a wavelength of 543 nm, the absorbance values reduced from 0.124 to 0.011 for the anode and from 0.157 to 0.001 for the cathode. The highest levels of color removal were attained at the 60-minute mark, reaching 94.81% removal for the anode and an impressive 99.91% removal for the cathode (as illustrated in Fig. 5b). Concurrently, the concentrations of RR 141 decreased from an initial 100–5.19 mg/L for the anode and from 100 mg/L to a mere 0.09 mg/L for the cathode. The percentage of COD reduction figures reflects how effectively organic pollutants are eliminated from the solution. However, it is crucial to note that negative values suggest an increase in COD, indicating the potential formation of byproducts or the creation of new compounds during the electrolysis process. Regarding the anode, the percentage of COD reduction ranges from −7.95% to −

Fig. 4 (**a**, **b**) Are for experimental run at 0.3 A; (**c**–**e**) Are for experimental run at 0. 5 A

46.05%, while for the cathode, it varied from −20.75% to 22.18% (as depicted in Fig. 5c). These results suggest that the electrodialysis process might face difficulties in efficiently eliminating organic pollutants and could even result in the generation of new compounds or byproducts. Further investigation and optimization efforts are necessary to enhance the efficiency of COD removal in this specific experimental setup. In a similar manner, during electrodialysis at pH 4.83, the data indicated a gradual decrease in absorbance and the concentration of the colored compound

Table 1 Data obtained for experimental run at pH 9.02 for raw waste sourced from Kothari Processors Pvt. Ltd.

Initial $I = 0.5$ A; Initial pH = 9.02; Distance between electrode = 30 mm c/c							
	Absorbance at 543 nm		Equilibrium pH			% color removal	
Time (min)	[a]A	[a]C	[a]A	[a]C	Voltage (V)	[a]A	[a]C
0	0.382	0.382	9.02	9.02	9.9	0	0
15	0.275	0.364	6.94	10.18	10.3	28.01	4.71
30	0.205	0.349	5.9	10.24	12.8	46.34	8.64
45	0.186	0.315	4.02	10.82	13.8	51.31	17.54
60	0.177	0.265	1.84	11.63	15.5	53.66	30.63
75	0.169	0.115	1.4	12.48	16.3	55.76	69.90
90	0.16	0.04	1.26	12.92	17.4	**58.12**	**89.53**

[a]Here A and C are Anode and Cathode respectively

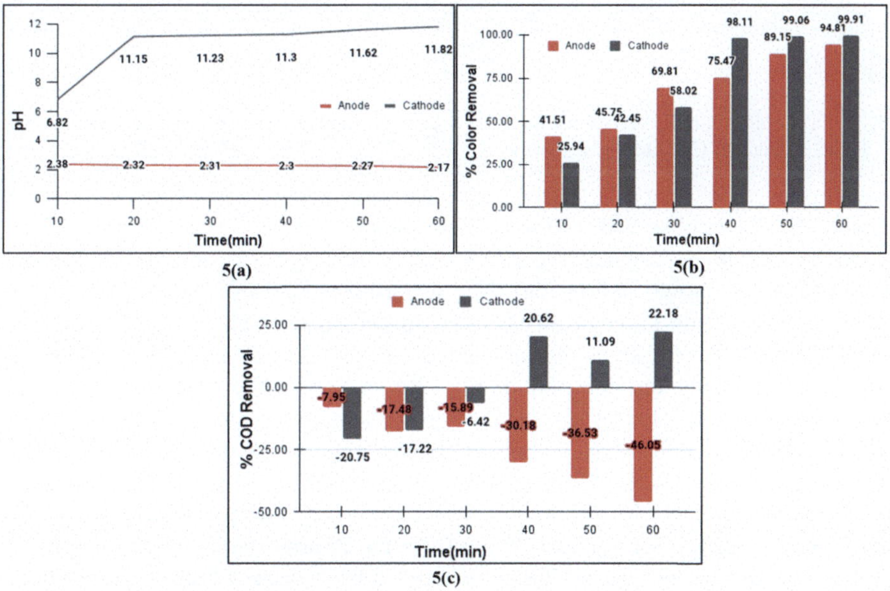

Fig. 5 (**a**) Equilibrium pH variation with time for catholyte and anolyte; (**b**) Percent color removal vs Time; (**c**) Percent COD removal vs time

(RR 141) over time. The equilibrium pH trend had remained the same as before, with an increase at the cathode and a decrease at the anode (as shown in Fig. 6a). Specifically, absorbance values at 543 nm had decreased from 0.314 to 0.153 for the anode and from 0.494 to 0.004 for the cathode. The highest percentages of color removal had been achieved at the 60-minute mark, with 38.21% removal for the anode and 99.62% removal for the cathode (as shown in Fig. 6b). The corresponding RR 141 concentrations had dropped from 100 to 14.43 mg/L for the anode and from 100 to 0.38 mg/L for the cathode. On the other hand, the maximum percent COD

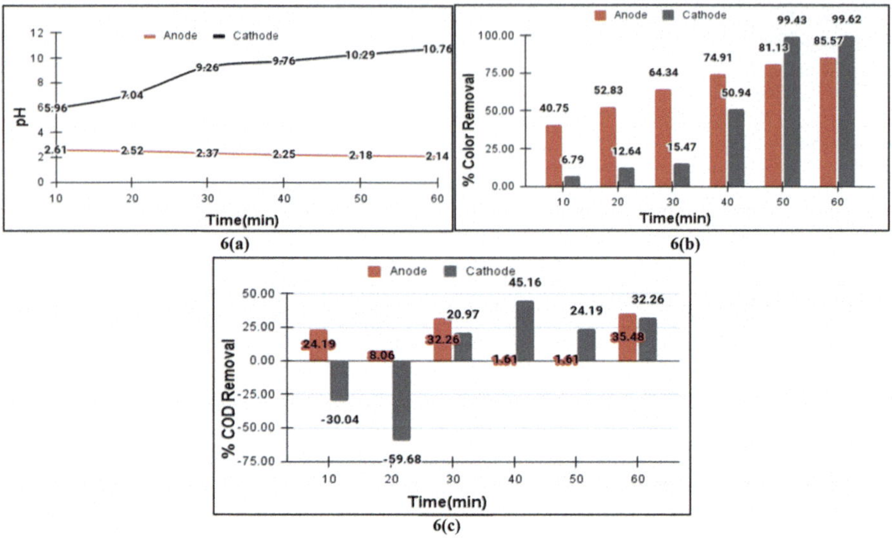

Fig. 6 (**a**) Equilibrium pH variation with time for catholyte and anolyte; (**b**) Percent color removal vs Time; (**c**) Percent COD removal vs time

removal had been observed for the cathode sample at 40 minutes, which had been approximately 45%, whereas the percent COD removal for the anode had been inconsistent (as shown in Fig. 6c). The experiment had also been conducted at pH 8.92, and all the parameters had been investigated in the same manner. The equilibrium pH had exhibited the same trend for both cathode and anode samples as observed previously (as depicted in Fig. 7a). A substantial percentage of color removal had been noted at the cathode just 30 minutes into the ED operation, reaching approximately 98%. In contrast, the maximum percentage of color removal at the anode had been 38.21%, observed after 60 minutes of ED operation (as illustrated in Fig. 7b). Following the experiment, the concentrations of RR 141 had decreased to 0.17 and 61.79 mg/L for the cathode and anode, respectively. Conversely, the percentage of COD removal had been notably high for the cathode sample, reaching approximately 97.17%. However, for the anode samples, it had decreased as the operating time increased (as shown in Fig. 7c).

3.2 *ATR-IR Spectroscopy*

To assess the functional groups within a liquid sample, we employed the attenuated total reflection (ATR) technique, following the methodology outlined [22]. For the analysis, we selected samples obtained from both anode and cathode chambers at two time points, namely, 30 and 60 minutes, from an experimental run conducted at a pH level of 8.92. These samples were subjected to ATR-IR spectroscopy at the

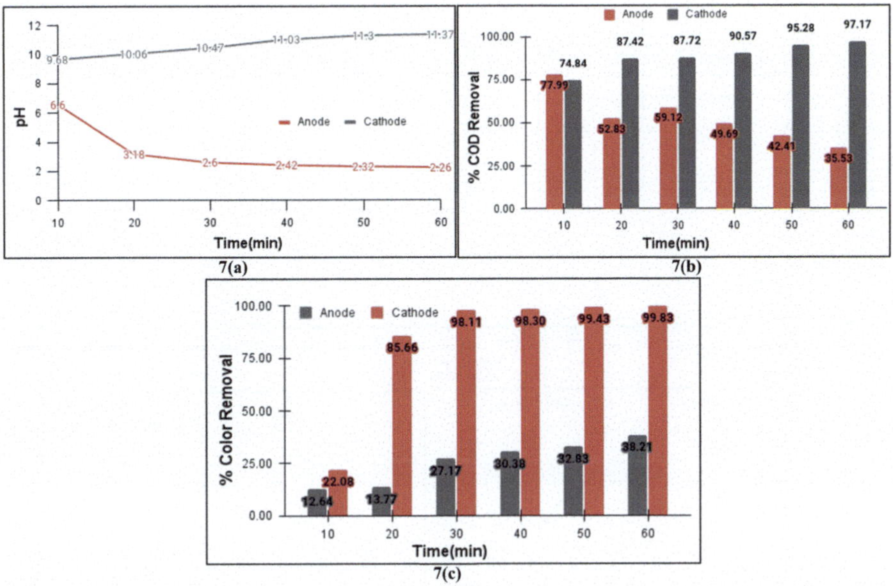

Fig. 7 (**a**) Equilibrium pH variation with time for catholyte and analyte; (**b**) Percent color removal vs Time; (**c**) Percent COD removal vs time

Indian Association for the Cultivation of Science, Kolkata-700032, alongside an untreated RR 141 sample. The PerkinElmer FT-IR C120947 instrument was used for this analysis. The ATR-IR spectrum of the untreated RR 141 sample, presented in Fig. 8a, revealed distinct peaks: one at 1234 cm^{-1} corresponding to strong C-O stretching alkyl aryl ether or medium C-N stretching amine; another at 1404 cm^{-1} corresponding to medium O-H bending carboxylic or alcohol; a peak at 1532 cm^{-1} indicative of strong N-O stretching nitro compounds; one at 1736 cm^{-1} denoting strong C=O stretching aldehyde groups; a peak at 2044 cm^{-1} associated with strong N=C=S stretching isothiocyanate; a 2880 cm^{-1} peak representing medium C-H stretching alkane; another at 3490 cm^{-1} representing medium N-H stretching alcohol primary amine; and, finally, a peak at 3652 cm^{-1} corresponding to strong, broad O-H stretching alcohol free group. Figure 8b illustrates the ATR-IR response for the anode sample collected at the 30-minute mark, exhibiting peaks within the same spectral range as the untreated RR 141 sample. Meanwhile, Fig. 8c displays the ATR-IR response for the cathode sample collected at 30 minutes, which, apart from most peaks mirroring earlier samples, notably lacks the C=O stretching aldehyde group. Instead, a peak at 1635 cm^{-1} emerges, corresponding to a medium C=C stretching alkene group, disubstituted or vinylidene, conjugate alkene, or cyclic alkene, as well as N-H bending amine. Figure 8d, e depicts the ATR-IR spectra for samples collected from the anode and cathode chambers at the 60-minute mark, respectively. These spectra exhibit the same functional groups as observed earlier. Importantly, it is worth noting that the intensity of these peaks diminishes with

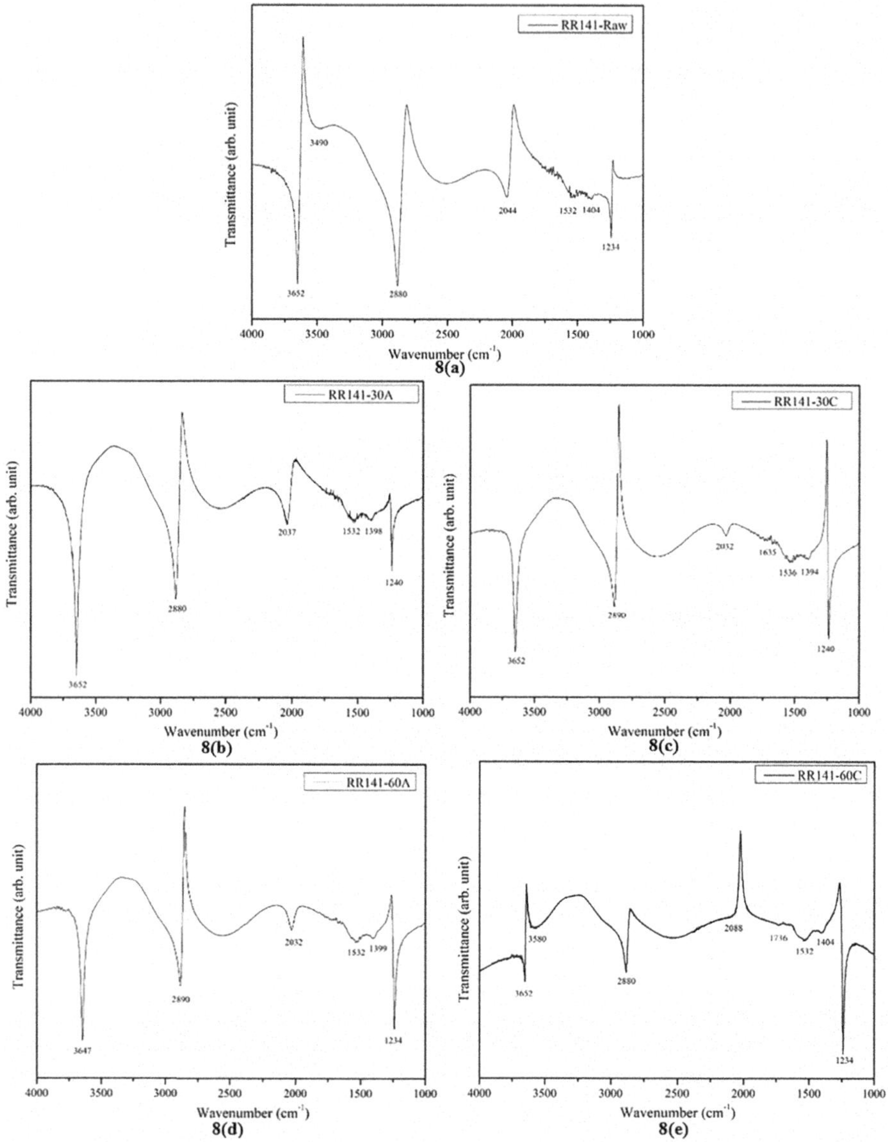

Fig. 8 ATR-IR spectroscopy for: (**a**) raw RR 141 dye sample; (**b**, **c**) 30 minutes sample from anolyte and catholyte, respectively; (**d**, **e**) 60 minutes sample from anolyte and catholyte, respectively

increasing time for the cathode samples, suggesting a reduction in the concentration of these functional groups over time.

3.3 Observations and Findings on the Degradation Study for Real Textile Wastewater

During electrodialysis at pH 9.02, the data showed a gradual decrease in absorbance and concentration of the colored compound (RR 141) over time. The equilibrium pH of cathode samples was increasing, whereas for anode samples, it was decreasing as the ED operation proceeded (as shown in Fig. 9a). The absorbance values at 543 nm decreased from 0.382 to 0.16 for the anode and from 0.382 to 0.04 for the cathode. The highest percentage of color removal was achieved at the 90-minute mark, with 58.12% removal for the anode and 89.53% removal for the cathode (shown in Fig. 9b).

4 Conclusion

In this comprehensive study of electrodialysis carried out at different pH levels (2.78, 4.83, and 8.92), we have observed intriguing trends in the removal of the colored compound RR 141 and the reduction in chemical oxygen demand (COD) over time. These results provide valuable insights into the effectiveness and challenges associated with this electrochemical process for wastewater treatment. At pH 2.78, we observed a substantial reduction in RR 141 concentration and absorbance, indicating effective color removal. Notably, the cathode outperformed the anode, achieving an impressive 99.91% color removal, whereas the anode achieved 94.81%. However, the COD reduction results presented concerns, with negative values suggesting the potential formation of byproducts during the process. Further optimization is essential to enhance COD removal efficiency. Similarly, at pH 4.83, we observed a significant decrease in RR 141 concentration and absorbance, with remarkable color removal efficiency at the cathode (99.62%). However, the anode showed inconsistent COD removal results, warranting further investigation. At pH 8.92, the cathode consistently outperformed the anode in color removal,

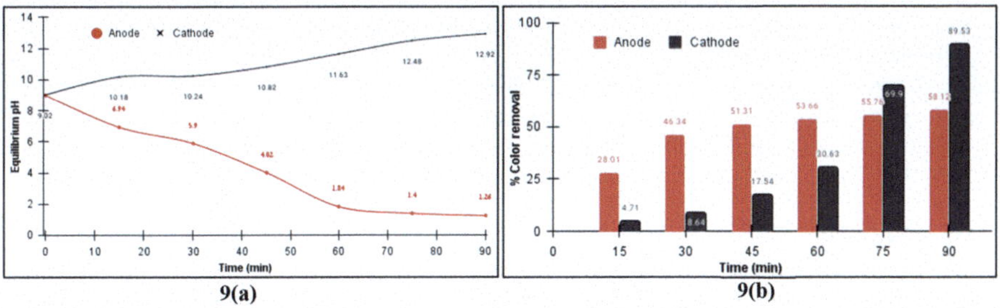

Fig. 9 (**a**) Equilibrium pH variation with time for catholyte and anolyte; (**b**) Percent color removal vs time

achieving nearly 98% removal within 30 minutes. The cathode also exhibited high COD removal (97.17%), while the anode's COD removal efficiency decreased over time.

In summary, electrodialysis demonstrates potential for effective color removal, especially at higher pH levels. However, challenges related to COD removal and potential byproduct formation require careful consideration and optimization for practical application. These findings lay the foundation for further research and process refinement in the realm of wastewater treatment and environmental remediation.

Acknowledgment The current research project is funded by the Ministry of Human Resource Development, Government of India, under the Research Fellowship Program.

References

1. M. Adnan, B. Xiao, P. Xiao, P. Zhao, S. Bibi, Heavy metal, waste, COVID-19, and rapid industrialization in this modern era—fit for sustainable future. Sustainability (Switzerland) **14** (2022). https://doi.org/10.3390/su14084746
2. P.V. Nidheesh, M. Zhou, M.A. Oturan, An overview on the removal of synthetic dyes from water by electrochemical advanced oxidation processes. Chemosphere **197**, 210–227 (2018). https://doi.org/10.1016/j.chemosphere.2017.12.195
3. M.R. Islam, M.G. Mostafa, Textile dyeing effluents and environment concerns-a review. J. Environ. Sci. Nat. Resour. **11**, 131–144 (2018)
4. F. Uddin, Environmental hazard in textile dyeing wastewater from local textile industry. Cellulose **28**, 10715–10739 (2021). https://doi.org/10.1007/s10570-021-04228-4
5. A. Khatri, M.H. Peerzada, M. Mohsin, M. White, A review on developments in dyeing cotton fabrics with reactive dyes for reducing effluent pollution. J. Clean. Prod. **87**, 50–57 (2015). https://doi.org/10.1016/j.jclepro.2014.09.017
6. S. Papić, N. Koprivanac, A.L. Božić, Removal of reactive dyes from wastewater using Fe(III) coagulant. Color. Technol. **116**, 352–358 (2000). https://doi.org/10.1111/j.1478-4408.2000.tb00013.x
7. F. Asjadi, E. Salahi, I. Mobasherpour, Removal of reactive red 141 dye from aqueous solution by titanium hydroxyapatite pellets. J. Dispers. Sci. Technol. **37**, 14–22 (2016). https://doi.org/10.1080/01932691.2015.1018423
8. H. Hafdi, M. Joudi, J. Mouldar, B. Hatimi, H. Nasrellah, M.A. El Mhammedi, M. Bakasse, Design of a new low cost natural phosphate doped by nickel oxide nanoparticles for capacitive adsorption of reactive red 141 azo dye. Environ. Res. **184**, 109322 (2020). https://doi.org/10.1016/j.envres.2020.109322
9. S. Al Arni, S. Ghareba, C. Solisio, M.S. Alves Palma, A. Converti, Methods of reactive red 141 dye decolorization, treatment, and removal from industrial wastewaters: a critical review. Environ. Eng. Sci. **38**, 577–591 (2021). https://doi.org/10.1089/ees.2020.0338
10. N.N. Rao, V. Chaturvedi, G. Li Puma, Novel pebble bed photocatalytic reactor for solar treatment of textile wastewater. Chem. Eng. J. **184**, 90–97 (2012). https://doi.org/10.1016/j.cej.2012.01.004
11. E. Yuksel, M. Eyvaz, E. Gurbulak, Electrochemical treatment of colour index reactive orange 84 and textile wastewater by using stainless steel and iron electrodes. Environ. Prog. Sustain. Energy **32**, 60–68 (2013). https://doi.org/10.1002/ep.10601

12. A. Paz, J. Carballo, M.J. Pérez, J.M. Domínguez, Biological treatment of model dyes and textile wastewaters. Chemosphere **181**, 168–177 (2017). https://doi.org/10.1016/j.chemosphere.2017.04.046
13. G.D. Saratale, S.D. Kalme, S.P. Govindwar, Decolorisation of textile dyes by *Aspergillus ochraceus* (NCIM-1146). Indian J. Biotechnol. **47**, 400 (2006)
14. S. Al-Amshawee, M.Y.B.M. Yunus, A.A.M. Azoddein, D.G. Hassell, I.H. Dakhil, H.A. Hasan, Electrodialysis desalination for water and wastewater: a review. Chem. Eng. J. **380**, 122231 (2020). https://doi.org/10.1016/j.cej.2019.122231
15. M. Paidar, V. Fateev, K. Bouzek, Membrane electrolysis—history, current status and perspective. Electrochim. Acta **209**, 737–756 (2016). https://doi.org/10.1016/j.electacta.2016.05.209
16. A. Siekierka, F. Yalcinkaya, Selective cobalt-exchange membranes for electrodialysis dedicated for cobalt recovery from lithium, cobalt and nickel solutions. Sep. Purif. Technol. **299**, 121695 (2022). https://doi.org/10.1016/j.seppur.2022.121695
17. P. Mandal, R. Mondal, P. Goel, E. Bhuvanesh, U. Chatterjee, S. Chattopadhyay, Selective recovery of carboxylic acid through PVDF blended anion exchange membranes using electrodialysis. Sep. Purif. Technol. **292**, 121069 (2022). https://doi.org/10.1016/j.seppur.2022.121069
18. L. Allat, R. Delimi, A. Mehellou, Mitigation of an anion exchange membrane fouling by coupling electrodialysis to anodic oxidation. Chem. Eng. Res. Des. **186**, 136–148 (2022). https://doi.org/10.1016/j.cherd.2022.07.042
19. M. Behera, P.S. Jana, T.T. More, M.M. Ghangrekar, Rice mill wastewater treatment in microbial fuel cells fabricated using proton exchange membrane and earthen pot at different pH. Bioelectrochemistry **79**, 228–233 (2010). https://doi.org/10.1016/j.bioelechem.2010.06.002
20. J. Winfield, I. Gajda, J. Greenman, I. Ieropoulos, A review into the use of ceramics in microbial fuel cells. Bioresour. Technol. **215**, 296–303 (2016). https://doi.org/10.1016/j.biortech.2016.03.135
21. T. Senasu, S. Nanan, Photocatalytic performance of CdS nanomaterials for photodegradation of organic azo dyes under artificial visible light and natural solar light irradiation. J. Mater. Sci. Mater. Electron. **28**, 17421–17441 (2017). https://doi.org/10.1007/s10854-017-7676-x
22. J. Fahrenfort, Attenuated total reflection: a new principle for the production of useful infra-red reflection spectra of organic compounds. Spectrochim. Acta **17**, 698–709 (1961). https://doi.org/10.1016/0371-1951(61)80136-7

Suspended Solids Removal in Combined Carbon Oxidation, Nitrification, and Denitrification of Wastewater by Moving Bed Hybrid Bioreactor: Membrane Bioreactor System

Tanmoy Bir and Debabrata Mazumder

Abbreviations

BF-MBR	Biofilm Membrane Bioreactor
CA	Cellulose Acetate
CMBR	Conventional Membrane Bioreactor
CPCB	Central Pollution Control Board
HMBR	Hybrid Membrane Bioreactor
MBHBR	Moving Bed Hybrid Bioreactor
MBR	Membrane Bioreactor
PE	Polyethylene
PP	Polypropylene
PS	Polysulfide
PTFE	Polytetrafluoroethylene
PVDF	Polyvinylidene Fluoride
UF	Ultrafiltration

T. Bir (✉) · D. Mazumder
Department of Civil Engineering, Indian Institute of Engineering and Science Technology, Shibpur, Howrah, West Bengal, India
e-mail: debabrata@civil.iiests.ac.in

D. Mazumder (ed.), *Sustainable Advanced Technologies for Environmental Management*, Springer Proceedings in Earth and Environmental Sciences,
https://doi.org/10.1007/978-3-031-64006-3_31

1 Introduction

Organic matter and nutrients like nitrogen and phosphorus are simultaneously present with considerable amount in several industrial effluents. Without adequate treatment, their discharge into waterbodies results in environmental problems like eutrophication, depleting dissolved oxygen. There are environmental regulations governing nitrogenous species in wastewater, and therefore, efficient removal of such contaminants from effluent has attracted a lot of attention. It is currently difficult to biologically extract nitrogen from wastewater with a high percentage of ammoniacal nitrogen. The maximum allowable concentration of ammoniacal nitrogen for discharge into public sewers (CPCB) is set to 50 mg/L. Numerous techniques have been developed to address nitrogen pollution, such as ion exchange, [1, 2] and electrodialysis [3–5]. However, the most commonly employed approach is biological method [6]. Nitrogen removal was also studied in denitrification along with carbon oxidation [7].

Membrane bioreactor (MBR) systems are particularly noteworthy for their ability to operate at high mixed liquor suspended solids (MLSS) levels, facilitating simultaneous nitrification and denitrification. It functions through the formation of an inner anoxic biomass zone, attributed to limited oxygen diffusion within the floc [8]. Moreover, the presence of carbon, which enhances denitrification capacity, is vital for achieving comprehensive nitrogen removal from wastewater [9]. This underscores the importance of an additional carbon source in single-stage reactor operation, minimizing the need for a larger footprint. A biofilm reactor combined with a conventional membrane bioreactor is a special configuration for the removal of organic matter along with suspended solids. A hybrid biofilm membrane bioreactor (HMBR) is a design in which both suspended and attached biomass are effective [10]. HMBR means the conventional MBR is attached with any kind of advanced process such as microfiltration, nanofiltration, or granular activated carbon-sponge fluidized bed bioreactor [11] or a nanofiltration unit [12].

In case of conventional MBR (CMBR), there are four types of materials used to make membrane: (i) ceramic, (ii) polymeric, (iii) composite, and (iv) metallic. Membranes of ceramic material are preferred because these membranes have good filtration ability for its high chemical resistivity, integrity, and also inert nature. Highly hydrophilic nature of ceramic membranes also makes them more resistant to fouling. However, they are not economically viable for use in MBR because of its high fabrication cost and brittle nature. On the other hand, membranes made of polymeric materials have good physical and chemical resistance, but they are mostly hydrophobic. Polymeric membrane materials include polyvinylidene fluoride (PVDF), polysulfide (PS), polyethylene (PE), cellulose acetate (CA), polypropylene (PP), polytetrafluoroethylene (PTFE), etc. Fouling occurs more easily in polymeric membranes because of their hydrophobic nature. Nevertheless, they are widely used nowadays due to the ease of fabrication of desirable pore sizes. Those membranes are called composite membranes which are made of two or more materials. These membranes are manufactured to achieve the advantages of all the parent materials in the produced membrane. In composite membranes, one material is used to construct the support layer while the other forms the active surface.

Composite membranes are made hydrophobic by coating with hydrophilic polymer to overcome fouling. Recent research has focused on this kind of modification to mitigate membrane fouling problems. Liu et al. [13] worked to modify a polypropylene membrane surface by using air plasma treatment, and the membrane was used to characterize antifouling properties in a submerged MBR. After study on this modified membrane, 35% higher flux recovery has been reported after cleaning with water and chemical. Another modification of polyether sulfone ultrafiltration (UF) membranes using different forms of nano-silver results to lower fouling in the membranes. Modification of membrane produces an innovative photocatalytic nanocomposite membrane which has great fouling resistance.

The aim of this study is to demonstrate the effective choice and application of a flat sheet membrane combined with HMBBR for the purpose of solid separation. The system effectively produces cleared effluent that meets the discharger limit for total solid content (100 ppm) as specified by the Central Pollution Control Board (CPCB, 1986). The final effluent quality in most cases also complied with requirement of permissible turbidity for potable purpose, which is 5 NTU (IS 10500, 2012). Moreover, it eliminates the need for a secondary clarifier and reduces large-scale unit operations into a smaller footprint.

2 Materials and Methods

2.1 Preparation of Synthetic Feed

The performance of hybrid bioreactor for combined carbon oxidation and SND process was studied using synthetic wastewater with C:N:P ratio 100:20:1 as the influent. The synthetic wastewater comprised of two different kinds of feed, i.e., heterotrophic (carbonaceous) feed containing carbon source and autotrophic (nitrogenous) feed containing nitrogen source. The composition of autotrophic and heterotrophic feed is given in Tables 1 and 2, respectively.

The two kinds of feed were given to the reactor as a combined feed for the purpose of acclimatization of biomass in the reactor. The optimal pH range was

Table 1 Composition of synthetic wastewater used as a carbonaceous feed

Constituents	Amount (g/L)
Dextrose	10
Ammonium sulfate	3.77
Potassium dihydrogen phosphate	0.175
Dipotassium hydrogen phosphate	0.2
Ferric chloride	0.02
Calcium chloride	0.2
Magnesium sulphate	0.22

Table 2 Composition of synthetic wastewater used as a nitrogenous feed

Constituents	Amount (g/L)
Ammonium sulphate	47.17
Sodium bicarbonate	17.7
Potassium dihydrogen phosphate	9.067
Ferric chloride	0.02
Calcium chloride	0.2
Magnesium sulphate	0.22

maintained within the range of (7 ± 0.1) to (8 ± 0.01) for good performance of combined carbon oxidation, nitrification, and denitrification process. It was observed sometimes that the pH dropped in the range of 5.8–6.8 at the end of batch operations, indicating occurrence of nitrification as well as carbon oxidation. Consequently, pH was maintained using 0.1 N NaOH or 0.1 N glacial acetic acid (CH_3COOH) depending on prevailing pH value.

Activated sludge floc development is affected by the physical and chemical characteristics of biomass, nutrient concentrations, and feed mix. The hybrid bioreactor consists of both suspended and attached biomass in the moving bed bioreactor (MBBR). Hybrid bioreactors (MBBR) use both suspended biomass (dispersed flocs) and attached biomass (biofilm) to remove carbon and nutrients from wastewater. The latter necessitates a substrate on the surface to grow. The main purpose of the membrane in the hybrid bioreactor is to remove suspended particles produced by autotrophic bacteria when breaking down organic matter like carbonaceous and nitrogenous compounds after eating stabilized organic matter. Biomass is cultivated on both the exterior and inner surfaces of the carrier. Bio flocs can be classified based on their sludge volume index as pin point, normal, or thickening sludge [14]. Considering the floc classification based on SVI, an approximate recommendation can be made for choosing a membrane with the suitable pore size.

2.2 Experimental Setup

Schematic diagram of setup of hybrid bioreactor with membrane module along with mechanism of combined carbon oxidation-nitrification-denitrification is shown on Fig. 1.

2.3 Selection of Membrane Material and Fabrication of Membrane Module

Membrane components are essential for both CMBR and HMBR. Combining a biofilm reactor with membrane separation of suspended solids (BF-MBR) is an alternative to the AS-MBR that may mitigate the effects of membrane contamination

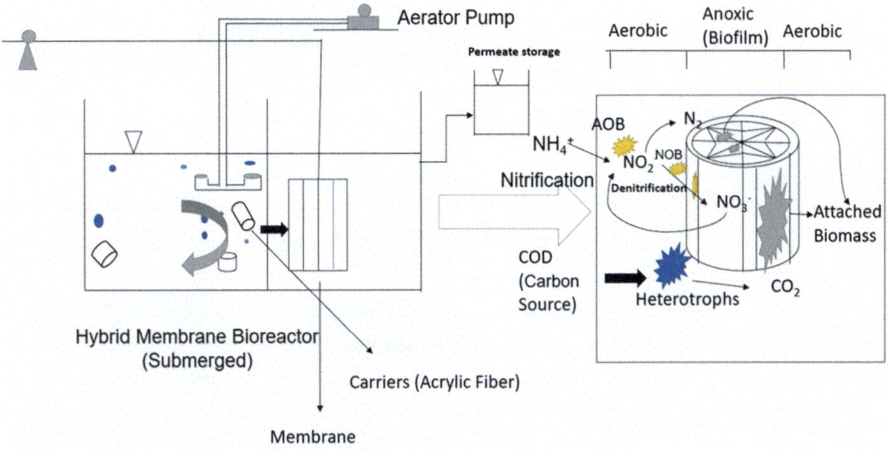

Fig. 1 Schematic diagram of hybrid membrane bioreactor

Table 3 Specification of membrane used in the present study

Type of membrane	Pore size (μ)	Material
Flat sheet	1	Polypropylene
	5	Polypropylene

caused by high biomass concentrations, according to Leikens and Ødegaard [15]. Based on available research outcomes, including that of Keckes et al. and Reis et al. [16, 17], the flat sheet membrane configuration has been chosen. A commercially available flat sheet membrane has been selected in accordance with the purpose of the present study. The benefits as listed below are the primary determinant in choosing a flat sheet membrane over a porous or multi-tube membrane [18].

1. The mechanical stability of the flat sheet membrane is commendable, as it prevents any wire breakage.
2. The cleansing cycle of the flat sheet membrane is extended and it is simpler to clean.
3. Flat sheet membranes are inexpensive to operate and have a lengthy lifespan. Ayala et al. [19] proposed that aerobic MBRs can maintain a membrane lifetime of 6–7 years without seeing a notable decrease in permeability.

Table 3 demonstrates the characteristics of membranes used in the present study. The flow chart of experimental plan is shown in Fig. 2.

After selection of membrane material, the membrane module has been fabricated. The fabricated external and internal membrane modules are shown in Figs. 3 and 4 respectively.

Membrane modules were integrated separately with the MBHBR, and the bioreactor effluent was passed through both of these modules. Total solids and turbidity were measured in order to evaluate the performance of membrane filtration.

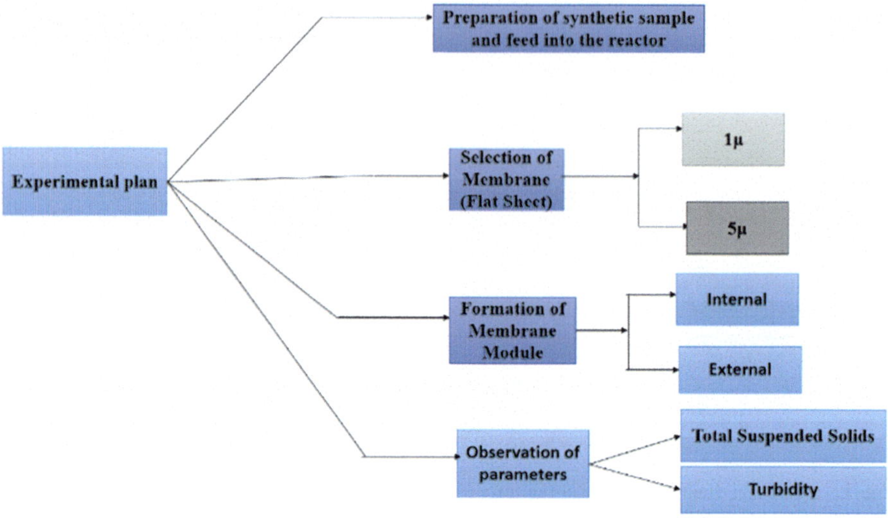

Fig. 2 Flow chart of experimental plan

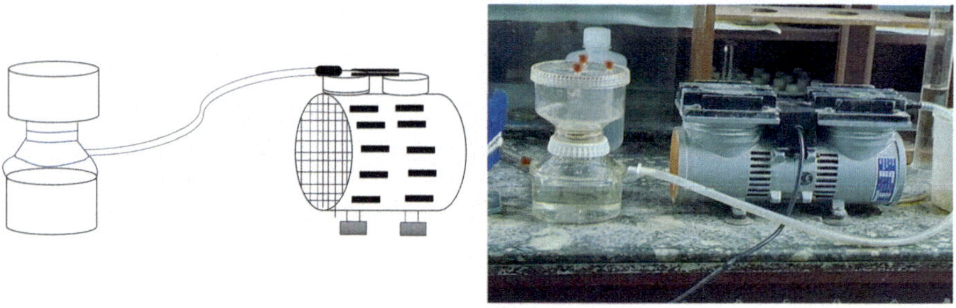

Fig. 3 External membrane module used in the present study

3 Results and Discussion

The membranes as mentioned in Table 3 undergone a number of observations on both internal and external conditions until they are subjected to concentration polarization and fouling. The primary goal of these observations was to determine whether membranes of both kinds could separate total solids and reduce turbidity, while placed after a hybrid bioreactor to produce a clear effluent. Furthermore, final filtration resistance was observed in order to facilitate the development of novel approaches to avoid fouling. Figures 5, 6, 7, 8, 9, and 10 exhibit the concentration profile of total solids varying with respect to the cumulative influent volume from the reactor for each type of membrane module comprising 1 and 5 μ flat sheet membrane.

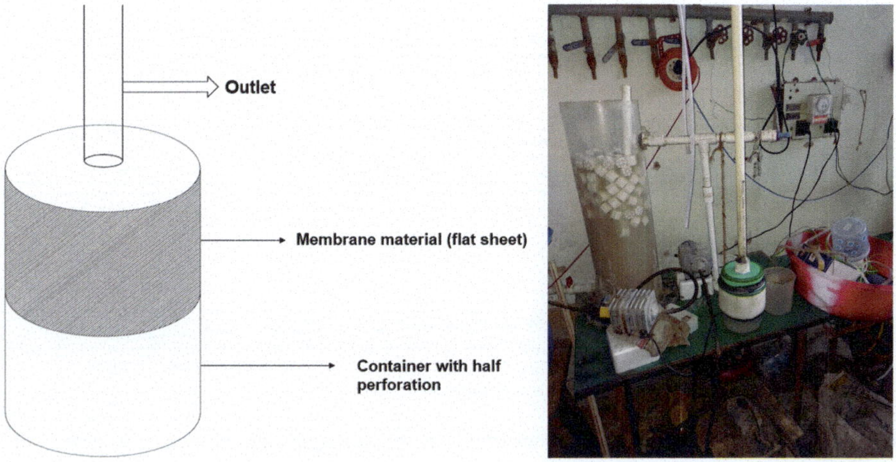

Fig. 4 Internal membrane module used in the present study

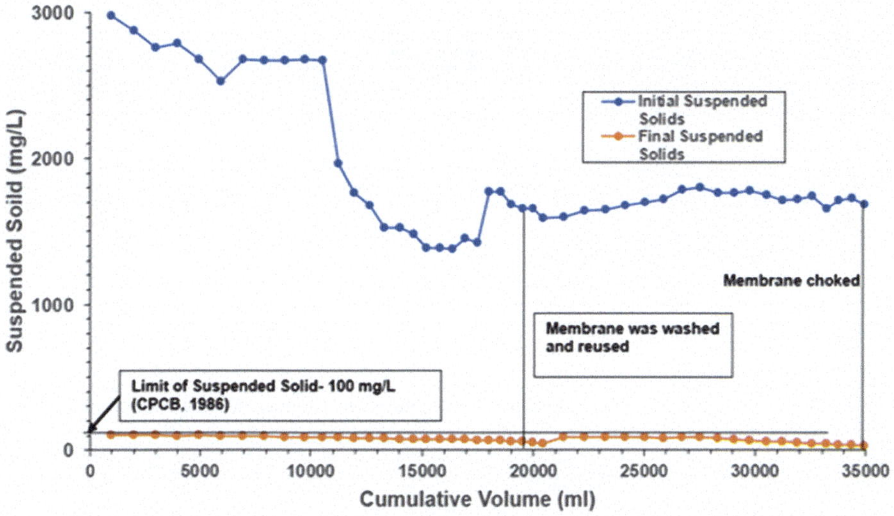

Fig. 5 Profile of initial and final concentration of total solid using internal membrane module (1 μ)

Figure 5 through Fig. 10 showed that the commercially available flat sheet membranes with pore sizes 1 and 5 μ were effective at separating the particles produced by the combined carbon oxidation nitrification and denitrification. The final concentration of total solids in the effluent from an internal module is 30 mg/L and 50 mg/L for membranes with pore sizes 1 and 5 μ, respectively. Similarly, the final turbidity is 1 and 2.8 NTU for external module with pore size 1 and 5 μ, respectively. This is to note that all the final total solids concentration and turbidity

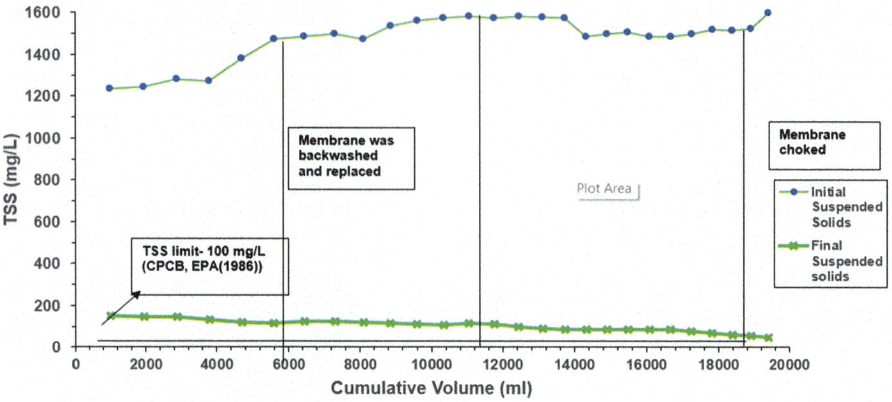

Fig. 6 Plot of initial and final concentration of total solid using internal membrane module (5 μ)

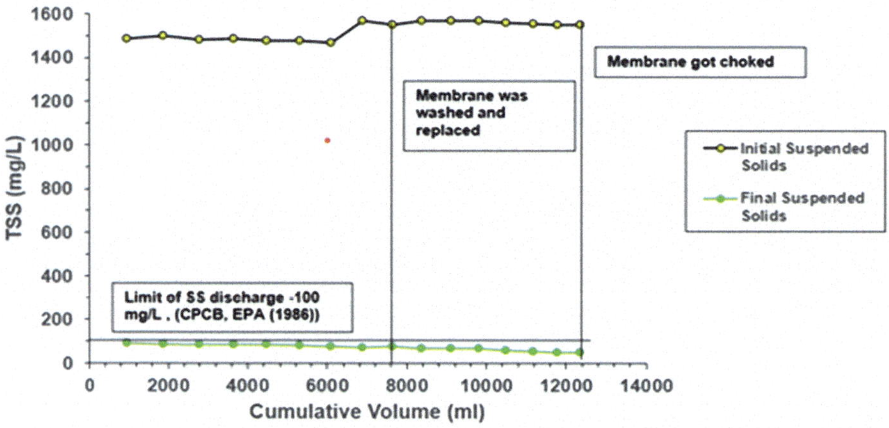

Fig. 7 Profile of initial and final concentration of total solid using combined internal membrane module with 1 and 5 μ together

are below 100 mg/L and 5 NTU, respectively, according to the standard discharge limits set by the CPCB (1986) and IS (10500-2012). Therefore, all the membranes satisfied the final effluent quality well below the desired limit ensuring the capability of solid separation.

However, as time passes, the filtration capacity of both membrane modules decreased, suggesting the development of membrane fouling. The intermediate vertical lines of each graph reflected the occurrence of concentration polarization of the membrane surface in both cases, but the filtering capacity was recovered immediately after backwashing. When filtration was stopped even after backwashing the membrane, fouling was presumed to occur. The primary goal of this study was to assess the efficiency of the membrane in making clear effluent, which was accomplished satisfactorily.

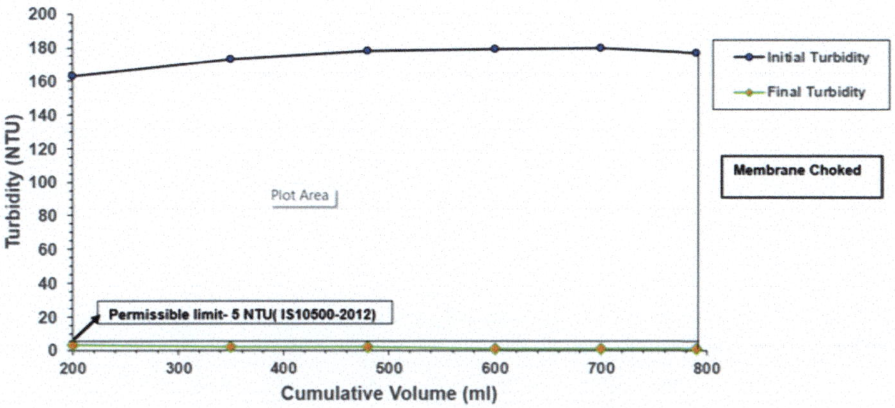

Fig. 8 Profile of initial and final concentration of turbidity using external membrane module (1 μ)

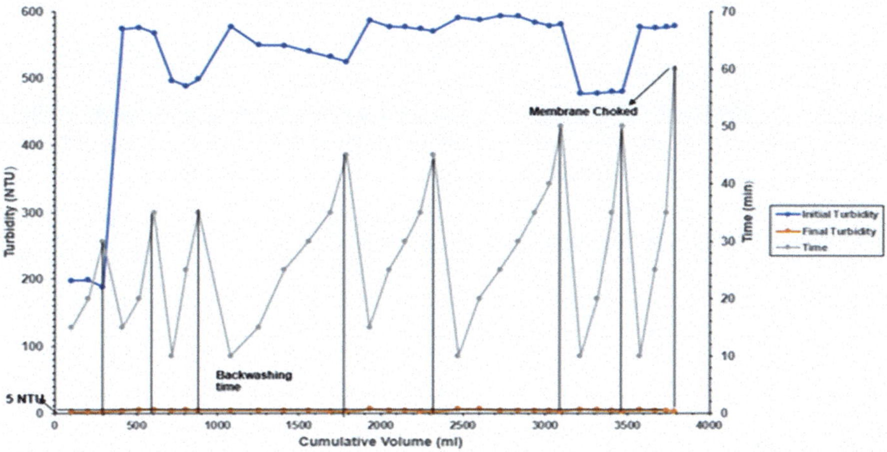

Fig. 9 Profile of initial and final concentration of total solid using combined external membrane module with 1 and 5 μ together

4 Conclusion

The integration of a membrane bioreactor (MBR) with a moving bed hybrid bioreactor (MBHBR) has proven to be a promising solution for enhancing the removal of suspended solids during combined carbon oxidation, nitrification, and denitrification in wastewater treatment. This hybrid system displayed remarkable efficiency in addressing suspended solids removal. Furthermore, the present study delved into the evaluation of commercially available membranes with different pore sizes (1 and 5 μ) and two membrane module configurations (external and internal) for suspended solids and turbidity removal. The results indicated that the membranes under consideration were efficient in suspended solids removal, but they were susceptible to

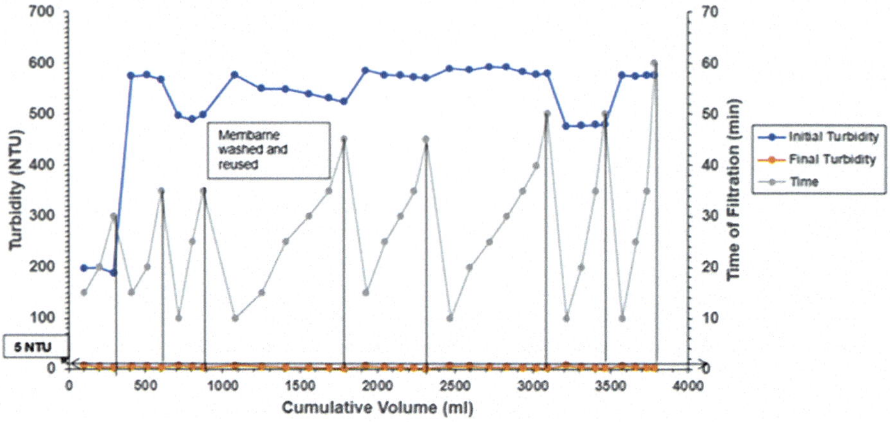

Fig. 10 Plot of initial and final concentration of total solid using external membrane module (5 μ)

concentration polarization and fouling, which could affect long-term performance. Considering the potential for future membrane advancements, it is essential to acknowledge the need for ongoing research and development in mitigating fouling and concentration polarization issues. This study also provides valuable insights into the efficiency of MBR-MBHBR systems and membrane filtration towards suspended solids removal, laying the foundation for further innovations in advanced wastewater treatment technologies.

Acknowledgments The present research work is supported under the Research fellowship program of Ministry of Education, Government of India.

References

1. X. Huang, S. Guida, B. Jefferson, A. Soares, Economic evaluation of ion-exchange processes for nutrient removal and recovery from municipal wastewater. NPJ Clean Water **3**(1), 7 (2020)
2. S. Leaković, I. Mijatović, Š. Cerjan-Stefanović, E. Hodžić, Nitrogen removal from fertilizer wastewater by ion exchange. Water Res. **34**(1), 185–190 (2000)
3. J.J. Schoeman, Nitrate-nitrogen removal with small-scale reverse osmosis, electrodialysis and ion-exchange units in rural areas. Water SA **35**(5), 721–728 (2009)
4. D. Vineyard, A. Hicks, K.G. Karthikeyan, P. Barak, Economic analysis of electrodialysis, denitrification, and anammox for nitrogen removal in municipal wastewater treatment. J. Clean. Prod. **262**, 121145 (2020)
5. C. Zhang et al., Recovery of ammonia nitrogen from landfill leachate using a biopolar membrane equipped electrodialysis system. Water Sci. Technol. **82**(9), 1758–1770 (2020)
6. Z. An, Q. Feng, R. Zhao, X. Wang, Bioelectrochemical methane production from food waste in anaerobic digestion using a carbon-modified copper foam electrode. Processes **8**(4), 416 (2020)
7. P.B. Dhamole, S.F. D'Souza, S.S. Lele, A review on alternative carbon sources for biological treatment of nitrate waste. J. Inst. Eng. (India): E. **96**, 63–73 (2015)

8. M. Sarioglu, G. Insel, N. Artan, D. Orhon, Model evaluation of simultaneous nitrification and denitrification in a membrane bioreactor operated without an anoxic reactor. J. Membr. Sci. **337**(1–2), 17–27 (2009)
9. L. Li, G. Yan, H. Wang, Z. Chu, Z. Li, Y. Ling, T. Wu, Denitrification and microbial community in MBBR using A. donax as carbon source and biofilm carriers for reverse osmosis concentrate treatment. J. Environ. Sci. **84**, 133–143 (2019)
10. I. Ivanovic, T.O. Leiknes, The biofilm membrane bioreactor (BF-MBR)—a review. Desalin. Water Treat. **37**(1–3), 288–295 (2012)
11. T.T. Nguyen, H.H. Ngo, W. Guo, Pilot scale study on a new membrane bioreactor hybrid system in municipal wastewater treatment. Bioresour. Technol. **141**, 8–12 (2013). https://doi.org/10.1016/j.biortech.2013.03.125
12. K. Choet, J. Cho, H.K. Shon, Fouling characteristics of a membrane bioreactor and nanofiltration hybrid system for municipal wastewater reclamation. Bioresour. Technol. **130**, 239–247 (2013). https://doi.org/10.1016/j.biortech.2012.12.007
13. L. Li, Y. Song, H. Liu, F. Qu, Recent developments in membrane materials and membrane fouling control for membrane bioreactors. RSC Adv. **6**(97), 95147–95172 (2016)
14. H.D. Park, I.S. Chang, K.J. Lee, *Principles of Membrane Bioreactors for Wastewater Treatment* (CRC Press, Boca Raton, 2015). https://doi.org/10.1201/b18368
15. T. Leiknes, H. Ødegaard, The development of a biofilm membrane bioreactor. Desalination **202**(1–3), 135–143 (2007)
16. S. Keskes, F. Hmaied, H. Gannoun, H. Bouallagui, J.J. Godon, M. Hamdi, Performance of a submerged membrane bioreactor for the aerobic treatment of abattoir wastewater. Bioresour. Technol. **103**(1), 28–34 (2012). https://doi.org/10.1016/j.biortech.2011.09.063
17. B.G. Reis, A.L. Silveira, Y.A.R. Lebron, V.R. Moreira, L.P.T. Teixeira, A.A. Okuma, M.C.S. Amaral, L.C. Lange, Comprehensive investigation of landfill leachate treatment by integrated Fenton/microfiltration and aerobic membrane bioreactor with nanofiltration. Process Saf. Environ. Prot. **143**, 121–128 (2020). https://doi.org/10.1016/j.psep.2020.06.037
18. I. Bodík, A. Blšt'Áková, L. Dančová, S. Sedlìček, Comparison of flat-sheet and hollow-fiber membrane modules in municipal wastewater treatment. Pol. J. Environ. Stud. **18**(3), 331–340 (2009)
19. D.F. Ayala, V. Ferre, S. Judd, Membrane life estimation in full-scale immersed membrane bioreactors. Energy Fuels **378**, 95–100 (2011). https://doi.org/10.1016/j.memsci.2011.03.013

Part V
Solid and Hazardous Waste Management

Utilization of Spent Mushroom Substrate for Biochar Production and Its Impact on Arsenic Accumulation in Rice Grain

Deepanjan Mridha and Tarit Roychowdhury

Abbreviations

As	Arsenic
AsIII	Arsenite
AsV	Arsenate
CK	Control
DAT	Days After Transplanting
ETC	Electron Transport Chain
FAO	Food and Agricultural Organization
GMB	Ganga-Meghna-Brahmaputra
iAs	Inorganic Arsenic
SDGs	Sustainable Development Goals
SMS	Spent Mushroom Substrate
SMSB	Spent Mushroom Substrate Biochar
TBARS	Thiobarbituric Acid Reactive Substances
WHO	World Health Organization

1 Introduction

Arsenic (As) is a hazardous element whose exposure can cause significant harm to living beings. As pollution in groundwater and soil offers a higher risk to mankind. The presence of increased levels of As in groundwater has a significant impact on a large portion of India's Ganga-Meghna-Brahmaputra (GMB) flood plain,

D. Mridha · T. Roychowdhury (✉)
School of Environmental Studies, Jadavpur University, Kolkata, West Bengal, India

D. Mazumder (ed.), *Sustainable Advanced Technologies for Environmental Management*, Springer Proceedings in Earth and Environmental Sciences,
https://doi.org/10.1007/978-3-031-64006-3_32

particularly in India and Bangladesh [1]. Rice is a key crop and the main supply of sustenance for over half of the global population. A considerable volume of rice is produced in polluted water and soil in the GMB plain of India [2]. Arsenic in the root soil might penetrate the root and translocate into the rice grain, eventually contaminating the food chain [3]. Therefore, the grain As is a challenge that requires an appropriate remedy. Arsenic is mostly found in extremely hazardous inorganic (arsenite: AsIII and arsenate: AsV) and less lethal organic (DMA and MMA) forms in nature [4]. In anaerobic condition, AsIII is the predominant species of As, and in aerobic condition, AsV is mainly present. These harmful inorganic As (iAs) species accumulate in rice crops and impair development and productivity [5]. The concentration of total As in grain also differs from cultivar to cultivar. However, people inhabiting in As-contaminated parts of the globe still consume a significant quantity of As via rice and rice-based products. Apart from water, grain also contributes a significant amount of As exposure to humans [6]. According to findings from a previous investigation, consuming As-enriched rice may result in an increase in As burden in the human body, even after drinking As-safe water [1]. The World Health Organization (WHO) and Food and Agricultural Organization (FAO) propose limiting the amount of As (iAs) in rice grains to 350 μg/kg for rough rice, 300 μg/kg for brown rice, and 250 μg/kg for polished rice. Countries are encouraged to reduce the amount of As in rice below this level [7, 8].

Until recently, experts all through worldwide have proposed several measures to reduce As availability, translocation, and finally accumulation in grain of rice. To lower rice grain As toxic effects, different remedial agents and methods were used in both indoor and field trials [3, 9, 10]. A carbon-rich byproduct of slowly burning biomass in an oxygen-restricted atmosphere is called biochar. Biochar's high C-content and environmental stability make it suitable for long-term sequestration of soil carbon [11]. Biochar is a good amendment since it has the capacity to immobilize toxic elements in soil while also improving soil health and plant development [12]. As a consequence, it has gained special interest as a potential remediating agent for As-contaminated soil owing to its varied functional groups, significant porosity, and unique surface area [11].

Biochar is a type of biological substance with high aromatization, stability, large surface area, rich pore, and great adsorption capability. Furthermore, biochar has a high capacity for ion exchange, which supports the growth and survival of microorganisms [11]. Applying biochar to the soil exhibited that the amount of organic matter in the soil had improved and that the organic carbon in the soil was breaking down more slowly. Another well-known property of biochar is its capacity to lower pesticides and toxic metal(loid)s, which has negative effects on crops, soil, and human health.

The residue leftover from cultivating mushrooms, also known as spent mushroom substrate (SMS), is frequently utilized as an agent in composting due to its uniform particle size, excellent degree of breathability, and water permeability. SMS features a lot of active binding sites that facilitate the adsorption of metal ions, high porosity, and a comparatively large surface area [13]. Moreover, the SMS might be utilized as a raw material to make biochar, ensuring that agricultural leftovers are recycled and repurposed. The biochar produced from the SMS has a more surface area and porosity than traditional biochar. It may be possible to improve the phytoremediation capabilities of polluted soils by using the SMS and biochar produced from SMS as amendment agents. SMS biochar (SMSB) is proven to be a good soil conditioner and potential agent for remediation of hazardous metal(loid)s. However, no such study has been focused on the effect of biochar As stress alleviation from rice plants. Rice grain contains a range of minerals, which are essential for human metabolic processes. Nevertheless, the exact role of biochar on nutrient accumulation in rice grain was unknown. Hence, importance should be given to the remediation of As from plot/field level to consumer plate. However, limited studies have been reported on the alleviation of As stress in rice plants using spent mushroom substrate biochar. Therefore, the present research sought to investigate the (i) impact of biochars [SMSB] on paddy growth and yield under As stress, (ii) effect of SMSB on As uptake and translocation in paddy plant, and (iii) effect of SMSB on accumulation of micronutrients in the harvested brown rice grains.

2 Materials and Methods

2.1 Synthesis of SMSB

SMSB was made from spent mushroom substrate (SMS). Initially, mushroom (*Pleurotus djamor*) was cultivated using rice straw as a substrate for 50 days, and thereafter, SMS was collected and air-dried for 10 days. SMS was chopped into small pieces and pyrolyzed in a furnace at a temperature of 500 °C for 1 h to produce SMS biochar (SMSB). The produced SMSB was sieved and stored for subsequent characterization and soil application.

2.2 Experimental Soil

The soil was randomly collected from 0 to 20 cm of As-contaminated rice field in Madhusudan Kathi village, West Bengal, India. The collected soil sample had been oven-dried (60 °C) before the soil get pulverized. For the pot experiments, the soil was passed through a sieve (<2 mm), and several soil physicochemical properties were assessed (Table 1).

2.3 Pot Experiment

This study used the rice cultivar Satabdi (IET 4786) for pot experiment. The rice seeds were disinfected with H_2O_2 solution (30% v/v) for 15 min and subsequent

Table 1 Soil physiochemical characteristics

EC (mS/cm)	1.84
pH	5.67
SOC (g/kg)	15.7
Available-N (mg/kg)	46.3
Available-P (mg/kg)	7.30
Available-K (mg/kg)	54.5
Total-As (mg/kg)	17.1
Available As (mg/kg)	1.10
Soil texture (%)	Clay: 55.2
	Silt: 21.3
	Sand: 23.5

washed with distilled water [3]. Rice seeds were seeded in a plastic seedbed in advance and kept there until transplanting. In this study, 10 inch bottom-sealed plastic containers were used to avoid the loss of nutrients and As. A total of 6 kg of dried soil was placed into each pot. A total of five experimental sets were used in this experiment and, each experimental set was triplicated. The details of the experimental sets consist of 0 (CK), 0.5% SMSB (w/w), and 1% SMSB. Prior to transplanting stage, basal doses of N (CH_4N_2O), P (Na_2HPO_4), and K (KCl) were amended as per the required dosage (N:P:K = 2:1:2). The required irrigation to individual pot was given through tap water (As: <3 μg/L) 2 days before the transplanting, and anaerobic condition was maintained till ripening stage. At the time of harvest, plant and soil samples were used to measure yields and for further assessment.

2.4 Chlorophyll, Lipid Peroxidation, H_2O_2, and Proline

The plant chlorophyll was extracted at 20 DAT. Briefly, 0.5 g of rice plant leaf tissue was extracted with acetone, and the OD of the acetone extract was measured at 663 and 645 nm [14]. Lipid peroxidation (TBARS: thiobarbituric acid reactive substances), H_2O_2, and proline osmolyte contents in leaf tissue were analyzed as per the protocol of Heath and Packer [15] and Velikova et al. [16], respectively.

2.5 Root Iron Plaque Extraction

In order to isolate Fe plaque from fresh root samples, a dithionite-citrate-bicarbonate (DCB) solution was used [19]. Roots of rice plant were initially washed to remove the dirt from the roots, and then the roots were incubated in 50 mL of DCB extraction solution (0.03 M $Na_3C_6H_5O_7$, 0.125 M $NaHCO_3$, and 0.8 g $Na_2S_2O_4$) at 25 °C for 1 h. After extraction, the rice roots were rinsed properly with distilled water and the final volume of the extracted solution was made up to 50 mL.

2.6 Plant Yield Measurement

Plant height, filled grain per panicle, and grain yield were measured after harvest of the rice plant (110 DAT) from each pot.

2.7 Analysis of Soil and Plant Samples

The extraction of different elements from soil (0.1 g) and plant (0.2 g) (Mn, Zn, Se, Cu, Fe and As) were done with HNO_3 and H_2O_2 at 2:1 ratio. The analysis of total As and Fe in soil and plant samples was measured utilizing atomic absorption spectroscopy (AAS) [17]. The concentration of Mn, Zn, Se, and Cu was determined by ICP-OES. The concentration of Si was determined with XRF instrument.

2.8 Quality Control

Periodic table mix 1 (ICP multielement standard solution IV, 1.11355, Sigma Aldrich) for ICP-OES was used as a standard for the quantitative determination of metal(loids) or micronutrients composition in biochar, soil, and plant using an ICP-OES. Standard reference materials (SRM), blank measurements, testing duplicates, and spiked samples were checked to ensure quality control of the generated analytical results. To assess the recovery of metal(loids) in soil and samples, rice flour 1568a (NIST, Gaithersburg, MD, USA), tomato leaf 1573a (NIST, Gaithersburg, MD, USA), and river sediment 1645 (NBS, Washington, DC) were used as reference materials. Metal(loids) recoveries in soil and plant samples varied from 89.3% to 101.6%.

3 Results and Discussion

3.1 Effect of SMSB Application on Plant Growth and Yield

The observations from the pot experiment showed that 0.5% and 1% SMSB treatments improved the paddy plant growth and yield compared to CK (Fig. 1a–c). The plant height remains the same in all treatments, which showed As stress does not pose threat to rice growth or rice plants have its own adaptability towards As stress (Fig. 1a). The number of filled grain was high in 0.5% and 1% SMSB treatment than CK. The 1% SMSB treatment has most filled grain counts (133 ± 5.51 filled grains/panicle) among all the treatment groups (Fig. 1b). SMSB has been shown to boost plant output by promoting growth and protecting plants

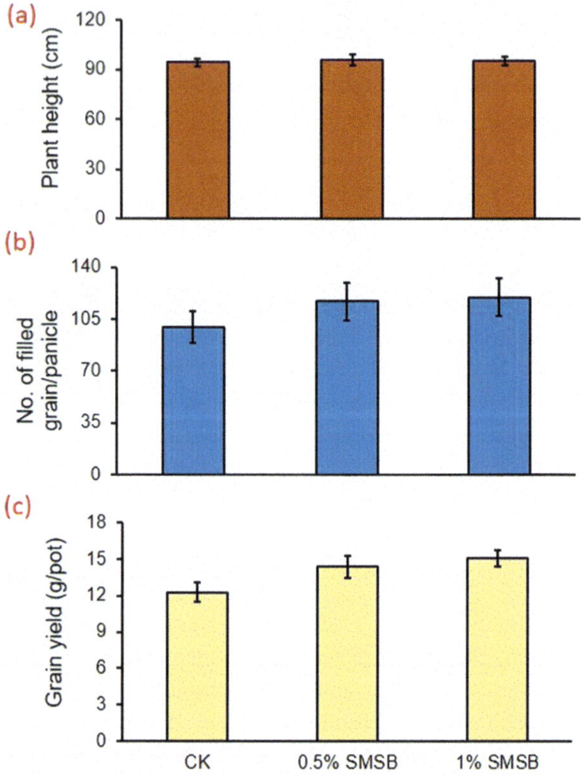

Fig. 1 Effect of SMSB treatment on (**a**) plant growth, (**b**) number of panicle, and (**c**) grain yield of rice

from As by enhancing soil physical qualities like increasing soil pH, CEC, and strengthening nutrient retention capacity [18]. Arsenic toxicity lowered the grain yield in CK. The rice grain yield in CK was decreased by 14.6–18.5% than SMSB treatments (Fig. 1c). The rice grain quantity was observed higher in 1% SMSB treatment by 4.6% than 0.5% SMSB treatment. Lin et al. [19] found that biochar improved the rice plant growth and grain output by enhancing soil CEC activities and reducing toxic As levels in rice. However, a proper dose selection for biochar is necessary prior to application since differing soil types and conditions may make it inappropriate in some situations.

3.2 *Effect of Biochars on Chlorophyll, TBARS, H_2O_2, and Proline Content in Plant*

Oxidative stress caused by metal(loid) stress reduces photosynthesis rate in plants. The overall amount of chlorophyll in paddy plants of CK was substantially reduced when compared to plants that were treated with SMSB (Fig. 2a). The 1% SMSB

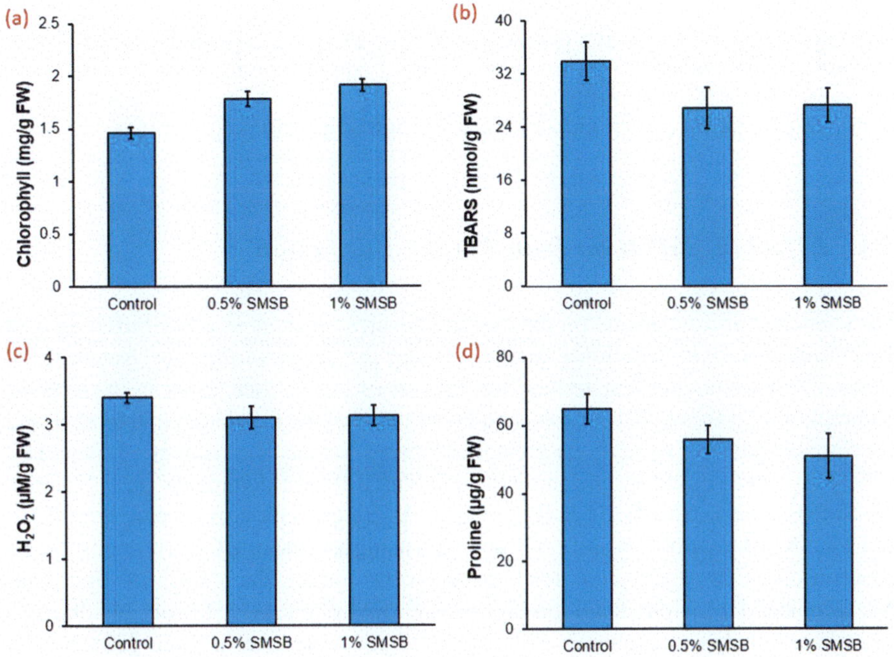

Fig. 2 Impact of biochar on (**a**) chlorophyll, (**b**) lipid peroxidation (TBARS), (**c**) H_2O_2, and (**d**) proline content of rice plant

treatment had the most chlorophyll concentration of all treatments. Compared to CK, the chlorophyll content of 0.5% and 1% SMSB was improved by 18% and 23.6%, respectively. A previous study found that biochar might increase photosynthetic system function under stressful conditions by enhancing chlorophyll synthesis, electron transport chain (ETC) activities, and PSI and PSII activity [20]. SMSB treatment improves photosynthesis in rice plants, leading to increased growth and production. Lipid peroxidation and peroxide are indications of oxidative stress in plants. Lipid peroxidation due to oxidative stress causes breakdown of cell membrane and electrolyte leakage.

Plants are harmed by stress caused by the production of reactive oxygen species, resulting into the destruction of cell membranes and exerting pressure on cellular processes [3, 5]. A high level of As accumulation in plants stimulates the synthesis of TBARS and H2O2 in rice plants. TBARS (lipid peroxidation) and H_2O_2 content were substantially high in CK as compared to SMSB (Fig. 2b, c). TBARS and H_2O_2 content in 0.5% and 1% SMSB were not significantly different. The TBARS and H_2O_2 content in plant were significantly increased after 0.5% and 1% SMSB treatments. The TBARS content in 0.5% and 1% SMSB was lowered by 21% and 19.7% than CK. However, in case of H_2O_2, 0.5% and 1% SMSB treatment reduced its level by 9% and 8%, respectively, as compared to CK. Proline is an important substance that helps to prevent membrane damage induced by oxidative stress caused by As. The elevated oxidative stress in CK increased the proline level in the rice plant (Fig. 2d). The difference in proline level between 0.5% and 1% SMSB was insignificant. However, the proline content in 1% SMSB was lowered by 22% when compared with CK. Our results corroborate earlier research showing that TBARS, H_2O_2, and proline levels in plants were reduced by biochar treatment [21].

3.3 Effect of SMSB on Iron Plaque Formation

The existence of iron-plaque on the rice roots' surface serves as a key buffer against the entrance of toxic metalloids such as As into the cortex of the root. Retention of met(loid) in Fe-plaque can reduce the uptake of As by rice roots as well as its accumulation in shoot [12]. The As and Fe contents of the Fe-plaque in the 0.5% and 1% SMSB additions were much greater compared to CK. The content of As and Fe in Fe-plaque was raised by enriching the soil with different treatments are as follows: 1% SMSB >0.5% SMSB >CK (Fig. 3a, b). As concentration in Fe-plaque of 0.5% and 1% SMSB treatment was 19.5% and 22.4% higher than the CK, respectively. Fe concentrations at 0.5% and 1% SMSB were higher than CK, in line with As. Increased Fe-plaque development on rice roots as a result of using biochar and amendments helps to sequester As and decreases its uptake in plant [22]. SMSB treatment increases the quantity of amorphous Fe-oxides in root soil, which stimulates Fe-plaque synthesis and leads to accumulation of As in Fe-plaques. Greater Fe-plaque growth on the outermost layer of the rice root caused by Fe^{2+} oxidation limited the buildup of As in rice tissues, especially the grains [23]. The immobilization of As by SMSB might potentially be attributed to the production of amorphous Fe^{3+}-arsenate complexes [24].

3.4 Arsenic Availability and As Concentration in Different Parts of Rice Plant

The administration of 0.5% and 1% SMSB did not reduce bioavailable fraction of As in the soil than the CK (Fig. 4a). The findings imply that biochar treatment had no effect on the As availability in soil. The SMS is basically rice straw which is enriched in silica (Si); thus, our synthesized SMSB is naturally enriched with Si. Similarly, Yang et al. [25] observed that treatment with of Si-enriched biochar did not affect the bioavailability of As.

In soil contaminated with As, SMSB additions had a noticeable influence on total As deposition in rice plants ($p \leq 0.05$). Both 0.5% and 1% SMSB treatment in As-contaminated soil was shown to diminish As levels in shoot of rice plants (Fig. 4b–d). Based on the findings, As accumulation pattern in 0.5% and 1% SMSB treatments was almost similar. The As content of aboveground biomass was found to be more effectively reduced by both SMSB treatments than CK.

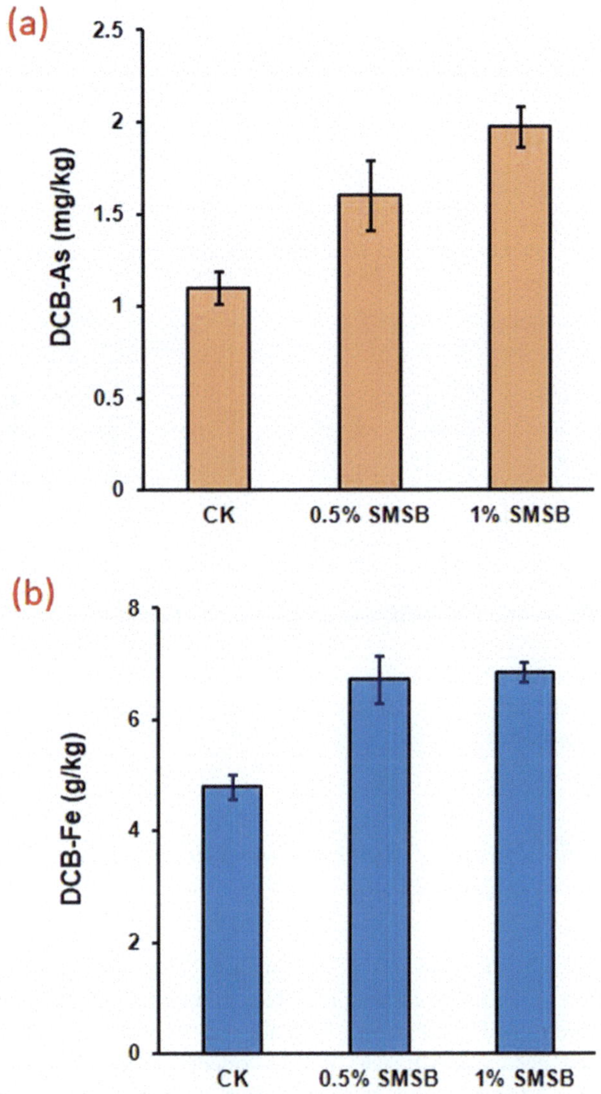

Fig. 3 Effect of SMSB on (**a**) Fe and (**b**) As concentration of root Fe-plaque

On the other hand, the 0.5% and 1% SMSB decrease rates were roughly comparable. Application of 1% SMSB decreased the As content in roots, leaf, and grains by 18%, 18.4%, 19%, and 27%, respectively, as compared to CK. In comparison to the CK plants, these data demonstrate a significant improvement in plant growth and health. The large amount of intrinsic Si present in SMSB may be the reason for rice plant's decreased As deposition. Biochar enhanced with Si lowered As accumulation in rice aboveground biomass, as reported by Herath et al. [26]. OsLsi1 and OsLsi2, which are the primary Si transporters in rice, also strictly regulate AsIII intake. High levels of Si in the rhizospheric zone cause rice to downregulate OsLsi1 and OsLsi2

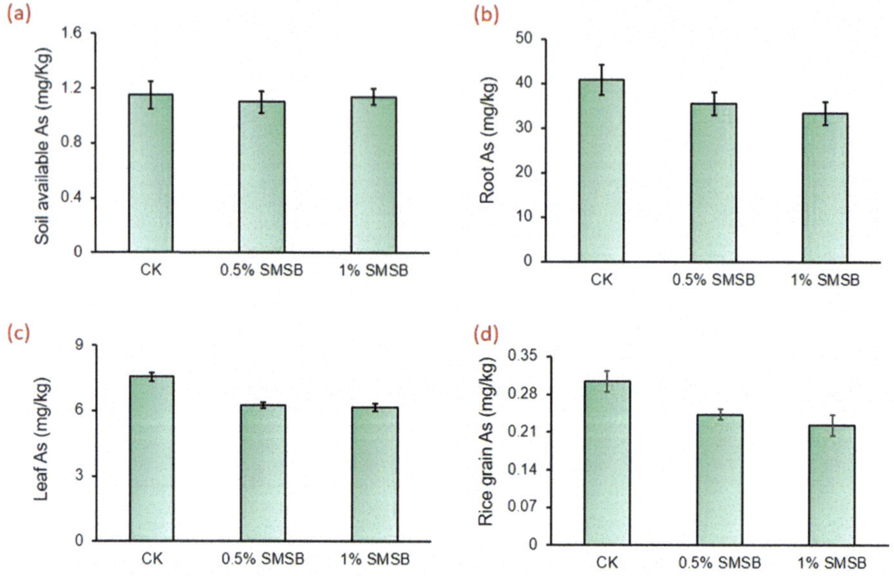

Fig. 4 Effect of SMSB on (**a**) available As, (**b**) root As, (**c**) leaf As, and (**d**) rice grain As

proteins. Additionally, when Si transporter is scarce, competition between Si and AsIII intensifies, restricting AsIII's capacity to enter rice roots through Si transporter [27].

3.5 *Effect of Biochars on Micronutrient Concentrations in Rice Grain*

In both emerging and affluent nations, controlling soil micronutrient insufficiency is one of the major issues [18]. Rice grain contents of Zn, Fe, Cu, Se, Si, and Mn were determined (Table 2). Overall, 0.5% and 1% SMSB treatments significantly enhanced Zn, Mn, Cu, and Si concentrations compared to CK. In comparison to CK, Fe and Se concentrations were not improved in rice grain obtained from 0.5% and 1% SMSB treatment. Similar findings were also noted in a previous study, which revealed that applying biochars increases the concentrations of micronutrients in rice and wheat grains [28]. Biochar binds micronutrients more effectively and slowly releases them into the rhizosphere [29]. A previous study found that rice and wheat plants with biochar-treated soil had less accumulation of Cd and Cu [30].

Other research found that applying biochar increased the concentrations of Zn and Cu in grains and vegetables, which contradicted previous findings [31, 32]. In countries where the majority of the population consumes rice and meals made from rice, foodstuff that has been supplemented with vital micronutrients can assist to enhance health [33].

Table 2 Different micronutrient content in rice grain

Treatments	Cu	Fe	Mn	Zn	Se	Si
CK	5.9	38.0	9.1	11.9	0.126	2760
0.5% SMSB	7.4	40.0	14.1	16.3	0.128	4130
1% SMSB	7.5	39.6	17.7	15.1	0.119	5570

4 Conclusion

The current investigation revealed that SMSB was unable to reduce plant available As; however, still it manages to reduce As uptake and accumulation in rice plants. Evidence suggests that using SMSB in the process of producing As-safe rice grain has benefits of its own. By reducing oxidative stress brought on by As in the plant, the increased Fe-plaque development in the aftermath of SMSB treatment encouraged As retention in Fe-plaque, inhibiting As access into roots and promoting growth. The content of micronutrients in rice grains was also enhanced through the application of SMSB. The effects of SMSB on As absorption in rice in field circumstances require more research, despite the encouraging results of this pot study. Long-term SMSB performance, its impact on soil microbiota, and other environmental regimes should all be carefully considered.

Acknowledgments Deepanjan Mridha, first author of this paper University Grant Commission (UGC), Government of India for providing fellowship, ICP-OES analyst at Calcutta University Ms. Upasana Kanji and field supervisor Mr. Pijush Sarkar.

References

1. N.R. Chowdhury, A. Das, M. Joardar, A. De, D. Mridha, R. Das, M.M. Rahman, T. Roychowdhury, Flow of arsenic between rice grain and water: Its interaction, accumulation and distribution in different fractions of cooked rice. Sci. Total Environ. **731**, 138937 (2020). https://doi.org/10.1016/j.scitotenv.2020.138937
2. D. Mridha, P.C. Gorain, M. Joardar, A. Das, S. Majumder, A. De, N.R. Chowdhury, U. Lama, R. Pal, T. Roychowdhury, Rice grain arsenic and nutritional content during post harvesting to cooking: A review on arsenic bioavailability and bioaccessibility in humans. Food Res. Int. **154**, 111042 (2022). https://doi.org/10.1016/j.foodres.2022.111042
3. D. Mridha, I. Ray, J. Sarkar, A. De, M. Joardar, A. Das, N.R. Chowdhury, K. Acharya, T. Roychowdhury, Effect of sulfate application on inhibition of arsenic bioaccumulation in rice (*Oryza sativa* L.) with consequent health risk assessment of cooked rice arsenic on human: A pot to plate study. Environ. Pollut. **293**, 118561 (2022). https://doi.org/10.1016/j.envpol.2021.118561
4. D. Mridha, I. Paul, A. De, I. Ray, A. Das, M. Joardar, N.R. Chowdhury, P.B.S. Bhadoria, T. Roychowdhury, Rice seed (IR64) priming with potassium humate for improvement of seed germination, seedling growth and antioxidant defense system under arsenic stress. Ecotoxicol. Environ. Saf. **219**, 112313 (2021). https://doi.org/10.1016/J.ECOENV.2021.112313
5. I. Ray, D. Mridha, J. Sarkar, M. Joardar, A. Das, N.R. Chowdhury, A. De, K. Acharya, T. Roychowdhury, Application of potassium humate to reduce arsenic bioavailability and

toxicity in rice plants (*Oryza sativa* L.) during its course of germination and seedling growth. Environ. Pollut. **313**, 120066 (2022). https://doi.org/10.1016/j.envpol.2022.120066
6. M. Joardar, A. Das, N.R. Chowdhury, D. Mridha, A. De, K.K. Majumdar, T. Roychowdhury, Health effect and risk assessment of the populations exposed to different arsenic levels in drinking water and foodstuffs from four villages in arsenic endemic Gaighata block West Bengal India. Environ. Geochem. Health **43**, 3027–3053 (2021). https://doi.org/10.1007/s10653-021-00823-3
7. FAO, *Codex Alimentarius Commission. 37th Session* (FAO, Geneva, 2014) https://www.fao.org/news/story/en/item/238558/icode/
8. FAO, *Thirty-Ninth Session Rome, Italy* (2016). http://www.fao.org/fao-who-codexalimentarius/roster/detail/en/c/421755/
9. P. Kumarathilaka, J. Bundschuh, S. Seneweera, A. Marchuk, Y.S. Ok, Iron modification to silicon-rich biochar and alternative water management to decrease arsenic accumulation in rice (*Oryza sativa* L.). Environ. Pollut. **286**, 117661 (2021). https://doi.org/10.1016/J.ENVPOL.2021.117661
10. I. Ray, D. Mridha, T. Roychowdhury, Waste derived amendments and their efficacy in mitigation of arsenic contamination in soil and soil–plant systems: A review. Environ. Technol. Innov. **24**, 101976 (2021). https://doi.org/10.1016/J.ETI.2021.101976
11. E. Wen, X. Yang, H. Chen, S.M. Shaheen, B. Sarkar, S. Xu, H. Song, Y. Liang, J. Rinklebe, D. Hou, Y. Li, Iron-modified biochar and water management regime-induced changes in plant growth, enzyme activities, and phytoavailability of arsenic, cadmium and lead in a paddy soil. J. Hazard. Mater. **407**, 124344 (2021). https://doi.org/10.1016/j.jhazmat.2020.124344
12. M.S. Islam, A.S.I.A. Magid, Y. Chen, L. Weng, J. Ma, M.Y. Arafat, Z.H. Khan, Y. Li, Effect of calcium and iron-enriched biochar on arsenic and cadmium accumulation from soil to rice paddy tissues. Sci. Total Environ. **785**, 147163 (2021). https://doi.org/10.1016/j.scitotenv.2021.147163
13. Z. Jia, J. Deng, N. Chen, W. Shi, X. Tang, H. Xu, Bioremediation of cadmium dichlorophen co-contaminated soil by spent *Lentinus edodes* substrate and its effects on microbial activity and biochemical properties of soil. J. Soils Sediments **17**, 315–325 (2017)
14. D.I. Arnon, Copper enzymes in isolated chloroplasts. Polyphenol oxidase in Beta vulgaris. Plant Physiol. **24**, 1–15 (1949). https://doi.org/10.1104/pp.24.1.1
15. R.L. Heath, L. Packer, Photoperoxidation in isolated chloroplasts. I. Kinetics and stoichiometry of fatty acid peroxidation. Arch. Biochem. Biophys. **125**, 189–198 (1968). https://doi.org/10.1016/0003-9861(68)90654-1
16. V. Velikova, I. Yordanov, A. Edreva, Oxidative stress and some antioxidant systems in acid rain-treated bean plants protective role of exogenous polyamines. Plant Sci. **151**, 59–66 (2000). https://doi.org/10.1016/S0168-9452(99)00197-1
17. A. Das, M. Joardar, A. De, D. Mridha, N.R. Chowdhury, M.T.B.K. Khan, P. Chakrabartty, T. Roychowdhury, Pollution index and health risk assessment of arsenic through different groundwater sources and its load on soil-paddy-rice system in a part of Murshidabad district of West Bengal, India. Groundw. Sustain. Dev. **15**, 100652 (2021). https://doi.org/10.1016/j.gsd.2021.100652
18. W. Hartley, P. Riby, J. Waterson, Effects of three different biochars on aggregate stability, organic carbon mobility and micronutrient bioavailability. J. Environ. Manag. **181**, 770–778 (2016). https://doi.org/10.1016/j.jenvman.2016.07.023
19. L. Lin, M. Gao, W. Qiu, D. Wang, Q. Huang, Z. Song, Reduced arsenic accumulation in indica rice (*Oryza sativa* L.) cultivar with ferromanganese oxide impregnated biochar composites amendments. Environ. Pollut. **231**, 479–486 (2017). https://doi.org/10.1016/j.envpol.2017.08.001
20. S. Lyu, G. Du, Z. Liu, L. Zhao, D. Lyu, Effects of biochar on photosystem function and activities of protective enzymes in *Pyrus ussuriensis* Maxim. under drought stress. Acta Physiol. Plant. **38**, 220 (2016). https://doi.org/10.1007/s11738-016-2236-1

21. A. Majumdar, M.K. Upadhyay, B. Giri, J. Karwadiya, S. Bose, M.K. Jaiswal, Iron oxide doped rice biochar reduces soil-plant arsenic stress, improves nutrient values: An amendment towards sustainable development goals. Chemosphere **312**(1), 137117 (2023). https://doi.org/10.1016/j.chemosphere.2022.137117
22. A. Sebastian, M.N.V. Prasad, Trace element management in rice. Agronomy **5**, 374–404 (2015). https://doi.org/10.3390/agronomy5030374
23. N.R. Chowdhury, R. Da, M. Joarda, S. Ghosh, S. Bhowmick, T. Roychowdhur, Arsenic accumulation in paddy plants at different phases of pre-monsoon cultivation. Chemosphere **210**, 987–997 (2018). https://doi.org/10.1016/j.chemosphere.2018.07.041
24. A. Mensah, B. Marschner, S.M. Shaheen, J. Wang, S. Wang, J. Rinklebe, Arsenic contamination in abandoned and active gold mine spoils in Ghana: Geochemical fractionation, speciation, and assessment of the potential human health risk. Environ. Pollut. **261**, 114–116 (2020). https://doi.org/10.1016/j.envpol.2020.114116
25. X. Yang, E. Wen, C. Ge, A. El-Naggar, H. Yu, S. Wang, E.E. Kwon, H. Song, S.M. Shaheen, H. Wang, J. Rinklebe, Iron-modified phosphorus-and silicon-based biochars exhibited various influences on arsenic, cadmium, and lead accumulation in rice and enzyme activities in a paddy soil. J. Hazard. Mater. **443**, 130203 (2023)
26. I. Herath, F.J. Zhao, J. Bundschuh, P. Wang, J. Wang, Y.S. Ok, K.N. Palansooriya, M. Vithanage, Microbe mediated immobilization of arsenic in the rice rhizosphere after incorporation of silica impregnated biochar composites. J. Hazard. Mater. **398**, 123096 (2020). https://doi.org/10.1016/j.jhazmat.2020.123096
27. J.F. Ma, N. Yamaji, N. Mitani, X.Y. Xu, Y.H. Su, S.P. McGrath, F.J. Zhao, Transporters of arsenite in rice and their role in arsenic accumulation in rice grain. Proc. Natl. Acad. Sci. **105**, 9931–9935 (2008). https://doi.org/10.1073/PNAS.0802361105
28. P.L. Korai, T.A. Sial, G. Pan, H. Abdelrahman, A. Sikdar, F. Kumbhar, S.A. Channa, E.F. Ali, J. Zhang, J. Rinklebe, S.M. Shaheen, Wheat and maize-derived water-washed and unwashed biochar improved the nutrients phytoavailability and the grain and straw yield of rice and wheat: A field trial for sustainable management of paddy soils. J. Environ. Manag. **297**, 113250 (2021). https://doi.org/10.1016/j.jenvman.2021.113250
29. S.S. Dhaliwal, R.K. Naresh, A. Mandal, R. Singh, M.K. Dhaliwal, Dynamics and transformations of micronutrients in agricultural soils as influenced by organic matter build-up: A review. Environ. Sustain. Indic. **1–2**, 100007 (2019). https://doi.org/10.1016/j.indic.2019.100007
30. L. Cang, J. Xing, C. Liu, Y. Wang, D. Zhou, Effects of different water management strategies on the stability of cadmium and copper immobilization by biochar in rice-wheat rotation system. Ecotoxicol. Environ. Saf. **2002**, 110887 (2020). https://doi.org/10.1016/j.ecoenv.2020.110887
31. J. Gartler, B. Robinson, K. Burton, L. Clucas, Carbonaceous soil amendments to biofortify crop plants with zinc. Sci. Total Environ. **465**, 308–313 (2013). https://doi.org/10.1016/j.scitotenv.2012.10.027
32. E. Moreno-Jimenez, J.M. Fernandez, M. Puschenreiter, P.N. Williams, C. Plaza, Availability and transfer to grain of As, Cd, Cu, Ni, Pb and Zn in a barley agri-system: Impact of biochar, organic and mineral fertilizers. Agric. Ecosyst. Environ. **219**, 171–178 (2016). https://doi.org/10.1016/j.agee.2015.12.001
33. A. Khare, S. Samudre, A. Arora, Sneak-peek into iron deficiency anemia in India: the need for food-based interventions and enhancing iron bioavailability. Food Res. Int. **162**(Part A), 111927 (2022). https://doi.org/10.1016/j.foodres.2022.111927

Implications of Green Solvents on Delignification and Saccharification of Neem Oil Seed Cake for Sustainable Biorefinery Applications

Sundaram Deepika Bharathi and Samuel Jacob

Abbreviations

AOAC	Association of Official Analytical Chemists
ChCl	Choline Chloride
CuCl	Copper Chloride
DES	Deep Eutectic Solvent
DNS	3,5-Dinitrosalicylic Acid
Glyc	Glyceline
HBA	Hydrogen Bond Acceptor
HBD	Hydrogen Bond Donor
ILs	Ionic Liquids
LCB	Lignocellulosic Biomass
NOC	Neem Oilseed Cake
OVAT	One Variable at a Time
Reln	Reline
S/L Ratio	Solid-to-Liquid Ratio

S. D. Bharathi · S. Jacob (✉)
Department of Biotechnology, School of Bioengineering, College of Engineering and Technology, Faculty of Engineering and Technology, SRM Institute of Science and Technology, Kattankulathur, Tamil Nadu, India
e-mail: ds7505@srmist.edu.in; samueljb@srmist.edu.in

D. Mazumder (ed.), *Sustainable Advanced Technologies for Environmental Management*, Springer Proceedings in Earth and Environmental Sciences,
https://doi.org/10.1007/978-3-031-64006-3_33

1 Introduction

Lignocellulosic biomass (LCB) offers a more promising future in producing alternative fuels and manufacturing value-added chemicals [1]. The demand for sustainable components in second-generation biorefineries can be fulfilled by modern and environmentally benign technologies based on bioprocess mechanisms [2]. The valorization of pentose (C5) and hexose (C6) sugars from LCB has gained popularity in recent decades to develop commercially relevant green chemicals [3]. The oil is extracted from the neem kernels and the leftover solid is known as neem oilseed cake (NOC). Despite the fact that the NOC is proteinaceous, it has a bitter taste and contains triterpenoids such as azadirachtin, nimbin, and salanin that are considered to be harmful ingredients including in cattle feed is unsafe compared to other oilseed cakes [4]. NOC comprises 24–50% carbohydrates and has not been investigated for valorization. As a result, NOC was chosen as a LCB for this study as a valuable resource for value-added chemical synthesis. Over the last few decades, several pretreatment processes have been developed to minimize biomass recalcitrance and fractionate the holocellulose component [2, 5]. The cellulose-hemicellulose-lignin matrix is structurally fractionated into three different streams by this processing [6]. Organosolv pretreatment can be done either with or without a catalyst (salt, acid, or base) with a range of organic or aqueous-organic solvents. This combination hydrolyses lignin-carbohydrate bonds and lignin linkages partially, thus the obtained solid residue majorly consisting of cellulose and some hemicellulose [7].

This study aims to further investigate the capability of NOC to separate high-quality lignin, hemicellulose, and cellulose via organosolv pretreatment under optimized conditions for enhanced saccharification using deep eutectic solvent (DES). DES is a modern technological breakthrough used to pretreat and hydrolyze lignocellulosic materials [8]. DES is comprised of a hydrogen bond donor (HBD) and hydrogen bond acceptor (HBA) that has a melting point lesser than its constituents [9]. Compared to conventional treatments and saccharification, DES has some advantages including its biodegradability, ease of synthesis, mild treatment conditions, non-flammability, and environment friendliness [10]. This study has been conducted by synthesizing HBA (choline chloride) and HBD (oxalic acid) to hydrolyze the NOC. Therefore, NOC is proven to be a suitable biomass for the recovery of sugars through green solvents that have not been reported elsewhere according to the authors' knowledge.

2 Materials and Methods

2.1 Collection and Processing of NOC

NOC was collected from the local neem oil extraction unit in Tamil Nadu, India. The NOC was powdered by pulverizing the solid lumps, and 0.2 mm sized particles were obtained after sieving. According to the Association of Official Analytical Chemists (AOAC) method, NOC was subjected to proximate analysis (total volatile solids, moisture, and ash content) [11]. The biochemical analysis including hemicellulose, cellulose, lignin, crude fiber, protein, and total carbohydrate [12–17] was determined. NOC was stored in an airtight container for further treatment.

2.2 Organosolv Pretreatment Process

The traditional organosolv method was carried out by the procedure described by Constant et al. [18]. 140 mL of aqueous ethanol (50%, v/v) with 3% (w/v) catalyst (CuCl) were added to 10.0 g of dried pulverized NOC in a round bottom flask with a solid-to-liquid (S/L) ratio of 1:14 (w/v). In a reflux flask, the mixture was stirred for 90 min at 60 °C at a constant speed of 250 rpm. Subsequently, vacuum filtration was used for the separation of solid and black liquor (liquid). The treated NOC (solid residue) was washed three times with aqueous ethanol of 10 ml (50, % v/v) at 60 °C, and the washes were mixed with the black liquor before oven drying at 70 °C. Water was added at three times the volume of the resulting black liquor in a separating funnel and left overnight to precipitate the lignin. The solid residues containing cellulose were further subjected to saccharification using DES.

2.3 Recovery of Lignin and Solvent

The precipitate defined as a lignin fraction was collected and dried at 45 °C until a consistent weight was attained. Following the treatment, the black liquor and solid residue washes were combined and evaporated using a rotary vacuum to separate the solvent. The recovered solvent (ethanol) can be reused for further organosolv pretreatment.

2.4 Estimation of Lignin by Iodometric Method

In this study, the iodometric method was followed to estimate the lignin content of NOC [13]. To 50 mg of NOC, 63 mL distilled water was added. To the mixture, 7.5 mL of 0.1 N potassium permanganate and 4 N sulfuric acid were added and incubated for 10 min in the dark condition. Following the incubation, 1.5 mL of 1 N potassium iodide and 1 mL of 0.2% starch were added, and the mixture was titrated against 0.1 N sodium thiosulphate solution until it turned to a colorless solution.

Lignin content was determined using Eq. (1)

$$\text{Lignin content }(\%) = (\text{Kappa number} \times 0.155) \times 100 \quad (1)$$

Therefore, 0.155 is the slope obtained from the linear plot between klason lignin and kappa number.

Delignification (%) was calculated using Eq. (2).

$$\text{Delignification}\%(\text{w/w}) = \\ ((\text{Lignin}\%\text{of raw NOC} - \text{lignin}\%\text{of treated NOC})/\text{lignin}\%\text{of raw NOC}) \times 100 \quad (2)$$

2.5 Selection of range for selected parameters of Organosolv Pretreatment Through OVAT Approach

Three significant process parameters viz., solid-to-liquid (biomass:organosolv) ratio (S:L), temperature, and process time were selected. The boundary conditions followed for one variable at a time (OVAT) are S/L ratio from 1:8 to 1:16, temperature from 55 °C to 80 °C, and time from 30 to 150 min [19]. Three organic solvents were chosen (acetone, ethanol, and diethylene glycol) based on their efficiency for pretreatment. After solvent screening, NOC was treated with 50% (v/v) of the selected organic solvent with 3% (w/v) of Lewis acid catalyst (CuCl). Further, the treated biomass was employed for saccharification.

2.6 DES Synthesis

The DES was synthesized by the combination of mixing HBA and HBD. In this study, choline chloride (ChCl) acts as HBA and oxalic acid acts as HBD prepared at 1:2 molar ratio leading to the formation of Glyc (glyceline) and Reln (reline), respectively. The prepared mixture was treated at 60 °C with a constant stirring at 200 rpm until the mixture became a homogenous transparent liquid. The freshly synthesized DES was subjected to cooling and placed in a vacuum desiccator at 28 ± 2 °C before use in saccharification [20].

2.7 DES-Based Saccharification and Optimization of Its Process Parameters

The DES saccharification process described by Li et al. [21] was carried out. In a round bottom flask, freshly synthesized DES (10 mL) was mixed with organosolv-treated solid residue (1.0 g) at a S/L ratio of 1:10 (w/v). The saccharification was performed for 6 h at 85 °C. Following saccharification, the mixture was cooled and 15 mL of acetone/water (1:1, v/v) mixture was added, followed by vacuum filtering of the saccharified hydrolysate. The saccharified solid residue was washed three times with 15 mL of acetone/water after filtration, and the washes were mixed with the hydrolysate. The DES saccharified liquid hydrolysate was subjected to reducing sugar (C6 sugar) estimation using the dinitro salicylic (DNS) method [22]. Through the OVAT approach, the range of variables was selected from the chosen parameters for DES saccharification. Different reaction time (2–6 h), ratios of NOC:DES (ranging from 1:5 to 1:25, g/g), and temperatures (70–90 °C) were selected for the saccharification process [23].

The saccharification of delignified NOC biomass was calculated using Eq. (3) [24].

$$\text{Saccharification}\ (\%) = [(\text{Reducing sugar} \times 0.9)/\text{Cellulose in substrate}] \times 100 \quad (3)$$

3 Results and Discussion

3.1 Proximate Analysis of NOC

The components of NOC were determined using proximate analysis as shown in Table 1. It contains volatile solids of 84.99 ± 5.18%, w/w, representing the presence of biologically accessible compounds indicating the valorizable organic content of the biomass. Nearly similar volatile solids (73.20%) composition has been reported by Mulimani and Navindgi [25] in NOC. Therefore, high volatile matter and less ash content (15.01 ± 5.18%, w/w) of NOC could be suitable substrate for value-added chemicals production.

3.2 Biochemical Analysis of NOC

The biochemical examination of NOC (Table 2) revealed a total carbohydrate content of approximately 54.50 ± 0.10%, w/w, of which hemicellulose and cellulose were found to be 18.90 ± 0.11% (w/w) and 24.90 ± 0.08% (w/w), respectively.

Table 1 Proximate analysis of NOC (neem oilseed cake) on wet[a] and dry[b] weight basis; Avg. ± S. D, $n = 3$ (No. of trials)

Characterization of NOC	Composition (%, w/w)
Ash content[b]	15.01 ± 5.18
Moisture content[a]	93.15 ± 1.45
Total volatile solid content[b]	84.99 ± 5.18
Total solids[a]	6.85 ± 1.45

Table 2 Biochemical analysis of NOC (neem oilseed cake) on dry weight basis; Avg. ± S. D, $n = 3$ (No. of trials)

Characterization of NOC	Composition (%, w/w)
Lignin	6.20 ± 0.90
Hemicellulose	18.90 ± 0.11
Cellulose	24.90 ± 0.08
Protein	3.03 ± 0.20
Total carbohydrate	54.50 ± 0.10
Crude fiber	12.62 ± 0.12

Thus, it can be considered as a suitable substrate for producing value-added chemicals through fermentation. The previously reported range aligns with this biochemical composition [26]. Similarly, the recalcitrant lignin (6.20 ± 0.90%, w/w) of NOC limits the holocellulose content, requiring pretreatment combining hydrolysis for lignin breakdown (delignification) and conversion of holocellulose components into fermentable sugars through hydrolysis. Despite the fact that NOC contains 3.03 ± 0.20 (%, w/w) of protein, its utilization as animal feed is limited due to the presence of a toxic compound, triterpenoids [4].

3.3 Effect of Delignification Through Organosolv Pretreatment Using Different Solvents

In comparison to conventional treatment, the organic solvent does not allow for extensive cellulose dissolution. Lignin is solubilized in the organic solvent during organosolv pretreatment and can be recovered as organosolv lignin by diluting the solvent with water [27]. As a result, organosolv under severe conditions causes substantial deterioration. However, by using milder pretreatment conditions and a lower concentration of Lewis acid as a catalyst (CuCl), organosolv lignin can be recovered with larger amounts of β-aryl ether linkages [28]. By cleaving the lignin-carbohydrate complex and generating cellulose-enriched solid residue, organosolv treatment effectively removes lignin and hemicellulose. Based on the literature [29], three organic solvents (ethanol, acetone, and diethylene glycol) were chosen for treating NOC in this study.

For the delignification of NOC, three important process parameters were chosen: the S:L (biomass to organic solvent) ratio, the process time, and the temperature. Initially, the biomass was treated with three different solvents with varied S:L ratio. As represented in Fig. 1, ethanol resulted in the highest delignification (60.00 ± 2.04%) than acetone (51.70 ± 1.66%) and diethylene glycol (57.50 ± 2.04%). It has been observed that delignification increased when the S:L ratio was increased gradually from 1:8 to 1:14 with ethanol and acetone, whereas for diethylene glycol the maximum delignification is attained at 1:12. This rise could be attributed to an increase in the S:L ratio, which results in more organic solvent available to react with NOC as represented in Fig. 1. Based on the delignification

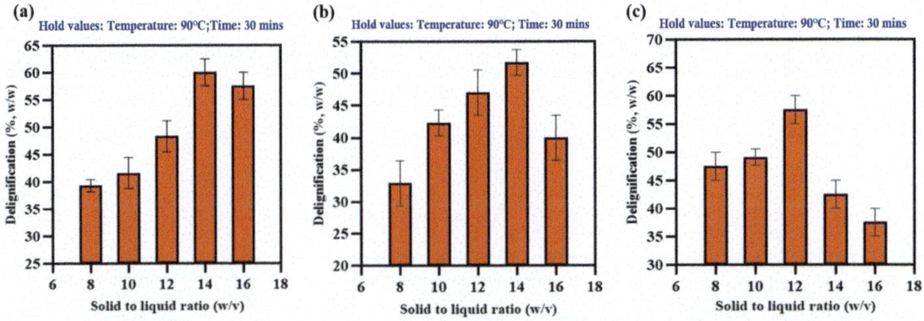

Fig. 1 Delignification of NOC through organosolv pretreatment using various solvents (**a**) Ethanol; (**b**) Acetone; (**c**) Diethylene glycol); Avg. ± S. D, $n = 3$ (No. of trials)

results, ethanol has been chosen as a suitable solvent for the treatment of NOC as it has a low boiling point, less cost and is highly efficient and easy to recover.

A similar delignification (72.80%) was observed when the almond shell was subjected to organosolv treatment using ethanol as a solvent [30]. Chen et al. [31] have reported delignification (57.80%) of wheat straw with a S:L ratio of 1:10. The reported delignification (64.00%) of wheat straw after treating it with ethanol and $CuCl_2$ [32] denotes the efficiency of solvent and catalyst in organosolv treatment. Therefore, this study achieved maximum delignification (60%) with an S:L ratio (1: 14) under mild temperature with less process time.

3.4 *Effect of Process Parameters for Organosolv Pretreatment of NOC*

Lignin is covalently bound to hemicellulose surrounding the cellulose microfibrils in lignocellulosic biomass [33], blocking cellulose access via enzymes or green solvents. When raw biomass was treated with cellulose hydrolysis, the cellulose yield was found to be decreased [34]. Organosolv pretreatment was proposed as one of the most efficient ways for increasing biomass saccharification because it could efficiently fractionate three key elements of lignocellulose into distinct fractions [35]. As a result, organosolv with ethanol and a Lewis acid catalyst generates a cellulose-enriched residue for subsequent DES hydrolysis [36]. Among the three important process parameters, delignification yielded 60.00 ± 2.04%, increasing gradually from 1:8 to a maximum of 1:14 (Table 3). Similarly, there was a rise in the delignification (70.23 ± 1.09%) at temperature (up to 60 °C) and time (up to 90 min) as shown in Fig. 2. From the obtained results, it is noted that there was a decrease in delignification percentage at higher time (150 min) and higher temperature (80 °C) which might be due to the lignin condensation that resulted in a decrease in its solubility [37].

Table 3 Optimization of organosolv pretreatment of NOC through one variable at a time (OVAT)

Process parameters	Optimized conditions through OVAT
Solid-to-liquid ratio (w/v)	1:14
Temperature (°C)	60
Time (min)	90

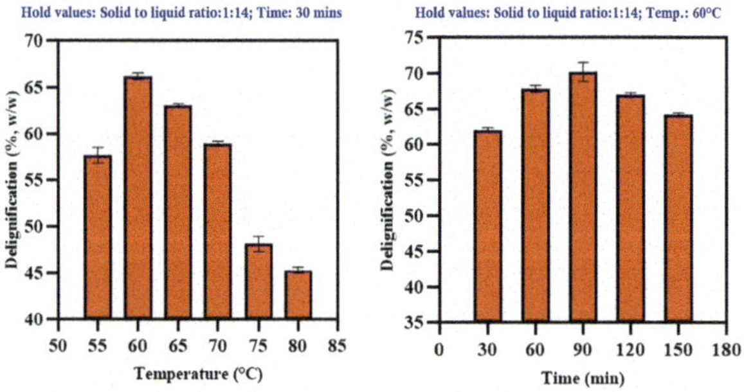

Fig. 2 Delignification of NOC through organosolv pretreatment using ethanol for various parameters (**a**) Temperature; (**b**) Time; Avg. ± S. D, $n = 3$ (No. of trials)

The previous study has reported that the ethanol/water ratio (1:1) provided the most effective delignification [38]. However, using 100% ethanol for delignification is not recommended. The presence of both ethanol and water (nucleophilic agent) results in a good solvent capable of promoting lignin cleavage and dissolving lignin particles. The nucleophilic agent accelerated lignin breakage with the addition of water but decreased the solvent's capacity to dissolve lignin during the process of delignification process [39]. Therefore, better delignification promoted between the solubility of water and ethanol mixtures and nucleophilicity [40–43].

The Lewis acid (CuCl) catalyst was chosen for this study which is found to be effective and gives the same degree of delignification compared to acids (H_2SO_4) [44]. Therefore, a mild concentration (3%, w/v) of catalyst was used in this study for effective delignification.

3.5 Saccharification of NOC Using DES (Chcl/Oxalic Acid)

In terms of physicochemical features, DES are regarded as a distinct class of green solvents that are comparable to ionic liquids (ILs). DESs have been widely recognized as IL analogues and developed as the new representation for green solvents [45]. DESs are frequently made up of two components held together by strong hydrogen bonds. The resulting eutectic mixture has a lower melting point than each component. According to Xu et al. [46], the acidic HBD of DES improves the overall effectiveness of delignification and hemicellulose dissolving.

For saccharification of cellulose using DES, three process parameters were selected, namely, biomass/DES ratio (g/g), process time (hr), and temperature (°C)

Table 4 Optimization of DES based saccharification of NOC through one variable at a time (OVAT)

Process parameters	Optimized conditions through OVAT
Solid-to-liquid ratio (w/v)	1:10
Temperature (°C)	80
Time (hours)	6

(Table 4). The maximum saccharification (68.67 ± 1.26%) was achieved at S/L ratio (1:10), temperature (up to 80 °C), and time (up to 6 h).

4 Conclusion

The recovery of fermentable sugars was employed in NOC by implicating green solvents for sustainable biorefinery applications. The mild organosolv pretreatment resulted in 60.00 ± 2.04% under optimized conditions, and the lignin was effectively recovered from the obtained black liquor. Similarly, the cellulosic fraction in the solid residue led to enhancement in saccharification with the help of DES. Hence, through this study, successful valorization of NOC was developed, and the recovered lignin would be upgraded into more valuable bio-products such as adsorbents, bioplastics and value-added chemicals leading to a sustainable green economy.

Acknowledgments The authors would like to acknowledge the Department of Biotechnology, SRM Institute of Science and Technology for the financial and infrastructure support to carry out the research work.

References

1. G. Singh, S. Sinha, K.K. Bandyopadhyay, M. Lawrence, D. Paul, Triauxic growth of an oleaginous red yeast Rhodosporidium toruloides on waste "extract" for enhanced and concomitant lipid and β-carotene production. Microb. Cell Factories **17**, 1–10 (2018)
2. P. Sannigrahi, A.J. Ragauskas, Fundamentals of Biomass Pretreatment by Fractionation, in *Aqueous Pretreatment of Plant Biomass For Biological and Chemical Conversion to Fuels And Chemicals*, ed. by C.E. Wyman, (John Wiley & Sons Ltd, Chichester, 2013), pp. 201–222
3. J. Wang, R. Ledesma-Amaro, Y. Wei, B. Ji, X.J. Ji, Metabolic engineering for increased lipid accumulation in *Yarrowia lipolytica*—a review. Bioresour Technol **313**, 123707 (2020)
4. A. Eifediyi, K.O. Remison, U.R. Samson, Effect of neem (*Azadirachta indica* L.) seed cake on the growth and yield of okra (Abelmoschus esculentus (L.) Moench). Poljoprivred **21**, 46–52 (2015)
5. A. Johansson, O. Aaltonen, P. Ylinen, Organosolv pulping—methods and pulp properties. Biomass **13**, 45–65 (1987)
6. N. Brosse, M.H. Hussin, A.A. Rahim, Organosolv Processes, in *Advances in Biochemical Engineering/Biotechnology*, (Springer, Berlin, Heidelberg, 2017)

7. J. Salapa, C. Katsimpouras, E. Topakas, D. Sidiras, Organosolv pretreatment of wheat straw for efficient ethanol production using various solvents. Biomass Bioenergy **100**, 10–16 (2017)
8. C. Li, C. Huang, Y. Zhao, C. Zheng, H. Su, L. Zhang, W. Luo, H. Zhao, S. Wang, L.J. Huang, Effect of choline-based deep eutectic solvent pretreatment on the structure of cellulose and lignin in bagasse. PRO **9**, 384 (2021)
9. R.D. Umai, S. Jacob, V. Kumar, Deep eutectic solvent pretreatment of water hyacinth for improved holocellulose saccharification and fermentative co-production of xylitol and lipids using Rhodosporidium toruloides NCIM 3547. Fermentation **8**, 591 (2022)
10. E.K. New, S.K. Tnah, K.S. Voon, K.J. Yong, A. Procentese, K.P. Yee Shak, W. Subramonian, C.K. Cheng, T.Y. Wu, The application of green solvent in a biorefinery using lignocellulosic biomass as a feedstock. J. Environ. Manag. **307**, 114385 (2022)
11. P. Cunniff (ed.), *Official Methods of Analysis of AOAC International*, vol 1, 16th edn. (AOAC International, Virginia, 1995)
12. J.A. Marlett, S.C. Lee, Dietary fiber, lignocellulose and hemicellulose contents of selected foods determined by modified and unmodified van Soest procedures. J. Food Sci. **45**, 1688–1693 (2006)
13. J.X. Sun, F. Sun, H. Zhao, R.C. Sun, Isolation and characterization of cellulose from sugarcane bagasse. Polym. Degrad. Stab. **84**, 331–339 (2004)
14. M.A. Hussain, M.E. Huq, S.M. Rahman, Estimation of lignin in jute by titration method. Pak. J. Biol. Sci. **9**, 521–522 (2002)
15. X. Yu, Y. Zheng, K.M. Dorgan, S. Chen, Oil production by oleaginous yeasts using the hydrolysate from pretreatment of wheat straw with dilute sulfuric acid. Bioresour. Technol. **102**, 134–140 (2011)
16. O.H. Lowry, N.J. Rosebrough, A.L. Farr, R.J. Randall, Protein measurement with the folin phenol reagent. J. Biol. Chem. **193**, 265–275 (1951)
17. R.S. Kirk, R. Sawyer, *Pearson's Composition and Analysis of Foods*, 9th edn. (Longman, Harlow, 1991)
18. S. Constant, C. Basset, C. Dumas, F. Di Renzo, M. Robitzer, A. Barakat, F. Quignard, Reactive organosolv lignin extraction from wheat straw: Influence of Lewis acid catalysts on structural and chemical properties of lignins. Ind. Crop. Prod. **65**, 180–189 (2015)
19. V.Z. Ong, T.Y. Wu, K.K.L. Chu, W.Y. Sun, K.P.Y. Shak, A combined pretreatment with ultrasound-assisted alkaline solution and aqueous deep eutectic solvent for enhancing delignification and enzymatic hydrolysis from oil palm fronds. Ind. Crop. Prod. **160**, 112974 (2021)
20. V.Z. Ong, T.Y. Wu, C.B.T.L. Lee, N.W.R. Cheong, K.P.Y. Shak, Sequential ultrasonication and deep eutectic solvent pretreatment to remove lignin and recover xylose from oil palm fronds. Ultrason. Sonochem. **58**, 104598 (2019)
21. X. Li, W. Tang, Y. He, Integrated understanding of acidic deep eutectic solvent choline chloride: Oxalic acid pretreatment to enhance the enzymatic hydrolysis of rape straw. Ind. Crop. Prod. **206**, 117691 (2023)
22. G.L. Miller, Use of dinitrosalicylic acid reagent for determination of reducing sugar. Anal. Chem. **31**, 426–428 (1959)
23. G. Rajeswari, S. Jacob, Co-fermentation of lactic acid and acetone-butanol-ethanol (ABE) from the deep eutectic solvent–pretreated Aloe vera leaf rind through sequential valorization of holocellulose. Biomass Conv. Bioref., 1–19 (2022). https://doi.org/10.1007/s13399-022-03688-3
24. S.D. Bharathi, S. Jacob, Comprehensive treatment strategy for Banana inflorescence bract to synthesize biodiesel and bioethanol through fungal biorefinery. Waste Biomass Valor. **15**(1), 417–436 (2023)
25. V.H. Mulimani, M.C. Tec, Analysis of physiochemical properties of de-oiled neem seed cake for their suitability in producing bio-oil. Int. J. Mech. Prod. Eng. **2320**, 2092 (2016)
26. R. Roychoudhury, *Ecofriendly Pest Management for Food Security* (Academic Press, Neem products, 2016), pp. 545–562

27. A. Eifediyi, K.O. Remison, Samson.U.R.: Effect of neem (*Azadirachta indica* L.) seed cake on the growth and yield of okra (Abelmoschus esculentus (L.) Moench). Poljoprivreda **21**, 1 (2015). https://doi.org/10.18047/poljo.21.1.8
28. R. Moreira, C.V.T. Mendes, M.B.F. Banaco, M.G.V.S. Carvalho, A. Portugal, New insights in the fractionation of Pinus pinaster wood: Sequential autohydrolysis, soda ethanol organosolv and acidic precipitation. Ind. Crop. Prod. **152**, 112499 (2020)
29. F. Weinwurm, T. Turk, J. Denner, K. Whitmore, A. Friedl, Combined liquid hot water and ethanol organosolv treatment of wheat straw for extraction and reaction modeling. J. Clean. Prod. **165**, 1473–1484 (2017)
30. D.W.K. Chin, S. Lim, Y.L. Pang, C.H. Lim, S.H. Shuit, K.M. Lee, C.T. Chong, Effects of organic solvents on the Organosolv pretreatment of degraded empty fruit bunch for fractionation and lignin removal. Sustain. For. **13**, 6757 (2021)
31. H. Chen, J. Zhao, T. Hu, X. Zhao, D. Liu, A comparison of several organosolv pretreatments for improving the enzymatic hydrolysis of wheat straw: Substrate digestibility, fermentability and structural features. Appl. Energy **150**, 224–232 (2015)
32. J. Quesada-Medina, F.J. Lopez-Cremades, P. Olivares-Carrillo, Organosolv extraction of lignin from hydrolyzed almond shells and application of the d-value theory. Bioresour. Technol. **101**, 8252–8260 (2010)
33. D. Gong, K.M. Holtman, D. Franqui-Espiet, W.J. Orts, R. Zhao, Development of an integrated pretreatment fractionation process for fermentable sugars and lignin: Application to almond (*Prunus dulcis*) shell. Biomass Bioenergy **35**, 4435–4441 (2011)
34. D.W.K. Chin, S. Lim, Y.L. Pang, L.K. Leong, C.H. Lim, Investigation of organosolv pretreatment to natural microbial-degraded empty fruit bunch for sugar based substrate recovery. Energy Procedia **158**, 1065–1071 (2019)
35. D.W.K. Chin, S. Lim, Y.L. Pang, M.K. Lam, Fundamental review of organosolv pretreatment and its challenges in emerging consolidated bioprocessing. Biofuels Bioprod. Biorefin. **14**, 808–829 (2020)
36. Z.K. Wang, S. Hong, J. Wen, C.Y. Ma, L. Tang, H. Jiang, J.J. Chen, S. Li, X.J. Shen, T.Q. Yuan, Lewis acid-facilitated deep eutectic solvent (DES) pretreatment for producing high-purity and antioxidative lignin. ACS Sustain. Chem. Eng. **8**, 1050–1057 (2020)
37. J.H. Choi, S.K. Jang, J.H. Kim, S.Y. Park, J.C. Kim, H. Jeong, H.Y. Kim, I.G. Choi, Simultaneous production of glucose, furfural, and ethanol organosolv lignin for total utilization of high recalcitrant biomass by organosolv pretreatment. Renew. Energy **130**, 952–960 (2019)
38. M. Michelin, S. Liebentritt, A.A. Vicente, J.A. Teixeira, Lignin from an integrated process consisting of liquid hot water and ethanol organosolv: Physicochemical and antioxidant properties. Int. J. Biol. Macromol. **120**, 159–169 (2018)
39. J.H. Ortega, Process design of lignocellulosic biomass fractionation into cellulose, hemicellulose and lignin by prehydrolysis and organosolv process. Master's thesis, Wageningen University, Wageningen, 2015
40. M.E. Vallejos, F.E. Felissia, A.A.S. Curvelo, M.D. Zambon, L. Ramos, M.C. Area, Chemical and physico-chemical characterisation of lignins obtained from ethanol-water fractionation of bagasse. Bio. Resources **6**, 1158–1171 (2011)
41. S. Bauer, H. Sorek, V.D. Mitchell, A.B. Ibáñez, D.E. Wemmer, Characterisation of Miscanthus x giganteus lignin isolated by ethanol organosolv process under reflux condition. J. Agric. Food Chem. **60**, 8203–8212 (2012)
42. J. Sun, T. Dutta, R. Parthasarathi, K.H. Kim, N. Tolic, R.K. Chu, N.G. Isern, J.R. Cort, B.A. Simmons, S. Singh, Rapid room temperature solubilization and depolymerisation of polymeric lignin at high loadings. Green Chem. **18**, 6012–6020 (2016)
43. M.G. Alriols, A. García, R. Llano-Ponte, J. Labidi, Combined organosolv and ultrafiltration lignocellulosic biorefinery process. Chem. Eng. J. **157**, 113–120 (2010)
44. N. Mahmood, Z. Yuan, J. Schmidt, C. Xu, Depolymerisation of lignins and their applications for the preparation of polyols and rigid polyurethane foams: A review. Renew. Sust. Energ. Rev. **60**, 317–329 (2016)

45. Y. Ye, Y. Zhang, J. Fan, J. Chang, Novel method for production of phenolics by combining lignin extraction with lignin depolymerisation in aqueous ethanol. Ind. Eng. Chem. Res. **51**, 103–110 (2012)
46. Y. Xu, K. Li, M. Zhang, Lignin precipitation on the pulp fibers in the ethanol-based organosolv pulping. Colloids Surf. A Physicochem. Eng. Asp. **301**, 255–263 (2007)

A Sustainable Solid Waste Management of *Aloe vera* Processing Industry Through Biorefinery Approach

Gunasekaran Rajeswari and Samuel Jacob

Abbreviations

AVR	*Aloe vera* rind
ChCl	Choline chloride
DES	Deep eutectic solvents
DNS	Dinitrosalicylic acid
FC	Folin–Ciocalteu
MW-NDES	Microwave-assisted natural deep eutectic solvents
N.D.	Negligible deviation
NDES	Natural deep eutectic solvents

1 Introduction

The foremost step in second-generation biorefineries is the renewable feedstock selection, where *Aloe vera* rind (AVR) waste procured from gel processing industries has found its prominence as a potent industrial feedstock in biofuel production owing to its high holocellulose component. Studies on AVR are mainly focused on the valorization of holocellulose component, where AVR lignin comprising around 16–20% (w/w) of the total biomass is often overlooked [1–3]. As a result of various treatment process opted for AVR biomass, lignin remains as a major recalcitrant by-product. Therefore, it necessitates a sustainable pretreatment strategy for the

G. Rajeswari · S. Jacob (✉)
Department of Biotechnology, School of Bioengineering, College of Engineering and Technology, Faculty of Engineering and Technology, SRM Institute of Science & Technology, Kattankulathur, Tamil Nadu, India
e-mail: rg5717@srmist.edu.in; samueljb@srmist.edu.in

D. Mazumder (ed.), *Sustainable Advanced Technologies for Environmental Management*, Springer Proceedings in Earth and Environmental Sciences,
https://doi.org/10.1007/978-3-031-64006-3_34

effective fractionation of AVR lignin and its conversion into lignin-derived value-added chemicals. Recently, pretreatment of biomass using natural deep eutectic solvents (NDES) has been considered as an alternative physiochemical strategy where DES possess high affinity towards lignin solubility on heating [4, 5]. NDES are synthesized by mixing the cellular metabolites (hydrogen bond acceptor, such as choline chloride, and hydrogen bond donors, such as alcohols, sugars, organic acids, and amino acids) to form a eutectic mixture which is categorized as Type III DES [6, 7]. The high abundance and superior nature of NDES make them a suitable alternative green solvent over other DES and ionic liquids [8, 9]. Among the NDES, polyols-based DES (choline chloride: glycerol) is highly preferred in lignocellulosic biomass fractionation owing to its interaction between the free –OH group of glycerol and etherified or free –OH groups of lignin which enhances the efficiency of selective lignin fractionation with >90% retention of glucan [10–12]. On the other hand, in the biodiesel industry, crude glycerol is primarily generated as a by-product which could constitute an inexpensive source for the glycerol-based NDES synthesis [13]. In addition, NDES is considered as a promising green solvent because of its unique properties such as ease of synthesis, renewable, biodegradable, non-toxic, tailorable, and environment friendly [14]. On the other hand, recent studies have revealed that combinative pretreatment strategy in biomass conversion could potentially enhance the depolymerization rate of lignin matrix thereby enabling the efficient valorization of holocellulose of lignocellulosic biomass [15]. In this regard, microwave-assisted pretreatment of lignocellulosic biomass has proven to be a feasible strategy as it able to enhance the lignocellulose degradation within a short duration [16]. Gaudino et al. [17] opted NDES pretreatment for the wheat straw delignification under microwave irradiation (120 °C, 30 min). Similarly, Muley et al. [18] have investigated the effect of microwave-assisted DES pretreatment of pine sawdust using ChCl and oxalic acid which resulted in a high yield of lignin. On comparing with the various conventional treatment strategy (pyrolysis, acid/alkali hydrolysis, and oxidation/reduction) opted for lignin valorization, microwave-assisted NDES (MW-NDES) pretreatment of lignocellulosic biomass is found to be more efficient owing to its several advantages, namely, short reaction time, selectively, and high lignin removal efficiency with retention of polysaccharides content (hemicellulose and cellulose) [16, 19, 20]. To the best of the authors' knowledge, this is the first study to inculcate MW-NDES for AVR lignin valorization using ChCl:glycerol at different microwave irradiation power levels (160 W, 320 W, and 480 W) (Fig. 1). The sugar monomers and lignin-derived hydrocarbons estimated in the pretreated liquor after MW-NDES pretreatment could be effectively tapped for the production of value-added chemicals.

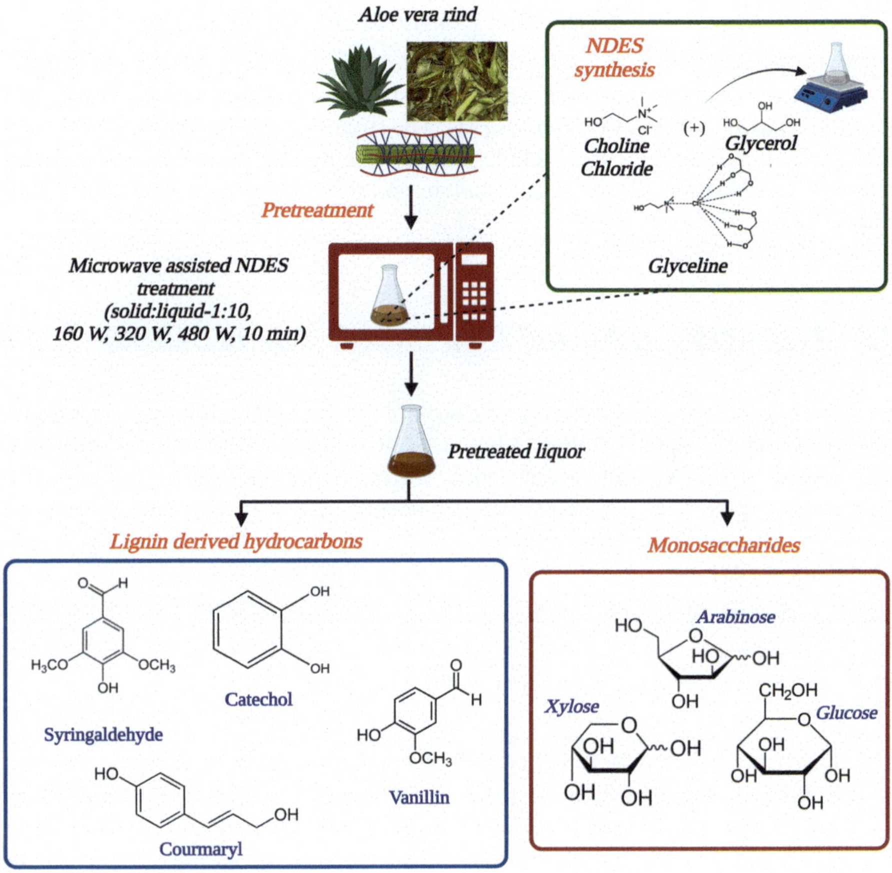

Fig. 1 A schematic representation of microwave-assisted NDES pretreatment of AVR for the production of various lignin-derived hydrocarbons and sugar monomers

2 Materials and Methods

2.1 Raw Substrate

Aloe vera rind (AVR) used in the study was collected from the gel extraction unit situated near Chennai, Tamil Nadu, India. The AVR was subjected to solar drying to remove the moisture content followed by milling. Then the sieved AVR feedstock of particle size ~0.2 mm was kept in an air-tight plastic bag at ambient temperature (28 ± 2 °C) for later use.

2.2 Proximate and Structural Compositional Analysis of AVR

A standard AOAC method was opted for estimating the volatile solid, ash, and moisture content of the AVR biomass [21]. The structural composition of AVR such as lignin, hemicellulose, and cellulose were determined based on the method reported by Hussain et al. [22], Marlett and Lee [23], and Updegraff [24], respectively.

2.3 Preparation of Natural Deep Eutectic Solvent (NDES)

NDES was prepared by heating the mixture of hydrogen bond donor (glycerol) and hydrogen bond acceptor (choline chloride—ChCl) at 1:2 molar ratio. The mixture was heated at 60 °C with a constantly stirring (150 rpm) for 1 h to form a homogenous glyceline. The synthesized glyceline was a transparent solvent and stored at ambient temperature (28 ± 2 °C) in a vacuum desiccator for further pretreatment [25].

2.4 Microwave-Assisted NDES Pretreatment of AVR

The microwave-assisted NDES (MW-NDES) pretreatment of AVR was performed in a laboratory microwave oven (M/S. LG, Model: MC2149BPB, India). The synthesized glyceline (10 mL) was added to the 50 mL Erlenmeyer flask containing 1 g of AVR and subjected to the microwave radiation at various power output (160 W, 320 W, and 480 W) for about 10 min. The solid to liquid ratio (1:10) and the reaction time for the pretreatment were set based on the earlier report [1, 26]. Based on the preliminary results, an increase in microwave irradiation (800 W) has resulted in no traces of residual liquid after pretreatment which in turn led to the breakage of reaction flask. Therefore, further increase in the power level of microwave irritation was not performed. The pretreated liquor was separated from the solid fraction by centrifugation at 5000 rpm for 15 min. Then the solid fraction was washed with acetone/water mixture (1:1) and subjected to drying in hot air oven at 60 °C until a constant weight is obtained. After drying, solid recovery (Eq. 1) and delignification percentage (Eq. 2) of MW-NDES pretreated AVR were determined using the following equations:

$$\text{Solid recovery percentage}\left(\frac{\text{w}}{\text{w}}\right) = \frac{\text{Dry AVR weight after pretreatment (g)}}{\text{Dry AVR weight before pretreatment (g)}} \times 100 \quad (1)$$

$$\text{Delignification percentage}\left(\frac{\text{w}}{\text{w}}\right) = \frac{\text{(Lignin content in untreated AVR} - \text{Lignin content in treated AVR) (g)}}{\text{Lignin content in untreated AVR (g)}} \times 100 \quad (2)$$

2.5 *Analysis of Lignin and Holocellulose Degraded Products in Pretreated Liquor*

2.5.1 Estimation of Lignin-Derived Hydrocarbons

The lignin-derived hydrocarbons such as catechol, gallic acid, coumaryl, vanillin, syringaldehyde, catechol, and 4-hydroxybenzoic acid were estimated based on the reported protocol [27, 28] using Folin–Ciocalteu (FC) reagent. To 3 mL of pretreated liquor, 0.5 mL of the FC reagent was added and incubated for about 3 min. After incubation, 2 mL of sodium carbonate (20%, w/v) was added to the test tube and kept in a water bath at 90 °C for 1 min. The phosphomolybdic acid in FC reagent reacts with the aromatic hydrocarbons derived from AVR lignin in the presence of sodium carbonate to produce molybdenum blue (blue colored complex) that was spectrometrically measured at 650 nm. The concentration of lignin-derived hydrocarbons in the pretreated liquor were calculated from the slope obtained from the linear equation of standard plots of the corresponding pure compounds.

2.5.2 Estimation of Monosaccharides

The concentration of hemicellulose and cellulose-derived monosaccharides such as xylose (C5), arabinose (C5), and glucose (C6) in the pretreated liquor was spectrometrically estimated by orcinol and dinitrosalicylic acid (DNS) method, respectively [29, 30]. The concentration of monosaccharides in pretreated liquor was determined based on the slope obtained from the linear equation of standard plots of pure xylose, arabinose, and glucose.

2.6 *Statistical Analysis*

All the statistical analysis were performed using GraphPad Prism 8.0.1 and MS Office Excel 2019, and the experimental values represented are the average of triplicates with the standard deviation (±) calculated from the three independent trials. The overall mass flow of lignin and its derivatives during MW-NDES pretreatment were visualized as an alluvial diagram using RAWGraphs 2.0.

3 Results and Discussion

3.1 *Characteristics of Cell Wall Components of AVR Biomass*

Proximate and cell wall components of AVR including volatile solid, ash content, lignin, hemicellulose, and cellulose were determined on the basis of weight percentage. The presence of significant amount of lignin (16.40 ± 0.67%, w/w), hemicellulose (15.57 ± 1.26%, w/w), and cellulose (38.39 ± 0.22%, w/w) in AVR biomass was found to be similar to the other reported studies [1–3, 31]. The proximate analysis of AVR constitutes total volatile solids (84.54 ± 1.16), moisture content (95.77 ± 1.25 (%, w/w)), and ash content (16.10 ± 1.44 (%, w/w)). The slight variation observed in the AVR cell wall components might be due to the AVR maturity and methods opted for analysis.

3.2 *Effect of MW-NDES Pretreatment on AVR Biomass*

3.2.1 Solid Recovery

After MW-NDES pretreatment of AVR, the solid recovery percentage was found to be decreased with an increase in the microwave power (Fig. 2). The highest solid recovery percentage (81.0%, w/w) was observed at 160 W, whereas at 480 W the recovery was decreased to 59.4% (w/w). Thus, at a higher microwave power, the AVR biomass was highly destructed and more solids are found to be solubilized in the pretreated liquor. Similarly, a significant amount of solid recovery percentage was decreased on increasing the temperature of microwave-assisted and conventional pretreatment of wheat straw (93.8−62.3%) and corn cob (85−55%), respectively [16, 32].

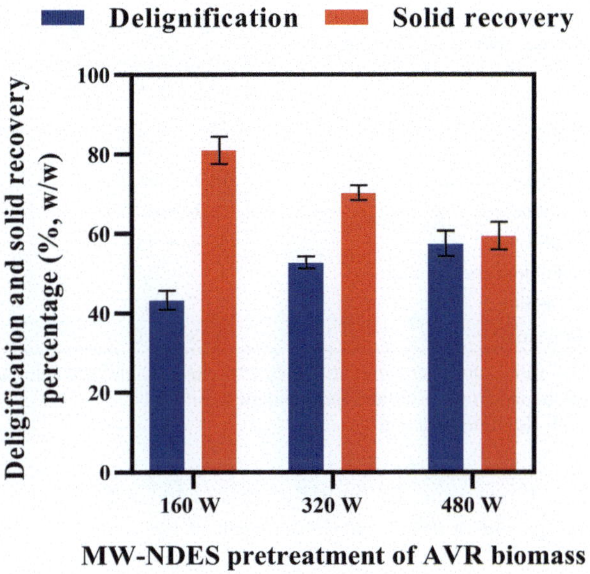

Fig. 2 Effect of MW-NDES pretreatment of AVR biomass on lignin degradation and solid recovery at different microwave irradiation power. The values are average of triplicate experiments and the standard deviation is represented as (±) calculated from mean and three independent trials

3.2.2 Lignin Removal

The influence of microwave intensity on removal of AVR lignin during pretreatment in the presence of synthesized DES (Chcl:glycerol) was studied. From the Fig. 2, it is observed that increasing the microwave irradiation intensity prominently reduces the AVR lignin content. In this regard, microwave irradiation at 480 W showed a maximum delignification of about 57.46% (w/w) with 6.97% (w/w) of residual AVR lignin in the solid residue. The microwave irradiation assisted with DES could lead to the cleavage of ester bonds between lignin and carbohydrates and thus results in the partial solubilization of AVR lignin [33]. Similarly, microwave irradiation has removed 45% of lignin in wheat straw (120 °C, 30 min) when the pretreatment is assisted with ChCl-based DES [17]. Recently, Ceaser et al. [34] has optimized the microwave-assisted pretreatment of softwood by ChCl-based DES (ChCl:formic acid) which resulted in the 90.1% of lignin removal at 800 W, 15% (w/w) solid loading for about 14 min. The microwave heating with non-ionizing radiation rapidly increases the non-targeted hydrolysis of cell wall components of lignocellulosic biomass. In addition, the ionic character and molecular polarity of synthesized DES are maximized with the microwave irradiation [20]. On the other hand, microwave-assisted acid pretreatment of oil palm empty fruit bunches, sugarcane bagasse, and palm bark has resulted in 59.6%, 27.0%, and 31.6% of lignin removal, respectively, which was found to be in the range of delignification percentage obtained in this study [35–37]. Therefore, NDES pretreatment could be substituted as a greener solvent for pretreatment of lignocellulosic biomass which

Table 1 Various lignin-derived hydrocarbons in the pretreated liquor obtained after MW-NDES pretreatment of AVR biomass

Lignin-derived hydrocarbons	MW-NDES pretreatment of AVR (mg/mL)		
	160 W	320 W	480 W
Gallic acid	0.38 ± 0.04	0.48 ± 0.11	0.66 ± 0.07
Syringaldehyde	0.36 ± 0.02	0.45 ± 0.08	0.63 ± 0.04
Coumaroyl	0.26 ± 0.05	0.32 ± 0.07	0.45 ± 0.01
Vanillin	0.34 ± 0.01	0.42 ± 0.04	0.59 ± 0.05
4-Hydroxybenzanoic acid	1.94 ± 0.71	2.41 ± 0.94	3.35 ± 1.23
Catechol	0.07 ± N.D.	0.08 ± N.D.	0.12 ± 0.04
Guaiacol	0.18 ± 0.05	0.23 ± 0.03	0.32 ± 0.02

The values are average of triplicate experiments, and the standard deviation is represented as (±) calculated from mean and three independent trials
N.D. Negligible deviation

could overcome the limitation associated with the conventional chemical treatment using acid or alkali (acid corrosion of the reactor and bio-incompatibility).

3.3 *Effect of MW-NDES Pretreatment on Lignin-Derived Hydrocarbons Formation*

In this study, we ascertained the efficacy of MW-NDES in AVR pretreatment for the efficient lignin fractionation. Table 1 represents the various lignin derived aromatic hydrocarbons obtained after MW-NDES pretreatment of AVR biomass. From Table 1, it is observed that the major phenylpropane monomers such as syringaldehyde, guaiacol, and coumaroyl are found to be formed in all the treatment sets. Among the different hydrocarbons, 4-hydroxybenzoic acid was found to be higher ranging from 1.94 to 3.35 mg/mL. An increase in the microwave irradiation power has significantly increased the formation of hydrocarbons by 1.24- (320 W) to 1.73-fold (480 W) as compared to the treatment set at 160 W. The maximum concentration of hydrocarbons at 480 W correlates with the highest delignification percentage (57.46%, w/w) of AVR biomass. The possible mechanism could be the free O–H group of glycerol interacts with the ether or β-O-4-aryl ester linkages between lignin and carbohydrate complex resulting in the lignin solubility by formation of new O–H bonds between lignin monomers and chloride anions [12, 38]. Further, short duration of microwave heating enhanced the lignin fractionation by hydrodeoxygenation or hydrogenation reaction through the cleavage of the C–C and C–O bonds that in turn results in the formation of lower molecular weight aromatic phenolic compounds [39]. Similarly, Toledano et al. [40] studied the effect of microwave-assisted nickel catalysis treatment for the depolymerization of lignin from olive tree pruning into phenolic monomers and bio-oil. Lignin-derived hydrocarbons possess some unique properties such as rigidity and hydrophobicity owing to their well-defined aromatic structure rather than lignin itself which aids in the production of renewable value-added chemicals [41]. For instance, vanillin is an industrially produced lignin derivative and finds applications as a flavoring agent in the food sector, a feed additive for poultry, and a fragrance ingredient [42, 43]. In recent times, vanillin has been a promising monomeric source for epoxy resins, polymer synthesis, and composite materials [44, 45]. Similarly, syringaldehyde is the precursor molecule of syringaresinol, a renewable and natural substitute for the bisphenols in epoxy-amine resins synthesis [46].

On the other hand, lignin monomers are utilized as the bioactive component in various sectors like pharmaceutical, nutraceutical, and food [47, 48]. Therefore, lignin-derived hydrocarbons would be a sustainable alternative for petroleum-based refinery products. Microwave-assisted extraction has been extensively applied in a variety of industrial and chemical processes such as drying, heating, coal treatment, oil reclamation, municipal waste treatment, sludge drying, biosolid treatment, and extraction [49–51]. In this regard, MW-NDES pretreatment of AVR biomass with efficient lignin removal in a short reaction time could be an alternative strategy for the conventional route. However, the high energy requirement and overall cost need to be addressed to enable the feasibility of its industrial applications.

3.4 Residual Monosaccharides in the Pretreated Liquor

After MW-NDES pretreatment of AVR, residual monosaccharides such as xylose, arabinose, and glucose were estimated in the pretreated liquor (Table 2). From the analysis, it is observed that as the microwave irradiation power is increased from 160 to 480 W, a gradual increase in the concentration of residual monosaccharides (xylose 0.19–0.3 mg/mL, arabinose 0.05–0.09 mg/mL, and glucose 0.18–0.30 mg/mL) was observed owing to the pronounced effect of degradation of holocellulose components of AVR biomass. Since NDES appears to be promising green solvent for the effective fraction of lignin from carbohydrate complex, the yield of residual monosaccharides in the pretreated liquor after MW-NDES pretreatment might be possibly due to the non-targeted hydrolysis of AVR by microwave irradiation.

Table 2 Various residual sugar monomers in the pretreated liquor obtained after MW-NDES pretreatment of AVR biomass

MW-NDES pretreatment	Xylose (mg/mL)	Arabinose (mg/mL)	Glucose (mg/mL)
160 W	0.19 ± 0.04	0.05 ± N.D.	0.18 ± 0.01
320 W	0.24 ± 0.07	0.07 ± N.D.	0.23 ± 0.03
480 W	0.30 ± 0.09	0.09 ± N.D.	0.30 ± 0.03

The values are average of triplicate experiments and the standard deviation is represented as (±) calculated from mean and three independent trials

N.D. Negligible deviation

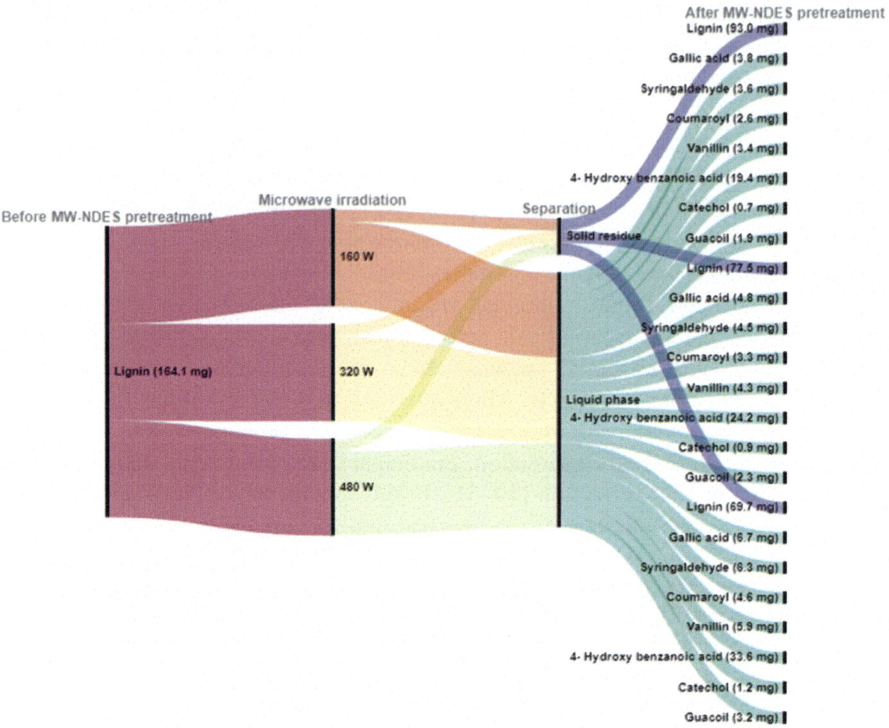

Fig. 3 Alluvial diagram representing the overall mass flow of the AVR lignin and its derivatives formed during the MW-NDES pretreatment

Similar kind of observation was reported during the microwave-assisted alkali pretreatment of AVR biomass by the author itself [3].

3.5 *Overall Mass Flow of Lignin and Its Derivatives During MW-NDES Pretreatment*

Overall mass flow for the conversion of AV rind lignin into various hydrocarbons after MW-NDES pretreatment was postulated as an alluvial diagram (Fig. 3). Herein, the cluster of nodes are represented as blocks, and stream fields represent the compositional changes over time. Based on the analysis of all the stream fields, a maximum of about 38.21% of AVR lignin conversion (i.e., 161.0 mg/g of untreated lignin into 61.53 air space mg/g of total hydrocarbons) was observed in the pretreated liquor obtained after MW-NDES treatment at 480 W leaving behind a 69.7 mg/g of residual lignin in the solids fraction.

In a biorefinery concept, the selective degradation of lignin and the recovery of lignin-derived hydrocarbons through MW-NDES enhances the prospect for AVR lignin valorization. In addition, MW-NDES aids in reducing the recalcitrance nature of AVR biomass wherein a significant amount of polysaccharides (hemicellulose and cellulose) retained in the biomass could be further tapped for carbohydrate-based biorefineries. In a nutshell, selective fractionation and conversion of each component of the AVR biomass into various biofuels and value-added chemicals unlocks a new paradigm for lignin and carbohydrate-based biorefineries.

4 Conclusion

In this study, MW-NDES of AVR has resulted in a 2.3-fold decrease in the lignin content at 480 W with a maximum delignification of about 57.46% (w/w). In addition, the solid recovery percentage was found to decrease with an increase in the microwave power, wherein AVR lignin is converted into various lignin-derived hydrocarbons in the pretreated liquor. Among the different sets, MW-NDES pretreatment at 480 W has shown a significant conversion (38.21%) of AVR lignin into various lignin-derived hydrocarbons in which 4-hydroxybenzoic acid was found to be highest (3.35 mg/mL). Thus, the combination of NDES with microwave irradiation showed a synergistic effect on AVR biomass by cleaving the linkage between lignin and carbohydrate complex and conversion of AVR lignin into monomers. Further, AVR lignin-derived hydrocarbons obtained after MW-NDES could be tapped for various commercial applications as value-added chemicals, additives, and copolymers. Overall, the short duration and rapid reaction rate of MW-NDES could be a benefit in establishing a sustainable strategy in the AVR biorefinery framework.

Acknowledgments The authors would like to acknowledge the Department of Biotechnology, School of Bioengineering, SRM Institute of Science and Technology, for the financial support provided for the research as well as infrastructure to carry out the research work.

References

1. G. Rajeswari, B. Arutselvy, S. Jacob, Delignification of Aloe vera rind by mild acid associated microwave pretreatment to persuade enhanced enzymatic saccharification. Waste Biomass Valori. **11**, 5965–5975 (2019)
2. G. Rajeswari, S. Jacob, Deciphering the aloe vera leaf rind as potent feedstock for bioethanol through enzymatic delignification and its enhanced saccharification. Ind. Crop. Prod. **143**, 111876 (2020)
3. G. Rajeswari, S. Jacob, A hybrid pretreatment strategy for delignification of Aloe vera processing waste and its effectiveness towards saccharification. J. Adv. Biotechnol. Exp. Ther. **5**, 537 (2022)
4. M. Jablonsky, A. Haz, V. Majova, Assessing the opportunities for applying deep eutectic solvents for fractionation of beech wood and wheat straw. Cellulose **26**, 7675–7684 (2019)

5. C. Li, C. Huang, Y. Zhao, C. Zheng, H. Su, L. Zhang, W. Luo, H. Zhao, S. Wang, L.-J. Huang, Effect of Choline-based deep eutectic solvent pretreatment on the structure of cellulose and lignin in bagasse. PRO **9**, 384 (2021)
6. A.P. Abbott, G. Capper, D.L. Davies, R.K. Rasheed, P. Shikotra, Selective extraction of metals from mixed oxide matrixes using choline-based ionic liquids. Inorg. Chem. **44**, 6497–6499 (2005)
7. V.Z. Ong, T.Y. Wu, C.B.T.L. Lee, N.W.R. Cheong, K.P.Y. Shak, Sequential ultrasonication and deep eutectic solvent pretreatment to remove lignin and recover xylose from oil palm fronds. Ultrason. Sonochem. **58**, 104598 (2019)
8. Y.H. Choi, J. van Spronsen, Y. Dai, M. Verberne, F. Hollmann, I.W. Arends, G.J. Witkamp, R. Verpoorte, Are natural deep eutectic solvents the missing link in understanding cellular metabolism and physiology? Plant Physiol. **156**(4), 1701–1705 (2011)
9. A. Paiva, R. Craveiro, I. Aroso, M. Martins, R.L. Reis, A.R.C. Duarte, Natural deep eutectic solvents–solvents for the 21st century. ACS Sustain. Chem. Eng. **2**(5), 1063–1071 (2014)
10. C.W. Zhang, S.Q. Xia, P.S. Ma, Facile pretreatment of lignocellulosic biomass using deep eutectic solvents. Bioresour. Technol. **219**, 1–5 (2016)
11. S. Wang, H. Li, L.P. Xiao, G. Song, Unraveling the structural transformation of wood lignin during deep eutectic solvent treatment. Front. Energy Res. **8**, 48 (2020)
12. M.A. Hossain, M.S. Rahaman, D. Yelle, H. Shang, Z. Sun, S. Renneckar, J. Dong, S. Tulaphol, N. Sathitsuksanoh, Effects of polyol-based deep eutectic solvents on the efficiency of rice straw enzymatic hydrolysis. Ind. Crop. Prod. **167**, 113480 (2021)
13. Y. Gu, F. Jérôme, Glycerol as a sustainable solvent for green chemistry. Green Chem. **12**, 1127–1138 (2016)
14. Z. Yang, Natural Deep Eutectic Solvents and Their Applications in Biotechnology, in *Application of Ionic Liquids in Biotechnology*, (Springer, Cham, 2019), pp. 31–59
15. Y.-L. Loow, E.K. New, G.H. Yang, L.Y. Ang, L.Y.W. Foo, T.Y. Wu, Potential use of deep eutectic solvents to facilitate lignocellulosic biomass utilization and conversion. Cellulose **24**, 3591–3618 (2017)
16. A. Isci, G.M. Erdem, S. Bagder Elmaci, O. Sakiyan, A. Lamp, M. Kaltschmitt, Effect of microwave-assisted deep eutectic solvent pretreatment on lignocellulosic structure and bioconversion of wheat straw. Cellulose **27**, 8949–8962 (2020)
17. E.C. Gaudino, S. Tabasso, G. Grillo, G. Cravotto, T. Dreyer, G. Schories, S. Altenberg, L. Jashina, G. Telysheva, Wheat straw lignin extraction with bio-based solvents using enabling technologies. C. R. Chim. **21**, 563–571 (2018)
18. P.D. Muley, J.K. Mobley, X. Tong, B. Novak, J. Stevens, D. Moldovan, J. Shi, D. Boldor, Rapid microwave-assisted biomass delignification and lignin depolymerization in deep eutectic solvents. Energy Convers. Manag. **19**, 1080–1088 (2019)
19. H. Rabemanolontsoa, S. Saka, Various pretreatments of lignocellulosics. Bioresour. Technol. **199**, 83–91 (2016)
20. Y.T. Tan, A.S.M. Chua, G.C. Ngoh, Deep eutectic solvent for lignocellulosic biomass fractionation and the subsequent conversion to bio-based products—A review. Bioresour. Technol. **297**, 122522 (2020)
21. AOAC, *Official Methods of Analysis*, 17th edn. (Association of Official Analytical Chemists, Maryland, 2000)
22. M.A. Hussain, M.E. Huq, S.M. Rahma, Estimation of lignin in jute by titration method. Pak. J. Biol. Sci. **5**, 521–522 (2002)
23. J.A. Marlett, S.C. Lee, Dietary Fiber, lignocellulose and hemicellulose contents of selected foods determined by modified and unmodified van soest procedures. J. Food Sci. **45**, 1688–1893 (1980)
24. D.M. Updegraf, Semimicro determination of cellulose in biological materials. Anal. Biochem. **32**, 420–424 (1969)

25. Z. Chen, X. Bai, A. Lusi, C. Wan, High-solid lignocellulose processing enabled by natural deep eutectic solvent for lignin extraction and industrially relevant production of renewable chemicals. ACS Sustain. Chem. Eng. **6**, 12205–12216 (2018)
26. C.Y. Ma, X. Gao, X.P. Peng, Y.F. Gao, J. Liu, J.L. Wen, T.Q. Yuan, Microwave-assisted deep eutectic solvents (DES) pretreatment of control and transgenic poplars for boosting the lignin valorization and cellulose bioconversion. Ind. Crop. Prod. **164**, 113415 (2021)
27. S. Sadasivam, A. Manickam, *A Biochemical Methods*, 2nd edn. (New age International, New Delhi, 1996)
28. K.P. Bastola, Y.N. Guragain, V. Bhadriraju, P.V. Vadlani, Evaluation of standards and interfering compounds in the determination of phenolics by Folin-Ciocalteu assay method for effective bioprocessing of biomass. Am. J. Anal. Chem. **8**, 416–431 (2017)
29. P.J. Pham, R. Hernandez, W.T. French, B.G. Estil, A.H. Mondala, A spectrophotometric method for quantitative determination of xylose in fermentation medium. Biomass Bioenergy **35**, 2814–2821 (2011)
30. G.L. Miller, Use of dinitrosalicylic acid reagent for determination of reducing sugar. Anal. Chem. **31**, 426–428 (1959). https://doi.org/10.1021/ac60147a030
31. S. Cheng, S. Panthapulakkal, M. Sain, A. Asiri, Aloe vera rind cellulose nanofibers reinforced films. J. Appl. Polym. Sci. **131**, 40592 (2014)
32. A. Procentese, E. Johnson, V. Orr, A.G. Campanile, J.A. Wood, A. Morzocchella, L. Rehmann, Deep eutectic solvent pretreatment and subsequent saccharification of corncob. Bioresour. Technol. **192**, 31–36 (2015)
33. Z. Zhu, D.J. Macquarrie, R. Simister, L.D. Gomez, S.J. McQueen-Mason, Microwave assisted chemical pretreatment of *Miscanthus* under different temperature regimes. Sustain. Chem. Process **3**, 15 (2015)
34. R. Ceaser, S. Rosa, D. Montané, M. Constantí, F. Medina, Optimization of softwood pretreatment by microwave-assisted deep eutectic solvents at high solids loading. Bioresour. Technol. **369**, 128470 (2023)
35. S. Ethaib, R. Omar, M.K. Siti Mazlina, A.B. Dayang Radiah, S. Syafie, Microwave-assisted dilute acid pretreatment and enzymatic hydrolysis of Sago Palm Bark. Bioresources **11**, 5687–5702 (2016)
36. W. Fatriasari, S.H. Anita, L. Risanto, Microwave assisted acid pretreatment of oil palm empty fruit bunches (EFB) to enhance its fermentable sugar production. Waste Biomass Valori. **8**(2), 379–391 (2016)
37. Z. Zhu, C.A. Rezende, R. Simister, S.J. McQueen-Mason, D.J. Macquarrie, I. Polikarpov, L.D. Gomez, Efficient sugar production from sugarcane bagasse by microwave assisted acid and alkali pretreatment. Biomass Bioenergy **93**, 269–278 (2016)
38. X.-J. Shen, J.-L. Wen, Q.-Q. Mei, X. Chen, D. Sun, T.-Q. Yuan, R.-C. Sun, Facile fractionation of lignocelluloses by biomass derived deep eutectic solvent (DES) pretreatment for cellulose enzymatic hydrolysis and lignin valorization. Green Chem. **21**, 275–283 (2019)
39. M. Zhou, J. Xu, J. Jiang, B.K. Sharma, A review of microwave assisted liquefaction of lignin hydrogen donor solvents: Effect of solvents and catalysts. Energies **11**(11), 2877 (2018)
40. A. Toledano, L. Serrano, A. Pineda, A.A. Romero, R. Luque, J. Labidi, Microwave-assisted depolymerisation of organosolv lignin via mild hydrogen-free hydrogenolysis: Catalyst screening. Appl. Catal. B Environ. **145**, 43–55 (2014)
41. N. Lee, Y.T. Kim, J. Lee, Recent advances in renewable polymer production from lignin-derived aldehydes. Polymers **13**(3), 364 (2021)
42. M.M. Gheisar, I.H. Kim, Phytobiotics in poultry and swine nutrition–a review. Ital. J. Anim. Sci. **17**(1), 92–99 (2018)
43. G. Banerjee, P. Chattopadhyay, Vanillin biotechnology: The perspectives and future. J. Sci. Food Agric. **99**(2), 499–506 (2019)
44. M. Fache, R. Auvergne, B. Boutevin, S. Caillol, New vanillin-derived diepoxy monomers for the synthesis of biobased thermosets. Eur. Polym. J. **67**, 527–538 (2015)

45. Z. Mirzakhanian, K. Faghihi, H.A. Geravi, M. Mahdieh, Semi-aromatic polyimide/Ag nanocomposite derived from vanillin. J. Appl. Polym. Sci. **133**(39), 44001 (2016)
46. M. Janvier, L. Hollande, A.S. Jaufurally, M. Pernes, R. Ménard, M. Grimaldi, J. Beaugrand, P. Balaguer, P.H. Ducrot, F. Allais, Syringaresinol: A renewable and safer alternative to bisphenol A for epoxy-amine resins. ChemSusChem **10**(4), 738–746 (2017)
47. R. Tinikul, P. Chenprakhon, S. Maenpuen, P. Chaiyen, Biotransformation of plant-derived phenolic acids. Biotechnol. J. **13**(6), 1700632 (2018)
48. L. Martínková, M. Grulich, M. Pátek, B. Křístková, M. Winkler, Bio-based valorization of lignin-derived phenolic compounds: A review. Biomol. Ther. **13**(5), 717 (2023)
49. Industrial Microwave Systems, Microwave Assisted Extraction (2024). https://industrialmicrowave.com/microwave-assisted-extraction/. Accessed 1 Mar 2024
50. Industrial Microwave Systems, Industrial Microwave Chemical Heating (2024). https://industrialmicrowave.com/industrial-microwave-applications/chemical-heating/. Accessed 1 Mar 2024
51. Ferrite Microwave Technologies, Industrial Microwave Processing (2024). https://industrialmicrowavesystems.com/applications/industrial-processing/. Accessed 1 Mar 2024

Elucidating Deoiled Rice Bran as an Effectual Feedstock for Fermentation Through Mild Acid-Mediated Combined Pretreatment and Saccharification

Ramalingam Kayalvizhi and Samuel Jacob

Abbreviations

DNS	3,5-Dinitro salicylic acid
DRB	Deoiled rice bran
HMF	Hydroxy methyl furfural
SEA	Solvent Extractors Association

1 Introduction

In the current economic scenario, many attempts are made to produce commercially important fuels and chemicals through a greener approach to minimize the negative impact on the environment caused by large-scale commercial production by conventional methods. In addition to it, these greener approaches also pave the way for the efficient utilization of lignocellulosic residues; otherwise might not have any commercial significance. Further fossil fuel-derived chemicals are non-renewable and pose a great threat to the environment [1]. In the first-generation biorefinery approach, food crops are utilized as feedstock for fuel and chemical production, and this approach competes with the food and fodder industry. To overcome this shortcoming, second-generation lignocellulosic biomasses are exploited as feedstock to produce fuels and chemicals. Commonly used lignocellulosic feedstocks

R. Kayalvizhi · S. Jacob (✉)
Department of Biotechnology, School of Bioengineering, College of Engineering and Technology, Faculty of Engineering and Technology, SRM Institute of Science and Technology, Kattankulathur, Tamil Nadu, India
e-mail: kr8416@srmist.edu.in; samueljb@srmist.edu.in

D. Mazumder (ed.), *Sustainable Advanced Technologies for Environmental Management*, Springer Proceedings in Earth and Environmental Sciences,
https://doi.org/10.1007/978-3-031-64006-3_35

are agricultural wastes, forest wastes, and domestic and industrial waste residues. These feedstocks are sustainable and cheaper making it a preferable choice for biorefinery [2].

India is an agriculture-based country and ranks second in rice production, after China [3]. Annual rice production in India is estimated at 115.2 million tons. Approximately, after processing 100 kg of rice straw, 5–15 kg of rice bran is generated and the major component of rice bran is oil (10–23%). The rice bran production is around 76 million tons annually on a global scale. Solvent Extractors Association (SEA) of India reported that in India the annual rice bran oil production is 9 lakh tons, and the production rate increases annually by 50,000 tons [4]. Once oil is extracted from rice bran, deoiled rice bran (DRB) is generated as a residue. This DRB residue accumulated in large amounts in the environment, as the bran oil production has been increasing in tremendous quantity every year. Initially, for a decade these residues are used as cattle feed, but not for a longer time as cattles show signs of disease conditions such as allergic dermatitis and gastrointestinal aliments due to the presence of elevated amount of anti-nutritional factors such as phytates, trypsin inhibitors, chymotrypsin inhibitors, subtilisin inhibitors, oxalates, saponins, tannins, and alkaloids. Due to these drawbacks, its application as animal feed is limited [5–7] and this solid residue is accumulated as waste without any applications. Despite being a holocellulose-rich residue, DRB contains substantial amount of carbohydrates that can be hydrolyzed into simple sugars, serving as a substrate for microorganisms to produce valuable chemicals and fuels. Recent advancements have demonstrated the utilization of sugar-rich DRB hydrolysate in the bioproduction of fuels and chemicals such as butanol [8], ethanol [9], methane [10], and microbial oil [11]. However, the potential of sugar-rich DRB extends beyond these applications, offering opportunities for the production of high-value products like organic acids and sugar alcohols using microbial cell factories. Leveraging DRB presents a sustainable alternative to conventional chemical methods, promoting environmental friendliness and contributing to the development of a circular bioeconomy. This current study aims to optimize mild acid-mediated pretreatment and saccharification conditions to maximize the extraction of fermentable sugars from DRB.

2 Materials and Methods

2.1 Biomass: Deoiled Rice Bran (DRB)

The substrate, DRB, has been procured from a local bran oil processing unit, Tiruchirapalli, Tamil Nadu, India. The biomass, DRB, was sun-dried for 8 h, then sieved to the particle size of 0.2 mm, and stored in a dry place at room temperature for further usage. The proximate analysis [12] and biochemical analysis were performed to estimate volatile solids, total solids, ash content, cellulose [13], hemicellulose [14], starch [15], lignin [16], and protein [17].

2.2 Parameters for Combined Pretreatment and Saccharification of DRB by Using Mild Acid

The monosaccharides present in the DRB were extracted through the process of combined pretreatment and saccharification using the action of a mild acid. DRB was treated with different strengths of H_2SO_4 (v/v) such as 1%, 1.5%, 2%, 2.5%, and 3% for the period of 6 h at 90 °C. In addition, the treatment process was varied by altering the ratios of DRB and H_2SO_4, (w/v) as 1:2; 1:4; 1:6; 1:8; and 1:10 respectively [11].

2.3 Detoxification of DRB-Derived Hydrolysate

The hydrolysate was detoxified to reduce the microbial growth inhibitory effects of lignin and sugar-derived compounds such as furfural, syringaldehyde, hydroxymethyl furfural, and acetic acid. After acid treatment, the solid residue and hydrolysate were separated through vacuum filtration. The pH of the hydrolysate was increased to 6.5 by the slow addition of calcium hydroxide with constant mixing. The inhibitors were allowed to precipitate and the depositions were removed using a vacuum filter [11]. The detoxified hydrolysate was stored at 4 ° C for further study.

2.4 Estimation of Sugars and Inhibitors

Pentose sugars (xylose and arabinose) present in the hydrolysate were estimated by Bial's method [18], and hexose sugars (glucose and mannose) were estimated by the 3,5-dinitro salicylic acid (DNS) method respectively [19]. The hydration reaction of pentose sugar with hydrochloric acid (key constituents of Bial's reagent) results in the production of 2-furaldehyde, and further, the presence of Fe^{3+} catalyst in Bial's reagent induces the formation of a colored complex, characterized by a peak absorption at 671 nm. The aldehyde group of glucose and mannose reduces the DNS to 3-amino-5-nitrosalicylic acid, resulting in a color change from yellow to orange-red with maximum absorption at 540 nm. The inhibitory compounds (furfural, syringaldehyde, hydroxymethyl furfural, and acetic acid) in the hydrolysate were estimated using the Folin–Ciocalteu reagent. The oxidoreduction reaction between phenolic compounds and components of the Folin–Ciocalteu reagent results in the formation of blue-colored metal oxide complexes which display maximum absorption at 650 nm [20].

3 Results and Discussion

3.1 Biochemical and Proximate Analysis of DRB Biomass

For any variety of paddy, the structural components may vary based on the geographical origin [21]. The proximate and biochemical composition analysis of DRB revealed that the starch (24.06%, w/w) and cellulose (23.55%, w/w) fractions are the major components followed by hemicellulose (15.54%, w/w) with a relatively small amount of lignin (3.93%, w/w). Thus, the presence of a high quantity of holocellulose components (39.09%, w/w) (cellulose and hemicellulose) paved the way for exploiting the DRB hydrolysate (rich in fermentable sugars) for fermentative-based product synthesis. According to recent literature, this compositional analysis was found to be within the range [22]. Lignin is a structural polymeric compound and it affects the process of saccharification. In the case of DRB, the lignin constitutes less than 4% (w/w), which makes DRB an efficient feedstock however it slightly impede with the saccharification process. Further, the presence of high volatile solid (86.73%, w/w) and total solids (98.4%, w/w) content followed by low moisture (1.6%, w/w) and ash (13.26%, w/w) content of DRB indicates that the holocellulose components are readily available for the biobased conversion process. The total protein and total reducing sugar content of DRB are estimated as 2.06% (w/w) and 0.01% (w/w), respectively.

3.2 Effect of Detoxification on Inhibitory Compounds

The chemical method of pretreatment is an efficient method to extract sugar from the biomass in a short period. However, it also results in the formation of inhibitory compounds. These compounds are produced during the process of pretreatment and are lignin- and sugar-derived compounds. Commonly observed inhibitors of acid treatment are furfural, hydroxymethyl furfural (HMF), syringaldehyde, and acetic acid. Table 1 represents the concentration of different inhibitory compounds before and after calcium hydroxide-mediated detoxification. These inhibitory compounds largely affect the growth of the microorganism if non-detoxified hydrolysate is used for fermentation without adopting an efficient detoxification strategy. Detoxification of the hydrolysate with the slow addition of calcium hydroxide and with

Table 1 Effect of detoxification on the concentration of inhibitory compounds

Inhibitory compounds	Before detoxification (g/L)	After detoxification (g/L)	Reduction percentage (w/w)
Furfural	0.19 ± 0.009	0.035 ± 0.006	81.58
HMF	0.14 ± 0.006	0.026 ± 0.005	81.43
Acetic acid	0.23 ± 0.005	0.051 ± 0.012	77.83
Syringaldehyde	0.34 ± 0.003	0.083 ± 0.007	75.56

simultaneous continuous stirring results in the precipitation of these inhibitory compounds [11]. The estimation of these inhibitory compounds using the Folin–Ciocalteu reagent indicated that the concentration were reduced significantly after detoxification.

3.3 *Effect of Mild Acid Treatment on Delignification*

Mild acid treatment is an efficient method to break down the lignin polymer present in the biomass. This aromatic polymer protects the underlying cellulose and hemicellulose from any sort of destruction. Lignin is a heterogenous polymer majorly composed of three components such as coniferyl alcohol, coumaryl alcohol, and sinapyl alcohol. Lignin content of DRB is estimated to be less than 4% (w/w) which is quite low that promotes DRB as an efficient second-generation lignocellulosic biomass. This mild acid treatment aids in removing the lignin polymer from the DRB biomass. The maximum delignification observed was 78.86% (w/w), representing a lignin content of 0.87% (w/w) in the residual DRB biomass after the mild acid treatment. Figure 1 represents the delignification efficiency of the mild acid treatment strategy for DRB biomass. It was observed that mild acid treatment proved to be an efficient delignification method. As a byproduct, some of the lignin degraded products formed are furfural, syringaldehyde, and HMF. These inhibitors at high concentrations affect the growth of microorganisms, during the processes of fermentative production of fuels and chemicals. Through effective detoxification methods, such as overliming with calcium hydroxide, these degraded products can be precipitated [23, 24].

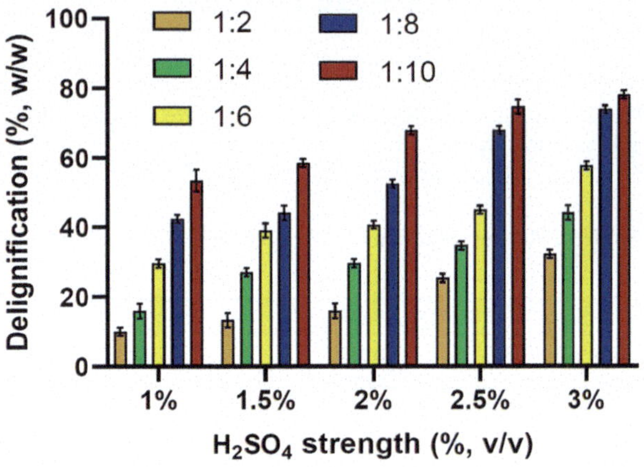

Fig. 1 Estimation of percentage of delignification in treated DRB after mild acid treatment

3.4 Effect of Mild Acid Treatment on Holocellulose Compounds

Cellulose is a hydrophilic polymer, composed of glucose units connected by β-(1,4)-glycosidic linkage, whereas hemicellulose is a heteropolymer that constitutes both pentose and hexose sugars. After the process of combined pretreatment and saccharification, the treated biomass was separated from the hydrolysate by vacuum filtration and dried. The residual cellulose and hemicellulose contents of the treated biomass were estimated. It is found that the treated biomass contains 4.89 and 2.55% (w/w) of residual cellulose and hemicellulose and the respective reduction percentage are 79.24 and 84.26 (%, w/w). Figure 2 and Table 2 represent the amount of cellulose and hemicellulose components remaining in the treated DRB biomass after mild acid treatment. Upon efficient saccharification, the hexose and pentose sugar can be extracted from the holocellulose components, and the crystallinity of cellulose is reduced significantly [25, 26]. The maximum saccharification observed was 87.72% (w/w). It indicates that the maximum holocellulose components were saccharified by this mild acid-mediated combined treatment and saccharification.

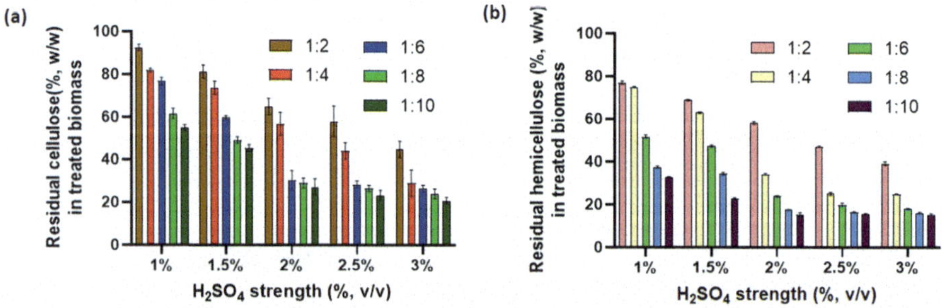

Fig. 2 Estimation of percentage of residual (**a**) cellulose (**b**) hemicellulose in treated DRB after mild acid treatment

Table 2 Percent reduction of polysaccharide and lignin in mild acid-treated DRB biomass

Structural component	Raw biomass % (w/w)	Treated biomass % (w/w)	% reduction (w/w)
Cellulose	23.55	4.89	79.24
Hemicellulose	15.54	2.55	84.62
Lignin	3.93	0.87	78.86

3.5 Effect of Mild Acid Treatment on Fermentable Sugar Yield

The strength of H_2SO_4 was varied as 1%,1.5%, 2%, 2.5%, and 3%, and the DRB: H_2SO_4 ratio was maintained as 1;2, 1:4, 1:6, 1:8, and 1:10, respectively. The treatment temperature and period were kept constant at 90 °C and 6 h, respectively, throughout the treatment [27]. Under optimized conditions, the predominant sugar obtained was glucose followed by xylose, mannose, and arabinose, respectively. Figure 3 represents the maximum monosaccharides hydrolysed from DRB biomass after acid treatment. Predominantly, many bacterial and fungal species are hexose carbon-utilizing organisms. For these hexose utilizers, glucose is the preferable carbon source. On the contrary, there are some yeasts such as *Candida* sp., [28] and *Pichia* sp. [29, 30] that are both pentose and hexose sugar-utilizing organisms. However, yeast like *Saccharomyces cerevisiae* [31] is only a hexose utilizer. Pentose sugars serve a crucial role in ATP metabolism and transport in eukaryotic organisms. It competes with the ATP binding site and restricts its transport across the cell membrane. It also modulates the production of ATP by allosterically inhibiting enzymes involved in its synthesis. Long-term exposure to high concentrations of pentose sugar can interfere with ATP-dependent cellular functions such as muscle contraction, neuronal transmission, and membrane transport in higher animals, and

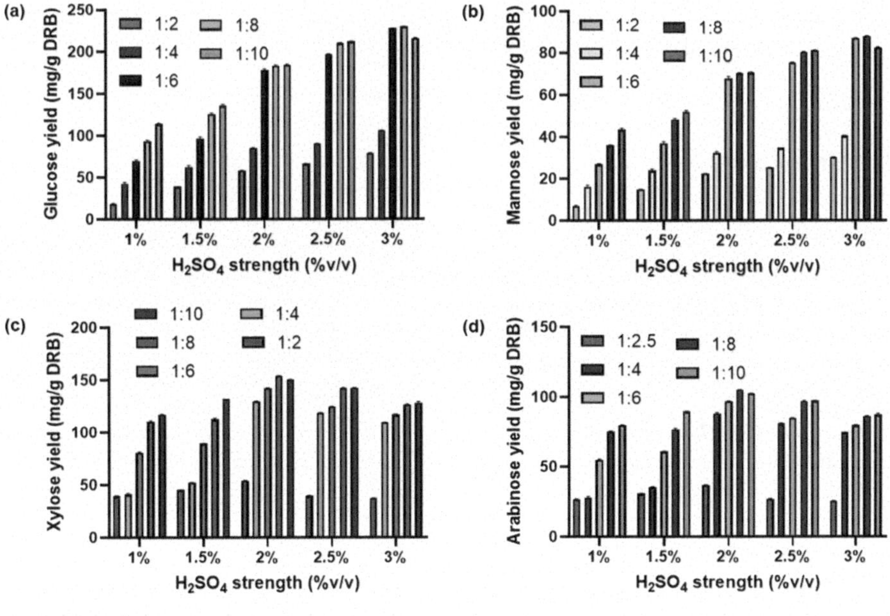

Fig. 3 Estimation of monosaccharides yield in mild acid-treated DRB hydrolysate with varied H_2SO_4 strength (v/v) and DRB:H_2SO_4 ratio (w/v) (**a**) Glucose (**b**) Mannose (**c**) Xylose (**d**) Arabinose

in yeast, it can affect cell growth and viability. For complete utilization of the lignocellulosic biomass, organisms are genetically engineered to increase their efficiency to utilize both hexose and pentose sugar and to overcome the feedback inhibition of the end products produced [32]. The maximum sugar yield after detoxification was estimated to be 529.64 mg/g of DRB.

3.6 Economic Analysis of Xylitol Production from DRB

An economic analysis of chemical production from biomass involves assessing the financial feasibility and profitability of its production from biomass feedstocks. The key components of this analysis include the cost of the feedstock, the cost of the conversion technology, capital investment, operating cost, and market demand for the product. This section delves into a case study of xylitol production from DRB. In the Indian scenario, the chemical-based production of xylitol is estimated at $ 9.0–10.5/kg xylitol with the emission of 3.83 Kg of CO_2/Kg xylitol. Conversely, xylitol production from DRB incurs an estimated cost of $15.36, encompassing expenses for DRB, sulfuric acid, calcium hydroxide, water, and energy consumption during the production process. The slightly increased cost is directly attributed to the elevated energy consumption during the process of pretreatment and saccharification. Since this approach involves the valorization of waste organic residue (DRB), it is highly recommended compared to the chemical method of xylitol production. However, the cost of production can be further reduced by implementing strategies such as enzymatic delignification and saccharification.

4 Conclusion

This work depict the efficiency of mild acid-mediated combined pretreatment and saccharification to extract fermentable monosaccharides from DRB biomass. This treatment showed significant lignin removal and thus exposed the cellulose and hemicellulose components of DRB for further saccharification process. However, the production of lignin and sugar-derived inhibitors was the major limitation of this treatment strategy. The process of detoxification through an overliming strategy using calcium hydroxide reduced the concentration of these inhibitors by a significant amount. The maximum sugar yield after detoxification was estimated to be 529.64 mg/g of DRB with the maximum saccharification of 87.72% (w/w). From the study, it has been shown that DRB-derived hydrolysate could be an efficient carbon source for the microbial production of fuels and chemicals. This inclusion marks one more lignocellulosic biomass as an efficient second-generation feedstock for the production of commercially important fuels and chemicals.

Acknowledgments The authors would like to acknowledge the Department of Biotechnology, SRM Institute of Science and Technology for the financial and infrastructure support to carry out the research work.

References

1. S.K. Maity, Opportunities, recent trends and challenges of integrated biorefinery: Part I. Renew. Sust. Energ. Rev. **43**, 1427–1445 (2015). https://doi.org/10.1016/j.rser.2014.11.092
2. S.S. Hassan, G.A. Williams, A.K. Jaiswal, Moving towards the second generation of lignocellulosic biorefineries in the EU: Drivers, challenges, and opportunities. Renew. Sust. Energ. Rev. **101**, 590–599 (2019). https://doi.org/10.1016/j.rser.2018.11.041
3. Rice Bran – IRRI Rice Knowledge Bank (2023). http://www.knowledgebank.irri.org/step-by-step-production/postharvest/rice-by-products/rice-bran. Accessed 11 Nov 2023
4. The Solvent Extractors' Association of India (2023). https://seaofindia.com/. Accessed 11 Nov 2023
5. S. Kumar, N.P. Sahu, N. Shamna, A. Ranjan, Feeding higher level of de-oiled rice bran causes stress to *Labeo rohita* fingerlings. Aquaculture **484**, 184–190 (2018). https://doi.org/10.1016/j.aquaculture.2017.11.029
6. N. Shridhar, Commercial cattle feed induced dermatosis in cattle and buffaloes of Karnataka: Role of defatted rice bran as cause. J. Pharm. Innov. **9**, 151–154 (2020)
7. A. Ranjan, S. Kumar, N.P. Sahu, et al., Strategies for maximizing utilization of de-oiled rice bran (DORB) in the fish feed. Aquac. Int. **30**, 99–114 (2022). https://doi.org/10.1007/s10499-021-00791-6
8. J. Lee, Fermentation of rice bran and defatted rice bran for butanol production using *Clostridium beijerinckii* NCIMB 8052. J. Microbiol. Biotechnol. **19**, 482–490 (2009). https://doi.org/10.4014/jmb.0804.275
9. A.K. Chandel, M.L. Narasu, R. Rudravaram, et al., Bioconversion of de-oiled rice bran (DORB) hemicellulosic hydrolysate into ethanol by *Pichia stipitis* NCM3499 under optimized conditions. Int. J. Food Eng. **5** (2009). https://doi.org/10.2202/1556-3758.1453
10. B. Jha, R. Chandra, V.K. Vijay, et al., Utilization of de-oiled rice bran as a feedstock for renewable biomethane production. Biomass Bioenergy **140**, 105674 (2020). https://doi.org/10.1016/j.biombioe.2020.105674
11. Y.A. Tsigie, C.-Y. Wang, N.S. Kasim, et al., Oil production from *Yarrowia lipolytica* Po1g using rice bran hydrolysate. J. Biomed. Biotechnol. **2012**, 1–10 (2012). https://doi.org/10.1155/2012/378384
12. A.D. Eaton, M.A.H. Franson, *Standard Methods for the Examination of Water and Wastewater*, 21st edn. (American Public Health Association, Washington, 2005)
13. D.M. Updegraff, Semimicro determination of cellulose in biological materials. Anal. Biochem. **32**, 420–424 (1969). https://doi.org/10.1016/S0003-2697(69)80009-6
14. J.A. Marlett, S.C. Lee, Dietary fiber, lignocellulose and hemicellulose contents of selected foods determined by modified and unmodified Van Soest procedures. J. Food Sci. **45**, 1688–1893 (1980). https://doi.org/10.1111/j.1365-2621.1980.tb07590.x
15. K.M. Clegg, The application of the anthrone reagent to the estimation of starch in cereals. J. Sci. Food Agric. **7**, 40–44 (1956). https://doi.org/10.1002/jsfa.2740070108
16. M.A. Hussain, H.M. Emdadul, S. Matiur Rah, Z. Ahmed, Estimation of lignin in jute by titration method. Pak. J. Biol. Sci. **5**, 521–522 (2002). https://doi.org/10.3923/pjbs.2002.521.522
17. O.H. Lowry, N.J. Rosebrough, A.L. Farr, R.J. Randall, Protein measurement with the folin phenol reagent. J. Biol. Chem. **193**, 265–275 (1951)

18. P.J. Pham, R. Hernandez, W.T. French, et al., A spectrophotometric method for quantitative determination of xylose in fermentation medium. Biomass Bioenergy **35**, 2814–2821 (2011). https://doi.org/10.1016/j.biombioe.2011.03.006
19. G.L. Miller, Use of dinitrosalicylic acid reagent for determination of reducing sugar. Anal. Chem. **31**, 426–428 (1959). https://doi.org/10.1021/ac60147a030
20. S. Sadasivam, A. Manickam, *Biochemical Methods*, 2nd edn. (New Age International, New Delhi, 1996)
21. A.G. Gopalakrishna, R.G. Raja Rajan, A.S. Bhatnagar, Rice bran: Chemistry, production and applications-a review. Beverage Food World **39**, 31–36 (2012)
22. P. Chugh, J. Kaur, R. Soni, et al., A low-cost process for efficient hydrolysis of deoiled rice bran and ethanol production using an inhouse produced multi-enzyme preparation from *Aspergillus Niger* P-19. J. Mater. Cycles Waste Manag. **25**, 359–375 (2023). https://doi.org/10.1007/s10163-022-01538-y
23. J. Wang, S.J. Fishwild, M. Begel, J.Y. Zhu, Properties of densified poplar wood through partial delignification with alkali and acid pretreatment. J. Mater. Sci. **55**, 14664–14676 (2020). https://doi.org/10.1007/s10853-020-05034-2
24. J. Gominho, R. Costa, A. Lourenço, et al., The effect of different pre-treatments to improve delignification of eucalypt stumps in a biorefinery context. Bioresour Technol. Rep. **6**, 89–95 (2019). https://doi.org/10.1016/j.biteb.2019.02.004
25. H.S. Hafid, F.N. Omar, J. Zhu, M. Wakisaka, Enhanced crystallinity and thermal properties of cellulose from rice husk using acid hydrolysis treatment. Carbohydr. Polym. **260**, 117789 (2021). https://doi.org/10.1016/j.carbpol.2021.117789
26. N. Chen, G. Zhang, P. Zhang, et al., Rice husk-based solid acid for efficient hydrolysis and saccharification of corncob. Bioresour. Technol. **292**, 121915 (2019). https://doi.org/10.1016/j.biortech.2019.121915
27. A. Lorenci Woiciechowski, C.J. Dalmas Neto, P. de Souza, L. Vandenberghe, et al., Lignocellulosic biomass: Acid and alkaline pretreatments and their effects on biomass recalcitrance–conventional processing and recent advances. Bioresour. Technol. **304**, 122848 (2020). https://doi.org/10.1016/j.biortech.2020.122848
28. W.G. Morais Junior, T.F. Pacheco, D. Trichez, et al., Xylitol production on sugarcane biomass hydrolysate by newly identified *Candida tropicalis* JA2 strain. Yeast **36**, 349–361 (2019). https://doi.org/10.1002/yea.3394
29. V. Narisetty, E. Castro, S. Durgapal, et al., High level xylitol production by *Pichia fermentans* using non-detoxified xylose-rich sugarcane bagasse and olive pits hydrolysates. Bioresour. Technol. **342**, 126005 (2021). https://doi.org/10.1016/j.biortech.2021.126005
30. T.M. Louie, K. Louie, S. DenHartog, et al., Production of bio-xylitol from d-xylose by an engineered *Pichia pastoris* expressing a recombinant xylose reductase did not require any auxiliary substrate as electron donor. Microb. Cell Factories **20**, 50 (2021). https://doi.org/10.1186/s12934-021-01534-1
31. K. Malik, E.-S. Salama, T.H. Kim, X. Li, Enhanced ethanol production by *Saccharomyces cerevisiae* fermentation post acidic and alkali chemical pretreatments of cotton stalk lignocellulose. Int. Biodeterior. Biodegradation **147**, 104869 (2020). https://doi.org/10.1016/j.ibiod.2019.104869
32. J. Sharma, V. Kumar, R. Prasad, N.A. Gaur, Engineering of *Saccharomyces cerevisiae* as a consolidated bioprocessing host to produce cellulosic ethanol: Recent advancements and current challenges. Biotechnol. Adv. **56**, 107925 (2022). https://doi.org/10.1016/j.biotechadv.2022.107925

Characterization of Construction and Demolition Waste Aiming to Recycle as Coarse and Fine Aggregate in Kolkata Municipal Corporation Area

Subhadip Pramanik and Debabrata Mazumder

Abbreviations

CDW	Construction and Demolition Waste
ITZ	Interfacial Transition Zone
KM	Kolkata Municipal Corporation
NA	Natural Aggregate
PSD	Particle Size Distribution
RA	Recycled Aggregate
RCA	Recycled Coarse Aggregate
RFA	Recycled Fine Aggregate

1 Introduction

The swift urban development and industrial growth in cities have caused a rising need for building materials, contributing to the exhaustion of natural resources and the generation of substantial quantities of construction and demolition waste (CDW). Since the CDW is typically dumped in landfills, it also influences the environment [1–3]. The materials from CDW emerge from the building of new structures or the remodeling/demolition of existing ones. These structures include buildings, bridges, roads, dams, etc. [4]. Half of the solid trash produced globally consists of building materials. Every stage of construction, from the extraction of raw materials to the

S. Pramanik (✉) · D. Mazumder
Department of Civil Engineering, Indian Institute of Engineering Science and Technology (IIEST), Shibpur, Howrah, West Bengal, India
e-mail: debabrata@civil.iiests.ac.in

D. Mazumder (ed.), *Sustainable Advanced Technologies for Environmental Management*, Springer Proceedings in Earth and Environmental Sciences,
https://doi.org/10.1007/978-3-031-64006-3_36

processing, manufacturing, shipping, and final disposal at the end of a building's useful life, has an impact on the environment that can be linked to CDW [5].

Regarding the environmental concerns associated with recycling facilities, numerous studies have been undertaken to evaluate the environmental impact of CDW recycling facilities. In a study conducted by Marzouk and Azab in 2014 [6], two types of experimental investigations were carried out. The initial study focused on assessing the greenhouse gas emissions from both primary and recycling aggregate units. The findings revealed that recycled aggregate is more environmentally sustainable compared to primary aggregate. In the second study, the researchers aimed to examine the environmental effects of landfilling versus recycling. Their conclusion suggested that recycling is a more environmentally friendly option than landfilling. The estimated proportions of CDW generated are as follows: 30% for India [7], 40% for the EU [8], 50% for the UK, 23% for Hong Kong, 41% for Malaysia [9], 30–40% for China, 26.1% for Germany [10], and 32% for Sweden [11], with respect to total generated solid waste. Laws and regulations that have created a market for recycled construction materials made from the CDW stream are now implemented by governments all over the world to minimize solid waste [12].

The CDW includes materials like concrete, bricks, metals, ceramic, tile, gypsum, asphalt, wood, glass, and plastics [13]. Concrete and clay brick waste make up around 80% of the CDW, and it is possible to turn these wastes into supplementary building materials [14]. The CDW waste mainly consists of a major amount of RCA [15] and RFA [16]. As it contains some different types of impurities, characterization is necessary before it is applied in the construction sector. The main problem with RA is that the recycled concrete aggregate comprises natural aggregate (NA) with a layer of fragile, adhered mortar on its surface, leading to the formation of a feeble interfacial transition zone (ITZ), and gives RA of inferior quality [17, 18]. In case of brick aggregate, it has higher water absorptivity and porosity, which directly affects the quality of concrete when it is used to produce new concrete [19, 20]. Due to the presence of ITZ and brick aggregate, RA shows higher water absorption and lower sp. gravity and also gives lower mechanical properties in terms of abrasion value and impact value [21]. The RA also contains some contaminants such as sulphate, chloride, oil, siliceous, and some other materials [22], It is mostly utilized in road sub-base, base, soil stabilization, and some others activities [23] due to lack of experience and standards for evaluating the quality of the material and shortage of a regular supply. This is not the most environmentally friendly method of recycling, either, as significant amounts of primary aggregates are still needed to produce fresh concrete.

The Kolkata Municipal Corporation (KMC) area, with its historical and contemporary construction activities, witnesses a substantial generation of CDW, posing an environmental threat and resource scarcity. In response to these concerns, the exploration and utilization of recycled materials from CDW have emerged as a critical component of sustainable urban development. Characterizing CDW, with a specific focus on RA, in the KMC area is not only technologically significant but also has practical implications for sustainable urban development. This understanding can guide policymakers, urban planners, and the construction industry in adopting more efficient and eco-friendly practices, reducing the environmental footprint, and harnessing the economic potential of recycled materials in construction projects within the city and beyond. This paper delves into some physical properties and mechanical properties of different sizes of aggregates and their limitations.

2 Materials and Methods

2.1 Materials

In this experimental investigation, the aim was to characterize various sizes of recycled fine and coarse aggregate from CDW. Samples were sourced from a fixed recycling facility situated in New Town, Kolkata. The process for handling CDW begins with the utilization of a mechanical jaw crusher, which is then followed by the separation process facilitated by vibrating screens. To remove metallic elements, magnetic conveyor belts are employed, and any non-metallic impurities, including wood, paper, plastics, glass, and gypsum, are manually extracted after the initial crushing phase. The recycling plant yields five distinct fractions of RA, with particle sizes finer than the sieve size of 4.75, 6.3, 10, 20, and 40 mm, as shown in Fig. 1.

Fig. 1 Various particle sizes of RA

2.2 Methods

Throughout the characterization process, a series of experiments conforming to various standard codes were carried out, as detailed in Table 1. Corresponding to each fraction, four samples were chosen for testing, and the average value was then evaluated. The physical properties of aggregate like particle size distribution (PSD), Sp. gravity, bulk density, water absorption, and percentage of void were checked for both RFA and RCA. Flakiness, elongation index, and combined flakiness elongation index and also some mechanical characterization tests were performed for RCA only. To check the PSD and some mechanical strength of RA, the samples were dried at 105 °C. Samples were cured under water for 24 h before checking the water absorption of RA.

3 Results and Discussion

3.1 Particle Size Distribution of Recycled Aggregate

Aggregate grading plays a pivotal role, in influencing several aspects such as the workability, density, porosity, degree of compaction, strength, and durability of concrete [30]. The particle size distribution (PSD) of each aggregate fraction was evaluated using the testing method specified in the ASTM C136-01 standard, which focuses on the geometric properties of aggregates. Figure 2 depicts the PSD of diverse RA spanning from 4.75 to 40 mm. The PSD graph displays varying percentages of aggregate passing for each size category. As an example, the fine aggregate (4.75 mm) was classified as Zone II fine aggregate, exhibiting a favorable fineness modulus of 3.04 based on sieving data, indicating its high quality. This observation reveals that the PSD in RFA is superior to that in RCA. A consistent distribution of particle sizes implies that the aggregate is uniformly graded across different dimensions. This uniform grading fosters improved particle interaction, resulting in heightened compactness and mechanical strength [22].

Table 1 Physical tests performed for recycled coarse and fine aggregate and the corresponding standard

Test methods	Standard code
Particle size distribution [24]	ASTM C136-01
Sp. gravity and water absorption of coarse aggregate [25]	ASTM C127-07
Sp. gravity and water absorption of fine aggregate [26]	ASTM C128-07
Bulk density and percentage of voids [27]	ASTM C29/C29M - 07
Flakiness and elongation index [28]	ASTM D 4791-99
Impact, crushing, and abrasion value of coarse aggregate [29]	IS 2386-IV

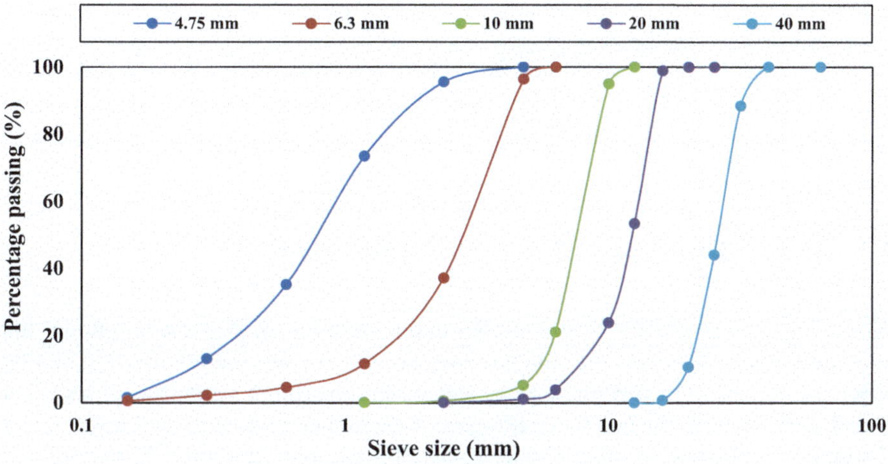

Fig. 2 Particle size distribution profile of different sizes of RA

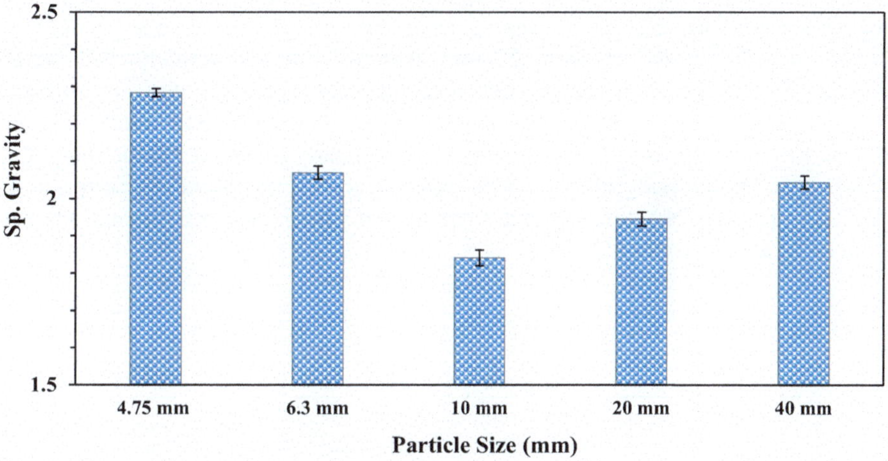

Fig. 3 Profile of sp. gravity of different sizes of RA

3.2 *Sp. Gravity of Recycled Aggregate*

Notably, the sp. gravity is an important factor in the composition of concrete mixtures. The sp. gravity of RA is subject to variation based on the type of recycled material and its origin. To ascertain the sp. gravity of RA, it is customary to adhere to established protocols outlined in applicable testing standards, such as ASTM C127 or ASTM C128 for fine and coarse aggregates. Figure 3 illustrates the specific gravity values for various sizes of RA. The highest specific gravity is linked to 4.75 mm aggregates, whereas the lowest is noted for 10 mm aggregates. According to Poon and Chan [31], the density of recycled concrete aggregate is higher as compared to recycled brick aggregate. Here, RCA contain higher amount of brick aggregate as compared to RFA, due to that reason the sp. gravity is higher in case of RFA as compared to RCA. The RCA with a size of 10 mm holds the highest quantity of brick aggregate (42.8%), resulting in the lowest sp. gravity for this aggregate size.

3.3 Water Absorption of Recycled Aggregate

Water absorption on different sizes of RA is shown in Fig. 4. Water absorption is the most important parameter for concrete mix design. An inverse relationship can be observed between the specific gravity and water absorption properties of recycled gravel. It is well-established that aggregates with higher sp. gravity and lower water absorption contribute to improved concrete strength and durability. Minimum and maximum water absorption is observed as 5.52% and 14.53% for 4.75 and 10 mm sizes, respectively. Water absorption is higher in the case of RCA as it contains a greater amount of recycled brick aggregate, and compared to NA, brick aggregate has higher porosity [32]. Brick aggregate exhibits weight percentages of 42.8%, 29.7%, and 30.1% for 10 mm, 20 mm, and 40 mm, respectively. The water absorption is elevated in the 10 mm size recycled aggregate due to a greater concentration of brick particles. In contrast, the water absorption is comparable for 20 and 40 mm sizes, as their fractional content is nearly identical. Furthermore, the presence of old adhering mortar on the NA surface increases water absorption in RCA. Compared to NA, this adhering mortar usually has higher porosity and a larger capacity to absorb water [33, 34].

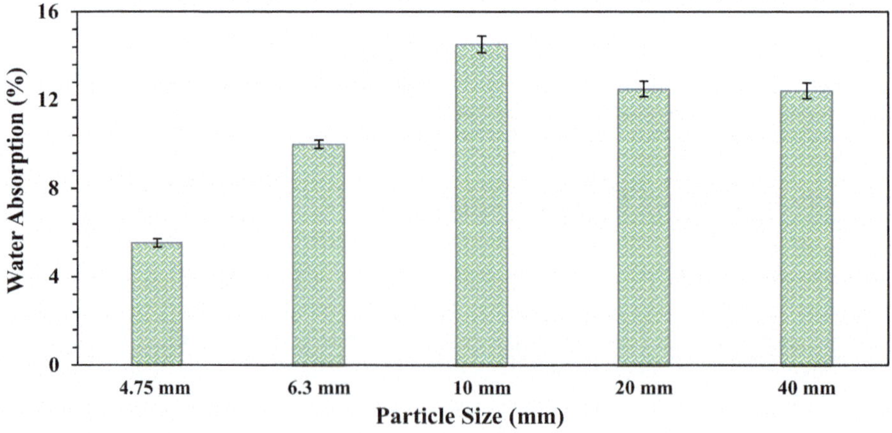

Fig. 4 Profile of water absorption of different sizes of RA

3.4 Bulk Density of Recycled Aggregate

One important consideration when designing a concrete mix is the aggregate's bulk density, as the unit weight of concrete is directly influenced by it. A lower concrete unit weight is the consequence of a decrease in aggregate bulk density. Figure 5 illustrates the variations in bulk density among different sizes of RA. This density is directly proportional to the specific gravity of the aggregate. The aggregate's bulk density is intricately linked to factors such as the crushing procedure, original material quality, particle shape, and size, as reported by Silva et al. [2]. The presence of ancient mortar, which is less dense than NA, further reduces the density of RCA, contributing to lower RCA density. The number of crushing stages also plays a role, with more stages resulting in higher aggregate density due to cumulative breakdown of the cement mortar paste [35]. The maximum bulk density is observed for 4.75 mm aggregate size as 1265 kg/m^3, while the minimum bulk density is recorded for 10 mm size as 1164 kg/m^3. Coarse aggregate shows a trend, where larger sizes lead to higher bulk density, ranging from 1164 to 1185 kg/m^3. When this lightweight aggregate is used in concrete preparation, it is referred to as lightweight concrete.

3.5 Percentage of Voids on Recycled Aggregate

The percentage of voids in aggregate is a crucial factor in the field of civil engineering construction. These voids represent the empty spaces or gaps between individual particles of the aggregate material. Figure 6 illustrates the percentage of voids in various sizes of RA, both in loose and compacted states. The aggregate's bulk density and specific gravity both affect the percentage of voids, as outlined by ASTM C29/C29M -07 [27]. This parameter directly affects how workable the

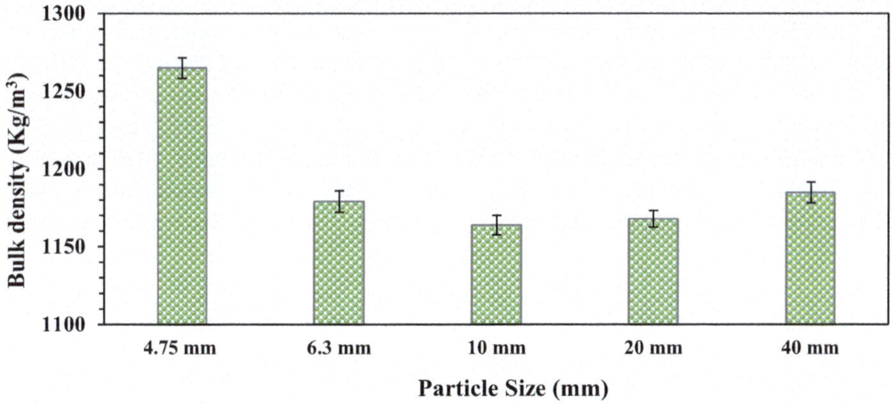

Fig. 5 Profile of bulk density of different sizes of RA

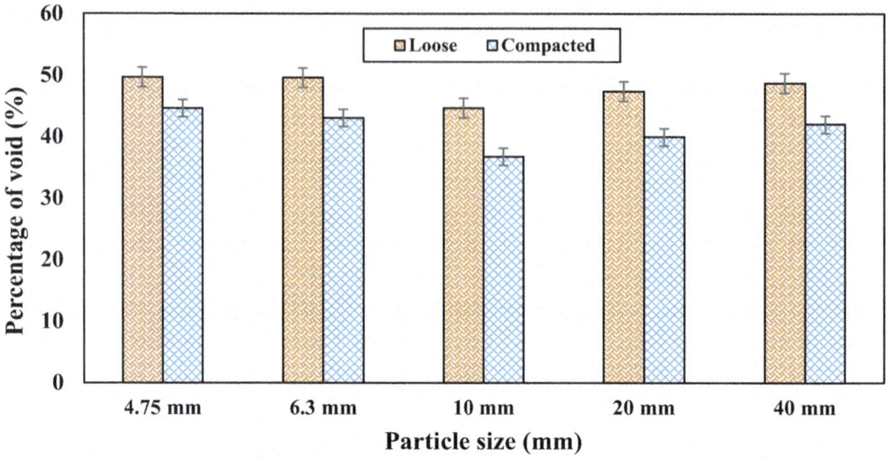

Fig. 6 Percent of voids of different sizes of RA

concrete is; a higher void percentage can result in a rigid mix, complicating the placement and finishing processes. Excessive voids may enhance permeability and reduce the concrete's durability. Figure 6 indicates that void percentages are higher in the loose state compared to the compacted state. The void percentage remains nearly the same for aggregate sizes 4.75 and 6.3 mm. However, the RA with larger aggregate sizes results in higher void percentages.

3.6 Flakiness and Elongation Index of Recycled Coarse Aggregate

The flakiness index, elongation index, and combined flakiness-elongation index for different RCA sizes are shown in Fig. 7. The workability of concrete mixes may be negatively impacted by excessive flakiness and elongation, making it more difficult to achieve a consistent and cohesive blend. Concrete's strength and durability may be reduced by the presence of flaky and elongated particles [36]. Therefore, it is essential to manage these characteristics to uphold the performance of the construction material. It is evident from Fig. 7 that the highest combined flakiness and elongation value is associated with 40 mm aggregates (43.32%), while the lowest value is observed for 20 mm aggregates (35%). This conforms to the standards outlined in IS 383-2016, which stipulate that value not exceeding 40% [37]. As per the standard for combined flakiness and elongation index, the values for 10 and 20 mm aggregate sizes fall within the acceptable range. However, for 40 mm aggregate size, the value exceeded the standard limit. The elevated flakiness and elongation values in RCA are attributed to a higher presence of bricks and ceramic tiles, a phenomenon also noted by Pacheco and de Brito in 2021 [38].

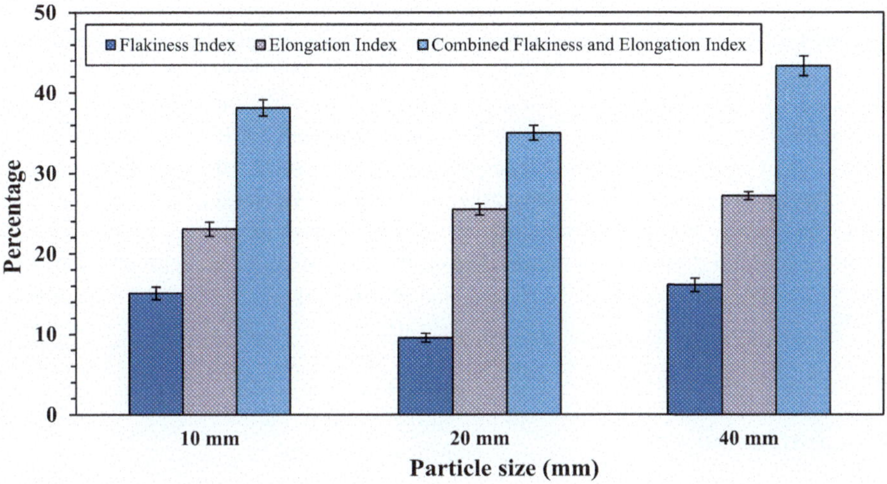

Fig. 7 Profile of flakiness and elongation index of different sizes of RCA

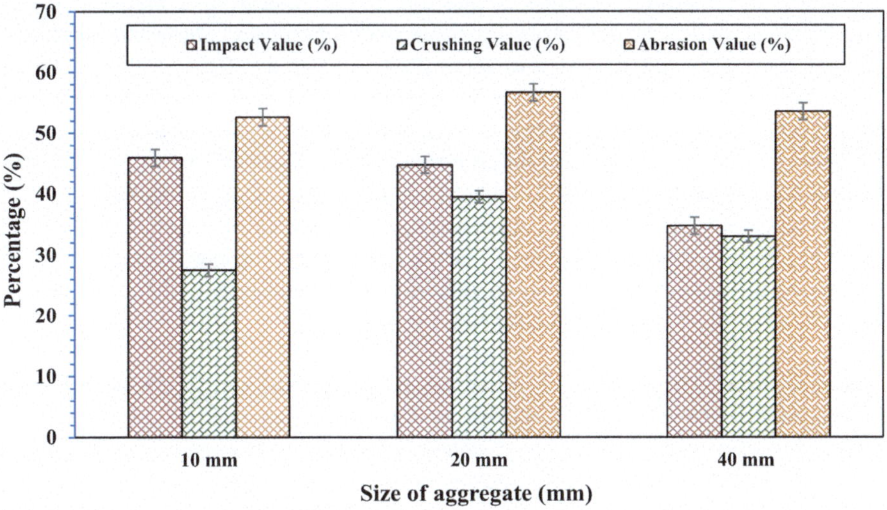

Fig. 8 Profile of impact, crushing, and abrasion value of different sizes of RCA

3.7 Mechanical Properties of Recycled Coarse Aggregate

In order to know the mechanical characteristics of coarse aggregate, tests were conducted by IS 2386-IV-1963 [29]. Figure 8 illustrates some mechanical properties of RCA for 10, 20, and 40 mm sizes. According to IS 383-2016, when concrete is used in wearing surfaces, the impact value, crushing value, and abrasion value should not surpass 30%. In applications other than wear surfaces, the impact value ought to be limited to 45%, the crushing value should remain the same as 30%, and the abrasion value should not go beyond 50% [37]. The results from Fig. 8 indicate that the impact value for only the 40 mm size aggregate is lower, at 34.68, meeting the standard for applications in concrete other than wearing surfaces. However, the values for 10 and 20 mm sizes surpass the standard limit. In terms of the crushing value, only the 10 mm size aggregate falls within the standard limit. The abrasion value for all sizes of RCA exceeds 50%, indicating a deviation from the standard limit. This higher value is attributed to the elevated quantum of brick material, and the presence of a greater amount of adhered mortar pores on the aggregate surface. The mechanical properties of RCA are deemed inadequate based on the available results.

4 Conclusion

Evaluating the properties of recycled aggregates is crucial for optimizing concrete mix design. Conducting a sieve analysis of recycled aggregates, the 4.75 mm-sized aggregates are classified as Zone II fine aggregate, indicating a higher fineness modulus value that is advantageous for construction purposes. The water sorptivity is higher and the specific gravity is lower in RCA compared to RFA, primarily due to a higher proportion of brick aggregate and also due to the presence of pores adhered mortar surface in the natural aggregate (NA). Among the recycled coarse aggregates, the 20 mm-sized aggregates exhibit superior quality based on sp. gravity, water absorption, and combined flakiness and elongation index values, compared to 10 and 40 mm aggregates. The mechanical properties of RCA data represent that it should not be applied on concrete wearing surfaces. After assessing five types of recycled aggregates, it becomes evident that recycled fine aggregates exhibit better quality in comparison to recycled coarse aggregates.

Acknowledgments The present research work is supported under the research fellowship program of Ministry of Education, Government of India.

References

1. S.K. Kirthika, S.K. Singh, Durability studies on recycled fine aggregate concrete. Constr. Build. Mater. **250**, 118850 (2020). https://doi.org/10.1016/j.conbuildmat.2020.118850
2. R.V. Silva, J. De Brito, R.K. Dhir, Properties and composition of recycled aggregates from construction and demolition waste suitable for concrete production. Constr. Build. Mater. **65**, 201–217 (2014). https://doi.org/10.1016/j.conbuildmat.2014.04.117
3. A.A. Nurhanim, State of art reviews on physico-chemical properties of waste concrete aggregate from construction and demolition waste. Iran. J. Energy Environ. **13**, 340–348 (2022). https://doi.org/10.5829/ijee.2022.13.04.03

4. A.A. Oyenuga, R. Bhamidiarri, Sustainable approach to managing construction and demolition waste: An opportunity or a new challenge? Int. J. Innov. Res. Sci. Eng. Technol. **4**, 10368–10378 (2015). https://doi.org/10.15680/IJIRSET.2015.0411007
5. S.K. Ghosh, H.S. Haldar, S. Chatterjee, P. Ghosh, An optimization model on construction and demolition waste quantification from building. Procedia Environ. Sci. **35**, 279–288 (2016). https://doi.org/10.1016/j.proenv.2016.07.008
6. M. Marzouk, S. Azab, Environmental and economic impact assessment of construction and demolition waste disposal using system dynamics. Resour. Conserv. Recycl. **82**, 41–49 (2014). https://doi.org/10.1016/j.resconrec.2013.10.015
7. CPCB. Guidelines on environmental management of construction & demolition (C & D) wastes. Central Pollution Control Board, Ministry of Environment, Forests & Climate Change, Government of India (2017).
8. K.R.A. Nunes, C.F. Mahler, Comparison of construction and demolition waste management between Brazil, European Union and USA. Waste Manag. Res. **38**, 415–422 (2020). https://doi.org/10.1177/0734242X20902814
9. C.M. Mah, T. Fujiwara, C.S. Ho, Construction and demolition waste generation rates for high-rise buildings in Malaysia. Waste Manag. Res. **34**, 1224–1230 (2016). https://doi.org/10.1177/0734242X16666944
10. M. Li, A. Kühlen, J. Yang, F. Schultmann, Improvement of the statutory framework for construction and demolition waste management exemplified in Germany and Australia, in *Urban Environment*, (Springer, Dordrecht, 2013), pp. 15–25. https://doi.org/10.1007/978-94-007-7756-9_2
11. A. Kadawo, M. Sadagopan, O. During, K. Bolton, Combination of LCA and circularity index for assessment of environmental impact of recycled aggregate concrete. J. Sustain. Cem.-Based Mater. **12**, 1–12 (2023). https://doi.org/10.1080/21650373.2021.2004562
12. C. Medina, W. Zhu, T. Howind, M.I. Sánchez De Rojas, M. Frías, Influence of mixed recycled aggregate on the physical-mechanical properties of recycled concrete. J. Clean. Prod. **68**, 216–225 (2014). https://doi.org/10.1016/j.jclepro.2014.01.002
13. R.V. Silva, J. De Brito, R.K. Dhir, Availability and processing of recycled aggregates within the construction and demolition supply chain: A review. J. Clean. Prod. **143**, 598–614 (2017). https://doi.org/10.1016/j.jclepro.2016.12.070
14. Z. Ma, Q. Tang, H. Wu, J. Xu, C. Liang, Mechanical properties and water absorption of cement composites with various fineness and contents of waste brick powder from C&D waste. Cem. Concr. Compos. **114**, 103758 (2020). https://doi.org/10.1016/j.cemconcomp.2020.103758
15. V. Radonjanin, M. Malešev, S. Marinković, A.E.S. Al Malty, Green recycled aggregate concrete. Constr. Build. Mater. **47**, 1503–1511 (2013). https://doi.org/10.1016/j.conbuildmat.2013.06.076
16. H. Hafez, R. Kurda, R. Kurda, B. Al-Hadad, R. Mustafa, B. Ali, A critical review on the influence of fine recycled aggregates on technical performance, environmental impact and cost of concrete. Appl. Sci. **10**, 1018 (2020). https://doi.org/10.3390/app10031018
17. R. Wang, N. Yu, Y. Li, Methods for improving the microstructure of recycled concrete aggregate: A review. Constr. Build. Mater. **242**, 118164 (2020). https://doi.org/10.1016/j.conbuildmat.2020.118164
18. Y. Ding, J. Wu, P. Xu, X. Zhang, Y. Fan, Treatment methods for the quality improvement of recycled concrete aggregate (RCA) – A review. J. Wuhan Univ. Technol. Mater. Sci. Ed. **36**, 77–92 (2021). https://doi.org/10.1007/s11595-021-2380-3
19. L. Berredjem, N. Arabi, L. Molez, Mechanical and durability properties of concrete based on recycled coarse and fine aggregates produced from demolished concrete. Constr. Build. Mater. **246**, 118421 (2020). https://doi.org/10.1016/j.conbuildmat.2020.118421
20. S. Hou, Z. Duan, J. Xiao, L. Li, Y. Bai, Effect of moisture condition and brick content in recycled coarse aggregate on rheological properties of fresh concrete. J. Build. Eng. **35**, 102075 (2021). https://doi.org/10.1016/j.jobe.2020.102075

21. F. Kazemian, H. Rooholamini, A. Hassani, Mechanical and fracture properties of concrete containing treated and untreated recycled concrete aggregates. Constr. Build. Mater. **209**, 690–700 (2019). https://doi.org/10.1016/j.conbuildmat.2019.03.179
22. M. Martín-Morales, M. Zamorano, A. Ruiz-Moyano, I. Valverde-Espinosa, Characterization of recycled aggregates construction and demolition waste for concrete production following the Spanish Structural Concrete Code EHE-08. Constr. Build. Mater. **25**, 742–748 (2011). https://doi.org/10.1016/j.conbuildmat.2010.07.012
23. E. Moreno-Pérez, J. Hernández-Ávila, Y. Rangel-Martínez, E. Cerecedo-Sáenz, A. Arenas-Flores, M.I. Reyes-Valderrama, E. Salinas-Rodríguez, Chemical and mineralogical characterization of recycled aggregates from construction and demolition waste from Mexico city. Minerals **8**, 237 (2018). https://doi.org/10.3390/min8060237
24. ASTM Standard C136-01. Standard test method for sieve analysis of fine and coarse aggregates. American society for testing and materials, West Conshohocken, Pennsylvania (2001).
25. ASTM Standard C127-07. Standard test method for density, relative density (Specific Gravity), and Absorption of coarse aggregate. American society for testing and materials, West Conshohocken, Pennsylvania (2007).
26. ASTM Standard C128-07. Standard test method for density, relative density (Specific Gravity), and Absorption of fine aggregate. American society for testing and materials, West Conshohocken, Pennsylvania (2007).
27. ASTM Standard C 29/C 29M – 07. Standard test method for bulk density ("Unit Weight") and Voids in aggregate. american society for testing and materials, West Conshohocken, Pennsylvania (2007).
28. ASTM Standard D 4791-99. Standard test method for flat particles, elongated particles, or flat and elongated particles in coarse aggregate. American society for testing and materials, West Conshohocken, Pennsylvania (1999).
29. Bureau of Indian Standards, *IS 2386: Part IV, Methods of Test for Aggregates for Concrete – Mechanical Properties*, vol 2366 (Bureau of Indian Standards, New Delhi, 1997), pp. 466–472. http://shop.bsigroup.com/en/ProductDetail/?pid=000000000010037736
30. M. Etxeberria, E. Vázquez, A. Marí, M. Barra, Influence of amount of recycled coarse aggregates and production process on properties of recycled aggregate concrete. Cem. Concr. Res. **37**, 735–742 (2007). https://doi.org/10.1016/j.cemconres.2007.02.002
31. C.S. Poon, D. Chan, Feasible use of recycled concrete aggregates and crushed clay brick as unbound road sub-base. Constr. Build. Mater. **20**, 578–585 (2006). https://doi.org/10.1016/j.conbuildmat.2005.01.045
32. C.S. Poon, D. Chan, Effects of contaminants on the properties of concrete paving blocks prepared with recycled concrete aggregates. Constr. Build. Mater. **21**, 164–175 (2007). https://doi.org/10.1016/j.conbuildmat.2005.06.031
33. J.V.M. Raman, V. Ramasamy, Various treatment techniques involved to enhance the recycled coarse aggregate in concrete: A review. Mater. Today Proc. **45**, 6356–6363 (2020). https://doi.org/10.1016/j.matpr.2020.10.935
34. A. Djerbi, Effect of recycled coarse aggregate on the new interfacial transition zone concrete. Constr. Build. Mater. **190**, 1023–1033 (2018). https://doi.org/10.1016/j.conbuildmat.2018.09.180
35. S. Nagataki, A. Gokce, T. Saeki, M. Hisada, Assessment of recycling process induced damage sensitivity of recycled concrete aggregates. Cem. Concr. Res. **34**, 965–971 (2004). https://doi.org/10.1016/j.cemconres.2003.11.008
36. F. Agrela, A. Barbudo, A. Ramírez, J. Ayuso, M.D. Carvajal, J.R. Jiménez, Construction of road sections using mixed recycled aggregates treated with cement in Malaga, Spain. Resour. Conserv. Recycl. **58**, 98–106 (2012). https://doi.org/10.1016/j.resconrec.2011.11.003
37. Bureau of Indian Standards, *IS 383: Specification for Coarse and Fine Aggregates From Natural Sources for Concrete* (Bureau of Indian Standards, New Delhi, 1970), pp. 1–24
38. J. Pacheco, J. de Brito, Recycled aggregates produced from construction and demolition waste for structural concrete: Constituents, properties and production. Materials (Basel) **14**, 5748 (2021). https://doi.org/10.3390/ma14195748

Biodegradation Study on Conventional Plastics Using Single-Stage Anaerobic Digestion

Penaganti Praveen and Debabrata Mazumder

Abbreviations

COD	Chemical oxygen demand
HDPE	High-density polyethylene
HRT	Horizontal retention time
LDPE	Low-density polyethylene
MLVSS	Mixed liquor volatile suspended solids
MSM	Mineral salt media
PCL	Polycaprolactone
PET	Polyethylene
PHA	Polyhydroxyalkanoate
PHB	Polyhydroxy butyrate
PHBV	Polyhydroxy butyrate-co-hydroxy valerate
PLA	Polylactic acid
PP	Polypropylene
PS	Polystyrene
PVC	Polyvinyl chloride
SEM	Scanning electron microscope
UASB	Up flow anaerobic sludge blanket reactor
VOC	Volatile organic compound

P. Praveen (✉) · D. Mazumder
Civil Engineering Department, Indian Institute of Engineering Science and Technology, Shibpur, Howrah, West Bengal, India
e-mail: debabrata@civil.iiests.ac.in

D. Mazumder (ed.), *Sustainable Advanced Technologies for Environmental Management*, Springer Proceedings in Earth and Environmental Sciences,
https://doi.org/10.1007/978-3-031-64006-3_37

1 Introduction

Plastics are a category of synthetic materials created from polymers, which consist of long chains of molecules. They are recognized for their versatility, durability, and wide range of applications. Plastics are derived from various organic compounds, primarily sourced from petroleum and natural gas [1]. However, there is an increasing trend of plastic production using renewable resources like corn and sugarcane as the sources [2]. In the modern society, plastics have become ubiquitous, finding applications in numerous products that include packaging materials, consumer goods, automotive parts, electronics, construction materials, and medical devices [3]. They are preferred due to their lightweight, cost-effective manufacturing processes, resistance to moisture, chemicals, and degradation. However, the extensive use of plastics (conventional plastics) has raised significant environmental concerns. Plastics are highly durable and non-biodegradable leading to their persistence in the environment for extended periods [4]. This persistence contributes to plastic pollution, which poses risks to the wildlife and the ecosystem. The accumulation of plastic waste in oceans and landfills has prompted global awareness regarding the urgent need for sustainable plastic management [2]. On the other side, the increased recycling efforts and the development of alternative materials are not meeting the need for sustainable plastic management.

Although alternative materials like biodegradable plastics exist, achieving comparable workability across all sectors poses a challenge due to variations in their characteristics and properties. Biodegradable plastics, derived from substances susceptible to microbial breakdown, undergo enzymatic degradation by microorganisms like bacteria and fungi [5]. These microorganisms release enzymes that decompose the plastic into smaller molecules, serving as nourishment. The speed of plastic biodegradation relies on factors like plastic type, environmental conditions, and microbial presence [6]. Certain plastics, including polylactic acid (PLA) and polyhydroxyalkanoate (PHA), readily biodegrade in diverse environmental settings. In contrast, plastics like polyethylene terephthalate (PET), high-density polyethylene (HDPE), low-density polyethylene (LDPE), polystyrene (PS), polypropylene (PP), and polyvinylchloride (PVC) exhibit limited biodegradability in natural conditions but may degrade under certain circumstances, such as the presence of specific microorganisms [3]. The environmental conditions that affect the biodegradation of plastic include temperature, moisture content, and oxygen concentration. The higher the temperature and the moisture content, the quicker the plastic biodegradation [7].

Anaerobic digestion (AD) is a promising technology for stabilizing the waste and converting it to energy in form of methane gas generation [8]. The variation in the plastic properties and operating methods of the AD system makes it difficult to compare the biodegradation studies from the literature, and sometimes, the study's results contradict. For instance, in a study by Vasmara and Marchetti [9], the biodegradation assessment of ground PLA cups during 90 days of mesophilic anaerobic digestion yielded no conclusive results. Another investigation by Itävaara et al. [10] observed a 60% biodegradation yield for PLA granules within 100 days under similar mesophilic anaerobic conditions. The degradation efficiency of polymers exhibits an improvement under thermophilic conditions compared to the mesophilic ones [8, 11]. Cazaudehore et al. [5] categorized biodegradable plastics in anaerobic digestion into three significant groups: polymers readily biodegradable under both mesophilic and thermophilic conditions, e.g., thermoplastic starch (TPS) and polyhydroxybutyrate (PHB); polymers displaying slow biodegradation under mesophilic conditions but quicker biodegradation under thermophilic conditions, e.g., polycaprolactone (PCL) and polylactic acid (PLA); and polymers meeting the biodegradable criteria of EN 13432 standards. However, not all notable biodegradable polymers have caught equal attention; existing studies predominantly focus on PHB, PCL, and PLA degradation in mesophilic conditions and PLA, PCL, and polybutylene succinate (PBS) degradation in thermophilic conditions [11, 12]. Conversely, conventional plastics like PP and PET have observed no biodegradation under mesophilic anaerobic digestion even after 56 days [1]. Only a few studies investigated the biodegradability of bioplastics through an anaerobic digestion system, while that of conventional plastics investigations are limited.

With this background, the present study was investigated to measure the biodegradation potential of conventional plastics like LDPE, HDPE, PS, PVC, PET, and PP in a single stage anaerobic suspended growth batch reactor under mesophilic condition. The experimental period of the study was also increased compared to the previous studies, in order to examine the ultimate biodegradation potential of each plastic. SEM (scanning electron microscope) was used to assess the path in which these plastic pellets biodegraded under anaerobic conditions. This research is expected to contribute in the development of a more sustainable approach for disposal of conventional plastics.

2 Materials and Methods

Laboratory experiment was conducted to study the biodegradability of plastics pellets using anaerobic suspended growth batch reactor system. The extent of biodegradation was assessed by mass loss, SEM analysis, and analytical methods.

2.1 *Substrates and Inoculum*

Plastic pellets (LDPE, HDPE, PS, PVC, PET, and PP) were used as solid substrates for anaerobic digestion for the degradation by the microbes in the digester. These pellets were purchased from the polymer manufacturing industry through respective suppliers. Xylene is a volatile organic compound (VOC) commonly used as a solvent. It is also a bio-degradant and can be degraded by microorganisms. Xylene was used as a supporting substrate to acclimatize the anaerobic digestion system for

Table 1 Properties of the inoculum

Properties	pH	MLSS (mg/L)	MLVSS (mg/L)	COD (mg/L)	VFA (mg/L)	Alkalinity (mg/L)	SVI (mL/g)
Inoculum	7.8	12,000	8400	450	150	1250	150

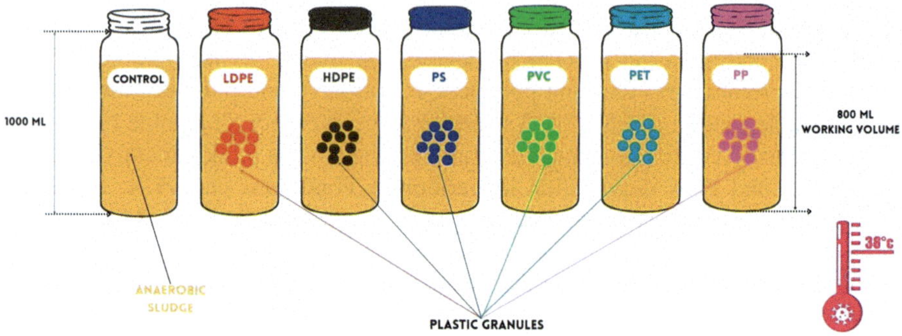

Fig. 1 Schematic arrangement of the experimental setup

the biodegradation of conventional plastics because it can help to enrich the microbial community, induce the production of enzymes that can degrade plastic, and increase the rate of biodegradation [13].

The inoculum used for anaerobic biodegradation system is from the UASB wastewater treatment unit. It is an anaerobic digester used to treat wastewater from the plastic manufacturing industry. It is a high-rate digester that can produce methane from similar kind of substrates. The inoculum from this unit contained a diverse microbial community capable of breaking down plastics and petroleum oil. The properties of the inoculum are shown in the Table 1.

2.2 *Experimental Setup*

The anaerobic digestion system for the biodegradation of plastic pellets was conducted in the small-scale laboratory units. The schematic diagram of the experimental setup is shown in Fig. 1. The anaerobic biodegradation was carried out in a single-stage anaerobic suspended growth batch reactor with a working volume of 0.8 L for each plastic individually. Therefore, 7 nos. of 1-L glass containers were considered for six different types of plastics (LDPE, HDPE, PS, PVC, PET, and PP) and seventh one for control. Temperatures were maintained at mesophilic conditions (37 ± 1 °C), and the study was conducted for 250 days. The reactors were initiated with substrate material (LDPE, HDPE, PS, PVC, PET, and PP) of 10 g of each plastic mixed with 500 mL of active methanogenic sludge collected from the UASB digestor of a wastewater treatment plant operated by a plastic manufacturing industry. Also, 300 mL of cow dung was added into the single-stage batch reactor to achieve the desired solids content for the experiment, and mineral salt medium (MSM) media [14] (K_2HPO_4 0.7 g/L; $NaNO_3$ 2.0 g/L; Na_2HPO_4 0.9 g/L; $CaCl_2$ 0.1 g/L; $MgSO_4{\cdot}7H_2O$ 0.4 g/L; $MnSO_4{\cdot}4H2O$ 1.5 g/L; $FeSO_4{\cdot}7H_2O$ 2.0 g/L; $(NH_4)_6Mo_7O_{24}{\cdot}4H_2O$ 0.6 g/L) was provided to the microbes.

Acclimatization was initially conducted using a supportive substrate such as xylene for 60 days. The feedstock was prepared for acclimatizing the microbes using xylene feed of 5000 mg/L, and feeding started with 50 mg/L to maintain all over 0.8 L working volume; thereafter feed was increased @ 50 mg/L every time up to 300 mg/L. The C:N:P ratio of 20:5:1 was maintained in the reactor by adjusting the feed using xylene (considering carbon content), ammonium nitrate, and potassium dihydrogen phosphate.

2.3 Analytical Procedure

The mass of the granules was measured using a weighing balance. The amount of mass loss is a reasonable indication about the extent of biodegradation. The cumulative mass loss was noted at the end of every week. Before measuring the mass of plastic pellets, the pellets were washed with distilled water and dried in a hot air oven at 60 °C for 1 h.

$$\text{Cummulative mass loss }\% = \frac{\text{Initial mass} - \text{residual mass}}{\text{Initial mass}} \times 100$$

Where:

Initial mass = mass of the sample at the beginning of the experiment,
Residual mass = mass of the sample on a particular measured day.

It is important to note that the cumulative mass loss may not be a linear function of time. The biodegradation rate may vary over time, depending on the environmental conditions and the type of microorganisms present. Hence, it is essential to measure the cumulative mass loss at regular intervals to estimate the extent of biodegradation. On the other side, the biological parameters like pH, COD, and MLVSS were monitored and analyzed to identify the biodegradation pathway. SEM analysis was conducted to study the surface morphology of each plastic and thereby to understand the nature of biodegradation.

3 Results and Discussion

The importance of studying the biodegradation of plastics using anaerobic digestion has become crucial and has increasingly received attention due to the toxicity of plastic waste in the environment, eventually affecting the ecosystem. As stated

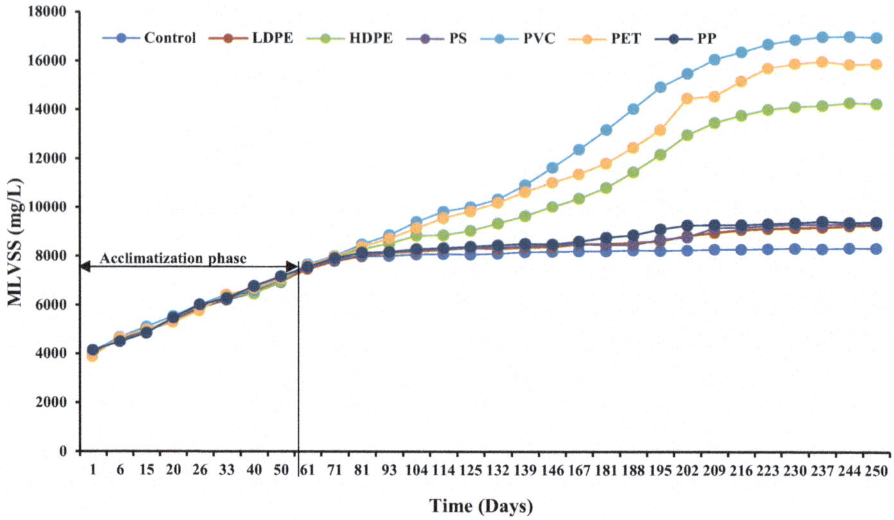

Fig. 2 Profile of MLVSS growth in various reactors containing individual plastic

earlier, the biodegradation of conventional plastics (LDPE, HDPE, PET, PVC, PS, and PP) was investigated using anaerobic digestion. The anaerobic digestion system consisted of a 1 L suspended growth batch reactor for all six plastics individually, along with one control operated at a temperature of 37 °C and a pH of 7.5. The batch period was adopted as 250 days. The biodegradation rates of these six plastics were assessed through the parameters, namely, COD, MLVSS, mass loss, and SEM images. During the acclimatization phase, the pH of all the reactors, including the control, fell below 7.0, and it was adjusted and maintained between 7.5 and 8.0.

Microbial biomass, measured as mixed liquor volatile suspended solids (MLVSS), increased in anaerobic reactors (PVC, PET, and HDPE) during plastic degradation due to the growth of microorganisms capable of metabolizing plastics as shown in Fig. 2. It reveals that PVC has the highest growth in the MLVSS growth profile, rising from 8000 to 17,000 mg/L after the acclimatization phase, followed by PET and HDPE, increased to 14,000 and 16,000 mg/L. This indicates successful acclimatization of microorganisms to plastic degradation byproducts, supporting efficient conversion and utilization of the plastic residuals. In contrary, for PS, PP, and LDPE reactors, MLVSS was enhanced from 4000 to 8000 mg/L during the acclimatization phase and, thereafter, including control reactor MLVSS was reduced to 7000 mg/L due to the lack of microbial activity and endogenous decay. Although PS, PP, and LDPE were slightly higher in the control reactor, the nominal differences were not statistically significant. In view of that, PVC, PET, and HDPE appear to be more readily available carbon sources for microbes to maintain growth in anaerobic digestion systems than PS, PP, and LDPE. This is also attributable to the chemical structure of these polymers. PVC, PET, and HDPE are relatively amorphous polymers, as they have a disordered crystal structure [15, 16], increasing their

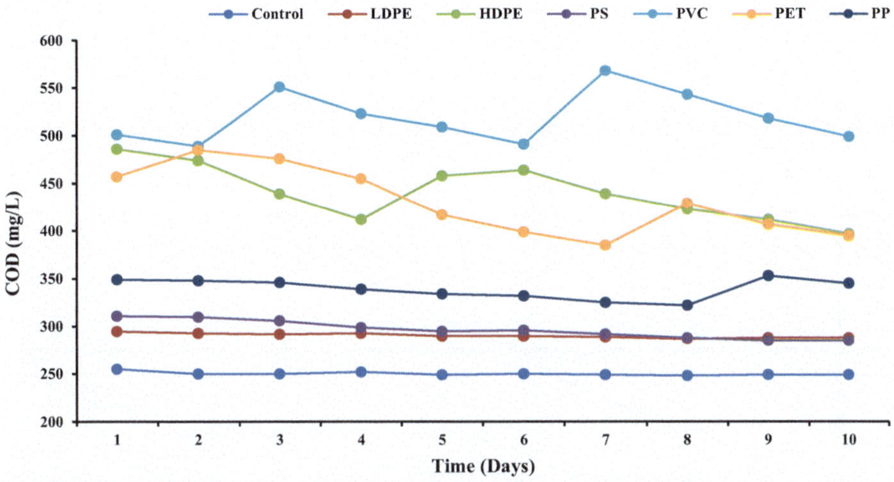

Fig. 3 Profile of COD in various reactors containing individual plastic

susceptibility to enzymatic attack. PS, PP, and LDPE are more crystalline polymers, implying that they have a more ordered structure [17], making them more resistant to enzymatic attack. Another possible explanation for the observed results is that different polymers attract different microbial communities for biodegradation. Hypothetically, PVC, PET, and HDPE may attract microbial communities that are more efficient in biodegrading these polymers.

The six plastics were subjected to anaerobic digestion, with the COD parameter monitored for biodegradation behavior. Figure 3 shows the COD profile and the control of each plastic-containing reactor. LDPE, PP, and PS displayed consistent COD values, indicating limited biodegradation. Their resistance to anaerobic microbial activities had contributed to this stability. Meanwhile, PVC, PET, and HDPE showed notable increase in COD values, suggesting potential biodegradation. It has been observed that plastic degradation in anaerobic reactors led to an initial increase in COD from 500 to 560 mg/L on the third day due to the breakdown of complex plastic polymers into simpler organic compounds. Thereafter, COD was decreased from 560 to 490 mg/L on the sixth day possibly due to the mineralization of the simple organic compounds. Similar phenomenon had taken place in the case of PET and HDPE. Microbial adaptation to plastic degradation caused a temporary surge in COD as microorganisms got acclimated to utilize plastic constituents as a carbon source. As the anaerobic microbial community becomes more efficient in breaking down plastics, the COD may subsequently decrease, reflecting the successful degradation of plastic compounds. The results of MLVSS growth and COD profile of PVC, PET, and HDPE are reflected in the mass loss profile as shown in Fig. 4.

The results of mass loss of all six plastics are shown in Fig. 4, indicating that PVC has the highest biodegradation rate, followed by PET and HDPE. The percent mass loss for PVC was 11.6%, while PET and HDPE exhibited a mass loss of 3.7% and 3.2%, respectively, after 250 days of digestion. The remaining plastics, PS, LDPE,

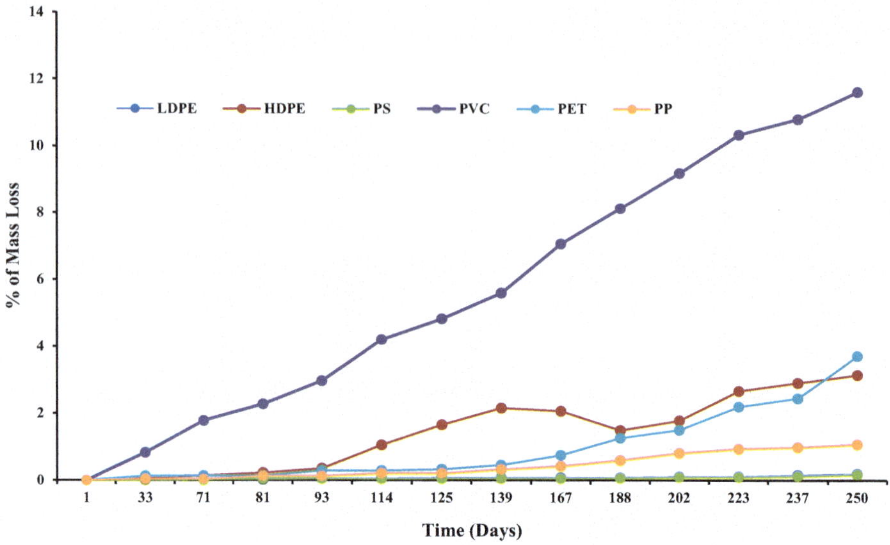

Fig. 4 Profile of mass loss in various reactors containing individual plastic over 250 days

and PP, do not have any considerable mass loss. SEM analysis was conducted for the six plastics after 250 days of suspended growth anaerobic digestion process. SEM images of plastics before and after mineralization showed substantial changes in the surface of the PVC material (Fig. 5a, b) and nominal degradation of the PET plastic (Fig. 5e, f) and HDPE (Fig. 5c, d). PP also has some surface changes (Fig. 5e, f) and showed 1% mass loss as shown in Fig. 4. On the other hand, the SEM images of LDPE (Fig. 5g, h) and PS (Fig. 5k, l) have not shown any qualitative surface changes.

The results of this study demonstrate that PVC, PET, and HDPE can be biodegraded using anaerobic digestion. Though the results do not appear to be perfectly corroborating, when compared to previous studies, it is considerable. One study investigated biodegradability under anaerobic digestion and reported no degradation observed for PET and PP even after 50 days of digestion in a mesophilic anaerobic digester [1]. Conventional plastics, exemplified by cellulose acetate [18] and PETE [19], exhibit minimal biodegradation. Successful biodegradation of PVC was achieved only when the plastic underwent a prior Fenton oxidation process [20]. The presence of HDPE, PP, and PS reportedly inhibits biogas production [21]. Incorporating additives that induce abiotic oxidation to convert them into oxo-degradable plastics does not result in the biodegradation of PE and PETE [22]. Conversely, PETE-PLA copolymers achieved mineralization ranging from 34% to 69%, based on the initial aromatic to aliphatic ratio [19]. Similarly, the introduction of starch in composites with PE selectively promoted the biodegradation of the bio-based polymer [23]. Consequently, conventional plastics, even when combined with pro-oxidant additives, do not exhibit substantial biodegradation. This

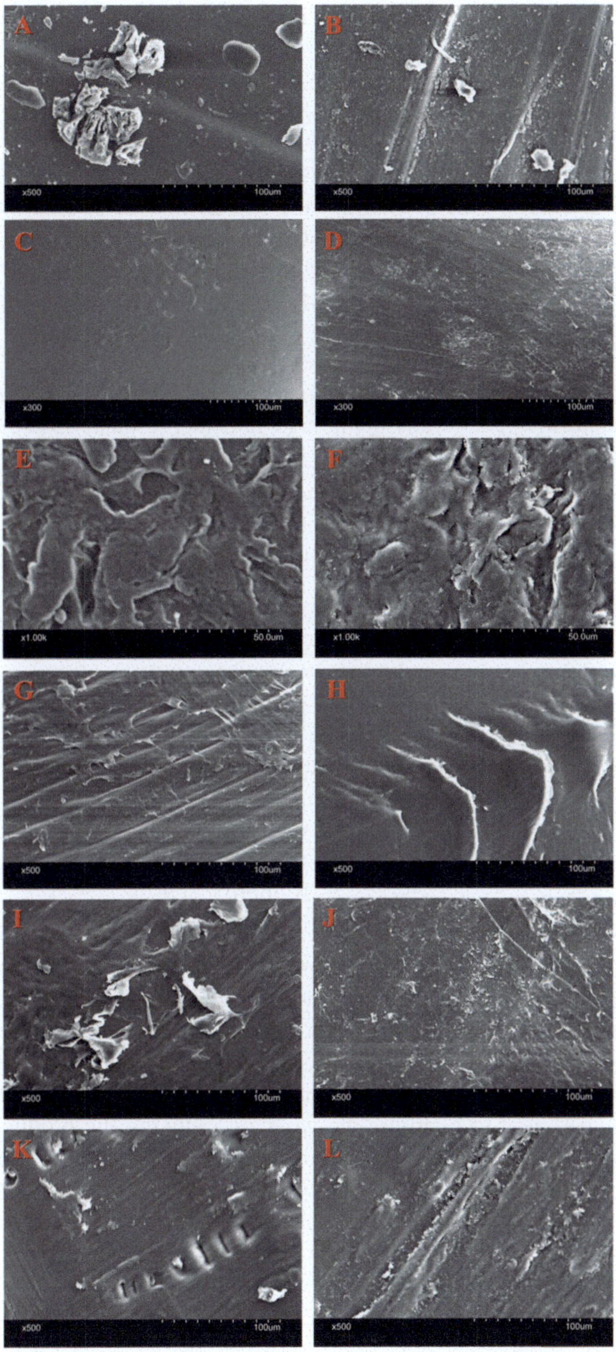

Fig. 5 Scanning electron micrographs of plastics during 1 year of single-stage anaerobic digestion reactor. From top to bottom: PVC (**a**: initial, **b**: final), HDPE (**c**: initial, **d**: final), PET (**e**: initial, **f**: final), LDPE (**g**: initial, **h**: final), PP (**i**: initial, **j**: final) and PS (**k**: initial, **l**: final)

could be due to the experimental conditions which led to negative results in literature.

The degradation of plastics within anaerobic digestion systems is a multifaceted process involving diverse microorganisms. The specific organisms engaged in plastic biodegradation vary, contingent on the plastic type and anaerobic digestion system conditions. Generally, plastic biodegradation in anaerobic digestion follows a two-step progression. Firstly, extracellular enzymes catalyze the hydrolysis of plastics into smaller molecules. Subsequently, acetogenic and methanogenic bacteria degrade the smaller molecules yielding methane and carbon dioxide. Biodegradation of plastics within anaerobic digestion systems presents several advantages for plastic waste management. Anaerobic digestion stands out as a simple, cost-effective, and sustainable technology. It not only generates renewable energy in the form of methane but also aids in mitigating greenhouse gas emissions by preventing methane release from landfills. This study's findings underscore anaerobic digestion as a promising technology for PVC, PET, and HDPE biodegradation.

4 Conclusion

This study demonstrates that the conventional plastics are not readily biodegradable under mesophilic anaerobic conditions. Only PVC, PET, and HDPE can be biologically degraded, with PVC being the most biodegradable and HDPE being the least. LDPE, PS, and PP showed no biodegradation even after 250 days. The biodegradation rates observed in this study are significantly lower than those reported for biodegradable plastics under similar conditions. This suggests that anaerobic digestion is less effective for the biodegradation of conventional plastics than bioplastics. Further research is needed to optimize the anaerobic digestion process for the biodegradation of conventional plastics. This may include investigating the effect of various operating parameters (such as temperature, pH, and retention time) on biodegradation rates and developing more efficient microbial strains for plastic degradation. Additionally, it is essential to note that although some conventional plastics showed nominal biodegradation in this study, the biodegradation rates were comparatively still low. This means that even if anaerobic digestion were optimized, it would likely be a partial solution for managing conventional plastic waste. Therefore, it is essential to continue the work on reducing the production and use of conventional plastics and developing more effective methods for their recycling and disposal. Further research is needed to determine the specific microorganisms involved in the biodegradation of different types of plastics in anaerobic digestion systems. Also, it is required to optimize the working conditions of anaerobic digestion systems for the biodegradation of conventional plastics.

Acknowledgments The present research work is supported under the research fellowship program of Ministry of Education, Government of India.

References

1. E.F. Gómez, F.C. Michel, Biodegradability of conventional and bio-based plastics and natural fiber composites during composting, anaerobic digestion and long-term soil incubation. Polym. Degrad. Stab. **98**, 2583–2591 (2013). https://doi.org/10.1016/j.polymdegradstab.2013.09.018
2. G. Gadaleta, S. De Gisi, Z.K. Chong, J. Heerenklage, M. Notarnicola, K. Kuchta, L. Cafiero, M. Oliviero, A. Sorrentino, C. Picuno, Degradation of thermoplastic cellulose acetate-based bioplastics by full-scale experimentation of industrial anaerobic digestion and composting. Chem. Eng. J. **462**, 142301 (2023). https://doi.org/10.1016/j.cej.2023.142301
3. P. Praveen, D. Mazumder, A simplistic mathematical model for two-stage anaerobic digestion of plastic wastes, in *Springer Proceedings in Earth and Environmental Sciences*, (Springer Nature, Cham, 2023), pp. 255–268. https://doi.org/10.1007/978-3-031-37596-5_19
4. Z. Shang, R. Wang, X. Zhang, Y. Tu, C. Sheng, H. Yuan, L. Wen, Y. Li, J. Zhang, X. Wang, G. Yang, Y. Feng, G. Ren, Differential effects of petroleum-based and bio-based microplastics on anaerobic digestion: A review. Sci. Total Environ. **875**, 162674 (2023). https://doi.org/10.1016/j.scitotenv.2023.162674
5. G. Cazaudehore, F. Monlau, C. Gassie, A. Lallement, R. Guyoneaud, Active microbial communities during biodegradation of biodegradable plastics by mesophilic and thermophilic anaerobic digestion. J. Hazard. Mater. **443**, 130208 (2023). https://doi.org/10.1016/j.jhazmat.2022.130208
6. D. Pangallo, A. Gelsomino, F. Fazzino, A. Pedullà, P.S. Calabrò, The fate of biodegradable plastic during the anaerobic co-digestion of excess sludge and organic fraction of municipal solid waste. Waste Manag. **168**, 98–106 (2023). https://doi.org/10.1016/j.wasman.2023.05.053
7. Z. Cai, M. Li, Z. Zhu, X. Wang, Y. Huang, T. Li, H. Gong, M. Yan, Biological degradation of plastics and microplastics: A recent perspective on associated mechanisms and influencing factors. Microorganisms **11**, 1661 (2023). https://doi.org/10.3390/microorganisms11071661
8. A. Abraham, H. Park, O. Choi, B.I. Sang, Anaerobic co-digestion of bioplastics as a sustainable mode of waste management with improved energy production—A review. Bioresour. Technol. **322**, 124537 (2021). https://doi.org/10.1016/J.BIORTECH.2020.124537
9. C. Vasmara, R. Marchetti, Biogas production from biodegradable bioplastics. Environ. Eng. Manag. J. **15**, 2041–2048 (2016) http://omicron.ch.tuiasi.ro/EEMJ/
10. M. Itävaara, S. Karjomaa, J.F. Selin, Biodegradation of polylactide in aerobic and anaerobic thermophilic conditions. Chemosphere **46**, 879–885 (2002). https://doi.org/10.1016/S0045-6535(01)00163-1
11. V. Bátori, D. Åkesson, A. Zamani, M.J. Taherzadeh, I. Sárvári Horváth, Anaerobic degradation of bioplastics: A review. Waste Manag. **80**, 406–413 (2018). https://doi.org/10.1016/J.WASMAN.2018.09.040
12. G. Cazaudehore, R. Guyoneaud, P. Evon, L. Martin-Closas, A.M. Pelacho, C. Raynaud, F. Monlau, Can anaerobic digestion be a suitable end-of-life scenario for biodegradable plastics? A critical review of the current situation, hurdles, and challenges. Biotechnol. Adv. **56**, 107916 (2022). https://doi.org/10.1016/J.BIOTECHADV.2022.107916
13. S. Wong, N. Ngadi, T.A.T. Abdullah, Study on dissolution of low-density polyethylene (LDPE). Appl. Mech. Mater. **695**, 170–173 (2014). https://doi.org/10.4028/www.scientific.net/AMM.695.170
14. L. Giacomucci, N. Raddadi, M. Soccio, N. Lotti, F. Fava, Polyvinyl chloride biodegradation by Pseudomonas citronellolis and Bacillus flexus. New Biotechnol. **52**, 35–41 (2019). https://doi.org/10.1016/J.NBT.2019.04.005
15. D. Sorolla-Rosario, J. Llorca-Porcel, M. Pérez-Martínez, D. Lozano-Castelló, A. Bueno-López, Study of microplastics with semicrystalline and amorphous structure identification by TGA and DSC. J. Environ. Chem. Eng. **10**, 106886 (2022). https://doi.org/10.1016/J.JECE.2021.106886
16. R. Nisticò, Polyethylene terephthalate (PET) in the packaging industry. Polym. Test. **90**, 106707 (2020). https://doi.org/10.1016/J.POLYMERTESTING.2020.106707

17. Z.O.G. Schyns, M.P. Shaver, Z.O.G. Schyns, M.P. Shaver, Mechanical recycling of packaging plastics: A review. Macromol. Rapid Commun. **42**, 2000415 (2021). https://doi.org/10.1002/MARC.202000415
18. C.J. Rivard, W.S. Adney, M.E. Himmel, D.J. Mitchell, T.B. Vinzant, K. Grohmann, L. Moens, H. Chum, Effects of natural polymer acetylation on the anaerobic bioconversion to methane and carbon dioxide. Appl. Biochem. Biotechnol. **34–35**, 725–736 (1992). https://doi.org/10.1007/BF02920592/METRICS
19. S. Hermanová, P. Šmejkalová, J. Merna, M. Zarevúcka, Biodegradation of waste PET based copolyesters in thermophilic anaerobic sludge. Polym. Degrad. Stab. **111**, 176–184 (2015). https://doi.org/10.1016/J.POLYMDEGRADSTAB.2014.11.007
20. T. Mackuak, A. Takáčová, M. Gál, M. Marton, J. Ryba, PVC degradation by Fenton reaction and biological decomposition. Polym. Degrad. Stab. **120**, 226–231 (2015). https://doi.org/10.1016/J.POLYMDEGRADSTAB.2015.07.005
21. J.W. Lim, D.W.Q. Ting, K.C. Loh, T. Ge, Y.W. Tong, Effects of disposable plastics and wooden chopsticks on the anaerobic digestion of food waste. Waste Manag. **79**, 607–614 (2018). https://doi.org/10.1016/J.WASMAN.2018.08.033
22. S. Selke, R. Auras, T.A. Nguyen, E. Castro Aguirre, R. Cheruvathur, Y. Liu, Evaluation of biodegradation-promoting additives for plastics. Environ. Sci. Technol. **49**, 3769–3777 (2015). https://doi.org/10.1021/ES504258U/SUPPL_FILE/ES504258U_SI_001.PDF
23. M. Day, K. Shaw, D. Cooney, Biodegradability: An assessment of commercial polymers according to the Canadian method for anaerobic conditions. J. Environ. Polym. Degrad. **2**, 121–127 (1994). https://doi.org/10.1007/BF02074780/METRICS

Comparing Efficiency of Stabilisation of Black Cotton Soil Mixed with Fly Ash Using Plastic, Bentonite, and Bio-cement

Chandrima Bhadra, Debarghya Sarkar, Sankhadip Rang, Soumyadip Das, Shalini Banerjee, Anit Kumar Das, and Supriya Pal

Abbreviations

BCS	Black Cotton Soil
CBR	California Bearing Ratio
FA	Fly Ash
HDPE	High-Density Polyethylene
IS	Indian Standard
MDD	Maximum Dry Density
MICP	Microbially Induced Calcite Precipitation
OMC	Optimum Moisture Content
PET	Polyethylene Terephthalate
UCS	Unconfined Compressive Strength

1 Introduction

This study delves into the formidable challenges posed by black cotton soil in India, recognised for its inadequacy in construction due to its high plasticity, low bearing capacity, and propensity to swell. Researchers have explored various waste-based admixtures as potential stabilising agents in response to these challenges. Among

C. Bhadra (✉) · D. Sarkar · S. Rang · S. Das · S. Banerjee · A. K. Das
Department of Civil Engineering, Heritage Institute of Technology, Kolkata, West Bengal, India
e-mail: chandrima.bhadra@heritageit.edu

S. Pal
Department of Civil Engineering, National Institute of Technology Durgapur, Durgapur, West Bengal, India

D. Mazumder (ed.), *Sustainable Advanced Technologies for Environmental Management*, Springer Proceedings in Earth and Environmental Sciences,
https://doi.org/10.1007/978-3-031-64006-3_38

these, fly ash, a byproduct of thermal power plants, stands out for its known pozzolanic properties and high shear strength. Despite environmental concerns, plastic has garnered attention for its potential in recycling and has been considered for soil stabilisation. Bio-cement, derived from *Bacillus* bacteria, has shown promise in improving soil properties by precipitating calcium carbonate and enhancing bearing capacity [19]. Furthermore, bentonite has been recognised for reducing soil permeability and promoting cohesion, making it a favourable option for stabilising black cotton soil.

Previous studies have highlighted the significance of curing in achieving optimal UCS for bentonite-treated soil. For instance, curing for 3 days at a temperature of 60 °C has been found to enhance UCS significantly, reaching up to seven times higher than that of natural soil [1]. Moreover, despite being categorised as hazardous solid waste, fly ash has been effectively utilised in various construction applications such as roads and highways [2]. Studies suggest that the properties of black cotton soil can be suitably improved by stabilising it with fly ash up to specific percentages [13].

The utilisation of plastic in various fields is extensive, but its environmental impact must be addressed [12]. However, incorporating plastic in soil stabilisation endeavours presents an opportunity to mitigate its adverse effects, especially considering its economic and engineering effectiveness [3]. Research indicates plastic scraps contribute to soil stabilisation and reduction of waste when added at specific concentrations. Similarly, bentonite clay has been identified as a viable alternative for soil stabilisation due to its ability to fill voids and reduce free water content [5].

Bio-cement, extracted from bacterial precipitation, has demonstrated its potential to enhance soil-bearing capacity when combined with fly ash [6]. Furthermore, studies on Microbially Induced Calcite Precipitation (MICP) treatments have shown that the strength of soil increases as the size of microbes decreases [21]. Additionally, the incorporation of PET granules has led to a substantial rise in Maximum Dry Density (MDD) and California Bearing Ratio (CBR) values while reducing swelling potential [9].

Elasticity and resistance to deformation have been significantly improved by incorporating high-density polyethylene (HDPE) strips into the soil [18]. Moreover, employing recycled plastic and demolition wastes in construction, to some extent, offers numerous environmental and economic advantages [16].

The primary objective of this study is to compare the efficacy of plastic, bentonite, and bio-cement when mixed with fly ash for stabilising black cotton soil, taking into account both economic and environmental considerations. The study will observe the effects of these admixtures on various soil properties, ultimately drawing conclusions on which admixture combinations yield the highest Unconfined Compressive Strength and lowest swell index value [20].

2 Materials and Methods

2.1 *Materials*

2.1.1 Black Cotton Soil

Expansive soil, like black cotton, has montmorillonite clay minerals [17]. It is found in dark grey to black. The soil samples from Jamnagar, Gujrat, in western India, were used to study the strength parameters with various admixtures. The basic properties of the soil are listed in Table 1.

2.1.2 Fly Ash

Due to the incineration of the pulverised coal in the thermal power plant, fly ash, as shown in Fig. 1, has been formed, an industrial waste with pozzolanic characteristics [28]. Fly ash is light grey in colour and has a high pH value. This study uses fly ash samples from TATA industries, Mithapur, Dwarka, and Gujrat. Specific gravity was 2.34.

2.1.3 Plastic

Plastic pellets (as shown in Fig. 2) acquired from the Dhapa landfill and collected from Chowbaga, Anandapur, and Kolkata are used in this study. The plastic is recyclable and can be used safely as a stabilisation agent in black cotton soil. We perform the sieve analysis on the collected sample passing from a 4.75 mm sieve and retained on a 2 mm sieve [15]. The type of plastic is HDPE-2 (high-density

Table 1 Properties of raw black cotton soil

Sl. no	Property	Result
1	Natural moisture content	1.76%
2	Specific gravity	2.58
3	Plastic limit	27%
4	Liquid limit	62.21%
5	Plasticity index	35.21%
6	Sand percentage	38.50%
7	Silt & clay percentage	61.50%
8	Free swell index	50.00%
9	Optimum moisture content	21%
10	Maximum dry density	1.364 g/cc
11	Unconfined compressive strength	179.50 KN/m^2
12	Cohesion	89.75 KN/m^2
13	California bearing ratio	11.22%

Fig. 1 Fly ash

Fig. 2 Plastic pellet

polyethylene-2). It has high tensile strength and can sustain up to 130 °C temperature. It is almost nonreactive [4].

2.1.4 Bentonite

A certain amount of bentonite (as shown in Fig. 3) is added to the soil as a soil stabiliser to improve the engineering properties of black cotton soil. Bentonite is alkaline in nature. It improves the binding property between the particles, which can help to increase the soil's bearing capacity and observe an increment in strength compared to natural soil.

2.1.5 Bio-cement

Bio-cement (as shown in Fig. 4) is prepared from cultured *Bacillus* genus bacteria (e.g. *Sporosarcina pasteurii*) via a process called Microbial Induced Calcite Precipitation (MICP). In this process, urea is hydrolysed into ammonia and carbonate, leading to precipitation of calcium carbonate ($CaCO_3$). When inoculated in a cementitious medium, these bacteria precipitate $CaCO_3$ at high pH (>10), producing

Fig. 3 Bentonite

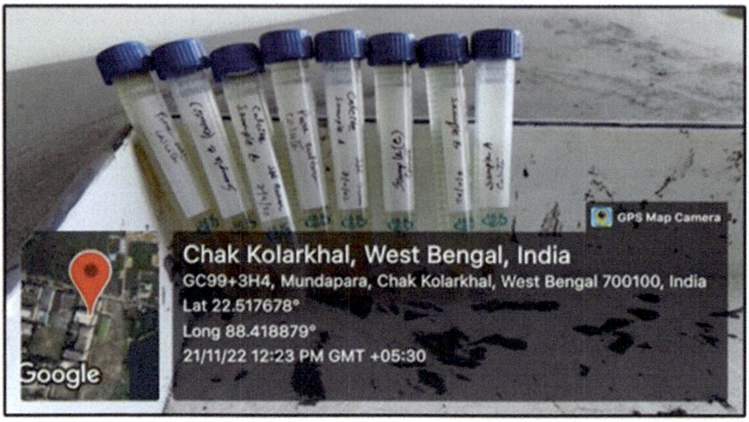

Fig. 4 Bio-cement extract

bio-cement. Bio-cement has self-healing capability. Bio-cement for the current study is collected from the biotech lab of Heritage Institute of Technology, Kolkata.

2.2 *Methodology*

2.2.1 Preparation of Bio-cement

The soil sample was obtained from a home garden located in Kolkata. A 10-g portion of the soil was introduced into 100 mL of Media-1. The composition of Media-1 consisted of 2% glucose, 2 g/L of peptone, and 5 g/L of NaCl. The mixture was then incubated in an incubator shaker at 37 °C and pH 7 for 24 h. Following this incubation period, the flasks were removed, and the supernatant was carefully collected. Subsequently, the supernatant was inoculated into fresh Media-1 and allowed to grow for 24 h. The resulting culture was then transferred into Media-2, comprising 0.5 M urea, 0.5 M calcium chloride, and 2.12 g of sodium bicarbonate per litre, in a total volume of 1 L of water. After 3 days, precipitation was observed to occur. This precipitate was collected and directly added to the soil and fly ash mixture as bio-cement.

2.2.2 Preparation of Test Specimen

Sieve analysis, specific gravity, liquid limit, plastic limit, free swell, CBR tests, OMC and MDD, and unconfined compression test were performed to identify the properties of raw black cotton soil [7]. Then, fly ash was added to the soil in various proportions. To find optimum fly ash content, combinations of (BC SOIL + 0%FA),

(BC SOIL + 10%FA), (BC SOIL + 25%FA), and (BC SOIL + 50%FA) were taken, and UCS compaction parameters were tested.

The optimum ash value was 25% (refer to Sect. 3.2). After that, 5% of different admixtures were added to prepare (BC SOIL + 25%FA + 5%PLASTIC), (BCSOIL + 25%FA + 5%BENTONITE), and (BC SOIL + 25%FA + 5%BIOCEMENT) samples for liquid limit, plastic limits, free swell test, OMC and MDD, and unconfined compression test (UCS).

2.2.3 Specific Gravity Test

A specific gravity test was conducted for raw black cotton soil using IS2720 (Part 3/Sec 1)-1987 specifications. 10 g of dry sample was used in a specific gravity bottle and the weight of the empty bottle, bottle + dry soil, bottle + distilled water, and bottle + distilled water + soil. The specific gravity was determined.

2.2.4 Liquid Limit

We conduct liquid limit tests by preparing 100 g of samples using IS:2720 (Part-5)-1985 specifications—Casagrande liquid limit device used with Casagrande tool for grooving. By turning the crank at two revolutions per second, the cup is lifted and dropped until the bottom of the groove is closed for a length of 12.5 mm. The number of blows was recorded and repeated five times to plot the graph, and the liquid limit was read at a water content of 25 blows.

2.2.5 Plastic Limit

This test was also conducted concerning IS:2720 (Part-5)-1985 by preparing a sample passing through a 425 mm sieve and rolling it on a glass plate until a 3 mm diameter is reached, repeating and watching when crumbling occurs, and then collecting and calculating the moisture content.

2.2.6 Compaction Test

The OMC and MDD test was performed according to Standard Proctor's Test procedure to determine the compaction factor. The proctor mould is filled in 3 layers, and 25 blows are given to each layer with a standard rammer of 2.6 kg, which can be dropped from a height of 31 cm. The mould with soil is weighted, and the bulk density is calculated. A small sample is collected from the compacted soil and weighed for moisture content determination. Then, the dry density is found based on the calculated bulk density and moisture content.

2.2.7 UCS Test

According to IS: 2720 (Part 10): 1991, a sample was prepared using a 38 mm diameter and 76 mm length mould to determine the compressive strength of cohesive soil. After trimming the sample to the correct size, the sample is placed on the bottom plate of the compression device, the upper plate is adjusted to make contact with the specimen, and the dial gauge of the proving ring and strain is set to zero. Compress the specimen until a crack appears on the specimen. Dial gauge reading should be taken for every deformation of the specimen. Two or three specimens are tested similarly, as shown in Fig. 5.

3 Results and Discussion

3.1 Characteristics of Black Cotton Soil

The properties of raw black cotton soil are shown in Table 1.

As seen from Table 1, the plasticity index and swell index, which are 35.21% and 50%, respectively, are on the higher side, so fly ash is added with bentonite, plastic pellets, and cement, respectively. Relevant tests were performed to ensure that the amended soil could be used in the field without excessive swelling and shrinkage and to increase its bearing capacity.

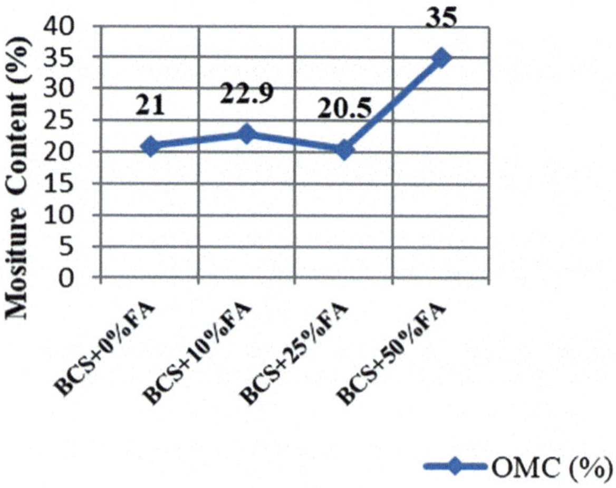

Fig. 5 Variation of OMC (BCS + FA)

3.2 Optimum Fly Ash Content

Compaction and UCS tests were conducted on black cotton soil amended with varying percentages of fly ash (10%, 25%, and 50%). Analysis revealed that the maximum Modified Proctor Density (MDD) value of 1.415 g/cc was achieved at a 25% fly ash content, as depicted in Fig. 6. Similarly, Fig. 7 illustrates that a 25% fly ash content resulted in the highest UCS value of 488.68 KN/m^2. Figure 5 displays the relationship between optimum moisture content (OMC) and fly ash content.

The initial rise in MDD value shown in Fig. 6 can be attributed to the finer fly ash particles infiltrating soil voids. However, beyond a certain threshold of fly ash content, MDD decreases. This decline occurs as the voids become fully saturated with fly ash particles, causing them to replace soil particles. Notably, fly ash possesses a lower specific gravity (2.34) than the soil sample (2.58), further facilitating its replacement of soil particles.

The identified optimal fly ash content aligns with findings from other scholarly references [14], which also reported maximum MDD and unsoaked California

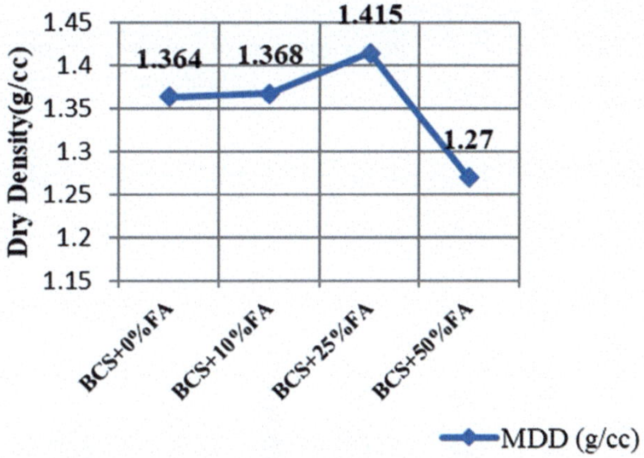

Fig. 6 Variation of MDD (BCS + FA)

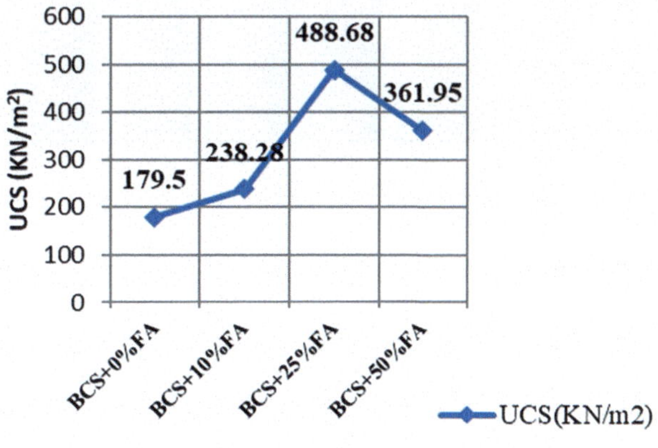

Fig. 7 Variation of UCS (BCS + FA)

Bearing Ratio (CBR) values at a 20% fly ash content. Beyond this point, CBR and MDD values gradually decline [8].

3.3 *Properties of the Soil Amended with Admixtures*

3.3.1 Atterberg Limit Tests

The Atterberg limit test (comprising liquid limit and plastic limit) followed the guidelines outlined in IS 2720 (Part-5): 1985. Black cotton soil blended with 25% fly ash underwent testing with 5% bentonite, 5% plastic, and 5% bio-cement, respectively. Analysis of the results presented in Table 2 and Fig. 8 reveals that the lowest values for liquid limit, plastic limit, and plasticity index were observed at 57.39%, 41.91%, and 15.48%, respectively, when the soil was amended with 25% fly ash in conjunction with cement. This indicates a reduced propensity for volume changes with alterations in moisture content. Notably, a similar trend was observed

Table 2 Atterberg limit of the amended soil

Sample composition	Pure BCS	BCS + 25% FA	BCS + 25%FA + 5%bentonite	BCS + 25% FA + 5%plastic	BCS + 25%FA + 5%bio-cement
Liquid limit (%)	62.21	57.94	68.11	67.78	57.39
Plastic limit (%)	27	42.46	51	44.20	41.91
Plasticity index (%)	35.21	15.48	17.11	23.58	15.48

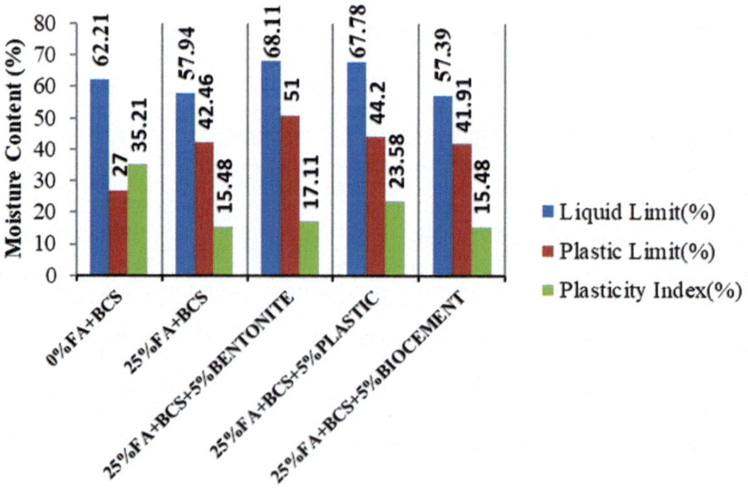

Fig. 8 Variation of Atterberg limits

in literature where the swelling potential of expansive soil was significantly diminished upon incorporating 25% fly ash with 5% bio-cement [6].

3.3.2 Compaction Parameters

The Standard Proctor's test was conducted to ascertain the OMC and MDD of the samples. Among the combinations tested, it was observed that the mixture comprising soil, 25% fly ash, and 5% plastic exhibited the most optimal enhancement of MDD, reaching 1.459 g/cc, as detailed in Table 3 and depicted in Fig. 9. This suggests that the inclusion of plastic pellets facilitates maximum compaction. The variation in OMC can be discerned from Fig. 10. Literature corroborates these findings, indicating that adding plastic granules consistently increases dry density to a certain percentage, typically around 6%. Moreover, the rise in OMC can be attributed to the expulsion of water under compaction efforts [9]. Similarly, studies have shown that incorporating glass and plastic results in a continuous increase in MDD with up to 6% admixture content [10].

Table 3 Compaction parameters of the amended soil

Sample omposition	Pure BCS	BCS + 25% FA	BCS + 25%FA + 5%bentonite	BCS + 25% FA + 5%plastic	BCS + 25%FA + 5%bio-cement
MDD (g/cc)	1.364	1.415	0.368	1.459	1.383
OMC (%)	21	20.5	28	23	21.45

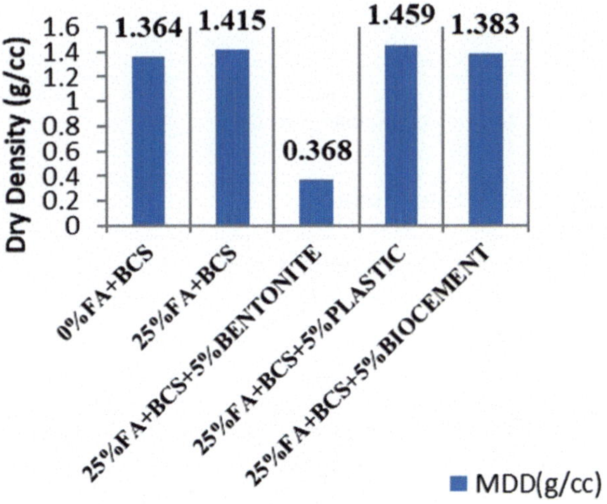

Fig. 9 Variation of MDD

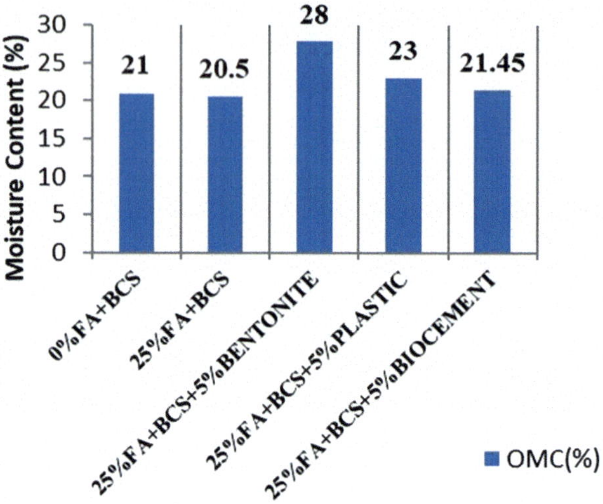

Fig. 10 Variation of OMC

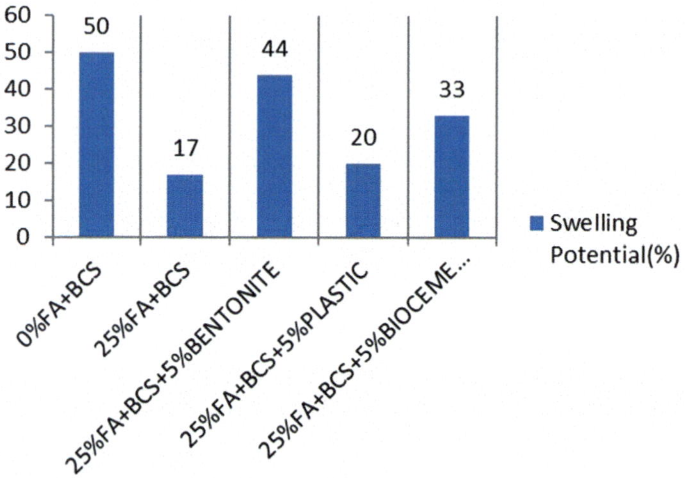

Fig. 11 Variation of swelling potential

3.3.3 Free Swell Index

The soil's swelling potential was evaluated per IS 2720 (Part-40): 1977, Reaffirmed 2021, for soil amended with various admixture combinations. A graphical representation of the swelling potential for all combinations is presented in Fig. 11. Analysis from Table 4 indicates that the most significant reduction in soil swelling tendency occurred when mixed solely with 25% fly ash. Consequently, adding fly ash alone may suffice when considering only the swelling potential.

However, plastic emerges as a more suitable option when both UCS and swelling potential parameters are considered for soil stabilisation [11]. Plastic, which

Table 4 Variation of free swell index

Sample composition	Pure BCS	BCS + 25% FA	BCS + 25%FA + 5%bentonite	BCS + 25% FA + 5%plastic	BCS + 25%FA + 5%bio-cement
Free swell index (%)	50	17	44	20	33

Table 5 Variation of UCS

Sample composition	Pure BCS	BCS + 25% FA	BCS + 25%FA + 5%bentonite	BCS + 25% FA + 5%plastic	BCS + 25%FA + 5%bio-cement
UCS (KN/m^2)	179.5	488.68	458.5	548.94	500.72
Cohesion (c) (KN/m^2)	89.75	244.34	229.25	274.47	250.36

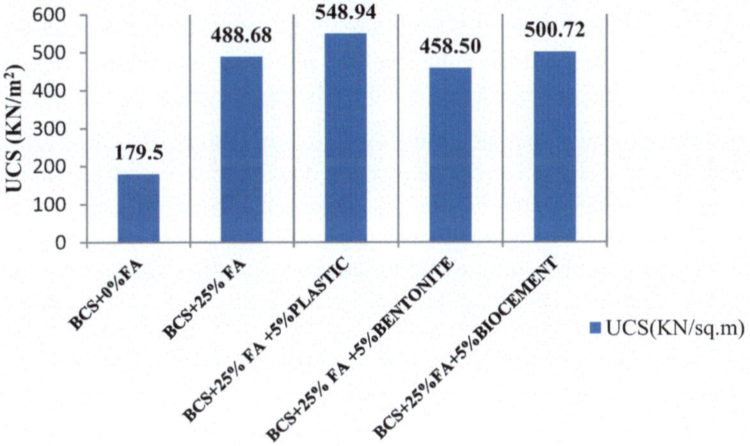

Fig. 12 Variation of UCS

demonstrated the second-best reduction in swelling potential, also exhibited the maximum enhancement in UCS value, as detailed in Table 5. Notably, research suggests that the free swelling value gradually decreases with increasing plastic content by up to 9% [9].

3.3.4 Unconfined Compressive Strength

UCS testing following IS 2720 (Part 10): 1991 on raw soil and amended soil samples yielded the results in Table 5 and Fig. 12. It can be observed that when black cotton soil is mixed with 25% fly ash and 5% plastic, the sample gains a maximum strength of 548.94 KN/m^2. The study reveals that initially, UCS value increases with increasing glass and plastic content, but strength decreases beyond the optimum percentage (6%) [10].

It was revealed through the test results that when raw black cotton soil was mixed with 25% fly ash, the plasticity index and swell index were reduced by 56% and 66%, respectively, as compared to the raw soil. Also, MDD and UCS values were increased by 3.74% and 172.24%, respectively.

The maximum increase in UCS value and MDD was observed for the combination of soil, 25% fly ash, and 5% plastic, which increased by 205.82% and 6.96%, respectively, to that of raw soil. A decrease in swelling potential and plasticity index of 60% and 33% was also observed for this specific combination.

4 Conclusion

Based on the findings of all the laboratory trials, the admixtures demonstrated satisfactory outcomes in reducing the swelling tendency and diminishing plasticity when compared to the raw black cotton soil. The key conclusions are as follows:

- The combination of bio-cement and fly ash proved to be the most effective in reducing the plasticity of the soil. However, due to the time-consuming nature of bio-cement production and the need for precise control over bacterial culture, it may not be sustainable for small-scale applications.
- While the addition of bentonite and fly ash successfully reduced plasticity to an acceptable degree, it did not adequately diminish the swelling tendency of the soil.
- Soil amended with plastic and fly ash exhibited the highest strength and maximum increase in MDD (Maximum Dry Density) value. Although it did not achieve the lowest swelling potential, it approached the minimum threshold of 17%. Adjusting plastic content or reducing the size of plastic pellets could further decrease the swelling potential.

The effectiveness of stabilisation with plastic was found to surpass that of other admixtures. Additionally, plastic is readily available and cost-effective, unlike other admixtures. Consequently, plastic emerges as a suitable and sustainable stabilising agent in construction projects where mitigating swelling tendency and enhancing soil strength are imperative.

Acknowledgments We, the authors, express our heartfelt gratitude to the professors, teachers, and lab assistants of HERITAGE INSTITUTE OF TECHNOLOGY, KOLKATA, for providing us with the relative data required to make this experiment a great success.

References

1. S. Sabrin, S. Siddiqui, N. Muhammad, Understanding the effect of heat treatment on subgrade soil stabilised with bentonite and magnesium alkalinization. Transp. Geotech. **21**, 100287., ISSN: 2214-3912 (2019). https://doi.org/10.1016/j.trgeo.2019.100287
2. P.M. Shanmugavadivu, M.M. Sindhu, G.K. Mukilan, Stabilization of black cotton soil using fly ash. Paripex Indian J. Res. **8**(2), 45–46 (2019)
3. P. Bhattarai, A.V.A. Bharat Kumar, K. Santoshi, T.C. Manikanta, K. Tejeswini, Engineering behaviour of soil reinforced with plastic strips. Int. J. Civil Struct. Environ. Infrastruct. Eng. Res. Dev. **3**(2), 83–88 (2013)
4. A. Vismaya, M. Simon, P.K. Jayasree, L. Peter, Strength characteristics of subgrade soil stabilized with plastic bag strips, in *Proceedings of the Indian Geotechnical Conference 2019*, vol. 136, (Springer, Singapore, 2021), pp. 485–493. https://doi.org/10.1007/978-981-33-6444-8_43
5. M.B. Baker, R. Abendeh, A. Sharo, A. Hanna, Stabilization of sandy soils by bentonite clay slurry at laboratory bench and pilot scales. Coatings **12**, 1922 (2022)
6. M. Li, C. Fang, S. Kawasaki, V. Achal, Fly ash incorporated with bio cement to improve the strength of expansive soil. Sci. Rep. **8**, 2565 (2018). https://doi.org/10.1038/s41598-018-20921-0
7. A.K. Agarwal, V. Rajurkar, P. Mokadam, Effect of waste synthetic bag pieces on the CBR value of expansive soil. J. Mater. Eng. Struct. **2**, 26–32 (2015)
8. A. Mehta, K. Parate, B.S. Ruprai, Stabilization of black cotton soil by fly ash. Int. J. Appl. Innov. Eng. Manag., 1–5 (2013) ISSN 2319-4847
9. V.A. Salimath, S.R. Kulkarni, Stabilization of black cotton soil by using plastic granules as a subgrade material. Int. Res. J. Eng. Technol. **8**(12), 1674–1679 (2021)
10. K. Subash, R. Sreerag, A.D. Deeraj, S. Sukesh, V. Dilna Sathian, J. John, Stabilization of black cotton soil using glass and plastic granules. Int. J. Eng. Res. Technol. **5**(4), 480–483 (2016)
11. S.K. Mohanty, P.K. Pradhan, C.R. Mohanty, Stabilization of expansive soil using industrial wastes. Geomech. Eng. **12**(1), 111–125 (2017)
12. A. Gupta, V. Saxena, V. Gaur, V. Kumar, T. Kumar, A review paper on stabilization of soil using plastic waste as an additive. Int. Res. J. Eng. Technol. **6**(5), 6435–6440 (2019)
13. P. Priya, R.V. Singh, Stabilization of black cotton soil using fly ash. Int. J. Recent Technol. Eng. **9**(5), 91–96 (2021)
14. Pallavi, T. Pradeep, P.D. Poorey, Stabilization of black cotton soil using fly ash and nylon fiber. Int. Res. J. Eng. Technol. **3**(11), 1283–1288 (2016) e-ISSN: 2395-0056
15. A.K. Rai, G. Singh, A.K. Tiwari, Comparative study of soil stabilization with glass powder, plastic, and e-waste: A review. Mater. Today Proc. **32**, 771–776 (2020). https://doi.org/10.1016/j.matpr.2020.03.570
16. A. Arulrajah, E. Yaghoubi, Y.C. Wong, S. Horpibulsuk, Recycled plastic granules and demolition wastes as construction materials: Resilient moduli and strength characteristics. Constr. Build. Mater. **147**, 639–647 (2017)
17. S.R. Soni, P.P. Dahale, R.M. Dobale, Disposal of solid waste for black cotton soil stabilization. Int. J. Adv. Eng. Sci. Technol. **8**(1), 113–120 (2011)
18. A.K. Choudhary, J.N. Jha, K.S. Gill, Utilization of plastic wastes for improving the sub-grades in flexible pavements, in *Paving Materials and Pavement Analysis*, (ASCE, Reston, 2010), pp. 320–326
19. S. Prasad, A.K. Mantripragada, Utilization of industrial waste in flexible pavement construction. Electron. J. Geotech. Eng. **13**, 1–12 (2009)
20. R. Dilrukshi, K. Nakashima, S. Kawasaki, Soil improvement using plant-derived urease-induced calcium carbonate precipitation. Soils Found. **58**(4), 894–910 (2018)
21. B. Soundara, P. Kulanthaivel, S. Nithipandian, V. Soundaryan, A critical review on soil stabilization using bacteria. IOP Conf. Ser.: Mater. Sci. Eng. **955**, 012065 (2020). https://doi.org/10.1088/1757-899X/955/1/012065

22. Bureau of Indian Standards, *Indian Standard 2720(Part-4): Grain Size Analysis* (Bureau of Indian Standards, New Delhi, 1985)
23. Bureau of Indian Standards, *Indian Standard 2720(Part-5): Determination of Liquid and Plastic Limit* (Bureau of Indian Standards, New Delhi, 1985)
24. Bureau of Indian Standards, *Indian Standard 2720(Part-10): Determination of Unconfined Compressive Strength* (Bureau of Indian Standards, New Delhi, 1991)
25. Bureau of Indian Standards, *Indian Standard 2720(Part-40): Determination of Free Swell Index of Soils* (Bureau of Indian Standards, New Delhi, 1977)
26. Bureau of Indian Standards, *Indian Standard 2720(Part-16): Laboratory Determination of CBR* (Bureau of Indian Standards, New Delhi, 1987)
27. Bureau of Indian Standards, *Indian Standard 2720(Part-3/Sec1): Determination of Specific Gravity* (Bureau of Indian Standards, New Delhi, 1980)
28. P. Kilabanur, M.S. Dharek, P. Sunagar, K.S. Sreekeshava, P. Thejasw, *Index and Strength Properties of Black CottonSoil using Combination of Geopolymer and Flyash*. IOP Conf. Series: Materials Science and Engineering 955 (2020)

Index

D. Mazumder (ed.), *Sustainable Advanced Technologies for Environmental Management*, Springer Proceedings in Earth and Environmental Sciences, https://doi.org/10.1007/978-3-031-64006-3

R

S

T

GPSR Compliance

The European Union's (EU) General Product Safety Regulation (GPSR) is a set of rules that requires consumer products to be safe and our obligations to ensure this.

If you have any concerns about our products, you can contact us on ProductSafety@springernature.com

In case Publisher is established outside the EU, the EU authorized representative is:

Springer Nature Customer Service Center GmbH
Europaplatz 3
69115 Heidelberg, Germany